OPTIMAL CONTROL SYSTEMS

MATHEMATICS IN SCIENCE AND ENGINEERING

A SERIES OF MONOGRAPHS AND TEXTBOOKS

Edited by Richard Bellman

1. Tracy Y. Thomas. Concepts from Tensor Analysis and Differential Geometry. Second Edition. 1965
2. Tracy Y. Thomas. Plastic Flow and Fracture in Solids. 1961
3. Rutherford Aris. The Optimal Design of Chemical Reactors: A Study in Dynamic Programming. 1961
4. Joseph La Salle and Solomon Lefschetz. Stability by Liapunov's Direct Method with Applications. 1961
5. George Leitmann (ed.). Optimization Techniques: With Applications to Aerospace Systems. 1962
6. Richard Bellman and Kenneth L. Cooke. Differential-Difference Equations. 1963
7. Frank A. Haight. Mathematical Theories of Traffic Flow. 1963
8. F. V. Atkinson. Discrete and Continuous Boundary Problems. 1964
9. A. Jeffrey and T. Taniuti. Non-Linear Wave Propagation: With Applications to Physics and Magnetohydrodynamics. 1964
10. Julius T. Tou. Optimum Design of Digital Control Systems. 1963
11. Harley Flanders. Differential Forms: With Applications to the Physical Sciences. 1963
12. Sanford M. Roberts. Dynamic Programming in Chemical Engineering and Process Control. 1964
13. Solomon Lefschetz. Stability of Nonlinear Control Systems. 1965
14. Dimitris N. Chorafas. Systems and Simulation. 1965
15. A. A. Pervozvanskii. Random Processes in Nonlinear Control Systems. 1965
16. Marshall C. Pease, III. Methods of Matrix Algebra. 1965
17. V. E. Beneš. Mathematical Theory of Connecting Networks and Telephone Traffic. 1965
18. William F. Ames. Nonlinear Partial Differential Equations in Engineering. 1965

MATHEMATICS IN SCIENCE AND ENGINEERING

19. J. Aczél. Lectures on Functional Equations and Their Applications. 1965

20. R. E. Murphy. Adaptive Processes in Economic Systems. 1965

21. S. E. Dreyfus. Dynamic Programming and the Calculus of Variations. 1965

22. A. A. Fel'dbaum. Optimal Control Systems. 1965

23. A. Halanay. Differential Equations: Stability, Oscillations, Time Lags. 1965

In preparation

Dimitris N. Chorafas. Control Systems Functions and Programming Approaches

David Sworder. Optimal Adaptive Control Systems

Namik Oğuztöreli. Time Lag Control Processes

Milton Ash. Optimal Shutdown Control in Nuclear Reactors

OPTIMAL CONTROL SYSTEMS

A. A. FEL'DBAUM

INSTITUTE OF AUTOMATICS AND TELEMECHANICS MOSCOW, USSR

TRANSLATED BY A. KRAIMAN

1965

ACADEMIC PRESS New York and London

ACADEMIC PRESS INC.
111 Fifth Avenue, New York, New York 10003

United Kingdom Edition published by
ACADEMIC PRESS INC. (LONDON) LTD.
Berkeley Square House, London W.1

LIBRARY OF CONGRESS CATALOG CARD NUMBER: 65-26397

PRINTED IN THE UNITED STATES OF AMERICA

OPTIMAL CONTROL SYSTEMS
WAS ORIGINALLY PUBLISHED AS:
OSNOVY TEORII OPTIMAL'NYKH AVTOMATICHESKIKH SISTEM
BY FIZMATGIZ, MOSCOW, 1963

Foreword to the Russian Edition

The rapid development of various branches of the theory of optimal automatic systems has made timely an attempt to cover the fundamental aspects of this theory from some unified point of view. Such an attempt was undertaken by the author in the chapter "Optimal systems" published in the collection "Disciplines and Techniques of Systems Control" under the editorship of Dr. Peschon (Luxembourg), and also in a course of lectures delivered in 1961-1962 to post-graduate students and employees of the Institute of Automation and Remote Control. The point of view presented there has been taken as the basis for the present book. It determined the sequence of presentation. The book is divided into six chapters. In the first chapter the statement of the problem is given, in the second a survey of the mathematical means applicable to its solution. The third chapter is devoted to so-called systems with complete information, and the fourth to those systems with maximal but partial information about the controlled object. In the fifth chapter the theory of systems with partial information about the object and with its passive storage is considered. Finally, in the sixth chapter systems with active information storage are studied.

It would have required an excessive increase in the size of the book to delve deeply into the details of the ramifications of the theory. Therefore the author limited himself to the discussion and illustration, by examples, of the fundamental aspects of the theory. Only the minimum information necessary for the investigator is given; moreover, this information is presented at the level of "engineering" rigor.

The book on the whole is theoretical, but it has been written, according to the widespread expression, "by the engineer for engineers."

Some very important aspects of research were not included, in particular, the Kolmogorov-Wiener theory and papers developing it were omitted since there is extensive literature in this field.

The author considers it his accepted duty to thank Ya. Z. Tsypkin and A. G. Butkovskii for discussing a number of points, A. V. Khram for help with the compilation of the bibliography and revising dates, and K. Man'chak, R. C. Rutman, V. N. Novosel'tsev, E. P. Maslov, V. P. Zhivoglyadov, and I. V. Tim for much help with the appearance and editing of the manuscript.

A. A. Fel'dbaum

August 14, 1962

Contents

CHAPTER I

Problem of the Optimal System

1. Significance of the Theory of Optimal Systems

The technology of information transmission and processing—in brief, cybernetic engineering—has grown during recent years at an exceptionally rapid rate. The development of its most important direction—automation—is characterized by the rapid spread of automatic systems and by an expansion of its range of application. But automation is growing not only "in breadth," but "in depth" as well. New principles of automatic control are emerging which solve all the more complex problems of control and replace man in all the more complicated spheres of his activity.

Automatic control systems are becoming complex. In the simple early types the measuring, controlling, and actuating units were frequently combined. In complicated contemporary automatic control systems these units are usually separate, sometimes very complex systems by themselves. The central part of the system—the control unit—is frequently realized in the form of a continuous- or discrete-operating controller. A definite law is imposed on this mechanism, or as it is otherwise called, a control policy. In contemporary controllers of universal or specialized type, the most complicated algorithms can be realized. At the present time computer technology permits computing speeds of the order of hundred thousands and millions of elementary operations per second. Therefore for many applications the control system can be considered inertialess.

Analogous trends appear in other branches of cybernetic engineering as well. The problems of ultra-long-range communication, telemetry and remote control, the separation of radar and television pictures from a background of natural or artificial noise, and the questions of image recognition are closely related to the realization of complex algorithms in information processing. Image recognition systems represent a typical example of systems of this sort. Systems of this type have at present found application basically in the recognition of letters and numbers printed and written by hand, entering a computer for example, and also in the recognition of commands given back by a man to a machine in the form of oral speech. But the perspectives of such systems are exceedingly wide; indeed, the recognition and classification of complex images and situations, masked by ambient conditions, noise and so forth, permit man to achieve a flexible orientation in the environment and take the proper decision. Algorithms modeling the comparatively complex functions of man's intellectual activity are complicated; therefore a system realizing them must also contain a sufficiently complex computer.

General control theory is also developing in parallel with the growth of technology, emerging as the foundation for the vast complex of branches of technology dealing with information transmission and processing. This general theory is called engineering cybernetics; it in turn represents the branch, direction, or section of general cybernetics (see [1.1]) which examines the processes of control and information processing both in engineering systems and in living organisms, and in collectives composed of living beings and machines.

Both cybernetics in general, and engineering cybernetics in particular, arose as a result of the long process of extending earlier disconnected theories, ideas, and principles developed in separate disciplines. This extension process is still continuing today. Until recent years control theory has been characterized by a series of comparatively weakly connected directions isolated from one another to a considerable extent. Only now has a clear tendency for the various directions to come together, and for the general concepts, ideas, methods, and theory to crystallize, begun to show.

At present there are several very important directions in the development of engineering cybernetics. One of them is the development of the theory and principles of designing complicated control systems consisting of a large number of elements, including complex interconnections of parts and complicated operating conditions.

Another important direction in engineering cybernetics is the development of the theory and principles of operating systems with automatic adaptation (self-adapting or adaptive systems). The process of automatic adaptation consists of the measurement of parameters, characteristics, and in general properties of the system or its parts, performed in an open loop, or by means of automatic regulation, or with the aid of automatic search. For example, the optimizer in a system with automatic optimization by means of automatic search thus varies the characteristics of the primary controller, so that the best operation is achieved when the properties of the controlled object change. The theories of learning, self-adjusting, and self-organizing systems are found along this direction.

The subject of this book is the third of the most important directions of engineering cybernetics. This direction is the theory of optimal processes, i.e., best in a known sense, and the theories of optimal control systems, i.e., optimal systems for information transmission and processing. The problems of optimal systems are becoming central by virtue of the following reasons.

(a) Any scientifically sound system is optimal, since in choosing some system by the same token we prefer it to another; hence, we regard it as better than other systems in some respect. The criteria with the aid of which the choice is made (below they are called optimality criteria) can be different. But for any choice, in the final result a criterion for optimality does exist. Otherwise the selection of a system would be impossible.

In former times, when industrial processes were not automated and technology was based to a considerable extent on experiment and practical people, when the means of measuring and computing engineering were not so developed as they are now, the attempts at a clear understanding and definition of optimality criteria, and especially the attempts to construct optimal systems, were often pointless. But now we are advancing into a new era, an age of the construction of scientifically sound and automated industrial processes. Therefore the importance of optimal control problems is growing. The solution of these problems will permit the use of manufacturing assemblies to be brought to maximum effectiveness, productivity to be increased, the quantity of production to be improved, the economy of electrical energy and valuable raw materials to be ensured, and so on.

(b) Any law of nature is a statement of limiting character, an assertion of what is possible and what is impossible in a certain field. Likewise the laws of the general type in cybernetics must give the possibility of making a judgment of what is attainable and what is impossible to achieve under definite real conditions. Therefore they can be formulated in the form of statements about the "ceiling of possibilities" for control. Indeed the finding of this "ceiling" among them is the problem of the theory of optimal systems.

Consequently, the theory of optimal systems must essentially help in the difficult matter of formulating the general laws of cybernetics. This is a matter for the future, but evidently not so distant.

The problems of optimal systems have arisen in many fields of cybernetic engineering. These are the problems of constructing automatic control systems which are optimal in speed of response, the problems of the best filtering of a signal from the noise mixed with it, the problems of constructing optimal signal detectors, optimal "predicting" devices, optimal methods of recognizing images, optimal strategy of automatic search, and so on. But between all these problems, so diverse at first glance, there is a deep intrinsic connection. In the following presentation this connection will be revealed and emphasized. Indeed, this connection is the basis for constructing a unified theory of optimal systems.

In engineering cybernetics there are, besides those enumerated above, other important directions as well—for example, the study of the stability and self-oscillations of systems. In addition, such very important directions as the general theory of systems (part of which is the theory of automata) and the general theory of signals (part of which is information theory) belong to general cybernetics to no lesser extent than the engineering aspect.

The indicated directions are not all independent of one another. On the contrary, a definite connection exists between them. For example, in the theory of finite automata, the determination of the simplest automatic machine realizing a prescribed algorithm is of great interest. Likewise, the problems of finding the most reliable automata for a given complexity are posed. These are systems which are optimal in reliability. The questions of automatic adaptation are even more closely connected with the theory of optimal systems. The most important domain of these questions consists of the theory of automatic optimization systems, connected with the theory of optimal systems by many channels. Let us enumerate some of these.

(a) The optimal system—this is the ideal at which the automatic optimization system aims (but which it does not always attain).

(b) For a sufficiently slow change in the characteristics of the controlled object, a primary controller can be constructed according to optimal system theory, by providing it, however, with variable parameters. A secondary controller—an automatic optimizer—by observing the operation of the system, changes the parameters of the primary device, such that on the whole the system remains close to optimal, in spite of an unexpected variation in the characteristics of the controlled object.

(c) When the characteristics of the object change comparatively rapidly, then the system, by operating according to the principle mentioned above, can turn out to be far from optimal. In this case the problem arises of finding the best search algorithm of optimal control methods, i.e., the problem arises of designing an optimal automatic search system or, in general, an optimal automatically adaptive system.

The theory of optimal systems is closely related to other directions of engineering cybernetics, if only because any problem being implemented by a controller can be solved by a best method which is optimal in some sense. Hence the problems of constructing optimal systems arise in any field of engineering cybernetics.

For example, let us touch briefly on the problem of constructing automatic control systems which are optimal in speed of response. This problem arises in the development of servomechanisms, automatic compensators, and tracking drives of engineering assemblies; in the design and operation of chemical and metallurgical reactors and stoves, in rocket control systems, and also in a number of other fields. We will consider as an example an automatic compensator (Fig. 1.1). The problem for this device is measuring and recording the voltage E, which can vary with time. In the compensator the difference ΔU is measured between the voltage E and the voltage U compensating it. The latter represents the potential difference between the moving indicator D of the potentiometer P and the lower point of the potentiometer. A stable constant voltage U_0 intentionally greater than E is supplied to the potentiometer. It is assumed that the potentiometer has a uniform winding. First let us assume that the position of moving indicator D is always such that compensation occurs, i.e., $E = U$. Then the difference $\Delta U = 0$. In this condition the coordinate of the moving indicator D is proportional to the measured voltage E. Therefore the pen attached to the moving indicator D can draw on a uniformly moving paper tape (not shown in the figure) a graph of the variation in E as a function of time.

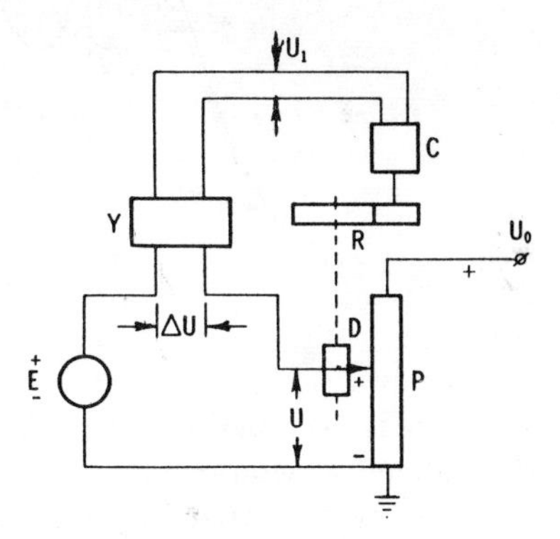

FIG. 1.1

Thus the problem of the automatic potentiometer consists of maintaining the equality $\Delta U = 0$ with sufficient accuracy. The voltage ΔU is fed to the input of the system Y, where it is amplified and transformed. The output voltage U_1 of the system Y acts on the servomotor C. If ΔU deviates from zero, then a voltage U_1 appears at the input to the servomotor C and the servomotor shaft starts to rotate, changing the position of the moving indicator D by means of the gearbox R, such that the equality $E = U$ is restored.

If the system is sufficiently precise and the voltage E varies slowly enough, then the condition $\Delta U = 0$ is maintained with the required accuracy. But in the case where the voltage E can vary with great rapidity, it would be required that the automatic compensator be "quick-acting." The most severe case corresponds to a jump in the voltage E—for example, from a value equal to zero up to E_{max} (Fig. 1.2,

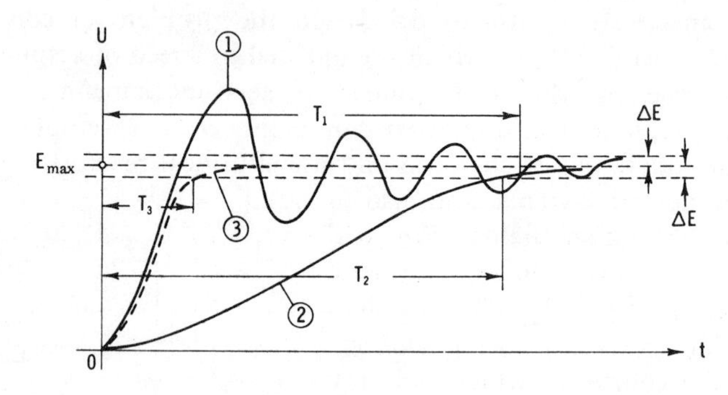

FIG. 1.2

where it is assumed that the jump occurs at $t = 0$). In an ideal system the voltage U would have to change from 0 to E_{max} in a jump as well. But it is evident that in a real compensator such a jump is impossible. In fact, the servomotor C cannot develop a moment (and hence an acceleration as well) larger than a certain maximally possible value. Usually there is also a limit imposed on the speed of rotation of the servomotor shaft. The latter likewise cannot exceed a maximally possible value. Therefore it could only be required that the curve $U = U(t)$ (Fig. 1.2) reach the region $E_{max} - \Delta E < U < E_{max} + \Delta E$ as rapidly as possible, i.e., for a minimal value of $t = T$, where ΔE is a sufficiently small quantity, and moreover U will be within this region for $t > T$ as well. The quantity T is called the "control time." It is thus required to construct a controller Y (or only its input, low-power part, if the power amplifier stages are given), so that the condition $T = \min$ would be ensured under the prescribed constraints imposed on the parameters of the servomotor C. Such a system will be the most rapid-operating, or in other words, optimal in speed of response.

The problem of constructing systems which are optimal in speed of response is very complex even in the simplest cases. In fact, let us assume at first that the system as a whole is described by a second-order linear differential equation with constant coefficients. For small damping coefficients (i.e., small coefficients of the first derivative of U in the equation) the curve of $U(t)$ has a sharply oscillatory character (curve 1 in Fig. 1.2). In this case the control time $T = T_1$ is large. If the damping coefficient is made large, then the process $U(t)$ acquires an aperiodic character (curve 2 in Fig. 1.2). In this case the control time is also large. By means of an optimal setting of the damping coefficient (it is usually chosen barely less than critical) the control time can be reduced. The corresponding curve is not shown on the figure. But very simple arguments show that the best results can be achieved by going over to a nonlinear system.

Let us make the damping coefficient depend on the difference ΔU between E and U. For large values of ΔU let it be small, and the curve of $U(t)$ denoted by 3 in Fig. 1.2 will go along curve 1. But when the difference ΔU becomes small, let the damping coefficient be sharply increased. Then the "tail" of curve 3 will obtain the same character as curve 2; curve 3 will approach the value E_{max} smoothly, and the control time T_3 will turn out to be considerably smaller than with any linear system. These very simple arguments are justified by theory and experiment. It turns out that a system which is optimal in speed of response must be nonlinear, even in the very simple case under consideration. Meanwhile the investigation of nonlinear systems is in general incomparably more difficult than the study of linear ones. But the practical and general theoretical value of the theory of optimal systems is too great, which fully justifies the big effort expended on its development.

The general theoretical significance of optimal system theory has been stressed above. As to the practical value of this theory, then one must distinguish two aspects here. In the first place, without it automatic systems which are optimal or close to it cannot be constructed. Indeed, often even in simple cases the engineer's intuition is quite insufficient for finding optimal operating laws for control systems. Meanwhile optimal modes of operation for assemblies can provide a very great economic effect; in defense technology, criteria of quite another type are suitable, but there as well the value of optimal systems is evident. On the other hand, the theory of optimal systems permits the "ceiling" to be estimated, which can be achieved in the best, optimal system and permits it to be compared with the indices of the operating non-optimal system. This comparison permits clarifying whether, in the case under consideration, one should be concerned with the development of an optimal system, or one can be satisfied with the existing one.

2. Classification of Optimal Systems

In Fig. 1.3 the block diagram of an automatic control system is depicted. The controller is denoted by the letter A, and the controlled object by the letter B. The nature of the object can be arbitrary. For example, a rolling mill or a chemical reactor can play the part of a controlled object. A shop business or an individual motor can be the controlled object. The "controlled variable" x appears at the output of

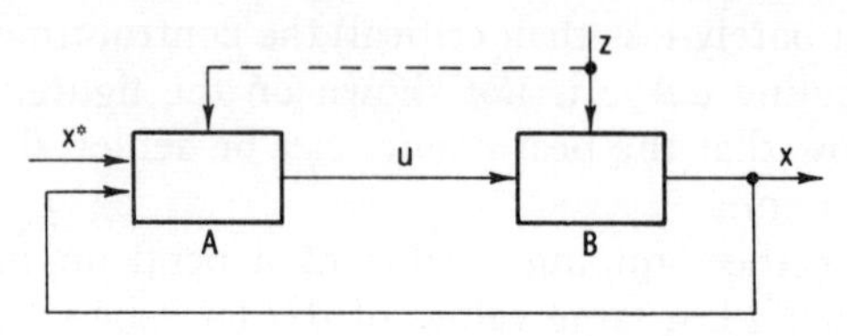

FIG. 1.3

the object B. By the controlled variable is meant the parameters characterizing the state of the controlled object. In the general case there are several such parameters $x_1, ..., x_n$. It is convenient to regard these variables as coordinates of a vector $\bar{x}$:

$$\bar{x} = (x_1, ..., x_n). \tag{1.1}$$

The vector $\bar{x}$ is also called the output vector or output variable of the object B.

The control action u of the controller A acts at the input to the object B. If there are several such actions $u_1, u_2, ..., u_r$, then they can be combined into the vector $\bar{u}$ with coordinates u_j $(j = 1, ..., r)$:

$$\bar{u} = (u_1, ..., u_r). \tag{1.2}$$

To the input of the controller A the "driving action" x^* is supplied, representing the instruction of what must be the output variable x of the object. This instruction must realize the purpose of control (the corresponding notion is made more precise in the next section). The instruction can be a collection of n variables $x_1^*, ..., x_n^*$, which we will regard as coordinates of a vector $\bar{x}^*$:

$$\bar{x}^* = (x_1^*, ..., x_n^*). \tag{1.3}$$

For example, it could be required that in the ideal case the conditions

$$x_i = x_i^* \qquad (i = 1, ..., n) \tag{1.4}$$

be satisfied, where the x_i^* are prescribed functions of time.

Automatic control systems are divided into two classes: open-loop and closed-loop systems. The latter are also called feedback systems. In open-loop systems the controller A does not receive information about the actual state of $\bar{x}$ of the object B. In closed-loop systems the controller A receives this information by the feedback path (at the bottom in Fig. 1.3). The principle of operation of a closed-loop system can be described briefly in the following way: if the quantity $\bar{x}$ does not conform to the demands $\bar{x}^*$, then the controller A renders such action $\bar{u}$ on the object B that $\bar{x}$ approaches these demands.

The deviation of the variable $\bar{x}$ from the demands can occur for various reasons:

(a) Improper, inaccurate or delayed use by the device A of the information contained in it or arriving at it about the characteristics and state of the object and about the purpose of control. This deficiency can in principle be corrected by improving the operating law (algorithm) of the controller A.

(b) A limitation in control resources, i.e., the impossibility, for that or other causes, of supplying to the object B those control actions $\bar{u}$ which would ensure the required behavior $\bar{x}$ of the object. In practice the control resources are always limited, and this circumstance must be taken into account.

(c) The cause of the deviation of $\bar{x}$ from the requirements could turn out to be some hitherto unforeseen and uncontrolled "perturbing action" z, acting on the object B and having an influence on its output variable $\bar{x}$. If the disturbances $z_1, ..., z_l$ act on various parts of the object B, then we will represent them in the form of a vector $\bar{z}$:

$$\bar{z} = (z_1, ..., z_l). \tag{1.5}$$

The perturbing action $\bar{z}$ is frequently called noise. Noise acting on the controlled object B can cause a hitherto unforeseen variation in its characteristics. The influence of load variation on the object can be considered as a special case of noise action.

We will assume that the algorithm of the controller A ensures the successful operation of the system for the defined characteristics of the object B. But with changes in them the operation of the system can deteriorate and the variable $\bar{x}$ begin to deviate considerably from the demands.

The principle of feedback creates in many cases the possibility of satisfying the demands presented to the variable $\bar{x}$, even in the presence of considerable noise $\bar{z}$ acting on the object B. However, if the characteristics of the object B are complex and vary rapidly over a wide range, then the control problem is made difficult. In such cases the obtaining of information about the noise $\bar{z}$ or even about some of its components $z_1, ..., z_{l'}$ $(l' < l)$ can render essential aid and improve the result of control. Let the noise be measured and the result of the measurement act (see the dashed line in Fig. 1.3) on the controller A. Then the latter can calculate and supply the control action $\bar{u}$ which compensates, neutralizes the influence of the noise $\bar{z}$, and will bring the output variable $\bar{x}$ of the object B into the best correspondence with the demands. This method bears the name of compensation. The compensation circuit in Fig. 1.1 is not a feedback path, since the value of the "input" is transmitted along it, and not the output variable of the object. Systems to which the principle of compensation is applied along with the feedback principle are sometimes called compound.

It should be noted that the range of application of the compensation principle is considerably narrower than the range of application of the feedback principle. This stems mainly from the fact that a large number of different noises $z_1, ..., z_l$ act on the object B. A significant part of these noises does not in general lend itself to measurement, and therefore cannot be compensated for with the aid of the circuit shown dashed in Fig. 1.3. Even if the fundamental possibility of measuring the set of noises z_i existed, then the calculation of the action $\bar{u}$ neutralizing them would be extremely complicated. Therefore the controller A would turn out to be too cumbersome, and the results of system operation would just be inadequately successful, since not all noise can be measured. Meanwhile the feedback principle permits just the deviation of the controlled variable $\bar{x}$ from the demands to be measured and permits the control action $\bar{u}$ to be formulated which brings $\bar{x}$ close to the required value. It is evident that the feedback principle is far more universal, and in general gives simpler control methods than the compensation principle. But in a number of cases, when the measurement of the basic perturbing effect is realized simply enough, the compensation method or its combination with the feedback principle proves to be more successful.

Usually the object B is given and its properties cannot vary. Meanwhile the algorithm of the controller A is not defined at all most of the time, and it can be chosen from a wide class of possible algorithms.†

† Frequently the high-capacity forcing part of the controller is prescribed; then it should relate to the controlled object and be regarded as part of it. Therefore sometimes the "controlled object" is replaced by the notion of the "invariable part of the system."

Thus the problem of constructing an optimal system reduces to the problem of developing a controller A which in a known sense controls the object B in the best way.

In practice a series of independent requirements not having a direct relation to the object B is usually presented to the device A. For example, it can be required that the device A be reliable enough, and also not too complicated. It can be required that its weight, size, or energy consumption not be too large. To facilitate the design or for other reasons the device A can be taken as linear or its block diagram even prescribed in advance, regarding only the parameters of the individual elements in it as unknown. But below our principal attention is turned to the consideration in which any requirements or limitations directly concerning the controller A are absent. We will assume that, if it is required, this device can be arbitrary—for example, as complex as desired, and also inertialess. Such an absence of limitations depends on the broad possibilities of modern computer technology (see the preceding section). Besides, the imposing of further limitations on the controller A can in general severely complicate the problem of finding the optimal system. Such a complication arises if, for example, it is required that the complexity or reliability, or the cost of the controller A not pass some upper bound. It is clear that if the choice of device A is limited to a known class of systems defined beforehand, or the circuit selected in advance and only its parameters regarded as unknown, then the problem is greatly simplified. But as a rule the value of its solution still diminishes to a large extent. Indeed, the very difficulty with the design of an optimum controller is this determination of a general form, a general structure for the algorithm. Unfortunately, here the engineer's intuition or mathematics can render aid only in the simplest cases and is powerless in any more complex ones. Therefore as a rule neither the general form of algorithm is known beforehand, nor even a restricted enough class of relations to which it belongs. Any unjustified *a priori* choice of a restricted class of relations therefore deprives the solution to the problem of the value which it would have in the absence of similar limitations.

If the limitations imposed on A are absent, then the algorithm of the optimal device A is determined only by the following factors, relating to the object B and the method of its combination with A:

(1) The characteristics of the object B;

(2) The demands presented to the object B;

(3) The character of the information about the object B entering the controller A.

A similar consideration of these factors is necessary with a view to a detailed statement of the problem. Each of the factors indicated above can be symbolically represented in the form of a certain direction orthogonal to the other, as shown in Fig. 1.4, and a point or region in such a three-dimensional space can be associated with each type of optimal system.

The directions indicated on Fig. 1.4 are the directions of classification for optimal systems [1.2]. Such a classification is useful in that it permits the place of every type of optimal system to be correctly determined among other types. The investigation of all possible types of optimal systems from general points of view reveals the unity of the basic aspects of the theory, in spite of essential differences in individual types of systems.

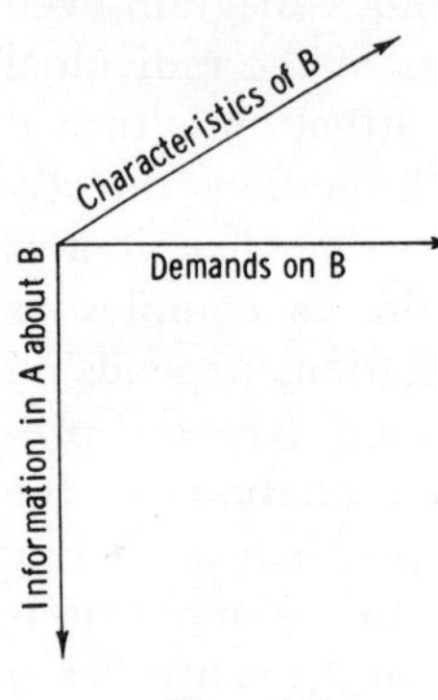

FIG. 1.4

The first direction indicated in Fig. 1.4 is the classification according to characteristics of objects. From Fig. 1.3 it is seen that the object B is characterized by the dependence of its output variable $\bar{x}$ on the input variables $\bar{u}$ and $\bar{z}$. We will represent this dependence symbolically in the following manner:

$$\bar{x} = \bar{F}(\bar{u}, \bar{z}). \tag{1.6}$$

In the general case the dependence $\bar{F}$ is an operator, i.e., a law of correspondence between two sets of functions. For example, in formula (1.6) the vector-function $\bar{x}$ depends on the form of the vector-functions $\bar{u}$ and $\bar{z}$. The operator $\bar{F}$ of the object can be prescribed in various ways—with the aid of formulas, graphs, or tables. Frequently this dependence is given in the form of differential equations, for example these:

$$\begin{aligned} dx_1/dt &= f_1(x_1, \ldots, x_n; u_1, \ldots, u_r; z_1, \ldots, z_l; t), \\ &\cdots \\ dx_n/dt &= f_n(x_1, \ldots, x_n; u_1, \ldots, u_r; z_1, \ldots, z_l; t). \end{aligned} \tag{1.7}$$

Here, in the general case, the f_i are nonlinear functions of $x_1, \ldots, x_n$; $u_1, \ldots, u_r$; $z_1, \ldots, z_l$, and time t. By introducing the vector notations

$$d\bar{x}/dt = (dx_1/dt, \ldots, dx_n/dt); \qquad \bar{f} = (f_1, \ldots, f_n), \tag{1.8}$$

Eq. (1.7) can be rewritten in a more compact and easily visible vector form:

$$d\bar{x}/dt = \bar{f}(\bar{x}, \bar{u}, \bar{z}, t). \tag{1.9}$$

In this expression the vector $\bar{f}$ is a vector function of the vectors $\bar{x}$, $\bar{u}$, $\bar{z}$ and the scalar t. For prescribed initial conditions, i.e., for a given vector $\bar{x}^{(0)} = (x_1^{(0)}, ..., x_n^{(0)})$, where

$$x_i^{(0)} = (x_i)_{t=0} \qquad (i = 1, ..., n), \tag{1.10}$$

Eq. (1.7) or (1.9) permits the vector $\bar{x}(t)$ to be found, if only the vector function $\bar{u}(t)$ is known and the noise $\bar{z}(t)$ given.

The operators of objects can be classified in very different ways. Here we will be concerned only with certain important lines of classification. First of all, we will consider the division into continuous, discrete-continuous, and discrete systems. In the first of these types of systems the variables are examined at any instant of time, where these variables can vary continuously and, in principle, their level can be arbitrary (the solid curve in Fig. 1.5). Thus, according to the definition accepted in communication and control engineering, these variables are not quantized in time or in level. Such, for example, are the solutions $x_1, ..., x_n$ of Eq. (1.7), which are functions of continuous time t. But not all types of systems are characterized by variables defined at any instant of time. In impulse and digital control systems, and also for the application of impulse modulation to the transmission of signals, only values of the variables at discrete instants of time are of interest. Here if arbitrary levels of the variables are permitted, then this means that the latter are quantized in time, but are not quantized in level. The corresponding systems are called discrete-continuous. The operator of a discrete-continuous system can be defined, for example, by finite difference equations.

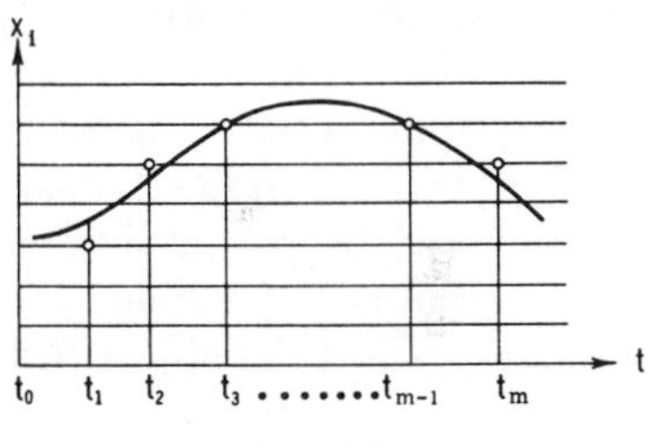

FIG. 1.5

We will denote by $x_i(m)$ the value of the variable x_i at the instant of time $t = t_m$ (see Fig. 1.5). Further, let us denote by

$$\bar{x}(m) = [x_1(m), ..., x_n(m)] \tag{1.11}$$

the vector $\bar{x}$ at the instant $t = t_m$, and by $\bar{u}(m)$ the vector $\bar{u}$ at $t = t_m$. Then the finite difference equations connecting subsequent values $x_i(m+1)$ with the preceding $x_i(m)$ can be written in the following form:

$$\begin{aligned} x_i(m+1) = g_i[&x_1(m), ..., x_n(m); \quad u_1(m), ..., u_r(m); \\ &z_1(m), ..., z_l(m); \quad m] \qquad (i = 1, ..., n), \end{aligned} \tag{1.12}$$

where in general the g_i are nonlinear functions of their arguments. Equations (1.12) can be written in the vector form

$$\bar{x}(m+1) = \bar{g}[\bar{x}(m); \bar{u}(m); \bar{z}(m); m]. \tag{1.13}$$

Here $\bar{g}$ is the vector with components $g_1, ..., g_n$.

In the third type of system only definite discrete levels of the variables are allowed (the grid of allowed levels is shown in Fig. 1.5). For example, a case is possible when $x_i(m) = aq$, where a is a constant and q an integer. Then the value of the variable x_i is represented by one of the allowed levels (circles in Fig. 1.5). Systems in which the variables are quantized both in time and in level are called discrete (or purely discrete). The operator of a discrete system can be characterized, for example, by Eq. (1.12); but all the variables in these equations must have only the allowed levels. In particular, the functions g_i as well can assume only the values permitted for the x_i.

A very large number of papers have been devoted to continuous optimal systems (see, for example, [3.2, 3.14, 3.16–3.18, 3.21–3.23]). Discrete-continuous optimal systems were examined in [3.19, 3.20, 3.24, 3.30, 3.31]. Paper [4.12] is devoted to purely discrete optimal systems.

Objects can also be distinguished according to the types of their equations. In a majority of papers on optimal systems objects are studied with lumped parameters, whose motion is characterized by ordinary differential equations. But a series of problems was posed in [3.42] and solved in [3.43, 3.44] for objects with distributed parameters, characterized by partial differential equations and integral equations.

"Constraints" of various forms likewise enter into a number of the characteristics of the object B. For example, the control actions $u_1, ..., u_r$ in the composition of the vector $\bar{u}$ (see Fig. 1.3) cannot have arbitrary values. On account of the physical properties of the object they cannot or—we will assume, because of the hazard in violating the normal operation of the object—must not exceed certain limits.

Very frequently the constraints have the form

$$|u_1| \leqslant U_1, ..., |u_r| \leqslant U_r, \tag{1.14}$$

where $U_1, ..., U_r$ are prescribed constants. A case is possible when the functions are bounded by certain control actions, for example:

$$\sum_{\nu=1}^{r} \lambda_\nu^2 u_\nu^2 \leqslant N, \tag{1.15}$$

where the λ_ν^2 and N are constants or prescribed time functions.

Let us consider the r-dimensional space of the vector $\bar{u}$ with Cartesian coordinates $u_1, \ldots, u_r$. Conditions (1.14) and (1.15) are special cases of the conditions restricting the location of the end of the vector $\bar{u}$ to a certain admissible region $\Omega(\bar{u})$ of this space. The expression "$\bar{u}$ belongs to the region $\Omega(\bar{u})$" is written symbolically in the following form:

$$\bar{u} \in \Omega(\bar{u}). \tag{1.16}$$

In the special case of conditions (1.14) the vector $\bar{u}$ is bounded by an r-dimensional parallelepiped; in the case of conditions (1.15), by an r-dimensional ellipsoid in $\bar{u}$-space.

Constraints can be imposed not only on the control actions u_j, but also on the coordinates x_i $(i = 1, \ldots, n)$ of the object B. For example, any given functions or functionals $H_\mu(\bar{x})$ of these coordinates must not exceed certain limits, which without restricting generality can be taken equal to zero:

$$H_\mu(x_1, \ldots, x_n) = H_\mu(\bar{x}) \leqslant 0 \qquad (\mu = 1, \ldots, m). \tag{1.17}$$

The functions or functionals $H_\mu(\bar{x})$ can be regarded as coordinates of an m-dimensional vector $\bar{H}(\bar{x})$. Conditions (1.17) impose constraints on the location of this vector. If the H_μ are single-valued functions of $\bar{x}$, then conditions (1.17) mean that the vector $\bar{x}$ is also bounded in n-dimensional $\bar{x}$-space by some admissible region $\Omega(\bar{x})$:

$$\bar{x} \in \Omega(\bar{x}). \tag{1.18}$$

In a more general case certain functionals L of $\bar{u}(t)$, $\bar{x}(t)$, and $\bar{z}(t)$ are bounded, i.e., quantities depending on the form of the functions $\bar{u}$, $\bar{x}$, $\bar{z}$ on some interval:

$$L_\mu[\bar{u}(t), \bar{x}(t), \bar{z}(t)] \in \Omega_\mu(L) \qquad (\mu = 1, \ldots, m), \tag{1.19}$$

where $\Omega_\mu(L)$ is the admissible domain of variation of the functional L_μ. A bound of the form

$$L = \int_0^T \left(\sum_{\nu=1}^{n} \alpha_\nu x_\nu^2 + \beta u^2\right) dt \leqslant N \tag{1.20}$$

can serve as an example, where T, α_ν, β, and N are positive constants.[†] In discrete-continuous or discrete systems analogous variables are bounded.

[†] In the formula for the constraint time t can also be introduced in explicit form.

Constraints are extremely important in the design of control systems. We will explain this statement by an example. Let it be required to construct a constant-current tracking system with minimal time for the transient process. In principle, by feeding arbitrarily large voltages to the input of the armature circuit of the servomotor, currents as large as desired, and arbitrarily large moments and accelerations of the motor shaft can be obtained in this circuit, which ensures as small a time for the transient process as desired. But only those processes are allowed in which the armature current and likewise the speed of the servomotor shaft will not fall outside certain definite limits. In particular, this circumstance does not permit the time of the transient process to be reduced without limit (see, for example, Fig. 1.2). In general, the existence of constraints gives meaning to the optimal system problem in many cases. The solution to this problem must answer the question: how are the best results obtained with limited resources?

The characteristics of the disturbance z acting on the object from an external medium can also be attributed to the characteristics of the object B—see Fig. 1.3 and Eqs. (1.7) and (1.9). Sometimes the characteristics of the disturbance are included in the composition of the object's operator. If the z_ν are known time functions ($\nu = 1, \ldots, l$), then their expressions can be substituted into the equations of the object B, for example, into Eq. (1.7) or (1.12). Then these equations will depend explicitly on time. For methodological purposes it is convenient to regard the unforeseen disturbances z_ν as external actions, applied from without to the object, while all the disturbances assumed to be known are included in the composition of the operator $\bar{F}$.

By being added to the other actions, for example, to the u_j, the disturbances z_ν can act on the inputs to the elements of the object B. These actions are called additive. But the z_ν can also act in a different way, by changing the coefficients of the equations of the elements or their parameters. Such actions are called parametric. In general, a clear distinction between these two types of actions is lacking in nonlinear systems.

The disturbances z_ν can be random variables or random processes, which are considered in Chapter II. In the first case the z_ν can be regarded as constant during one individual process in the system; in the second case the z_ν are random time functions, whose variation during the time of a single process cannot be neglected.

Sometimes the random disturbances z_ν do not appear explicitly in the conditions of the problem. But if $\bar{z}$ is random, then according to (1.9) for a given $\bar{u}$, the output variable $\bar{x}$ of the object B will be a random process. Instead of the characteristics of $\bar{z}$ the conditional probabilistic characteristics of the process $\bar{x}$ can be assigned directly, depending on $\bar{u}$ and the initial conditions $\bar{x}^0$, which at once replaces the representation of the operator $\bar{F}$ and the characteristics of the random disturbance $\bar{z}$. The object has been defined in such a manner, for example, in [4.12].

3. Optimality Criteria

The demands made on the behavior of the object B represent the second direction of classification of optimal systems (see Fig. 1.4). The assignment of a definite "control purpose" enters into the composition of these demands. In any case, the control purpose can be considered as the attainment of an extremum of some quantity Q—the optimality criterion. Either a maximum or a minimum of the quantity Q is necessary, depending on the demands. In the general case an optimality criterion depends both on the driving action $\bar{x}^*$ and on the output variable $\bar{x}$; it can also depend on $\bar{u}$ and $\bar{z}$, as well as on time t. For definiteness let it be required that the quantity Q be minimal:

$$Q(\bar{x}, \bar{x}^*, \bar{u}, \bar{z}, t) = \min. \tag{1.21}$$

This condition is the analytical formulation of the control purpose. We will note that Q is a functional, i.e., a number depending on the form of the functions $\bar{x}$, $\bar{x}^*$, $\bar{u}$, $\bar{z}$. For example, in a special case Q has the form

$$Q = \int_0^T [x(t) - x^*(t)]^2 \, dt, \tag{1.22}$$

where T is fixed. From formula (1.22) it is clear that Q depends on the form of the functions $x(t)$ and $x^*(t)$ on the interval $0 < t < T$.

Various engineering and economic indices can be selected as the criterion Q—for example, the output of an object or the quality of production, or the expenditure of raw materials or electrical energy, and so forth. The basis for choosing the optimality criterion Q, determined by specific engineering-economic conditions, is outside the scope of the theory of optimal systems and is not considered in this theory.

From formula (1.21) for Q not only can the minimal value $Q_{\min}$ be recognized, but also the degradation in system operation estimated by its deviation from the ideal. A measure of deterioration can be the difference $Q - Q_{\min}$ or some monotonic function of this difference, vanishing for $Q = Q_{\min}$.

Different lines of classification are possible according to the types of criteria Q. Thus, optimality criteria can be separated depending on whether they relate to a transient or steady-state process in the system. For an example let us consider integral criteria for processes in linear systems. Let the motion of some linear system with input variable x^* and output variable x (Fig. 1.6) be characterized by a linear differential equation with constant coefficients, relating the input variable x^* to the output x:

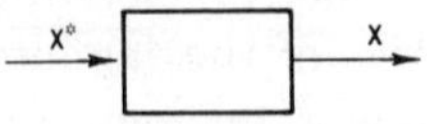

FIG. 1.6

$$a_0 \frac{d^n x}{dt^n} + a_1 \frac{d^{n-1} x}{dt^{n-1}} + \cdots + a_n x = b_0 \frac{d^m x^*}{dt^m} + \cdots + b_m x^*. \tag{1.23}$$

As is known, the solution of this equation has the form

$$x(t) = x_s(t) + x_d(t), \tag{1.24}$$

where $x_s(t)$ is a particular solution of the equation with a right-hand side, and $x_d(t)$ is the general solution of the equation without a right-hand side:

$$a_0 \frac{d^n x_d}{dt^n} + a_1 \frac{d^{n-1} x_d}{dt^{n-1}} + \cdots + a_n x_d = 0. \tag{1.25}$$

The physical meaning of formula (1.24) is that under definite auxiliary conditions $x_s(t)$ represents the steady-state process in the system, and $x_d(t)$ the transient process. If the system is stable, which will be assumed later on as well, then

$$x_d(t) \to 0 \quad \text{as} \quad t \to \infty. \tag{1.26}$$

In order to find the expression for $x_d(t)$, it is necessary, as is known, to solve as a preliminary the characteristic equation of the system

$$a_0 p^n + a_1 p^{n-1} + \cdots + a_n = 0, \tag{1.27}$$

and to find its roots $p_1, p_2, \ldots, p_n$. Then, by assuming all roots to be distinct without restricting generality, we will obtain

$$x_d(t) = C_1 e^{p_1 t} + \cdots + C_n e^{p_n t}, \tag{1.28}$$

where the constants C_i $(i = 1, \ldots, n)$ are determined from the initial conditions

$$(d^k x_d/dt^k)_{t=0} = (d^k x/dt^k)_{t=0} - (d^k x_s/dt^k)_{t=0}\,. \tag{1.29}$$

In order to clarify the character of the transient process, it is necessary to solve the characteristic equation (1.27), and, after finding its roots, to construct the graph of $x_d(t)$ according to Eq. (1.28). But the character of the solution can be determined more simply by calculating, for example, the integral:

$$I_1 = \int_0^\infty x_d(t)\, dt. \tag{1.30}$$

This integral is determined in general form as a function of the coefficients of Eq. (1.25) and the initial conditions without the need for finding the function $x_d(t)$ beforehand. If $x_d(t)$ is invariable in sign, for example, $x_d(t) > 0$ for any $t \geqslant 0$, then in general a decrease in the integral I_1 corresponds to an acceleration of the transient process. Therefore sometimes I_1 is taken as a criterion of "quality" for the transient process. But for processes with a change in the sign of $x_d(t)$, it can turn out that even a weakly damped process having a sharply oscillatory character possesses a small value of I_1.† Therefore the range of applicability of the criterion I_1 is limited. In [1.3] another criterion has been proposed:

$$I_2 = \int_0^\infty x_d^2(t)\, dt. \tag{1.31}$$

By selecting the parameters or algorithm of the controller A with the aim of minimizing the integral I_2, a satisfactory character for the transient process can frequently be achieved. The criterion I_2 has been applied to automatic control systems [1.4, 1.5]. But frequently the application of this criterion nevertheless brings an unnecessary oscillatory character to the transient process [1.6]. Therefore in [1.6] the so-called generalized integral criterion was proposed:

$$I_V = \int_0^\infty V\, dt, \tag{1.32}$$

where V is a quadratic form of the transient components x_{di} of the coordinates $x_1, \ldots, x_n$ of the system:

$$V = \sum_{i,j=1}^{n} a_{ij} x_{di} x_{dj}\,. \tag{1.33}$$

We will explain the geometrical meaning of the generalized integral

† Sometimes, however, the criterion I_1 can be used even when the processes have a clearly exhibited oscillatory character. See the paper by A. A. Voronov in *Automation and Remote Control*, No. 6, 1963 (Russian).

criterion by the simplest example, in which x_{d1} is the transient component of the error in the system, $x_{d1} = x_1$, $dx_{d1}/dt = x_2$. Let

$$I_V = \int_0^\infty (x_1^2 + T^2 x_2^2)\, dt = \int_0^\infty [x_{d1}^2(t) + T^2(dx_{d1}/dt)^2]\, dt, \qquad (1.34)$$

where $T = \text{const}$.

By choosing the parameters of the system such that the integral I_V be minimized, we suppress the prolonged existence of significant deviations x_{d1} (otherwise the component $\int_0^\infty x_{d1}^2\, dt$ of the integral I_V will be large), but likewise we suppress the prolonged existence of large values of the derivatives dx_{d1}/dt [otherwise the component $\int_0^\infty (dx_{d1}/dt)^2\, dt$ of the integral I_V will be large]. Thus, not only a quick, but also a smooth transient process without sharp oscillations is obtained.

The integral I_V differs from I_1 and I_2 principally in that it gives the possibility of making a rigorous judgement about the character of the transient process, according to the magnitude of I_V. This question is considered in greater detail in Chapter II.

Criteria (1.30)–(1.32) are used to estimate the transient process $x_d(t)$. To estimate the steady-state process $x_s(t)$ criteria of another type are used, for example,

$$x_{\text{mn sq}}^2 = \lim_{T\to\infty} \frac{1}{T} \int_0^T x^2(t)\, dt = \lim_{T\to\infty} \frac{1}{T} \int_0^T x_s^2(t)\, dt$$

$$+ \lim_{T\to\infty} \frac{1}{T} \int_0^T x_d^2(t)\, dt + \lim_{T\to\infty} \frac{2}{T} \int_0^T x_s x_d\, dt. \qquad (1.35)$$

The second term on the right side of (1.35) is equal to zero, since the integral $\int_0^T x_d^2(t)\, dt$ remains finite as $T \to \infty$. It is not difficult to see that the last term also vanishes. Therefore only the first term remains, corresponding to the steady-state process.

Other types of optimality criteria for transient and steady-state processes have been described in [1.7–1.9].

Frequently the control time or the magnitude of the maximal deviation of the process from some given value or time function is considered as the optimality criterion. In the latter case it is required that in an optimal system the minimum of the maximal deviation be attained, the so-called minimax.

It is important to emphasize that we cannot pose the problem of simultaneously achieving an extremum for "two" or more functions of one or several variables. In fact, generally the extrema for different functions or functionals do not correspond to the same value of the collection of arguments. Therefore in the general case there do not exist values of the arguments corresponding simultaneously to the extrema of two or more functions and functionals. Only the problem of attaining the extremum of a "single" function or functional can be posed, but here any number of auxiliary conditions can be imposed for restricting the other functions or functionals. These same constraints can bear a complicated character. For example, a choice of the vector $\bar{x}$ can be required such that the function $Q_1(\bar{x})$ attain a minimum, but here the values of the other functions $Q_2(\bar{x})$ and $Q_3(\bar{x})$ will not deviate in percentage ratio from their own extrema by more than ϵ_2 and ϵ_3, respectively. The problem of the existence of a value $\bar{x}$ satisfying these conditions can only be solved by considering a specific system.

Sometimes combined criteria are applied. Let us assume that a vector $\bar{x}$ must be chosen such that the function $Q_1(\bar{x})$ be minimal, while $Q_j(\bar{x}) \leqslant 0$ $(j = 2, ..., m)$. The latter inequalities bound a certain admissible region in the space of the vector $\bar{x}$, outside whose limits it is impossible to fall. But formally the constraints can be removed by applying the criterion

$$Q(\bar{x}) = Q_1(\bar{x}) + \sum_{j=2}^{m} \beta_j(Q_j)Q_j(\bar{x}), \tag{1.36}$$

where the functions β_j have the form

$$\beta_j = \begin{cases} 0 & \text{for} \quad Q_j \leqslant 0 \\ \gamma^2 \geqslant 1 & \text{for} \quad Q_j > 0 \end{cases} \qquad (j = 2, ..., m). \tag{1.37}$$

If the quantity γ^2 is sufficiently large, then the minimum point of the function Q either coincides with the minimum of Q_1, if the latter is inside the admissible region, or it lies practically on its boundary, not falling outside its limits. The functions $\beta_j(Q_j)$ can also be constructed in the form $(1 + Q_j)^{\alpha_j}$, where the numbers $\alpha_j \gg 1$. But usually the construction of formula (1.36) considerably complicates the analytical investigation. Large values of the coefficients can be avoided if formula (1.36) is replaced by the following:

$$Q(\bar{x}) = \beta_1(Q_2, ..., Q_m) \cdot Q_1(\bar{x}) + \sum_{j=2}^{m} \beta_j(Q_j)Q_j(\bar{x}), \tag{1.38}$$

where

$$\beta_1(Q_2, ..., Q_m) = \begin{cases} 1, & Q_j \leqslant 0 \quad (j = 2, ..., m), \\ 0, & \text{if just one of the } Q_j > 0; \end{cases}$$
$$\beta_j(Q_j) = \begin{cases} 1, & Q_j > 0, \\ 0, & Q_j \leqslant 0. \end{cases} \tag{1.39}$$

Here the construction of formula (1.38) is still more complicated. But for machine solution this formula is acceptable and is applied in such a form to certain automatic optimizers [3.25, 6.6].

Depending on the character of the optimality criterion the following types of optimal systems can be distinguished:

(a) uniformly optimal systems,

(b) statistically optimal systems,

(c) minimax optimal systems.

In the first type of system each individual process is optimal. For example, whatever the initial conditions or the driving actions would be (here the latter must belong to some prescribed class of admissible actions), in systems which are optimal in speed of response (see, for example, [3.1–3.24]) the state of the object arrives at that demanded in minimal time. And so, any uniformly optimal system copes with its own problem in the best way for each individual case.

In the second type of system it is not required nor is it possible to ensure the best system behavior for each individual process. In this type of system the optimality criterion Q has a statistical character. Such systems must be best on the basis of the average. Statistical criteria are applied to systems in which random factors are present in one form or another. A simple particular example is the problem of choosing the parameters $a_1, \ldots, a_k$ of a controller A whose circuit is given. We will assume that the primary criterion of quality is some scalar function

$$Q_1 = Q_1(a_1, \ldots, a_k; x_1^{(0)}, \ldots, x_n^{(0)}) = Q(\bar{a}, \bar{x}^0). \tag{1.40}$$

Here $\bar{x}^{(0)}$ is the vector of the initial conditions $x_i^{(0)}$ $(i = 1, \ldots, n)$ of the object B, while $\bar{a}$ is the parameter vector with coordinates a_j $(j = 1, \ldots, k)$.

The criterion Q_1 cannot be used directly for selecting the parameters a_j, since the values of $\bar{a}$ which are best for one set of initial conditions $\bar{x}^{(0)}$ generally do not prove best for another type. But if the *a priori* probability density $P(\bar{x}^{(0)})$ is known for the initial condition vector, then Q, the mathematical expectation, can serve as the criterion, or as is accepted in physics and engineering for designating it, the "mean value" of the quantity Q_1. We will denote by the letter M the mathematical expectation. Then according to Chapter II,

$$Q = M\{Q_1\} = \int_{\Omega(\bar{x}^0)} Q_1(\bar{a}, \bar{x}^0) P(\bar{x}^0)\, d\Omega(\bar{x}^0). \tag{1.41}$$

In this formula $\Omega(\bar{x}^{(0)})$ is the domain of variation of the vector $\bar{x}^{(0)}$, and $d\Omega(\bar{x}^{(0)})$ its infinitesimally small element.

As is known, the physical meaning of this estimate is the fact that for the bulk of trials which are carried out with the system, the quantity Q practically coincides with the arithmetic mean of the values of Q_1 obtained for each of the trials.

For such a formulation of the problem we will regard as the optimal system the one whose parameters a_i ensure a minimum for the quantity Q (in the general case with still additional constraints taken into account).

In the example considered the random initial values of the transient process can be represented as the result of the action of short random impulses on the object at the initial instant. Thus in this case there is the special case of the action of random noise $\bar{z}$ on the object. Another simple example can also be given of a system with a primary optimality criterion of the type $Q_1(\bar{x}, \bar{x}^*, \bar{u}, \bar{z})$, where the noise $\bar{z}$ is a random variable with probability density $P(\bar{z})$. Then as the optimality criterion the mean value Q of the quantity Q_1 can be chosen:

$$Q = M\{Q_1\} = \int_{\Omega(\bar{z})} Q_1(\bar{x}, \bar{x}^*, \bar{u}, \bar{z}) P(\bar{z})\, d\Omega(\bar{z}), \tag{1.42}$$

where $\Omega(\bar{z})$ is the domain of variation of the noise vector $\bar{z}$, and $d\Omega(\bar{z})$ its infinitesimally small element.

Optimal systems with statistical optimality criteria have been considered in a number of books (see, for example, [1.10–1.13]). This question is presented in greater detail in Chapters IV–VI.

Most frequently the mean value of some primary criterion appears as the statistical criterion. In some papers the statistical criteria are the probabilities of the variables x_i falling outside certain prescribed limits or the probabilities of an accident in the system.

Systems of the third type, called minimax optimal, ensure the best result compared to another system only in the worst case. In other words, the worst result in a minimax optimal system is better than the worst result in any other system. Such a formulation of the problem properly occurs in the case of the absence of *a priori* probability distributions. It will be considered in Chapter V.

The character of the requirements for the system is determined to a considerable extent by the form of the function $\bar{x}^*$ in formula (1.21) for Q. Everything unforeseen in advance for the control purpose (if the effect of the noise $\bar{z}$ is not considered) enters into this function, called the driving action, as stated above. If $\bar{x}^*$ is a regular function known in advance, then it can be included in the composition of the functional Q, and it will not appear in explicit form. But in practice a great deal is often not determined in advance for the control purpose.

For example, such is the situation for a servomechanism with tracking as its aim, whose future motion is unknown. In such a system a small value of the difference $[x^*(t) - x(t)]$ must be maintained, where $x^*(t)$ is a random time function not known beforehand. If the optimality criterion Q_1 depends on the random function, then most often the mathematical expectation $M\{Q_1\} = Q$ is taken as the best possible criterion.

4. The Information Input into the Controller about the Controlled Object

An important direction of classification is the division according to the character of the information about the controlled object B entering the controller A or accumulated in it until the beginning of the control process.

First, the division into systems with complete or partial information about the object must be introduced. From what was stated above (see also Fig. 1.3) it is clear that the information about the controlled object is combined from:

(a) Information about its operator, i.e., about the dependence $\bar{F}$ (1.6);

(b) Information about the disturbance $\bar{z}$ acting on the object B;

(c) Information about the state of the object B, for example, about all the variables $x_1, \ldots, x_n$ of the object, whose motion is characterized by Eqs. (1.7);

(d) Information about the control purpose, i.e., about the functional Q (1.21);

(e) Information about the driving action $\bar{x}^*$.

Complete information about any function means an absolutely precise knowledge of it. Thus, complete information about some time function $f(\tau)$ means that it is known, or in case of need its exact value can be determined at any time instant for $-\infty < \tau < \infty$. For example, if the function $f(\tau)$ is given by a formula, then complete information about the function means that all the coefficients in the composition of the formula are defined. If all the forms of information about the object B mentioned above are known in advance for the controller or are provided by the current information acting on it, then the system under consideration is a system with complete information about the object. A particular case is when *a priori* information is put into the controller A about the operator $\bar{F}$ of the object in the form of Eq. (1.7) and about the control purpose Q, when $\bar{z}$ and $\bar{x}^*$ are known beforehand and current information about the state $\bar{x}$ of the object is introduced into the controller A by the feedback circuit. In this case we have a system with complete information about the object B. In fact, whenever $\bar{z}$ and $\bar{F}$ are known, then for a given state $\bar{x}$ and action $\bar{u}$ controlled by the controller A, the entire behavior of the object B is predetermined for the future. But the problem cannot, by far, always be idealized so as to allow us to consider that there exists complete information about the object. For example, in open-loop systems the feedback circuit is missing, and hence information about the actual state of the object does not enter the controller.

Actually, the information about the object cannot be regarded as complete in any automatic control system, and frequently the lack of one or another form of information is very substantial. Let us return again to Fig. 1.3; it indicates a number of channels through which uncertainty penetrates into the automatic system. First, there is the channel of the driving action $\bar{x}^*$, which in many cases is not known in advance. We will assume that it is required to ensure the equality $\bar{x} = \bar{x}^*$. A requirement of this type is encountered, for example, in problems about the pursuit of a "dog" after a "hare." Various automatic machines, e.g., rockets, can play the part of the "dog" and "hare." What is the optimal behavior of the "dog" pursuing the "hare"? It is clear that he must run to intercept the "hare," and for this it is necessary to predict the future trajectory of the "hare" based on the analysis of its behavior in the past. But no prediction can be completely accurate. Hence, here is a channel through which uncertainty penetrates, as do the statistical methods of approach connected with it. Another channel is the noise $\bar{z}$, representing random variations in the characteristics of the object B unpredictable in advance, which most often do not lend themselves to direct measurement. Likewise, here the uncertainty and randomness account for the necessity of applying statistical methods. This type of uncertainty frequently proves to be the most important. In a majority of cases just the presence of uncertainty of the types mentioned above causes the necessity for complex forms of control. If everything about the object were known in advance, then an open-loop system could be realized in which the controller A is provided with a control program developed beforehand. In this case the necessity would hardly arise in general in the special theory of control. Among other things, in such a hypothetical system there would be no need for a feedback circuit.

However, the feedback circuit, being a powerful means for increasing the noise stability of a system, is itself a channel through which new forms of noise penetrate into the system. It is necessary to transmit along this circuit to the controller A data about the state of the object B, for example, the coordinates $x_1, ..., x_n$ of the object with Eq. (1.7), or the quantities equivalent to them, x, dx/dt, ..., $d^{n-1}x/dt^{n-1}$, for the object, characterized by Eqs. (1.23). Often the variable x can be measured with sufficient accuracy. But either the first and particularly the higher derivatives cannot be measured at all, or, if measurement is possible, the data are obtained with significant error. Here the repeated differentiation of the function $x(t)$ cannot help. In fact, on the one hand all differentiators create characteristic errors. On the other hand, small high-frequency noise, inevitably adding to the function $x(t)$, creates large distortions in the derivatives, generally larger the larger the order of the derivative.

Hence the data about the state of the object are obtained by the controller A with errors, sometimes very significant ones. Thus in practice the information about the state of the object is incomplete. An equivalent scheme can be conceived in which the data about the state of the object penetrate to the controller, by passing beforehand through some channel with random noise and being mixed with it. This is the third channel through which uncertainty penetrates into the system.

It should be pointed out that random noise exists inside any actual controller A. In continuous-acting devices many factors play this role—amplifier drift and noise, induction noise, and so on. In discrete-acting devices, for example, digital ones, round-off errors play this role, arising because of the presence of a finite number of digits in those numbers which are the results of elementary operations. But in what follows we will neglect errors in the controller. First, in the majority of cases these errors can be made small. Second, their calculation would greatly complicate the theory. Third, a rough calculation of controller errors can be performed in a number of cases by reducing them to its input or output. In these cases the controller errors become identical to the errors in the feedback channel or in the channel of the control action (or in the object B).

The incompleteness of the information about the object gives rise to the necessity for the study in time of the control process itself. Therefore in the general case the controller in an automatic system solves two problems that are closely related but different in character. First, on the basis of the arriving information, it ascertains the characteristics and state of the controlled object B. Second, on the basis of the data obtained on the object it determines what actions are necessary for successful control. The first problem is that of studying the object; the second is that of bringing the object to the required state. In the simplest types of systems the solution of one of these problems can be lacking or may be produced in primitive form. In complex cases the controller must solve both of the stated problems.

An analogy can be drawn between the operation of the controller and a man who interacts with his environment. The man studies these surroundings in order to influence them in a direction useful to himself. But in order to direct his own actions better, he must understand his environment better. Therefore sometimes he acts on the environment not in order to obtain a direct advantage, but only with the aim of understanding it better. Thus the influence on the environment and the study of it are closely linked to each other.

The process of studying the controlled object B has a definite similarity to the process of a man obtaining new information and knowledge. This information can be obtained in ready form, as from another man; it can be gained by observation; it can be found as a result of experiment. Optimal systems can also be separated into three types, corresponding to these methods of obtaining information:

(a) Optimal systems with complete information about the controlled object or with maximal possible information (see Chapters III and IV);

(b) Optimal systems with partial information about the object and its independent (or passive) storage in the control process (see Chapter V);

(c) Optimal systems with partial information about the object and its active storage in the control process (dual control) (see Chapter VI).

Below we assume that complete *a priori* information is established in the controller A about the operator $\bar{F}$ of the object and about the control purpose, i.e., about the functional Q. If, further, there is complete information about the driving action $\bar{x}^*$ (i.e., complete information about this variable in the past *and* in the present and future), complete information about the noise $\bar{z}$ (including, therefore, a precise knowledge of its future), and finally, complete "current" information about the state $\bar{x}$ of the object at a given instant of time t (and this information gives the possibility of finding out for a prescribed $\bar{u}(t)$ the complete behavior of the object in the future), then we will call such a system one with complete information (in the controller) about the controlled object.

Until recently the development of theories of systems with complete information about the object proceeded quite independently of and isolated from the development of another group of theories, appearing roughly simultaneously with the first group. In the second group of optimal system theories the same influences are not known *a priori*, but only the statistical characteristics of the random input actions. Thus there is not complete information about the object here. The basic problem under consideration in theories of the second group relates to the system whose block diagram is shown in Fig. 1.7a. From the beginning every system under consideration is taken as some filter F. The driving action x^* does not act on this filter directly, but through a communication channel, or generally through some prescribed system H^*, where it is mixed with a random fluctuation or noise h^*. Thus the mixture y^* of signal and noise is fed to the input of the filter F. The problem of the filter consists of delivering to the output a variable x which is closest, in some statistical sense, to x^* or to the result of some known transformation of x^*.

After solving the problem stated above, we can solve the subsequent problem of partitioning the filter F into the object B and the controller A, usually without relating to the former. These parts of F can be connected in succession, as in Fig. 1.7b, by forming an open-loop system, or by some other means. For example, in Fig. 1.7c there is an internal feedback circuit in the filter F. Usually the object B is prescribed beforehand,

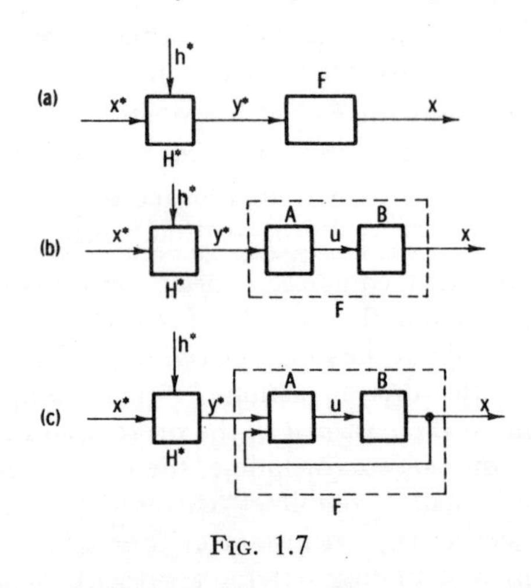

FIG. 1.7

and the algorithm of the controller A must be determined. If the optimal filter F is determined from the class of linear systems, then its subsequent partitioning into parts A and B gives rise to only comparatively small and usually not fundamental difficulties.† But if the optimal filter F is nonlinear, then the partition problem is incredibly complicated. In

† Nevertheless, in this case as well the problem is not always solved.

this case it is still suitable up to the start of design to prescribe the block diagram of the filter F for which the algorithm of part A is defined. But in the field of investigation of nonlinear filters, only the first steps have been taken for the present (see [1.10, 1.11, 1.17, 1.18]). On the basis of the bulk of papers in this field, starting with the classical investigations of A. N. Kolmogorov [1.14] and N. Wiener [1.15], the problems of obtaining optimal linear systems have been posed and solved (see, for example, [1.10–1.13, 1.16, 1.20–1.26]).

The second group of optimal systems mentioned above is characterized by the fact that the process of storing information about the action x^* does not depend on the algorithm, or in other words on the strategy of the controller A. Actually, the storage of information consists of "observing" the values of y^* and constructing a hypothesis about the process x^* according to them. By itself the observation process does not depend on the decisions taken by the device A about the character of the process x^*. The information obtained from observation can only be used properly, but it cannot be increased no matter what the strategy of the controller is. Therefore such systems can be called optimal systems with passive or independent (of the strategy of the controller) information storage. It is useful to distinguish the following variants in prescribing the characteristics of the action x^*, if it is irregular.

(a) $x^*(t)$ is a function belonging to some known class, for example:

$$x^*(t) = \sum_{i=1}^{\nu} C_i \psi_i(t), \tag{1.43}$$

where the $\psi_i(t)$ are known functions, and the C_i are random variables with given probabilistic characteristics. Further, $h^*(t)$ is a random function whose probabilistic characteristics are known. In such a case, the longer the observation of the variable y^* at the output of the channel H^* is carried out, the more accurately the future behavior of $x^*(t)$ can be predicted, for example, on the basis of refining the values of the coefficients C_i in formula (1.43).

(b) $x^*(t)$ is a random function, for which probabilistic characteristics are prescribed such that an exact knowledge of its past only gives the possibility of making predictions on its future. In this case these predictions can only be refined by observing $y^*(t)$ and refining the past and present values of $x^*(t)$ from the results of the observations. But the prediction of the behavior of the function $x^*(t)$ can never become as accurate as desired.

(c) $x^*(t)$ is a random function with unknown probabilistic characteristics, completely or partially. Here the problem reduces to that of ascertaining or refining the probabilistic characteristics of $x^*(t)$ by observing the variable $y^*(t)$, which permits the prediction of the behavior of $x^*(t)$ in the future to be refined (see, for example, [5.32, 5.33]).

In papers [6.1, 6.2] certain problems of a third group of optimal system theories are considered. This group has features similar to the theories of the first and second groups. But it also has its own special features, not inherent in the theories of the first two groups.

The block diagram considered in [6.1, 6.2] is shown in Fig. 1.8. The control action $\bar{u}$ acts on the object B through the communication channel G, where it is mixed with a random fluctuation (noise) $\bar{g}$.

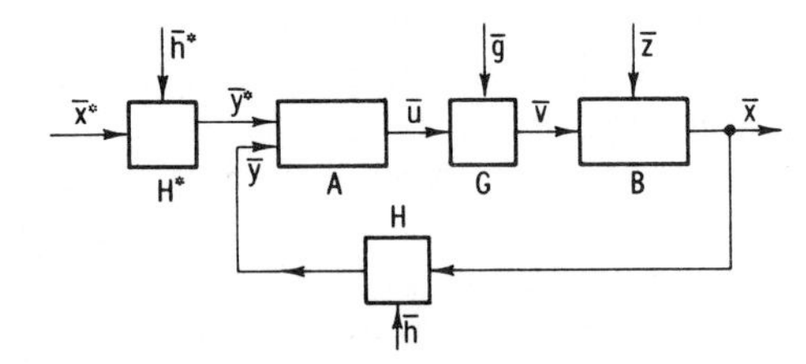

FIG. 1.8

Therefore in general the action $\bar{v}$ at the input to the object B is not equal to the variable $\bar{u}$. Further, the information about the state of the object passes through the communication channel H, where it is mixed with a random fluctuation (noise) $\bar{h}$, and after being transformed into the variable $\bar{y}$ acts on the input to the controller A. The external action $\bar{x}^*$ passes through the channel H^* as in Fig. 1.7.

The communication channel G with the noise $\bar{g}$ can be included in the structure of the object B. Then $\bar{g}$ becomes a component of the noise vector $\bar{z}$ of the object. The remaining blocks are essentially independent.

In the closed-loop circuit of Fig. 1.8, processes for which there is no analog in open-loop systems are possible. A study of the disturbance $\bar{z}$, i.e., by essentially varying the characteristics of the object B in an unexpected manner, can be made in this circuit not by passive observation but by an active method, by means of rational "experiments." The object would be "sensed" by the actions $\bar{u}$, having a trial perceptive character, and the results $\bar{y}$ of these actions analyzed by the controller A. The purpose of such actions is to further the more rapid and precise "study" of the characteristics of the object B, which will help to choose the best control law for the object.

However, the control action is necessary not only for the study of the object, but also for its "being brought" to the required state. Therefore in the diagram of Fig. 1.8 the control actions must have a reciprocal, "dual" character: to a certain extent they must be of a probing nature, but also controlling to a known degree. Therefore in [6.1] the theory of this type of system has been called the theory of dual control. Indeed, the duality of control is the fundamental physical fact distinguishing the third group of theories from the first two. In the first, dual control is not necessary, since without it the controller has complete information about the object. In the second, dual control is impossible, because information is accumulated by means of only one observation, and its rate of storage does not depend at all on the strategy of the controller.

The third direction of classification of optimal systems considered in this section (see Fig. 1.4) is closely related to the problems of obtaining information necessary for control. Just at this decisive point statistical methods penetrate into the theory of optimal systems. Such a penetration is not accidental. Yet it started not long ago with statistical control problems being only one of many chapters in the theory of automatic systems. However, the linking of probabilistic and regular aspects in the theory of automatic control is dictated by the fundamental problems of this theory. The probabilistic aspect is not a "strange" element or an addition to the basic "regular" theory. It introduces constraints into the structure of automatic control theory, being an essential part of it and being present invisibly behind the scenes even where nothing is explicitly referred to.

In fact, the fundamental problem for the automatic control system can be formulated thus: the attainment of a definite control purpose, i.e., providing a determinate behavior for the system in a definite manner in the presence of disturbances acting on it which are unforeseen beforehand, i.e., random. If there were no random disturbances, then the problem of control theory would be simplified so much that, perhaps, the need would not arise for a separate science. On the other hand, if the requirement for organizing a purposeful, determinate behavior of the system did not exist, then the situation would cease to differ essentially from the problems of physics, where the influence of random factors is also considered.

It is natural that in the first period of development of control theory, the simpler "regular" methods reigned supreme. But the "shadow" of random factors had already left its mark on the methods of the theory and modes of constructing automatic devices. Thus, for example, the general statement of the problem of stability of motion and "coarseness" of systems presupposes the presence of small noises which are unforeseen in advance and are essentially random. If it were not for them, there would be no stability problems either. Further, the most powerful principle of constructing automatic systems—the feedback principle—arose just because open-loop systems were weakly protected from the influence of random factors. Systems with feedback have considerably more noise stability. The appearance of self-tracking and self-adapting systems signifies a new stage in the increase of noise stability. In such systems a random law of motion within wide limits for the controlled object, as the result of a successful strategy of the controller, does not significantly influence the determinate behavior of the system as a whole. Sometimes, as will be shown in Chapter V, it is reasonable even to make the strategy of the controller random. The development of the theory of compound systems apparently still further increases the necessity for a statistical description of processes and systems and for statistical methods of solving problems. In general, the statistical approach involves wider and more general conformities than a regular one, and in this lies its large-scale practical value. Therefore a further increase in its role in automatic control theory should be expected in the future, in cases where random factors and the uncertainty of the situation are a basic consideration. It can only be added that by itself the statistical approach is not the only means of investigating indeterminate situations. Therefore it is quite natural that other methods of approach as well—game-theoretic methods, the method of inductive probability, and other directions—will begin to find greater and greater application in the theory of automatic systems.

5. Statement of Problems in the Theory of Optimal Systems

From the foregoing account it is seen that the kaleidoscopic variety of optimal systems does not prevent their systematic arrangement within the scope of a comparatively economic classification. By the same token a unity of approach to the problems of optimal systems is created. At the present time the construction of a unified general theory of optimal systems is becoming possible, including both the formulation of general problems and the methods of solving them. The general problems of this theory are formulated below. The concrete definition of problems for various directions will be made in subsequent chapters.

We will assume that the operator $\bar{F}[\bar{u}, \bar{z}, t]$ and the control purpose in the form of the functional Q are given. If the external actions operating on the system (Fig. 1.8) are random, then we will assume that its probabilistic characteristics are given. The prescribing of regular functions is a special case of assigning probabilistic characteristics. Then the mean value is given as a function of time and the dispersion (a measure of the scattering of the value of the random variable from trial to trial) is equal to zero. Let the operators of the elements H^*, H, and G in Fig. 1.8 also be assigned in the general case. Likewise we will assume that constraints are given, for example, of the type (1.16), (1.17), or (1.19). Usually $\bar{u}(t)$ is regarded as belonging to the class of piecewise-continuous functions with a finite number of discontinuity points of the first kind on an arbitrary finite interval.†

The problem consists in finding, for the conditions prescribed above, an algorithm, or as is sometimes said a strategy for the controller A, such that the optimality criterion Q takes on the least possible value. Such a strategy is called optimal.

In the general case the optimal strategy can turn out to be random. This means that the controller admits a random decision and can deliver to its own output a random variable $\bar{u}$ at the instant t. But the probabilistic characteristic $\Gamma\{\bar{u}(t)\}$ of this variable (for example, its probability density distribution) depends in some optimal way on all the information obtained earlier by the controller A and on the actions undertaken by it earlier, i.e., on the character of the functions $\bar{y}(\tau)$, $\bar{y}^*(\tau)$, and $\bar{u}(\tau)$ (Fig. 1.8) during the time τ elapsing from some initial instant t_0 to the present instant t, i.e., $t_0 \leqslant \tau < t$ (in a particular case $t_0 = -\infty$). The dependence $\Gamma\{\bar{u}(t)\}$ on the functions mentioned above and also on time t is expressed symbolically as

$$\Gamma\{\bar{u}(t)\} = \Gamma\{\bar{u}(t) | \bar{y}^*(\tau), \bar{y}(\tau), \bar{u}(\tau), t\} \qquad (t_0 \leqslant \tau < t), \tag{1.44}$$

and is read thus: the function Γ of $\bar{u}(t)$, "provided that" $\bar{y}^*$, $\bar{y}$, and $\bar{u}$ are prescribed on the interval from t_0 to t. Expression (1.44) is called the "conditional" probability characteristic. Γ is a functional depending on the form of the functions $\bar{y}^*$, $\bar{y}$, and $\bar{u}$.

In a particular case the optimal strategy is found to be regular if the probability characteristic is such that one of the possible values of $\bar{u}(t)$ has a probability equal to unity, and the probabilities of all the other values are equal to zero. The optimal regular strategy is expressed by the dependence

$$\bar{u}(t) = K[\bar{y}^*(\tau), \bar{y}(\tau), \bar{u}(\tau), t] \qquad (t_0 \leqslant \tau < t); \tag{1.45}$$

K is also a functional, i.e., a number depending on the form of the functions $\bar{y}^*(\tau)$, $\bar{y}(\tau)$, and $\bar{u}(\tau)$ on the interval $t_0 \leqslant \tau < t$.

† In some papers where so-called "sliding conditions" are considered, this number is infinite [2.28, 2.29].

In the particular case of systems with complete information about the object, when the elements H^*, H, and G in Fig. 1.8 are missing, and $\bar{z}$ and $\bar{x}^*$ are regular and can be included in the structure of the operator $\bar{F}$ and the criterion Q respectively, the general expression (1.45) for the algorithm of the optimal system is simplified still further and takes the form

$$\bar{u}(t) = \bar{K}[\bar{x}(\tau), \bar{u}(\tau), t] \qquad (t_0 \leqslant \tau < t). \tag{1.46}$$

If the state of the object characterized by the vector $\bar{x}$ determines its entire future behavior, independent of "previous history," i.e., of the values of $\bar{x}(\tau)$ and $\bar{u}(\tau)$ for $\tau < t$; if, for example, the object is characterized by equations of the type (1.7) and does not contain delays and ambiguous dependences; then at a given instant of time t, $\bar{u}(t)$ is a function of only the value of $\bar{x}$ at the same time instant:

$$\bar{u}(t) = \bar{K}[\bar{x}(t), t]. \tag{1.47}$$

Finally, if the equations of motion do not contain time t explicitly (such systems are often called stationary), then the optimal algorithm must be found in the form of the function

$$\bar{u}(t) = \bar{K}[\bar{x}(t)], \tag{1.48}$$

or, abbreviated,

$$\bar{u} = \bar{K}[\bar{x}]. \tag{1.49}$$

In the special case when there is only one control action u, the formula for the optimal algorithm acquires the form

$$u = K[\bar{x}]. \tag{1.50}$$

Thus in this case it is required to find the optimal scalar function K of the vector argument $\bar{x}$, i.e., of the position of a representative point—the end of the vector $\bar{x}$—in the phase space of the system. In other words, K is a function of the n variables $x_1, \ldots, x_n$.

The problem of determining the optimal strategy is sometimes called the problem of synthesizing the optimal system. This usage is not accurate at all, since a set of various possible systems will realize the same algorithm.

The formulation presented above of determining the optimal strategy or algorithm of the controller is fundamental to the theory of optimal systems, and has been worked out in its various ramifications since 1940. But there is still another problem—that of determining optimal processes, i.e., of finding the processes $\bar{u}(t)$ and $\bar{x}(t)$ as functions of time under the prescribed initial conditions $\bar{x}^0$. This problem is not essential, and it is advanced usually as a springboard; from its solution we may arrive at a solution of the problem of finding the optimal algorithm of the controller A. In fact, by eliminating the time t from the dependence of $\bar{u}(t)$ and $\bar{x}(t)$, the function $\bar{u}[\bar{x}]$ can be recognized under certain auxiliary conditions, i.e., the algorithm of the optimal controller.

In view of the fact that the problem of determining the optimal process has a definite independent value as well, we will give a statement of this problem for a definite class of systems with complete information about the object. We will assume that the motion of the object is described with the aid of n first-order equations of the type (1.7) for the coordinates $x_1, ..., x_n$, or the one vector equation

$$d\bar{x}/dt = \bar{f}(\bar{x}, \bar{u}, t), \tag{1.51}$$

where the f_i are functions which are continuous and differentiable in their arguments.

At the instant $t = t_0$ let the initial value of $\bar{x}$ be equal to $\bar{x}^{(0)}$. Figure 1.9 shows the point with coordinates

$$\bar{x}^{(0)} = (x_1^{(0)}, ..., x_n^{(0)}) \tag{1.52}$$

in the n-dimensional phase space of the system with Cartesian coordinates $x_1, ..., x_n$. As an example a three-dimensional phase space is depicted in Fig. 1.9. But all subsequent arguments are valid for arbitrary n.

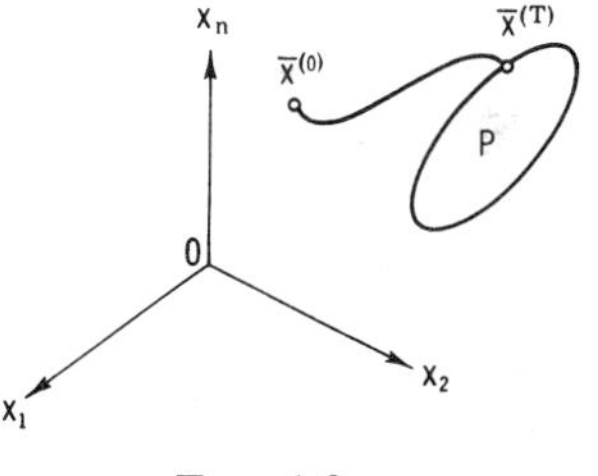

FIG. 1.9

As a result of the application of the control action $\bar{u}(t)$, the motion of the system takes place along a certain trajectory in phase space, and at an instant of time $t = T$ the representative point reaches the position $\bar{x}^{(T)}$. Let $\bar{x}^{(T)}$ belong to some subset P of points of phase space. In a particular case this subset can become a single point. For example, the subset P can be one-dimensional, i.e., it can be a curved line in n-dimensional space. In a particular case the subset P can generally coincide with the phase space. Then the problem is called the problem with a free end of the trajectory. Below we will consider chiefly only two cases: (a) the case when P degenerates into the fixed point $\bar{x}^{(T)}$, while the time T itself is not fixed beforehand; (b) the case of the problem with a free end of the trajectory, when T is fixed, i.e., established in advance. However, general methods of solving the problems are applicable to a considerably more general case, analyzed in Chapter II.

Let the constraints imposed on the vector $\bar{u}$ be prescribed as of the type (1.16):

$$\bar{u} \in \Omega(\bar{u}). \tag{1.53}$$

We will call an "admissible" control action a piecewise-continuous function $\bar{u}(t)$ which satisfies the constraint (1.53). In the problem of the optimal process it is required to find an admissible control action $\bar{u}(t)$ and the corresponding motion $\bar{x}(t)$ of the object, such that the trajectory of the representative point $\bar{x}$ in phase space passing from the initial position $\bar{x}^{(0)}$ to the position $\bar{x}^{(T)}$ belonging to the subset P will give a minimum of a certain functional Q. We will usually characterize the latter by the integral (for $t_0 = 0$)

$$Q = \int_0^T G[\bar{x}(t), \bar{u}(t), t]\, dt, \tag{1.54}$$

where G is a finite and usually positive scalar function of $\bar{x}$, $\bar{u}$, and t. We will note that from a formal point of view the explicit dependence on t in expressions (1.51), (1.54) can be removed by introducing the additional coordinate x_{n+1}, where $(x_{n+1})_{t=0} = 0$ and

$$\frac{dx_{n+1}}{dt} = 1. \tag{1.55}$$

Since here $x_{n+1} = t$, then x_{n+1} can be written everywhere instead of t. Therefore the new system of equations (1.51), with Eq. (1.55) added to it, does not contain the argument t in explicit form, but on the other hand is characterized by the $n + 1$ coordinates $x_1, \ldots, x_n, x_{n+1}$. Hence consideration can be limited to equations without an explicit dependence on t. We will treat these often later on.

In the particular case of the problem with a fixed end point $\bar{x}^{(T)}$ but the time T not fixed beforehand, if $G = 1$ is set in formula (1.54) we will obtain $Q = T$. This means that the condition $Q = \min$ is converted into $T = \min$. In this case we obtain the problem of "maximal speed of response," in which it is required to find a control law $\bar{u}(t)$ such that during a minimal time T the representative point $\bar{x}$ transfer from one fixed position $\bar{x}^{(0)}$ to another fixed position $\bar{x}^{(T)}$. The problem of maximal speed of response plays a large role in the formation of a general theory of optimal systems (see [3.25]).

We will introduce in addition to the coordinates $x_1, \ldots, x_n$ still another coordinate x_0, where $(x_0)_{t=0} = 0$ and

$$dx_0/dt = G[x_1, \ldots, x_n; u_1, \ldots, u_r; t]. \tag{1.56}$$

Comparing expressions (1.54) and (1.56), it is not difficult to be convinced that

$$Q = (x_0)_{t=T} = x_0^{(T)}, \tag{1.57}$$

i.e., the optimality criterion Q is equal to the value of x_0 at the final time instant $t = T$. Geometrically this means the following (see Fig. 1.10). In $(n + 1)$-dimensional space with coordinates $x_0, x_1, \ldots, x_n$ the phase trajectory of motion passes from the point $\bar{x}^{(0)}$ situated in the hyperplane $(x_1, \ldots, x_n)$, to the point M at which the coordinates $x_1, \ldots, x_n$ are fixed, but now the coordinate x_0 is "not" fixed. Hence M lies on the perpendicular erected from the point $\bar{x}^{(T)}$ of the hyperplane $x_0 = 0$ and parallel to the x_0 axis. It is required to find a control $\bar{u}(t)$ such that the final value $x_0^{(T)}$ of the coordinate x_0 be minimized.

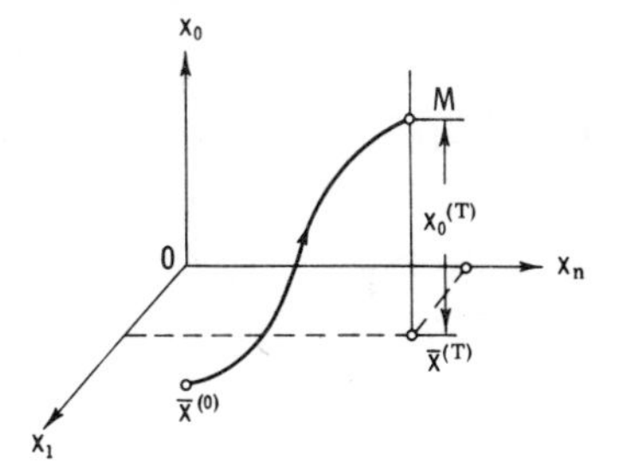

FIG. 1.10

In the particular cases of problems in the theory of optimal systems with incomplete information about the object, expression (1.45) for the functional K can be simplified by various means. We will assume that the feedback path is missing. Then $\bar{y}$ does not appear in formula (1.45) and

$$\bar{u}(t) = K[\bar{y}^*(\tau), \bar{u}(\tau), t] \qquad (t_0 \leqslant \tau < t). \tag{1.58}$$

Since $\bar{u}(\tau)$ at previous time instants is defined as a functional only of $\bar{y}^*(\tau)$, then the algorithm (1.58) can be written thus:

$$\bar{u}(t) = K[\bar{y}^*(\tau), t] \qquad (t_0 \leqslant \tau < t). \tag{1.59}$$

If we consider the problem in which it is required to find the algorithm of the filter F depicted in Fig. 1.7a, or if in the circuit of Fig. 1.7b the operator of the object B can be assumed to be unity (i.e., one in which the output variable equals the input), then $\bar{x}$ can be put in place of $\bar{u}$ in formula (1.59). If further the filter F is stationary here and $t_0 = -\infty$, then the formula takes the form

$$\bar{x}(t) = K[\bar{y}^*(\tau)] \qquad (-\infty < \tau < t). \tag{1.60}$$

In the special case of a single-channel filter x and y^* are scalar. Then

$$x(t) = K[y^*(\tau)] \qquad (-\infty < \tau < t). \tag{1.61}$$

For example, if it is given that the filter F belongs to the class of linear systems with constant coefficients, then the formula for K can be written in the form of a convolution

$$x(t) = \int_{-\infty}^{t} \varphi(t - \tau) y^*(\tau) \, d\tau, \tag{1.62}$$

where $\varphi(t)$ is the unit impulse response of the filter, or as it is often called, the weighting function. Let $h^*(t)$ and $x^*(t)$ be stationary random processes (see Chapter II) and

$$y^*(t) = h^*(t) + x^*(t), \tag{1.63}$$

while the optimality criterion takes the form

$$Q = \lim_{T\to\infty} \frac{1}{2T} \int_{-T}^{T} [x^*(t) - x(t)]^2 \, dt, \tag{1.64}$$

i.e., it is the mean square error. Then the problem of the optimal system takes the form of the problem of determining the weighting function $\varphi(t)$ of a physically realizable filter, i.e., a function satisfying the condition

$$\varphi(t) \equiv 0 \qquad (t < 0), \tag{1.65}$$

such that the optimality criterion be minimal. We have arrived at the problem of optimal linear filtering solved by N. Wiener [1.15]. It turned out that for solving the problem it is not required to know in an exhaustive manner the probabilistic characteristics of the processes $h^*(t)$ and $x^*(t)$. It suffices only to know the so-called correlation functions of these processes, which are determined from experiment in comparatively simple ways. The papers of A. N. Kolmogorov [1.14], A. Ya. Khinchin [1.19], and N. Wiener [1.15] laid the basis for the correlation theory of linear optimal systems. We will not analyze this theory—mainly because there already exists a collection of monographs and educational textbooks in this field. Furthermore, optimal problems with auxiliary constraints imposed on the controller A (in this case the requirement of linearity) are hardly considered in this book. Those interested can acquaint themselves with the various aspects of this theory in the literature ([1.10–1.13, 1.16–1.18, 1.20–1.26]).

CHAPTER II

Mathematical Methods Applicable to the Theory of Optimal Systems

1. Probability Theory

Chapter I emphasized the role of statistical methods in the theory of optimal systems. Therefore, to become acquainted with the theory, a prior understanding of at least certain basic elementary concepts in the theory of probability and mathematical statistics is necessary. In this section a brief summary is given of those notions and formulas which will be required in the subsequent presentation. A more detailed knowledge of probability theory and statistical methods can be found in a number of reference books and monographs (see, for example, [2.1–2.4, 1.10–1.13]). In addition, certain supplementary information of a statistical character will be reported in Chapters IV–VI as necessary.

In probability theory three classes of random phenomena are analyzed. Very simple random phenomena—the so-called "random events"—relate to the first class. The second, more complex class is called "random variables." Finally, the most complex class is "random processes."

The random event A is characterized by a certain number—the probability $p(A)$, where $0 \leqslant p(A) \leqslant 1$. In a mass production of trials, in the process of which the given random event A appears with the probability $p(A)$, the frequency, or how often the event appears (i.e. the ratio of the number N_A of appearances of the event to the overall number of trials N), differs little in practice from $p(A)$, if N is sufficiently large. For a certain event, $p(A) = 1$; for an impossible one, $p(A) = 0$.

If the events $A_1, A_2, \ldots, A_m$ are disjoint, i.e., any two of them cannot occur in the same trial, then the probability $p(A)$ of an event being the appearance of just one of them is expressed by the formula

$$p(A) = \sum_{i=1}^{m} p(A_i). \tag{2.1}$$

Frequently the notation $A = A_1 + \cdots + A_m$ is used for the event A. Here the "plus" sign replaces the word "or." If one of the events

A_i ($i = 1, ..., m$) occurs for sure, then $p(A) = 1$ and

$$\sum_{i=1}^{m} p(A_i) = 1. \tag{2.2}$$

The events $A_1, ..., A_m$ whose probabilities satisfy the equality (2.2) form a perfect group, as it is accepted to say.

We will assume that under definite conditions a series of trials takes place, to which is related the appearance or nonappearance of the random event A with probability $p(A)$, which we will now call absolute.

Let the appearance or nonappearance of another random event B with absolute probability $p(B)$ likewise be related to the trials. We will select from all the N trials only the portion N_B in which the event B appears. From all these N_B trials let only part of them, equal to $N_{A|B}$, be characterized by the appearance of the event A as well. We will call the ratio of $N_{A|B}$ to N_B the frequency of the event A "provided that" the event B appeared, or more briefly, the "conditional frequency" of the event A (provided that B appeared). For a large number of events N_B the ratio $N_{A|B}/N_B$ differs little in practice from a certain number, which we will agree to denote by $p(A \mid B)$ and call the "conditional probability" of the event A (provided that the event B appeared).

From what was stated above it is clear that $N_{A|B}$ is the number of trials in which both events A and B appeared. For a large number of trials N the ratio $N_{A|B}/N$ differs little in practice from the probability of the event C, consisting of the simultaneous occurrence of both events A and B. Frequently the notation $C = AB$ is used, where the operation of multiplication corresponds to the conjunction "and." Thus $p(C) = p(AB)$. Since

$$\frac{N_{A|B}}{N} = \frac{N_{A|B}}{N_B} \cdot \frac{N_B}{N}, \tag{2.3}$$

then by analogy the expression, not for the frequencies but for the probabilities corresponding to them, can be written:

$$p(AB) = p(A|B) \cdot p(B). \tag{2.4}$$

In view of the equivalence of the events A and B, by reasoning analogously we can obtain a formula in which A and B have interchanged places:

$$p(AB) = p(B|A) \cdot p(A). \tag{2.5}$$

Thus,

$$p(AB) = p(A|B)p(B) = p(B|A)p(A). \tag{2.6}$$

The events A and B are "independent" if the appearance of one of them does not affect the probability of the other appearing. If the probability of the event A occurring does not depend on the appearance of B, then $p(A \mid B) = p(A)$. In this case formula (2.6) takes the form

$$p(AB) = p(A)p(B). \tag{2.7}$$

This formula can be considered as the definition of the independence of the events A and B. From (2.6) and (2.7) it follows that if $p(A \mid B) = p(A)$, then also $p(B \mid A) = p(B)$; i.e., if A is independent of B, then B is independent of A as well.

The events $A_1, \ldots, A_m$ are called "pairwise independent" if any two of them are independent. But if, in addition, arbitrary products of these events $A_{i1}A_{i2} \cdots A_{il}$ and $A_{j1}A_{j2} \cdots A_{jq}$ not containing common factors are also independent, then the events $A_1, \ldots, A_m$ are called "independent." Let us emphasize that the concepts of pairwise independence and independence cannot coincide.

For arbitrary events, by applying formula (2.6) successively we obtain:

$$p(A_1A_2 \cdots A_m) = p(A_1)p(A_2|A_1) \cdots p(A_m|A_1, A_2, \ldots, A_{m-1}). \tag{2.8}$$

But if the events are independent, then expression (2.8) is simplified and takes the form

$$p(A_1A_2 \cdots A_m) = p(A_1)p(A_2) \cdots p(A_m). \tag{2.9}$$

Let $A_1, \ldots, A_m$ be disjoint events forming a perfect group. Then $A = A_1 + \cdots + A_m$ is a certain event and the probability of some other event B can be expressed in the following way:

$$\begin{aligned} p(B) &= p(BA) = p[B(A_1 + \cdots + A_m)] \\ &= p(BA_1 + BA_2 + \cdots + BA_m) = \sum_{i=1}^{m} p(BA_i), \end{aligned} \tag{2.10}$$

since the events BA_i and BA_j are disjoint for $i \neq j$. On the other hand, from (2.5),

$$p(BA_i) = p(A_i)p\,(B|A_i). \tag{2.11}$$

Substituting this expression into (2.10), we obtain

$$p(B) = \sum_{i=1}^{m} p(A_i)p(B|A_i). \tag{2.12}$$

Hence, on the basis of (2.6) and (2.12) the conditional probability $p(A_i \mid B)$ can be expressed as

$$p(A_i \mid B) = \frac{p(A_i)p(B \mid A_i)}{p(B)} = \frac{p(A_i)p(B \mid A_i)}{\sum_{i=1}^{m} p(A_i)p(B \mid A_i)}. \tag{2.13}$$

This formula, first discovered in 1784 by the Englishman Bayes, bears his name. Bayes' formula is widely used for calculating so-called *a posteriori* probabilities, i.e., probabilities obtained as a result of some trial. We will assume that the probabilities of the events A_i up to a certain trial—*a priori* probabilities—are denoted by $p(A_i)$. If as a result of the trial the event B appears, then after the experiment the probabilities of the events A_i must be redefined for the conditions when the new information appeared. These new probabilities will now be the conditional probabilities $p(A_i \mid B)$, since they must be calculated "under the condition that" the event B happened. But they are called *a posteriori* probabilities. Bayes' formula serves just as well for calculating the *a posteriori* probabilities $p(A_i \mid B)$ with respect to the given *a priori* probabilities $p(A_i)$. As is seen from (2.13), the conditional probabilities $p(B \mid A_i)$ must be known for such a calculation.

The random variable represents a more complex construction than the random event. This is a quantity that, as a result of a trial, assumes one and only one value from a set of possible values. Therefore, in order to describe a random variable, both the set of its possible values and their probabilities must be prescribed. Let the possible values $x_1, \ldots, x_n$ of the random variable ξ be discrete (where n can be finite or infinitely large). Then n probabilities of the form $p_i = p(x_i)$ must be given, where p_i is the probability of the random event being the appearance of the value x_i of the random variable ξ. It is evident that

$$\sum_{i=1}^{n} p_i = 1, \tag{2.14}$$

since the events $\xi = x_i$ for different i are disjoint by definition, and on the other hand form a perfect group, again from the definition of a random variable. But if the random variable ξ can assume arbitrary values on some interval, then another form of representation for its probabilistic characteristic is necessary. The most general form is the "distribution function" $F(x)$, or in other words, the integral distribution law. This is the probability of the random event $\xi < x$, that the quantity ξ was found to be less than some fixed level x:

$$F(x) = p(\xi < x). \tag{2.15}$$

Knowing $F(x)$ for arbitrary x, it is easy to find the probability of ξ occurring in the interval $a \leqslant \xi < b$. In fact, since the events $\xi < a$ and $a \leqslant \xi < b$ are disjoint and their sum is the event $\xi < b$, we can write:

$$p(\xi < b) = p(\xi < a) + p(a \leqslant \xi < b). \tag{2.16}$$

Hence it follows that:

$$p(a \leqslant \xi < b) = p(\xi < b) - p(\xi < a) = F(b) - F(a). \tag{2.17}$$

It is obvious that $F(a) = +1$ for $a = \infty$ and $F(a) = 0$ for $a = -\infty$; it is also evident that $F(x)$ is a monotonic nondecreasing function of x.

If $F(x)$ is continuous and differentiable in the entire interval $-\infty < x < \infty$, then the corresponding random variable ξ is called a "continuous" random variable. We will set

$$P(x) = \frac{dF(x)}{dx} = F'(x). \tag{2.18}$$

The function $P(x)$ is called the "probability density" or the differential distribution law of the random variable ξ. Since

$$P(x) = \frac{dF(x)}{dx} = \lim_{\Delta x \to 0} \frac{F(x + \Delta x) - F(x)}{\Delta x} = \lim_{\Delta x \to 0} \frac{p(x \leqslant \xi < x + \Delta x)}{\Delta x}, \tag{2.19}$$

then $P(x)\,\Delta x$ is, to within small terms $O\,(\Delta x)$ of higher order, the probability of the random variable being found in the infinitesimally small interval $x \leqslant \xi < x + \Delta x$. Further, the probability of the variable ξ being found in the interval $a \leqslant \xi < b$ is defined by the expression

$$p(a \leqslant \xi < b) = F(b) - F(a) = \int_a^b P(x)\,dx. \tag{2.20}$$

Hence it follows that:

$$\int_{-\infty}^{\infty} P(x)\,dx = F(\infty) - F(-\infty) = 1. \tag{2.21}$$

Important although not exhaustive characteristics of the random variable ξ are its so-called moments. A certain number representing the integral

$$\alpha_k = \int_{-\infty}^{\infty} x^k P(x)\,dx \tag{2.22}$$

is called the moment of order k. The first-order moment α_1 has a particularly important significance, and is called the "mathematical expectation" or the "mean value" of the random variable. It is often denoted by m_ξ or $M\{\xi\}$:

$$m_\xi = M\{\xi\} = \alpha_1 = \int_{-\infty}^{\infty} xP(x)\,dx. \tag{2.23}$$

The concept of the mean value arises from the arithmetic mean. If a sufficiently large number N of trials is carried out, then the arithmetic mean ξ_{am} of the values of the random variable ξ obtained by experiment differs little in practice from the mean value m_ξ (in the sense that the probability of significant deviations of ξ_{am} from m_ξ is small enough).

The kth-order moment of the difference $(\xi - m_\xi)$ is called the kth order "central moment" μ_k :

$$\mu_k = M\{(\xi - m_\xi)^k\} = \int_{-\infty}^{\infty} (x - m_\xi)^k P(x)\,dx. \tag{2.24}$$

The second-order central moment has a particular significance and is called the "dispersion." It is denoted by $D(\xi)$ or D_ξ :

$$D_\xi = D\{\xi\} = M\{(\xi - m_\xi)^2\} = \int_{-\infty}^{\infty} (x - m_\xi)^2 P(x)\,dx. \tag{2.25}$$

The dispersion characterizes to a known degree the scattering of values of the random variable ξ around its mean value m_ξ. The quantity $\sqrt{D_\xi}$ is called the standard deviation and is denoted by σ_ξ :

$$\sigma_\xi = \sqrt{D_\xi}\,. \tag{2.26}$$

In the general case this quantity must differ from the root mean square value of ξ, which we will denote by ξ_{rms} and define by the formula

$$(\xi_{\text{rms}})^2 = \alpha_2 = \int_{-\infty}^{\infty} x^2 P(x)\,dx. \tag{2.27}$$

Only in the case when $m_\xi = 0$ do expressions (2.25) and (2.27) coincide.

The mathematical expectation of an arbitrary function $f(\xi) = \eta$ of the random variable ξ is defined by the expression

$$M\{\eta\} = M\{f(\xi)\} = \int_{-\infty}^{\infty} f(x)P(x)\,dx. \tag{2.28}$$

In fact, the probability $P(y)\,dy$ that the function η will be between $y = f(x)$ and $y + dy = f(x + dx)$ is equal to the probability that the argument ξ of the function is between the values x and $x + dx$, and this probability is equal to $P(x)\,dx$. Thus, the mathematical expectation of the quantity η is defined by a general formula of the type (2.28) as the integral over infinite limits of $yP(y)\,dy = yP(x)\,dx = f(x)P(x)\,dx$, and this is the integrand on the right side of (2.28).

The most extensive distribution law for random variables is the "normal" or "Gaussian" law, for which

$$P(x) = \frac{1}{\sigma_x\sqrt{(2\pi)}} \exp\left\{-\frac{(x - m_x)^2}{2\sigma_x^2}\right\}. \tag{2.29}$$

Here m_x and σ_x are certain constants that are equal to the mean value and standard deviation, respectively. The correctness of this assertion can be verified from formulas (2.23), (2.25), and (2.26).

The distribution function for the normal law is obtained from the expression

$$\begin{aligned} F(x) &= \int_{-\infty}^{x} P(x)\,dx = \frac{1}{\sigma_x\sqrt{(2\pi)}} \int_{-\infty}^{x} \exp\left\{-\frac{(x - m_x)^2}{2\sigma_x^2}\right\} dx \\ &= \frac{1}{\sqrt{(2\pi)}} \int_{-\infty}^{(x-m_x)/\sigma_x} \exp\left\{-\frac{t^2}{2}\right\} dt. \end{aligned} \tag{2.30}$$

The function

$$\Phi(u) = \frac{1}{\sqrt{(2\pi)}} \int_{0}^{u} \exp\left\{-\frac{t^2}{2}\right\} dt \tag{2.31}$$

is introduced into probability theory. This integral is not expressed in terms of elementary functions. The function $\Phi(u)$ has been tabulated. It is evidently odd, i.e., $\Phi(-u) = -\Phi(u)$. By comparing (2.30) and (2.31) we see that for the normal distribution law

$$F(x) = \Phi\left(\frac{x - m_x}{\sigma_x}\right). \tag{2.32}$$

Let us consider a nonlinear inertialess element, depicted in Fig. 2.1, for which the relation between the output variable y and the input variable x is expressed by the formula

Fig. 2.1

$$y = f(x). \tag{2.33}$$

If x is a random variable, then y will also be random.

Let the probability density $P(x)$ for the random variable x be known. We will find the probability density $P(y)$ for the random variable y. Here the notation P is used for the probability density of both x and y. However, $P(y)$ denotes a function which in general differs from the function $P(x)$.

We will first assume that it is required to determine the probability of the event that y belongs to a certain subregion Ω_y of its own possible values: $y \in \Omega_y$. For example, this region can be the interval $a \leqslant y < b$. Let the values of x belonging to a certain region Ω_x correspond to the points y of the region Ω_y according to Eq. (2.33). Thus, $x \in \Omega_x$. Then the probability of the event $y \in \Omega_y$ is equal to the probability of the event $x \in \Omega_x$. Hence

$$p(y \in \Omega_y) = p(x \in \Omega_x) = \int_{\Omega_x} P(x)\, d\Omega_x , \tag{2.34}$$

where $d\Omega_x$ is an infinitesimally small element of the region Ω_x, and the integration is carried out over the entire region Ω_x. Using expression (2.34), we can find $P(y)$. For example, let $P(x)$ be the normal distribution law with a mean value m_x equal to zero:

$$P(x) = P_0(x) = \frac{1}{\sigma_x \sqrt{(2\pi)}} \exp\left\{ -\frac{x^2}{2\sigma_x^{\,2}} \right\}. \tag{2.35}$$

The curve of $P_0(x)$ is shown in Fig. 2.2a. Further, let

$$y = A + Bx, \tag{2.36}$$

where A and $B > 0$ are constants (Fig. 2.2b). Let us find the region Ω_x corresponding to some prescribed infinitesimally small interval of values of y between y_1 and $y_1 + dy$. Evidently, the region Ω_x is defined by the condition $x_1 \leqslant x < x_1 + dx$, where

$$x_1 = \frac{y_1 - A}{B} \tag{2.37}$$

and

$$dx = dy/B. \tag{2.38}$$

From (2.34) we find:

$$\begin{aligned} p(y_1 \leqslant y < y_1 + dy) &= P(y_1)\, dy = p(x_1 \leqslant x < x_1 + dx) \\ &= P_0(x_1)\, dx = P_0 \left(\frac{y_1 - A}{B} \right) \frac{dy}{B}. \end{aligned} \tag{2.39}$$

Hence it follows that if y_1 is replaced by y,

$$\begin{aligned} P(y) &= \frac{1}{B} P_0 \left(\frac{y - A}{B} \right) = \frac{1}{B} \cdot \frac{1}{\sigma_x \sqrt{(2\pi)}} \exp\left\{ - \frac{[(y - A)/B]^2}{2\sigma_x^{\,2}} \right\} \\ &= \frac{1}{B\sigma_x \sqrt{(2\pi)}} \exp\left\{ - \frac{(y - A)^2}{2(B\sigma_x)^2} \right\}. \end{aligned} \tag{2.40}$$

Let us set

$$\sigma_y = B\sigma_x, \qquad m_y = A. \tag{2.41}$$

Then formula (2.40) takes the form

$$P(y) = \frac{1}{\sigma_y\sqrt{(2\pi)}} \exp\left\{-\frac{(y - m_y)^2}{2\sigma_y^2}\right\}. \tag{2.42}$$

This expression conforms exactly to formula (2.29). Hence it follows that y has a normal distribution law with mean value A and standard deviation $B\sigma_x$ (Fig. 2.2c). The smaller B is, the more the curve of $P(y)$ is "concentrated" around the mean value $y = A$.

The nonlinear element can have a non-single-valued characteristic as well, when not one but several values of x correspond to one value of y. For example, let

$$y = Ax^2, \tag{2.43}$$

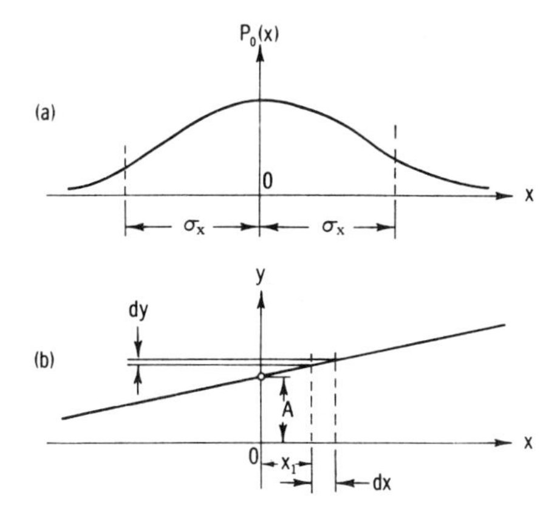

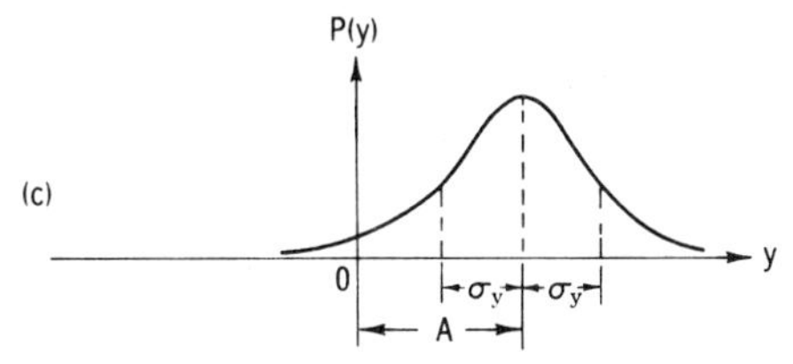

FIG. 2.2

where $A > 0$ (Fig. 2.3b). We will assume that the random variable x acts on the input to the element with the normal distribution $P_0(x)$

expressed by formula (2.35). We will find the probability of y being found between the values $w > 0$ and $w + dw$, i.e., the quantity

$$p(w \leqslant y < w + dw) = P(w)\, dw, \tag{2.44}$$

where the letter P denotes the probability density of y. From Figs. 2.3a and 2.3b it is seen that this probability is the sum of the probabilities of x being found in the intervals AB and CD:

$$\begin{aligned} P(w)\, dw &= p(w < y < w + dw) \\ &= p\left[+\sqrt{\left(\frac{w}{A}\right)} < x < +\sqrt{\left(\frac{w+dw}{A}\right)}\right] \\ &\quad + p\left[-\sqrt{\left(\frac{w+dw}{A}\right)} < x < -\sqrt{\left(\frac{w}{A}\right)}\right] \\ &= 2p\left[+\sqrt{\left(\frac{w}{A}\right)} < x < +\sqrt{\left(\frac{w+dw}{A}\right)}\right]. \end{aligned} \tag{2.45}$$

The latter transformation is valid since the distribution $P_0(x)$ is symmetrical. Therefore the two infinitesimally small areas shaded in Fig. 2.3a are equal to each other. Further, since to within small terms of higher order

$$\sqrt{\left(\frac{w+dw}{A}\right)} \approx \sqrt{\left(\frac{w}{A}\right)} + \frac{dw}{2\sqrt{(Aw)}}, \tag{2.46}$$

then (2.45) can be rewritten thus:

$$P(w)\, dw = 2P_0\left[x = +\sqrt{\left(\frac{w}{A}\right)}\right] \frac{dw}{2\sqrt{w}} \frac{1}{\sqrt{A}} \qquad (w > 0), \tag{2.47}$$

from which

$$P(w) = \frac{1}{\sqrt{(Aw)}} P_0\left[+\sqrt{\left(\frac{w}{A}\right)}\right] \qquad (w > 0). \tag{2.48}$$

Replacing w by y here and substituting $P_0(x)$ from (2.35), we will find:

$$\begin{aligned} P(y) &= \frac{1}{\sqrt{(Ay)}} \cdot \frac{1}{\sigma_x\sqrt{(2\pi)}} \exp\left\{-\frac{y}{2\sigma_x^2 A}\right\} \\ &= \frac{1}{\sigma_0\sqrt{(2\pi y)}} \exp\left\{-\frac{y}{2\sigma_0^2}\right\}, \end{aligned} \tag{2.49}$$

where

$$\sigma_0 = \sigma_x\sqrt{A}. \tag{2.50}$$

Formula (2.49) is valid only for $y > 0$. Values of $y < 0$ are impossible (Fig. 2.3b). Therefore $P(y) = 0$ for $y < 0$. This curve is depicted in Fig. 2.3c as a function of the variable $z = y/\sigma_0^2$. From (2.49) it follows that for $z > 0$

$$P(z) = P\left(\frac{y}{\sigma_0^2}\right) = \frac{1}{\sigma_0^2\sqrt{(2\pi z)}} \exp\left\{-\frac{z}{2}\right\}. \tag{2.51}$$

This distribution differs considerably from a normal one.

For several random variables $\xi_1, ..., \xi_n$ both the integral and differential characteristics can likewise be constructed. For example, the joint distribution density of the continuous random variables $\xi_1, ..., \xi_n$ is a function $P(x_1, ..., x_n)$, where $P(x_1, ..., x_n)\, dx_1 \cdots dx_n$ is the probability of the event consisting of the occurrence of ξ_1—the first of the random variables—in the interval $x_1 \leqslant \xi_1 < x_1 + dx_1$, and at the same time of ξ_2 in the interval $x_2 \leqslant \xi_2 < x_2 + dx_2$, and so on up to the occurrence of the last random variable ξ_n in the interval $x_n \leqslant \xi_n < x_n + dx_n$.

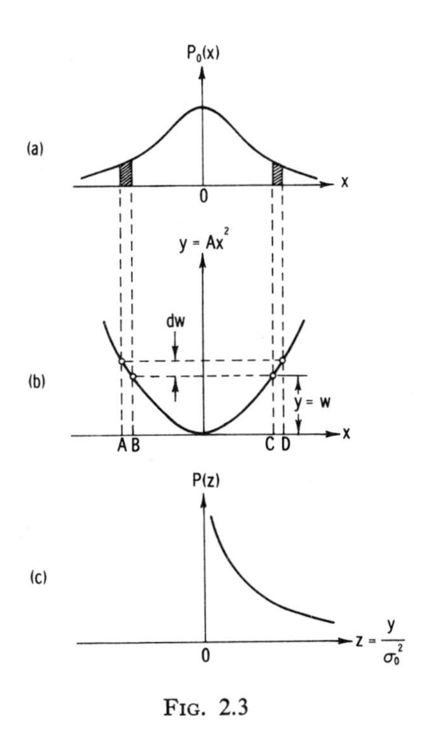

FIG. 2.3

The collection of random variables $\xi_1, ..., \xi_n$ can be considered as the Cartesian coordinates of a point or as components of a random vector $\bar{\xi} = (\xi_1, ..., \xi_n)$ in n-dimensional space. The probability $p(A)$ of the event A, consisting of the end of the vector $\bar{\xi}$ falling in the region Ω_A in this space, is evidently equal to the sum of the probabilities of occurrences in an infinitesimally small volume $d\Omega_A = dx_1 \cdots dx_n$ of this region:

$$p(A) = \int_{\Omega_A} P(x_1, ..., x_n)\, dx_1 \cdots dx_n = \int_{\Omega_A} P(\bar{x})\, d\Omega_A. \tag{2.52}$$

Here the abbreviated notation $P(\bar{x})$ has been introduced for $P(x_1, ..., x_n)$.

Since the probability of the end of the vector $\bar{\xi}$ falling somewhere in the n-dimensional space is equal to unity, then

$$\int_{-\infty}^{\infty} \cdots \int_{-\infty}^{\infty} P(x_1, ..., x_n)\, dx_1 \cdots dx_n = 1. \tag{2.53}$$

To find the probability density $P_i(x_i)$ for one variable ξ_i, $P(x_1, ..., x_n)$ can be integrated over the entire domain of variation of the other variables. For example,

$$P_1(x_1) = \int_{-\infty}^{\infty} \cdots \int_{-\infty}^{\infty} P(x_1, ..., x_n)\, dx_2 \cdots dx_n. \tag{2.54}$$

If the variables $\xi_1, ..., \xi_n$ are independent, then according to expression (2.9) the total probability $p(A)$ of the event A stated above is equal to the product of the probabilities for the random variables ξ_i $(i = 1, ..., n)$ taken separately. Therefore in this case

$$P(x_1, ..., x_n) = P_1(x_1) \cdot P_2(x_2) \cdots P_n(x_n). \tag{2.55}$$

Of course, the probability densities P_i for the various ξ_i can be different.

If the random variables ξ and η are related, then assigning a fixed value to one of them affects the probability distribution of the other. Let $P(y \mid x)\, dy$ be the probability of the random variable η being found in the fixed interval $y \leqslant \eta < y + dy$, provided that the random variable ξ has a certain fixed value x. We will call the probability density $P(y \mid x)$ for η with a fixed $\xi = x$ the "conditional probability density." Since the probability of η having some value is in general equal to unity,

$$\int_{-\infty}^{\infty} P(y \mid x)\, dy = 1. \tag{2.56}$$

By knowing the joint distribution density $P(x, y)$ of the variables ξ and η, it is not difficult to find the conditional distribution density

$P(y \mid x)$. In fact, on the basis of the multiplication theorem of probabilities we can write:

$$P(x, y)\, dx\, dy = [P(x)\, dx] \cdot [P(y \mid x)\, dy], \tag{2.57}$$

where $P(x)$ is the absolute probability density of ξ. Hence it follows that:

$$P(y \mid x) = P(x, y)/P(x). \tag{2.58}$$

It is not difficult to generalize formulas (2.56) and (2.58) to the case of vector variables $\bar{\xi}$ and $\bar{\eta}$, i.e., the collections $\xi_1, \ldots, \xi_n$ and $\eta_1, \ldots, \eta_n$.

The notions of mean values can also be introduced for the collection of random variables $\xi_1, \ldots, \xi_n$. For example, the mean value of the variable ξ_k can be computed from the formula

$$M\{\xi_k\} = m_{\xi_k} = \int_{-\infty}^{\infty} \cdots \int_{-\infty}^{\infty} x_k P(x_1, \ldots, x_n)\, dx_1 \cdots dx_n$$
$$= \int_{\Omega(\bar{x})} x_k P(\bar{x})\, d\Omega(\bar{x}), \tag{2.59}$$

where $\Omega(\bar{x})$ is the entire n-dimensional space of end points of the vector $\bar{\xi}$, and $d\Omega(\bar{x}) = dx_1 \cdots dx_n$ is an infinitesimally small element of this space.

The second-order central moments of the collection of variables $\xi_1, \ldots, \xi_n$ are defined by the formulas

$$M_{jk}\{\xi_j \xi_k\} = \int_{-\infty}^{\infty} \cdots \int_{-\infty}^{\infty} [x_j - M\{\xi_j\}]$$
$$\times [x_k - M\{\xi_k\}] P(x_1, \ldots, x_n)\, dx_1 \cdots dx_n. \tag{2.60}$$

For $j = k$ we obtain the dispersion of the random variable ξ_j, while for $j \neq k$ the corresponding quantity bears the name of the covariance of the random variables ξ_k and ξ_j. If these variables are independent, then, as is not difficult to show, the covariance is equal to zero (the converse in general is false: if the covariance is equal to zero, then it is still not possible to state that the variables ξ_j and ξ_k are independent). The dimensionless ratio

$$R_{jk} = \frac{M_{jk}}{\sqrt{(M_{jj} M_{kk})}} \tag{2.61}$$

is called the "correlation coefficient" between the random variables ξ_j and ξ_k.

The notions of conditional mean values are also formulated in a quite analogous manner. For example, the conditional mean value $M\{\eta \mid \xi\}$ of the random variable η for a fixed value x of the other random variable ξ, and for the conditional density $P(y \mid x)$ of η for a fixed ξ is obtained from a general formula of the type (2.58):

$$M\{\eta \mid \xi\} = M_{\eta \mid \xi} = \int_{-\infty}^{\infty} yP(y \mid x)\, dy$$
$$= \frac{1}{P(x)} \int_{-\infty}^{\infty} yP(x, y)\, dy. \tag{2.62}$$

The latter expression was obtained by substituting $P(y \mid x)$ from formula (2.58).

Let us consider an inertialess element of an automatic system (Fig. 2.4) for which the output variable x_k is a function of two input variables h_k and x_{k-1}:

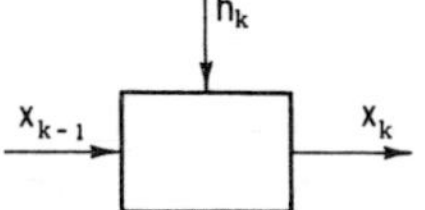

FIG. 2.4

$$x_k = f_k(x_{k-1}, h_k). \tag{2.63}$$

For example, x_{k-1} is the element's input variable and h_k the noise.

Let x_{k-1} be fixed and h_k a random variable with a distribution density $P(h_k)$. Then from formula (2.63) the conditional probability density $P(x_k \mid x_{k-1})$ can be found for x_k provided that some fixed value x_{k-1} is assigned, which in this case is regarded as a parameter.

Particular examples of determining $P(x_k)$ have been given above (see Fig. 2.2 and 2.3). The determination of the conditional probability density is carried out in essentially the same way.

Now the conditional mean value of x_k for a fixed x_{k-1} can be found:

$$M\{x_k \mid x_{k-1}\} = M_{x_k \mid x_{k-1}} = \int_{-\infty}^{\infty} x_k P(x_k \mid x_{k-1})\, dx_k\,. \tag{2.64}$$

If there is a chain of successively joined elements of such a type (Fig. 2.5), then for the last element the conditional mean value is

$$M\{x_n \mid x_{n-1}\} = M_{x_n \mid x_{n-1}} = \int_{-\infty}^{\infty} x_n P(x_n \mid x_{n-1})\, dx_n\,. \tag{2.65}$$

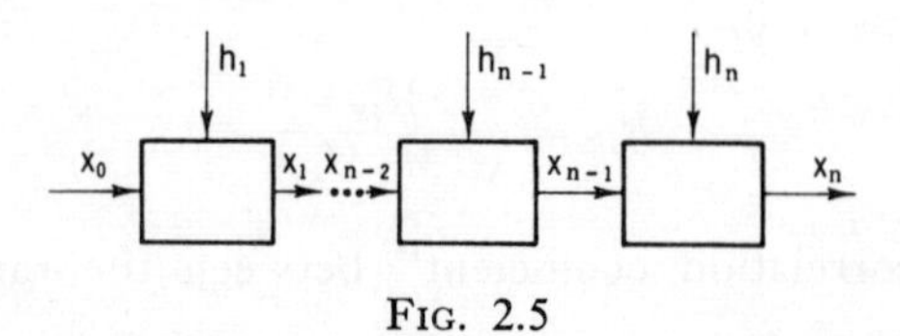

FIG. 2.5

But the quantity x_{n-1} is random even for a fixed value of the input x_{n-2} because of the presence of random noise h_{n-1}. By knowing the properties of the next-to-last element, the conditional density $P(x_{n-1}|x_{n-2})$ can be found. Here, with a view to economy of notations, the same letter P is used as for the function $P(x_n \mid x_{n-1})$, but in the general case these functions may be different. If x_{n-1} is a random variable, then the mean (2.65) must be regarded as a random variable as well. In turn the mean value of this random variable is equal to

$$\begin{aligned} M\{x_n \mid x_{n-2}\} &= M_{x_n|x_{n-2}} = M_{x_{n-1}|x_{n-2}}\{M_{x_n|x_{n-1}}\} \\ &= \int_{-\infty}^{\infty} P(x_{n-1} \mid x_{n-2}) \left[\int_{-\infty}^{\infty} x_n P(x_n \mid x_{n-1})\, dx_n\right] dx_{n-1} \\ &= \int_{-\infty}^{\infty}\int_{-\infty}^{\infty} x_n P(x_n \mid x_{n-1}) P(x_{n-1} \mid x_{n-2})\, dx_n\, dx_{n-1} \\ &= \int_{\Omega(x_n, x_{n-1})} x_n P(x_n \mid x_{n-1}) P(x_{n-1} \mid x_{n-2})\, d\Omega(x_n, x_{n-1}), \end{aligned} \tag{2.66}$$

where $\Omega(x_n, x_{n-1})$ is the notation for the entire domain of variation of the variables x_n and x_{n-1}, and $d\Omega(x_n, x_{n-1}) = dx_n\, dx_{n-1}$ for its infinitesimally small volume. By reasoning analogously for the chain of elements of Fig. 2.5 from end to beginning, we finally arrive at the formula

$$\begin{aligned} M\{x_n \mid x_0\} &= M_{x_n|x_0} \\ &= \int_{\Omega(x_1, x_2, \ldots, x_n)} x_n P(x_n \mid x_{n-1}) \cdot P(x_{n-1} \mid x_{n-2}) \cdots \\ &\qquad\qquad P(x_1 \mid x_0)\, d\Omega(x_1, \ldots, x_n) \\ &= \int_{\Omega(\bar{x})} x_n \left[\prod_{k=1}^{n} P(x_k \mid x_{k-1})\right] d\Omega(\bar{x}), \end{aligned} \tag{2.67}$$

where $\Omega(\bar{x})$ is the domain of variation of the vector $\bar{x} = (x_1, \ldots, x_n)$, and $d\Omega(\bar{x})$ its infinitesimally small volume.

We pass to the consideration of a more complex class of random phenomena, called random processes. A random, or as it is sometimes called, a "probabilistic" or "stochastic" process, is a random time function, i.e., a function which at each time instant is a random variable. Therefore the random process can be defined as a set of random variables $\xi(t)$ depending on the real argument t.

The individual observations of the random process $\xi(t)$ passing through single-type systems, i.e., under fixed controlled conditions of experiment, will give different functions $x(t)$ each time—different "samples" or "realizations" of the random process. The simplest of the probabilistic characteristics of this process is the one-dimensional distribution law $P_1(x_1, t_1)$, i.e., the probability density of the value $\xi(t_1)$ of the process at the time instant $t = t_1$. The expression $P_1(x_1, t_1)\, dx_1$ is the probability of the event that $x_1 < \xi(t_1) < x_1 + dx_1$. The more complicated function $P_2(x_1, t_1; x_2, t_2)$—the two-dimensional distribution function—is the joint probability distribution density of two random variables—the value of the process $\xi(t_1)$ at the time instant $t = t_1$ and the value of the process $\xi(t_2)$ at the time instant $t = t_2$. The expression $P_2(x_1, t_1; x_2, t_2)\, dx_1\, dx_2$ is the probability of the conditions $x_1 < \xi(t_1) < x_1 + dx_1$ and $x_2 < \xi(t_2) < x_2 + dx_2$ being satisfied. In general, any n values $\xi(t_1), \xi(t_2), ..., \xi(t_n)$ of the random process at the time instants $t_1, ..., t_n$ can be considered as n random variables. The collection of these variables is characterized by the joint probability density $P_n[x_1, t_1; x_2, t_2; ...; x_n, t_n]$, where $P_n\, dx_1 \cdots dx_n$ is the probability of satisfying the conditions

$$x_1 < \xi(t_1) < x_1 + dx_1, ..., x_n < \xi(t_n) < x_n + dx_n.$$

The probabilistic characteristics of the random process are completely known if the function P_n is known for arbitrary n.

The simplest type of random process is characterized by the independence of values of $\xi(t)$ at different time instants. Therefore for such a process the function

$$P_n[x_1, t_1; x_2, t_2; ...; x_n, t_n] = P_1(x_1, t_1) \cdot P_1(x_2, t_2) \cdots P_1(x_n, t_n). \tag{2.68}$$

Here P_1 is the one-dimensional distribution law, from which any n-dimensional law can be constructed according to (2.68).

Another example is the Markov random process, named after the famous mathematician A. A. Markov, who first investigated processes of this type. The complete probability density P_n for the Markov process is obtained from $P_2(x_1, t_1; x_2, t_2)$. Let $P(x_n, t_n \mid x_{n-1}, t_{n-1})$ be the conditional probability density of $\xi(t_n)$ at the time instant $t = t_n$ provided that at the time instant t_{n-1} preceding the instant t_n $(t_{n-1} < t_n)$, the value of $\xi(t_{n-1})$ was equal to x_{n-1}. For the Markov process this conditional probability density does not change, even if the values $x_{n-2}, ..., x_1$ of the process $\xi(t)$ at the preceding time instants $t_{n-2}, ..., t_1$ $(t_n > t_{n-1} > t_{n-2} > ... > t_1)$ become known. If $t = t_{n-1}$ is regarded as the given time instant, all values of $t > t_{n-1}$ refer to the future, while those of $t < t_{n-1}$ refer to the past. Then it can be considered that the probabilistic characteristics of the Markov process evaluated for future time instants are determined by the value x_{n-1} of the process at the given time instant t_{n-1} and do not depend on the "previous history" of this process, i.e., on its values in the past, for $t < t_{n-1}$.

We will show that in this case all the P_n can be expressed in terms of P_2. First of all, the one-dimensional density distribution P_1 is obtained from P_2 with the aid of formula (2.54):

$$P_1(x_1, t_1) = \int_{x_2=-\infty}^{x_2=\infty} P_2(x_1, t_1; x_2, t_2)\, dx_2. \tag{2.69}$$

In general, any "lower" densities P_i can be obtained from "higher" P_k $(k > i)$ by integrating with respect to those variables which do not appear in P_i.

In view of the independence of the conditional probability density of the "previous history" of the process we can write (taking $n > 2$):

$$P(x_n, t_n \mid x_{n-1}, t_{n-1}; \ldots; x_1, t_1) = P(x_n, t_n \mid x_{n-1}, t_{n-1}). \tag{2.70}$$

Here the same letter P has been used for two different probability densities on the left and right sides, which, however, are equal to one another for a Markov process.

According to theorem (2.6) for the probability of a compound event, the equality

$$P_2(x_n, t_n; x_{n-1}, t_{n-1}) = P(x_n, t_n \mid x_{n-1}, t_{n-1}) \cdot P_1(x_{n-1}, t_{n-1}) \tag{2.71}$$

holds. Therefore formula (2.70) can be rewritten thus:

$$P(x_n, t_n \mid x_{n-1}, t_{n-1}; \ldots; x_1, t_1) = \frac{P_2(x_n, t_n; x_{n-1}, t_{n-1})}{P_1(x_{n-1}; t_{n-1})}. \tag{2.72}$$

But by the probability theorem for a compound event

$$\begin{aligned} P_n(x_1, t_1; x_2, t_2; \ldots; x_n, t_n) &= P_{n-1}(x_1, t_1; \ldots; x_{n-1}, t_{n-1}) P(x_n, t_n \mid x_{n-1}, t_{n-1}; \ldots; x_1, t_1) \\ &= P_{n-1}(x_1, t_1; \ldots; x_{n-1}, t_{n-1}) \cdot \frac{P_2(x_n, t_n; x_{n-1}, t_{n-1})}{P_1(x_{n-1}, t_{n-1})}. \end{aligned} \tag{2.73}$$

Hence it is seen that for known P_1 and P_2, in this case any function P_n reduces to P_{n-1}, and this in turn to P_{n-2}, and so on, as a final result, to the function P_2. Thus for a Markov process any function P_n is expressed in terms of P_2.

The partition of random processes into stationary and nonstationary ones is very important. A random process is called stationary if all the distribution densities P_n do not depend on the shift of all points $t_1, t_2, \ldots, t_n$ along the time axis by the same magnitude t_0. For a stationary process the equality

$$P_n(x_1, t_1; x_2, t_2; \ldots; x_n, t_n) = P_n(x_1, t_1 + t_0; x_2, t_2 + t_0; \ldots; x_n, t_n + t_0) \tag{2.74}$$

holds. Thus the statistical character of a stationary random process remains invariant with time. The stationary random process is the analog of the steady-state process. From (2.74) it follows that for the one-dimensional distribution density P_1 the relation

$$P_1(x_1, t_1) = P_1(x_1, t_1 + t_0) \tag{2.75}$$

holds. But from this it follows that P_1 does not depend on t at all, i.e.,

$$P_1(x_1, t_1) = P_1(x_1). \tag{2.76}$$

For the two-dimensional distribution density, Eq. (2.74) takes the form

$$P_2(x_1, t_1; x_2, t_2) = P_2(x_1, t_1 + t_0; x_2, t_2 + t_0). \tag{2.77}$$

This condition means that P_2 does not depend on the time instants t_1 and t_2 themselves, but on their difference $t_2 - t_1 = \tau$:

$$P_2(x_1, t_1; x_2, t_2) = P_2(x_1, x_2, \tau). \tag{2.78}$$

The joint distribution densities of the variables $\xi(t_i)$ and $\eta(t_j)$ serve as the characteristics of the random processes $\xi(t)$ and $\eta(t)$ as related to one another. For example, $P_2(x_1, t_1; y_2, t_2)\, dx_1\, dy_2$ is the probability that $x_1 < \xi(t_1) < x_1 + dx_1$ and $y_2 < \eta(t_2) < y_2 + dy_2$.

If all such functions P_n do not depend on the shift of the time instants $t_1, t_2, \ldots$ by the same magnitude t_0, then the processes $\xi(t)$ and $\eta(t)$ are called "stationary dependent."

The mean value or mathematical expectation of the random process $\xi(t)$ at the instant $t = t_1$ is defined as the mathematical expectation of the random variable $\xi(t_1)$. To reduce the amount of notation, $\xi(t_1)$ can be replaced by $x(t_1)$. Then

$$M\{x(t_1)\} = \int_{-\infty}^{\infty} x_1 P_1(x_1, t_1)\, dx_1. \tag{2.79}$$

The mean square is defined in an analogous manner:

$$M\{x^2(t_1)\} = \int_{-\infty}^{\infty} x_1{}^2 P_1(x_1, t_1)\, dx_1 . \tag{2.80}$$

In view of condition (2.76), we can write for a stationary random process,

$$M\{x(t_1)\} = \int_{-\infty}^{\infty} x_1 P_1(x_1)\, dx_1 \tag{2.81}$$

and

$$M\{x^2(t_1)\} = \int_{-\infty}^{\infty} x_1{}^2 P_1(x_1)\, dx_1 . \tag{2.82}$$

The mean value of the product of the variables $x(t_1) = x_1$ and $x(t_2) = x_2$ is denoted by $K_x(t_1, t_2)$ and is called the "correlation" (or autocorrelation) function. By the definition of the mean value

$$\begin{aligned} K_x(t_1, t_2) &= M\{x(t_1)x(t_2)\} \\ &= \int_{-\infty}^{\infty}\int_{-\infty}^{\infty} x_1 x_2 P_2(x_1, t_1; x_2, t_2)\, dx_1\, dx_2 . \end{aligned} \tag{2.83}$$

In view of condition (2.78), formula (2.83) can be simplified for a stationary random process, since K_x depends only on $\tau = t_2 - t_1$:

$$\begin{aligned} K_x(\tau) &= M\{x(t_1)x(t_1 + \tau)\} \\ &= \int_{-\infty}^{\infty}\int_{-\infty}^{\infty} x_1 x_2 P_2(x_1, x_2, \tau)\, dx_1\, dx_2 . \end{aligned} \tag{2.84}$$

In the particular case of $\tau = 0$, we find from (2.84), with (2.82) taken into account,

$$K_x(0) = [K_x(\tau)]_{\tau=0} = M\{x^2(t_1)\}. \tag{2.85}$$

The cross-correlation function of the processes $x(t)$ and $y(t)$ is defined in a manner analogous to (2.83):

$$\begin{aligned} K_{xy}(t_1, t_2) &= M\{x(t_1)y(t_2)\} \\ &= \int_{-\infty}^{\infty}\int_{-\infty}^{\infty} x_1 y_2 P_2(x_1, t_1; y_2, t_2)\, dx_1\, dy_2 . \end{aligned} \tag{2.86}$$

If the processes $x(t)$ and $y(t)$ are stationary and moreover stationary dependent, then

$$P_2(x_1, t_1; y_2, t_2) = P_2(x_1, y_2, \tau), \tag{2.87}$$

where $\tau = t_2 - t_1$. Then K_{xy} depends only on τ:

$$K_{xy}(\tau) = M\{x(t_1)y(t_1 + \tau)\}$$
$$= \int_{-\infty}^{\infty} \int_{-\infty}^{\infty} x_1 y_2 P_2(x_1, y_2, \tau)\, dx_1\, dy_2 . \tag{2.88}$$

The correlation function, in the same way as the correlation coefficient (2.61) or the moment (2.60), is an estimate of the connection between values of the random process at different time instants.

From the very definition of the correlation function $K_x(t_1, t_2)$ it follows that $K_x(t_1, t_2) = K_x(t_2, t_1)$, while for a stationary random process $K_x(\tau) = K_x(-\tau)$.

There exists a subclass of stationary random processes called "ergodic," for which the "ensemble average" (i.e., the mathematical expectation) with probability equal to unity, is equal to the "time average." For example, the time average of the variable $x(t)$, defined by the expression

$$\hat{x} = \lim_{T\to\infty} \frac{1}{2T} \int_{-T}^{T} x(t)\, dt,$$

is equal to the ensemble average (2.81). Further, the time average of the square of the function $x(t)$

$$\hat{x}^2 = \lim_{T\to\infty} \frac{1}{2T} \int_{-T}^{T} x^2(t)\, dt$$

is equal to the ensemble average (2.82). An analogous equality also holds for the averages of the product:

$$K_x(\tau) = \int_{-\infty}^{\infty} \int_{-\infty}^{\infty} x_1 x_2 P_2(x_1, x_2, \tau)\, dx_1\, dx_2$$
$$= \lim_{T\to\infty} \frac{1}{2T} \int_{-T}^{T} x(t)x(t+\tau)\, dt. \tag{2.89}$$

In particular, setting $\tau = 0$, we find:

$$K_x(0) = \lim_{T\to\infty} \frac{1}{2T} \int_{-T}^{T} x^2(t)\, dt = \hat{x}^2(t). \tag{2.90}$$

This expression is sometimes called the "power" in the signal $x(t)$. If $x(t)$ is a voltage applied to a resistance of 1 ohm, then $\hat{x}^2(t)$ is equal to the mean value of the power extracted by this resistance.

The Fourier transform of the autocorrelation function $K_x(\tau)$ of a stationary random process

$$S_x(\omega) = \int_{-\infty}^{\infty} K_x(\tau) e^{-j\omega\tau} \, d\tau \tag{2.91}$$

is called the "spectral density" of the random process $x(t)$. The Fourier transform of the function $K_{xy}(\tau)$ of (2.88), i.e.,

$$S_{xy}(\omega) = \int_{-\infty}^{\infty} K_{xy}(\tau) e^{-j\omega\tau} \, d\tau, \tag{2.92}$$

is called the "cross-spectral density" of the processes $x(t)$ and $y(t)$. The physical meaning of the function $S_x(\omega)$ can be explained by substituting into the formula for the inverse Fourier transform

$$K_x(\tau) = \frac{1}{2\pi} \int_{-\infty}^{\infty} S_x(\omega) e^{j\omega\tau} \, d\omega \tag{2.93}$$

the value $\tau = 0$. Then we will obtain

$$K_x(0) = \frac{1}{2\pi} \int_{-\infty}^{\infty} S_x(\omega) \, d\omega. \tag{2.94}$$

On the left side of this expression is the power in the signal $x(t)$ [see (2.90)]. Hence the right side of (2.94) also represents the power, but expressed in the form of an integral with respect to the frequency ω. Here the differential $[S_x(\omega) \, d\omega]/2\pi$ has the meaning of that fraction of the signal power which falls in the infinitesimally narrow interval of the frequency spectrum from ω to $\omega + d\omega$.

The random signal for which $S(\omega) = S_0 = \text{const}$ is called "white noise". To this Fourier transform corresponds the pre-image

$$K_x(\tau) = S_0 \cdot \delta(\tau), \tag{2.95}$$

where $\delta(\tau)$ is the so-called unit impulse function, or the Dirac function, defined by the expression

$$\delta(\tau) = 0, \qquad \tau \neq 0, \qquad \int_{-\infty}^{\infty} \delta(\tau) \cdot d\tau = 1. \tag{2.96}$$

The function $\delta(\tau)$ represents an "infinitely high" and "infinitely narrow" impulse, appearing at the instant $\tau = 0$.

From formulas (2.95) and (2.96) it is seen that the autocorrelation function of white noise is equal to zero for $\tau \neq 0$. Hence it follows that correlation between values of white noise at different time instants is absent.

In the subsequent presentation we frequently replace the continuous random process $x(t)$ by a sequence of related random variables $x(t_1)$, ..., $x(t_n)$, which are its values at discrete time instants t_1, ..., t_n. This usually can be done if the frequency spectrum of all realizations of the random process is bounded by an upper limit ω_c, and the duration is bounded by a time T. Strictly speaking, for a process of finite duration the frequency spectrum is infinite; but a frequency ω_c can be found such that for $\omega > \omega_c$ the "tail" of the frequency spectrum is sufficiently small in intensity and does not have any substantial influence. Therefore according to the theorem of V. A. Kotel'nikov [5.16], instead of the function $x(t)$, without loss of information about it only a series of its discrete values $x(t_1)$, $x(t_2)$, ... can be considered, the so-called "samples" —separated by a distance in time not larger than $\Delta t = \pi/\omega_c$. The total number of samples is obtained by equating $T/\Delta t = T\omega_c/\pi = 2Tf_c$, where $f_c = \omega_c/2\pi$.

Let us denote $x(t_i)$ by x_i $(i = 1, ..., k)$ and introduce the vector

$$X = (x_1, x_2, ..., x_k). \tag{2.97}$$

The coordinates of the vector X are the "successive values" of the process $x(t)$ at the instants t_1, t_2, ..., t_k. The vector X can be considered in a k-dimensional space with Cartesian coordinates x_1, x_2, ..., x_k

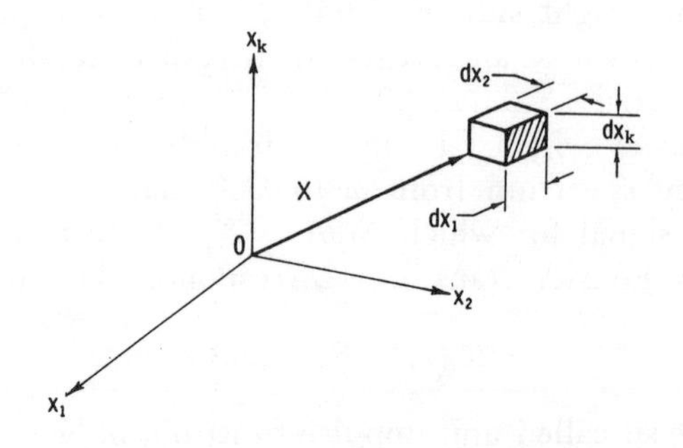

FIG. 2.6

(Fig. 2.6). The vector X is random since its coefficients are random variables. Let us denote by

$$P(X) = P(x_1, x_2, ..., x_k) \tag{2.98}$$

the probability density of this vector; this is not like that other one, the joint distribution density of the coordinates x_1, ..., x_k. The expression

$P(X)\,d\Omega(X)$, where $d\Omega(X) = dx_1, ..., dx_k$, is the probability of the end of the vector X occurring in the infinitesimally small volume $d\Omega(X)$ shown in Fig. 2.6. Since the end of the vector X is always found somewhere, then

$$\int_{\Omega(X)} P(X)\,d\Omega(X) = 1. \tag{2.99}$$

Here $\Omega(X)$ is the entire k-dimensional space of the vector X. In this case the random process is replaced by the random vector X.

2. Variational Methods

The finding of the extremum of some functional relates to a series of so-called variational problems whose role in the theory of optimal systems is evident.

Various groups of methods are applicable to the solution of variational problems. In 1696, when Johann Bernoulli posed the problem of the curve of steepest slope (brachistochrone), the so-called classical calculus of variations began to be developed. In the eighteenth century general methods of solving variational problems were given by Euler and Lagrange. A series of brilliant papers of the nineteenth century completed the structure of the classical calculus of variations. Its elementary foundations are examined in this section. For a more detailed acquaintance we refer the reader to the literature [2.4–2.8].

In the twentieth century the so-called direct methods of solving problems, also taking their basis from Euler, began to be applied. In recent time these methods have found application in physics and engineering. They are mentioned briefly below (see also [2.7, 2.8]).

The new problems arising in the middle of the twentieth century, among which the problems in the theory of optimal automatic control systems are not the newest, have caused the appearance of new methods of solving variational problems: the method of dynamic programming developed by the American mathematician R. Bellman and his coworkers, and also the maximum principle proposed and justified by the Soviet mathematician Academician L. S. Pontryagin and his pupils. These methods, which deserve particular consideration, will be presented in subsequent sections of this chapter.

In order to formulate the simplest problem in the calculus of variations, let us consider the functional I depending on the function $y(x)$:

$$I = \int_{x_0}^{x_1} F(x, y, y')\,dx, \tag{2.100}$$

where F is a prescribed function with the arguments x, y, and $y' = dy/dx$. We first consider that the limits of integration x_0 and x_1 are given constants. Let the function F be single-valued and continuous, together with its partial derivatives, up to the third order inclusively, for all values of x and y belonging to some region R of the (x, y) plane. Let us stipulate that the function $y = f(x)$ is single-valued and continuous on the segment (x_0, x_1) and has a continuous first derivative on this segment—in brief, that it belongs to the class $C^{(1)}$. We will call the curves of $f(x)$ admissible if they belong to the class $C^{(1)}$, lie entirely in the region R, and pass through the given points (x_0, y_0) and (x_1, y_1), where $y_0 = f(x_0)$ and $y_1 = f(x_1)$.

The problem is posed thus: to find among the admissible curves $f(x)$ the one for which the integral (2.100) has the least value.

It is not difficult to obtain a necessary condition that the curve must satisfy; this is the solution of the problem posed above. Let $f(x)$ be such a curve. Let us replace this function in the integral (2.100) by some other "close" function

$$y = f(x) + \alpha\eta(x), \tag{2.101}$$

where $\eta(x)$ is an arbitrary function of class $C^{(1)}$, vanishing at the ends of the segment,

$$\eta(x_0) = \eta(x_1) = 0, \tag{2.102}$$

and α is some small number. Then the integral I becomes a function $I(\alpha)$ of this number:

$$I(\alpha) = \int_{x_0}^{x_1} F[x, f(x) + \alpha\eta(x), f'(x) + \alpha\eta'(x)]\, dx. \tag{2.103}$$

If α is sufficiently small, then $I(\alpha)$ can be expanded in a series in powers of α:

$$I(\alpha) = [I(\alpha)]_{\alpha=0} + \frac{\alpha}{1!}\left(\frac{dI}{d\alpha}\right)_{\alpha=0} + \frac{\alpha^2}{2!}\left(\frac{d^2I}{d\alpha^2}\right)_{\alpha=0} + \cdots. \tag{2.104}$$

The expressions $\alpha(dI/d\alpha)_{\alpha=0}$ and $\alpha^2(d^2I/d\alpha^2)_{\alpha=0}$ are called the first and second variations of the integral I and are denoted by δI and $\delta^2 I$, respectively.

If the function $f(x)$ gives a minimum for the integral I, then

$$(dI/d\alpha)_{\alpha=0} = 0, \tag{2.105}$$

where this condition must be fulfilled for any function $\eta(x)$ belonging to the class $C^{(1)}$ and satisfying the boundary conditions (2.102).

Let us expand the expression for δI. From (2.103), by differentiating under the integral sign with respect to α and then setting $\alpha = 0$, we find:

$$\left[\frac{dI(\alpha)}{d\alpha}\right]_{\alpha=0} = \int_{x_0}^{x_1} \left[\frac{\partial F}{\partial y} \eta(x) + \frac{\partial F}{\partial y'} \eta'(x)\right] dx. \tag{2.106}$$

If we assume that the function $f(x)$ has a continuous second derivative $f''(x)$ (with the aid of more detailed arguments, it is shown that we can also manage without this assumption), then the second term in (2.106) can be integrated by parts:

$$\int_{x_0}^{x_1} \frac{\partial F}{\partial y'} \eta'(x)\, dx = \left[\eta(x) \frac{\partial F}{\partial y'}\right]_{x_0}^{x_1} - \int_{x_0}^{x_1} \eta(x) \frac{d}{dx}\left(\frac{\partial F}{\partial y'}\right) dx. \tag{2.107}$$

The first term in this expression vanishes by virtue of condition (2.102). Therefore, by combining (2.106) and (2.107) we arrive at the formula

$$\left[\frac{dI(\alpha)}{d\alpha}\right]_{\alpha=0} = \int_{x_0}^{x_1} \eta(x) \left[\frac{\partial F}{\partial y} - \frac{d}{dx}\left(\frac{\partial F}{\partial y'}\right)\right] dx. \tag{2.108}$$

If the left side of (2.108) is equal to zero according to (2.105), then the integral on the right side of (2.108) must also be equal to zero, and this must be true, moreover, for "arbitrary" functions $\eta(x)$ of class $C^{(1)}$. It is not difficult to prove [2.5–2.8] that this is possible only under the condition that the bracket under the integral is equal to zero:

$$\frac{\partial F}{\partial y} - \frac{d}{dx}\left(\frac{\partial F}{\partial y'}\right) = 0. \tag{2.109}$$

This is also the required "necessary" condition. It is a differential equation. By determining its solutions—the so-called "extremals"—we obtain the curves among which the solution to the problem must be sought. Equation (2.109) bears the name of Euler's equation.

For example, let it be required to find the curve $y = f(x)$ of class $C^{(1)}$ passing through the points M_0 and M_1 in the (x, y) plane (Fig. 2.7) with the prescribed coordinates

$$\begin{aligned} x_0 &= 0, & y_0 &> 0, \\ x_1 &> 0, & y_1 &= 0, \end{aligned} \tag{2.110}$$

and minimizing the integral

$$I = \int_0^{x_1} [y^2 + T^2(dy/dx)^2]\, dx, \tag{2.111}$$

where $T^2 = \text{const}$. In this case

$$F = y^2 + T^2(dy/dx)^2 = y^2 + T^2(y')^2. \tag{2.112}$$

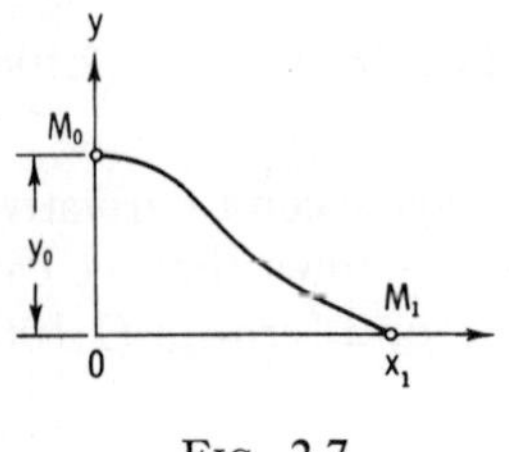

FIG. 2.7

Therefore

$$\partial F/\partial y = 2y, \qquad \partial F/dy' = 2T^2 y'. \tag{2.113}$$

Hence Euler's equation (2.109) takes the form

$$2y - 2T^2 \frac{d}{dx} y' = 0, \tag{2.114}$$

or

$$y - T^2\, d^2y/dx^2 = 0. \tag{2.115}$$

The solution to this equation has the form

$$y = C_1 e^{x/T} + C_2 e^{-x/T}, \tag{2.116}$$

where C_1 and C_2 are constants. Substituting the boundary conditions (2.110) into expression (2.116), we find the values of the constants

$$C_1 = \frac{y_0}{1 + e^{2x_0/T}}, \qquad C_2 = \frac{y_0}{1 + e^{-2x_0/T}}. \tag{2.117}$$

Thus the solution of Euler's equation is given by the function

$$y = y_0 \left(\frac{e^{x/T}}{1 + e^{2x_0/T}} + \frac{e^{-x/T}}{1 + e^{-2x_0/T}} \right). \tag{2.118}$$

This is the unique solution of Eq. (2.115) satisfying the boundary conditions (2.110). Meanwhile, it was shown above that if there exists a solution to the variational problem in the class $C^{(1)}$ of curves, then it must be sought among the solutions of Euler's equation. Therefore (under the hypothesis that a solution exists in the class $C^{(1)}$ of curves) the curve of (2.118) gives the integral (2.111) a stationary value. In other words, the integral can assume either a maximal or a minimal value, or the function $I(\alpha)$ has a point of inflection for $\alpha = 0$. Only additional arguments permit it to be established that (2.118) actually gives a minimum for the integral (2.111) and is the solution to the problem. But in general, it should not at all be assumed that a solution of Euler's equation, even if one exists, means a solution to the corresponding variational problem. Sometimes a solution of Euler's equation does not generally exist. For example, if the function F in (2.100) depends only on x and y, then Euler's equation takes the form

$$\frac{\partial F(x, y)}{\partial y} = 0. \tag{2.119}$$

This is not now a differential equation. For example, if $F(x, y) = xy$, then Eq. (2.119) has the form $x = 0$. In the general case, by solving Eq. (2.119) we can find a single one or several curves $y = f(x)$. But in general they do not pass through the required points (x_0, y_0) and (x_1, y_1). Therefore in this case an extremal of class $C^{(1)}$ satisfying the conditions of the problem can only be found for exceptional values of the coordinates of the boundary points.

In other cases it can turn out that a solution of the variational problem in the class $C^{(1)}$ does not exist in general. For example, let us find the minimum of the integral

$$I = \int_{-1}^{1} y^2[1 - (dy/dx)]^2 \, dx \tag{2.120}$$

for the boundary conditions $y(-1) = 0$, $y(1) = 1$. It is not difficult to see that the integrand $F \geqslant 0$, where the absolute minimum $F = 0$ of this function (and hence the minimum of the integral I as well) is attained on the broken line $y = 0$ for $x \leqslant 0$, $y = x$ for $x > 0$ (Fig. 2.8). But this broken line does not belong to the class $C^{(1)}$. By rounding off the line near the break point, it can be seen that the integral (2.120) is greater than zero for any curves of class $C^{(1)}$, although its value can also be made arbitrarily small.

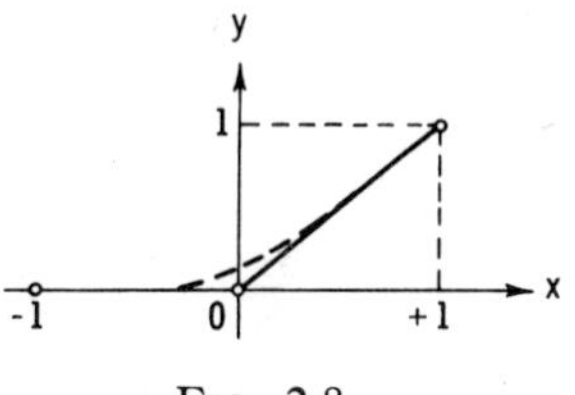

FIG. 2.8

What was presented above shows that the expression "a solution of the variational problem reduces to a solution of Euler's equation" must be accompanied by substantial reservations even for the restricted class considered.

If the integral I depends on several functions of the same variable x, then by a method analogous to that presented above, a necessary condition can be found for curves solving the variational problem in the form of Euler's equation. For example, let

$$I = \int_{x_0}^{x_1} F(x, y, z, y', z') \, dx, \tag{2.121}$$

where $y' = dy/dx$, $z' = dz/dx$. Here the boundary values $y(x_0)$, $y(x_1)$, $z(x_0)$, $z(x_1)$ are prescribed. Regarding z temporarily as a fixed function, we obtain instead of (2.121) an integral of the previous type (2.100), depending only on the form of the function $y(x)$. In such a case the function realizing an extremum must satisfy the same equation as (2.109):

$$\frac{\partial F}{\partial y} - \frac{d}{dx}\left(\frac{\partial F}{\partial y'}\right) = 0. \qquad (2.122)$$

Fixing $y(x)$ and reasoning analogously with respect to z, another equation can be obtained:

$$\frac{\partial F}{\partial z} - \frac{d}{dx}\left(\frac{\partial F}{\partial z'}\right) = 0. \qquad (2.123)$$

As a result it turns out that the functions y and z must satisfy the set of Eqs. (2.122) and (2.123). By solving these equations simultaneously, the desired functions $y(x)$, $z(x)$ can be sought among their solutions.

The variational problem can be generalized to the case when derivatives of higher orders enter into the integrand. We will stipulate that the function $y = f(x)$ belongs to the class $C^{(n)}$, if it is single-valued and continuous together with its derivatives up to nth order inclusively. Let the function $y = f(x)$ of class $C^{(n)}$ together with its derivatives up to $(n - 1)$th order assume prescribed values for $x = x_0$ and $x = x_1$, i.e.,

$$\begin{aligned} y &= y_0, \quad y' = y_0', \ldots, y^{(n-1)} = y_0^{(n-1)} \quad \text{for} \quad x = x_0, \\ y &= y_1, \quad y' = y_1', \ldots, y^{(n-1)} = y_1^{(n-1)} \quad \text{for} \quad x = x_1. \end{aligned} \qquad (2.124)$$

We will find a function $y = f(x)$ of this type that gives a minimum for the integral

$$I = \int_{x_0}^{x_1} F(x, y, y', \ldots, y^{(n)})\, dx. \qquad (2.125)$$

Let the function F have continuous partial derivatives up to the $(n + 2)$th order.

We replace y by a function $y + \alpha\eta(x)$ close to the required one, where $\eta(x)$ is a function of class $C^{(n)}$, vanishing with its $n - 1$ derivatives at the boundary points. Substituting into the integral (2.125), we obtain the expression

$$\left[\frac{dI(\alpha)}{d\alpha}\right]_{\alpha=0} = \int_{x_0}^{x_1} \left[\frac{\partial F}{\partial y}\eta(x) + \frac{\partial F}{\partial y'}\eta'(x) + \cdots + \frac{\partial F}{\partial y^{(n)}}\eta^{(n)}(x)\right] dx. \qquad (2.126)$$

We integrate the terms on the right side of (2.126) k times by parts:

$$\int_{x_0}^{x_1} \frac{\partial F}{\partial y^{(k)}} \eta^{(k)}(x)\, dx = \left[\frac{\partial F}{\partial y^{(k)}} \eta^{(k-1)}(x) - \frac{d}{dx}\left(\frac{\partial F}{\partial y^{(k)}}\right) \eta^{(k-2)}(x) + \cdots + (-1)^{k-1} \frac{d^{(k-1)}}{dx^{(k-1)}}\left(\frac{\partial F}{\partial y^{(k)}}\right) \eta(x)\right]_{x_0}^{x_1} + (-1)^k \int_{x_0}^{x_1} \frac{d^k}{dx^k}\left(\frac{\partial F}{\partial y^{(k)}}\right) \eta(x)\, dx. \tag{2.127}$$

The first term on the right side is equal to zero by virtue of the conditions imposed on $\eta(x)$ at the boundary points. Substituting the expression obtained into (2.126), we find:

$$\left[\frac{dI(\alpha)}{d\alpha}\right]_{\alpha=0} = \int_{x_0}^{x_1} \eta(x) \left[\frac{\partial F}{\partial y} - \frac{d}{dx}\left(\frac{\partial F}{\partial y'}\right) + \cdots + (-1)^n \frac{d^n}{dx^n}\left(\frac{\partial F}{\partial y^{(n)}}\right)\right] dx. \tag{2.128}$$

Since $(dI/d\alpha)_{\alpha=0} = 0$ for any function $\eta(x)$ of the type stated above, then as can be shown (see [2.8]) the bracket under the integral must be equal to zero:

$$\frac{\partial F}{\partial y} - \frac{d}{dx}\left(\frac{\partial F}{\partial y'}\right) + \cdots + (-1)^n \frac{d^n}{dx^n}\left(\frac{\partial F}{\partial y^{(n)}}\right) = 0. \tag{2.129}$$

Equation (2.129) is called the Euler-Poisson equation. If a function of class $C^{(n)}$ gives a minimum for the integral (2.125), then it must satisfy Eq. (2.129).

According to [1.9], let us consider the determination of a function minimizing the integral

$$I_V = \int_0^\infty V\, dt = \int_0^\infty \left[x^2 + \gamma_1 \left(\frac{dx}{dt}\right)^2 + \cdots + \gamma_{n-1}\left(\frac{d^{n-1}x}{dt^{n-1}}\right)^2\right] dt, \tag{2.130}$$

where x is a transient process in a stable linear system with the prescribed initial conditions

$$(x)_{t=0} = x(0), \quad \left(\frac{dx}{dt}\right)_{t=0} = x^{(1)}(0), \ldots, \left(\frac{d^{n-1}x}{dt^{n-1}}\right)_{t=0} = x^{(n-1)}(0). \tag{2.131}$$

Since the system is stable,

$$x \to 0 \quad \text{and} \quad x^{(i)} = d^i x/dt^i \to 0 \quad \text{as} \quad t \to \infty \quad (i = 1, \ldots, n-1). \tag{2.132}$$

Therefore

$$x(\infty) = x^{(1)}(\infty) = \cdots = x^{(n-1)}(\infty) = 0. \tag{2.133}$$

Conditions (2.131) and (2.133) can be regarded as the boundary values for $t = 0$ and $t = \infty$.

With the replacement of y by x and x by t, the Euler-Poisson equation (2.129) takes the form

$$\frac{\partial F}{\partial x} - \frac{d}{dt}\left(\frac{\partial F}{\partial x^{(1)}}\right) + \cdots + (-1)^n \frac{d^n}{dt^n}\left(\frac{\partial F}{\partial x^{(n)}}\right) = 0. \tag{2.134}$$

In the example considered

$$F = x^2 + \gamma_1(x^{(1)})^2 + \cdots + \gamma_{n-1}(x^{(n-1)})^2. \tag{2.135}$$

Substituting (2.135) into (2.134), we arrive at the differential equation

$$x - \gamma_1 x^{(2)} + \gamma_2 x^{(4)} + \cdots + (-1)^{n-1}\gamma_{n-1}x^{(2[n-1])} = 0. \tag{2.136}$$

This is a $2(n-1)$th-order linear equation with constant coefficients. For it we construct the characteristic equation

$$H(p) = 1 - \gamma_1 p^2 + \gamma_2 p^4 + \cdots + (-1)^{n-1}\gamma_{n-1}p^{2(n-1)} = 0. \tag{2.137}$$

We will assume that this equation has a root $p_1 = \alpha_1 + j\beta_1$. Since only even powers of p appear in equation (2.137), it also has the root $-p_1 = -\alpha_1 - j\beta_1$. Hence it follows that all the roots of Eq. (2.137) are arranged in pairs which are symmetric with respect to the origin (Fig. 2.9). Therefore, if the equation does not contain purely imaginary roots, then one-half the roots, i.e., $n - 1$ roots, are situated in the left half-plane, and the other one-half of the roots in the right. Hence the characteristic polynomial can be expressed in the form of a product

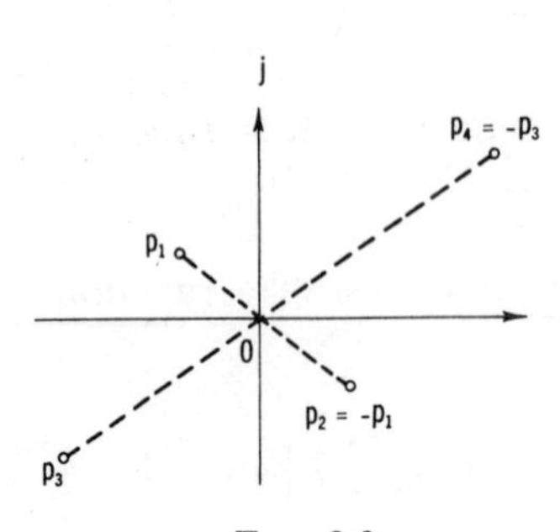

FIG. 2.9

$$H(p) = M(p)N(p), \tag{2.138}$$

where the roots of the $(n-1)$th-degree polynomial $M(p)$ are situated in the left half-plane, while the roots of the polynomial $N(p)$—also of $(n-1)$th degree— are in the right. Evidently, just the roots of the polynomial $M(p)$ must be taken into account in writing the solution for the extremal $x^*(t)$, while the roots of $N(p)$ lying in the right half-plane must be discarded in view of the boundary conditions (2.133). In other words, the terms in the solution of the form $C_i \exp(p_i t)$, where

p_i is in the right half-plane, must be absent (i.e., the C_i corresponding to them must be equal to zero); otherwise the boundary conditions (2.133) will be violated.

From certain considerations the kind of quadratic form in (2.130) can be chosen and the corresponding extremal sought. But as was done in [1.9], the equation of the extremal $x^*(t)$ can be assigned and the integral (2.130) corresponding to it sought. Let $x^*(t)$ be used as the solution of the differential equation

$$M(D)x^*(t) = 0, \tag{2.139}$$

where

$$D = d/dt \tag{2.140}$$

is used as the symbolic notation for the differential operator. Assigning Eq. (2.139) is equivalent to prescribing the roots or the coefficients of the characteristic equation

$$M(p) = 0. \tag{2.141}$$

Hence, let it be given that

$$M(p) = 1 + \vartheta_1 p + \vartheta_2 p^2 + \cdots + \vartheta_{n-1} p^{n-1}, \tag{2.142}$$

where $\theta_i > 0$. Since the roots of $N(p)$ are opposite to the roots of $M(p)$, then the polynomial $N(p)$ can be written by knowing $M(p)$ (the expression for $N(p)$ is easily derived from Viet's formula):

$$N(p) = 1 - \vartheta_1 p + \vartheta_2 p^2 + \cdots + (-1)^{n-1}\vartheta_{n-1} p^{n-1}. \tag{2.143}$$

Now $H(p)$ can be found from Eq. (2.138). From this equation the coefficients γ_i in formula (2.137) for $H(p)$ can be obtained after substituting (2.142) and (2.143) into its expression. Hence the form of the integral I_V (2.130) can be found, for which the extremal is the given curve $x^*(t)$ (or, with variable parameters for this curve, a family of extremals).

For example, let us consider the integral whose extremal is the solution of the second-order equation

$$\frac{d^2x^*}{dt^2} + 2d_0\omega_0 \frac{dx^*}{dt} + \omega_0^2 x^* = 0, \tag{2.144}$$

where $d_0 > 0$, $\omega_0 > 0$. As is known, for $d_0 < 1$, the solution $x^*(t)$ of this equation has the form of damped oscillations, while for $d_0 > 1$ the process is aperiodic. The quantity ω_0 is called the characteristic frequency of the oscillations and is close to the actual frequency of the damped oscillations for small values of d_0. Let us set

$$\omega_0 = \frac{1}{T_0}, \qquad d_0 = \sqrt{\left(\frac{\sigma+1}{2}\right)}. \tag{2.145}$$

Then Eq. (2.144) takes the form

$$T_0^2 \frac{d^2x^*}{dt^2} + 2T_0\sqrt{\left(\frac{\sigma+1}{2}\right)}\frac{dx^*}{dt} + x^* = 0. \tag{2.146}$$

Hence the coefficients of the polynomial $M(p)$ are

$$\vartheta_1 = 2T_0\sqrt{\left(\frac{\sigma+1}{2}\right)}, \qquad \vartheta_2 = T_0^2. \tag{2.147}$$

From the identity (2.138) it follows in this case that

$$(1 + \vartheta_1 p + \vartheta_2 p^2)(1 - \vartheta_1 p + \vartheta_2 p^2) = 1 - \gamma_1 p^2 + \gamma_2 p^4.$$

Removing the parentheses on the left side and equating coefficients in p^2 and p^4 on the left and right sides, we find

$$\gamma_1 = \vartheta_1^2 - 2\vartheta_2; \qquad \gamma_2 = \vartheta_2^2. \tag{2.148}$$

From this,

$$\gamma_1 = 4T_0^2\frac{\sigma+1}{2} - 2T_0^2 = 2\sigma T_0^2, \qquad \gamma_2 = T_0^4. \tag{2.149}$$

Hence the integral I_V, for which the solution of Eq. (2.144) or Eq. (2.146) is an extremal, has the form

$$I_V = \int_0^\infty [x^2 + \gamma_1(x^{(1)})^2 + \gamma_2(x^{(2)})^2]\,dt$$

$$= \int_0^\infty [x^2 + 2\sigma T_0^2(x^{(1)})^2 + T_0^4(x^{(2)})^2]\,dt. \tag{2.150}$$

A large group of problems in the calculus of variations contains auxiliary conditions imposed on the solution. The extremum of a functional, defined under such auxiliary conditions, is called a conditional extremum. Usually auxiliary conditions are considered in the form of equalities. For example, let it be required to find the curves $y_1(x)$, ..., $y_n(x)$ giving a minimum for the integral

$$I = \int_{x_0}^{x_1} F(x; y_1, y_2, \ldots, y_n; y_1', y_2', \ldots, y_n')\,dx \tag{2.151}$$

in the presence of the auxiliary conditions

$$\varphi_i(x; y_1, y_2, ..., y_n) = 0 \qquad (i = 1, ..., m; m < n). \tag{2.152}$$

Equations (2.152) are assumed to be independent.

For the solution the method of Lagrange multipliers is used. The integral

$$I^* = \int_{x_0}^{x_1} \Big[F + \sum_{i=1}^{m} \lambda_i(x)\varphi_i\Big]\, dx = \int_{x_0}^{x_1} F^*\, dx \tag{2.153}$$

is constructed, where

$$F^* = F + \sum_{i=1}^{m} \lambda_i(x)\varphi_i \tag{2.154}$$

and the $\lambda_i(x)$ are still unknown functions (Lagrange multipliers). The integral I^* is now investigated for an absolute extremum, i.e., the system of Euler's equations analogous to the system of Eqs. (2.122) and (2.123) is solved:

$$F^*_{y_j} - \frac{d}{dx} F^*_{y_j'} = 0 \qquad (j = 1, ..., n). \tag{2.155}$$

If the system of m equations (2.152) is added to this system of n equations, then in general the number $m + n$ of equations suffices for determining the $m + n$ unknown functions $y_1, ..., y_n, \lambda_1, ..., \lambda_m$, and the boundary conditions $y_j(x_0) = y_{j0}$ and $y_j(x_1) = y_{j1}$ $(j = 1, ..., n)$, which must be combined with the equations of the relations (2.152), give the possibility of determining the $2n$ arbitrary constants in the general solution of the system of Euler's equations.

The auxiliary conditions can have the character of differential equations (the general Lagrange problem):

$$\varphi_i(x, y_1, ..., y_n, y_1', ..., y_n') = 0 \qquad (i = 1, ..., m). \tag{2.156}$$

In this case the solution procedure remains the same. The auxiliary conditions can have the form of integral equations (the isoperimetric problem)

$$\int_{x_0}^{x_1} F_i(x, y_1, ..., y_n, y_1', ..., y_n')\, dx = l_i \qquad (i = 1, ..., m), \tag{2.157}$$

where the l_i are constants, and m can be less than, equal to, or greater than n. This problem can be reduced to the preceding one by introducing new coordinates, on which we will not dwell. Here the procedure for introducing Lagrange multipliers is simplified, since the λ_i turn out to be constants, i.e., the absolute extremum is determined for the integral

$$I^* = \int_{x_0}^{x_1} \left(F + \sum_{i=1}^{m} \lambda_i F_i\right) dx. \tag{2.158}$$

In the problems considered earlier, the curves whose ends were at two fixed points were taken for the admissible curves $y(x)$. In a wider class of problems the boundary points are not fixed, but it is required that they be situated on definite curves or surfaces G_0 and G_1 (Fig. 2.10).

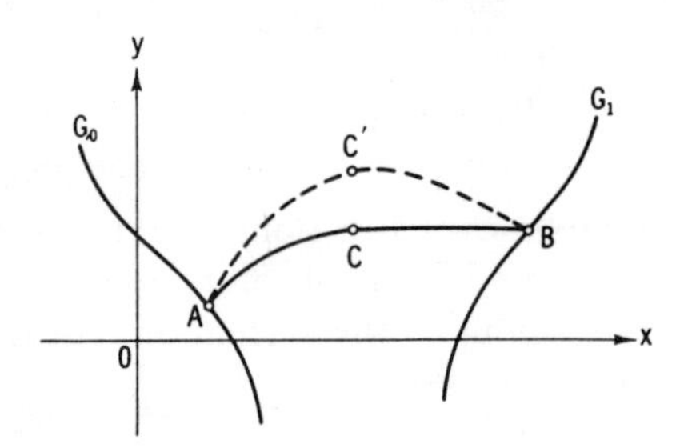

FIG. 2.10

If the curve ACB is the desired one, then after mentally fixing its two boundary points A and B, it can be compared with any other curves $AC'B$ passing through the same points. Since the curve ACB gives the integral I a value smaller than any other one $AC'B$ close to it, it must satisfy Euler's equation. But the solution of Euler's equation contains arbitrary constants. For example, in Fig. 2.10 there are four coordinates of the points A and B. Meanwhile, the very conditions for finding the points A and B on the boundary curves G_0 and G_1 give only two equations for determining their coordinates. A detailed consideration of the variations in the boundary points, which we omit, results in the conclusion that the so-called "transversality conditions" must still be satisfied at these points. By writing out these conditions, the missing relations can be found and the constants in the solutions of Euler's equations determined.

From Fig. 2.8 it is seen that the solution of the variational problem can sometimes be attained only by extending the class of admissible functions and including in the analysis, for example, piecewise-smooth functions. But in this case the analysis is made considerably more complex. It turns out that at break points (for example, the point 0 in Fig. 2.8) additional conditions must be satisfied, the so-called Erdman-

Weierstrass conditions. The problem is still further complicated if the solution is a function with a finite number of discontinuity points of the first kind. In this case the expression for the variation δI is so complex that the analysis of these questions is omitted even in the usual college mathematics textbooks on the calculus of variations. This analysis becomes particularly complicated if the number and location of discontinuity points is not known beforehand.

It was shown above that Euler's equations do not always give the necessary condition for an extremum of a functional in a convenient form. In the process of developing the calculus of variations, various more or less general necessary conditions have been worked out—the Weierstrass condition, the Clebsch condition, the Jacobi condition. By itself the problem of the calculus of variations has been generalized; as a result very general formulations of the problem have appeared—Bolza's problem, Mayer's problem. We will not consider these questions, since they are touched on when examining the maximum principle.

In the twentieth century the direct methods of the calculus of variations have attained considerable prevalence (see [2.7, 2.8]). For illustration only, we will briefly examine here the idea of Ritz's method—one of the simple variants of these methods. In this method functions of the following form are taken as admissible:

$$y_n = \sum_{i=1}^{n} \alpha_i P_i(x), \tag{2.159}$$

where the α_i are constant coefficients, and the P_i are certain prescribed functions. If expression (2.159) is substituted into the formula for the integral I, then the latter is converted into a function of the coefficients α_i:

$$I = I(\alpha_1, \ldots, \alpha_n). \tag{2.160}$$

Now the coefficients α_i can be chosen such that I will be minimized, for example, by solving the system of equations ("not" differential)

$$\partial I/\partial \alpha_i = 0 \qquad (i = 1, \ldots, n). \tag{2.161}$$

If this procedure can be carried out for arbitrary n, then as $n \to \infty$, if the limit exists, we obtain a function $y = \lim y_n$, which under certain additional restrictions is the exact solution of the variational problem. For n large enough the function y_n is an approximate solution of the problem.

We will now relate the possibilities of the methods described above for solving variational problems in the theory of optimal systems.

By comparing these problems with the problems considered in the classical calculus of variations, the following features of problems in optimal system theory can be noted:

(1) Both in the integral being minimized or the functional Q and in the equations of the object and the conditions of the constraints, not only do the coordinates x_i of the object appear, but also the control actions u_j ($j = 1, \ldots, r$).

(2) The constraints usually have the form of inequalities, for example, $|u_j| \leqslant U_j$, where the vector $\bar{u}$ may not only be situated "inside," but also "on the boundary" of the admissible region $\Omega(\bar{u})$ for it.

(3) The solution of the optimal problem often appears as piecewise-continuous functions $u_j(t)$ with a finite number of discontinuity points of the first kind, where the time instants when the jumps u_j occur are not determined in advance (frequently the determination itself of these time instants represents essentially the solution to the problem; see Chapter III).

By itself the first of these features does not present any difficulties. The u_j must only be included as functions taken equally with the x_i. In this case the $(n + r)$-dimensional space with the coordinates $x_1, \ldots, x_n, u_1, \ldots, u_r$ takes the place of the n-dimensional phase space of the vectors $\bar{x}$.

The equations of the system can now be regarded as the limiting conditions

$$\varphi_i = x_i' - f_i(x_1, \ldots, x_n, u_1, \ldots, u_r; t) = 0, \tag{2.162}$$

of the type of conditions (2.156). Sometimes $\bar{u}$ can be eliminated after substituting it from Eq. (2.162) into the integral for Q and the conditions of the constraints, or $\bar{x}$ can be eliminated if $\bar{x}$ can be expressed in terms of $\bar{u}$.

The second feature is already associated with greater difficulties. It is true that the constraints in the form of inequalities can be formally reduced to constraints in the form of equalities. In fact, other functions v_j which are related to the u_j by the equations

$$u_j = \Phi_j(v_j) \tag{2.163}$$

can be introduced in place of the u_j, where the Φ_j are chosen such that for any v_j the functions u_j will not leave the required intervals. If it is required to satisfy the conditions

$$|u_j| \leqslant U_j, \tag{2.164}$$

then we can choose, for example,

$$\Phi_j = U_j \sin v_j \tag{2.165}$$

or select the Φ_j as shown in Fig. 2.11. The substitution (2.165) has been used for continuous systems by Desoer [2.36], while a function of the type in Fig. 2.11 has been used by Miele [2.32]. Ya. Z. Tsypkin [3.30, 3.31] has applied the function (2.165) to discrete-continuous systems.

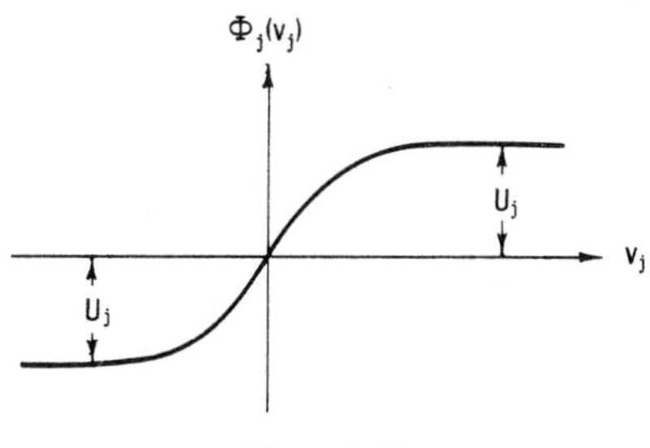

FIG. 2.11

If now $u_j = \Phi_j(v_j)$ is substituted into Eq. (2.162), then the new variables v_j can be arbitrary and the constraints (2.164) need not be considered in explicit form. Certain problems lend themselves to this solution. In general, a condition in the form of an inequality $L \leqslant 0$ can be replaced by the condition $M = 0$, where M is zero when L is negative, and M differs from zero for positive L. But in the general case the introduction of nonlinear functions related to these transformations can complicate the solution considerably.

The circumstance, at first glance unimportant, that according to (2.164) the u_j may not only be inside but also "on the boundary" of the allowed closed region, can sometimes be the cause of serious difficulties. In order to clarify their character, let us consider the simplest example. Figure 2.12a shows a continuous and differentiable function $\varphi(u)$, attaining a minimum "inside" the interval $|u| \leqslant 1$. Evidently, the minimum can be found among those points for which the condition

$$d\varphi/du = 0 \tag{2.166}$$

holds. The solution of Eq. (2.166) aids in finding the value $u = u^*$ minimizing $\varphi(u)$.

Meanwhile, if the values of u at the boundaries of the segment also have to be considered, then it can turn out (Fig. 2.12b) that the minimum will be on the boundary (in Fig. 2.12b the value of u corresponding to the minimum is equal to $+1$). Then the minimum point cannot be characterized by the condition (2.166).

Finally, as is seen from the preceding account, the third feature stated above complicates the calculations to a considerable degree by sometimes making it impossible in practice to overcome the difficulties of the usual classical methods. Indeed this feature renders a decisive influence, since in many cases the optimal control $\bar{u}$ has discontinuities of the first kind.

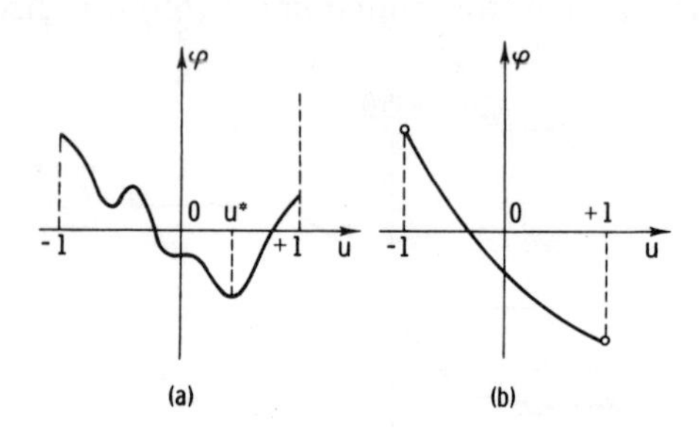

FIG. 2.12

As a result it turns out that only a restricted range of problems in optimal system theory with sufficiently "smooth" solutions admits an effective application of the methods of the classical calculus of variations described above.

In addition, the important direct methods are also far from able always to ensure the solution of problems in optimal control. Actually, in many problems of optimal system theory it is not known how to choose in advance, still not knowing the solution, the functions P_i in Ritz's method [see (2.159)] or in the methods related to it.

Therefore new methods—dynamic programming and the maximum principle—will be most frequently applied below, because they are more adequate for problems in optimal system theory than the classical methods of the calculus of variations.

It should be mentioned that the most general problems in the classical calculus of variations—the problems of Mayer and Bolza—are closely related to dynamic programming and the maximum principle (see [2.33, 2.34, 2.44]).

3. Dynamic Programming

Beginning in 1950, the learned American R. Bellman and a number of his coworkers developed a new general method for solving variational problems, which they called dynamic programming (see [2.9]). Later on (see [2.9, 2.10] and the bibliography for the following chapter) the method of dynamic programming was applied to a wide class of problems in the theory of optimal automatic control systems.

Let us consider once again the problem of the control of an object with the equation

$$d\bar{x}/dt = \bar{f}(\bar{x}, \bar{u}), \tag{2.167}$$

where $\bar{x}$ is an n-dimensional vector with coordinates $x_1, \ldots, x_n$, and $\bar{u}$ is an r-dimensional vector with coordinates $u_1, \ldots, u_r$. Let

$$\bar{u} \in \Omega(\bar{u}) \tag{2.168}$$

and let it be required to minimize the integral

$$Q = \int_0^T G[\bar{x}(t), \bar{u}(t)]\, dt, \tag{2.169}$$

where for the example T will be considered fixed. In Chapter I it was noted that the case with an explicit dependence of G and $\bar{f}$ on time can be reduced to expressions of the type (2.167) and (2.169).

The principle of optimality underlies the method of dynamic programming. This principle was formulated by R. Bellman for a wide range of systems, whose future behavior is completely or statistically determined by their state at the present. Therefore it does not depend on the character of their "previous history," i.e., the system behavior in the past, whenever the system is in a prescribed state at a given instant. For illustration let us consider the optimal trajectory in n-dimensional phase space (Fig. 2.13) with initial and final values of the vector $\bar{x}$ equal

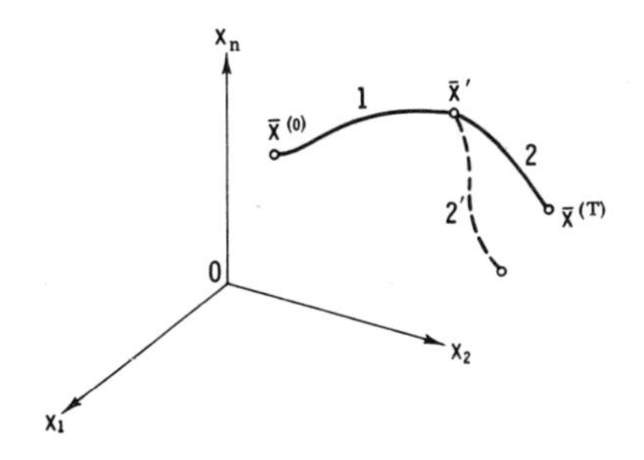

FIG. 2.13

to $\bar{x}^{(0)}$ at $t = t_0$ (usually $t_0 = 0$) and $\bar{x}^{(T)}$ at $t = T > t_0$. Let the initial conditions $\bar{x}^{(0)}$ be given; in general the value of $\bar{x}^{(T)}$ is not known. Let us mark off some intermediate point $\bar{x}'$ on the trajectory corresponding to $t = t'$, where $t_0 < t' < T$, and call the section of the trajectory from $\bar{x}^{(0)}$ to $\bar{x}'$ the first, and from $\bar{x}'$ to $\bar{x}^{(T)}$ the second. The second section corresponds to the part of the integral (2.169) equal to $\int_{t'}^{T} G[\bar{x}, \bar{u}]\, dt$. The second section of the trajectory can be regarded as an independent trajectory. It will be optimal if the integral corresponding to it is minimal. The first and second sections are marked by the numbers 1 and 2 in Fig. 2.13. The principle of optimality can be formulated thus:

The second section of an optimal trajectory is in turn an optimal trajectory.

This means that in the case when the initial state of the system is $\bar{x}'$, and the initial time instant $t = t'$, then, "independent of how the system arrived at this state," the trajectory 2 will be its subsequent optimal motion. In fact, let us assume the opposite. Then the criterion (2.169) considered for the interval of time from t' to T will be smallest, not for the trajectory 2 but for some other trajectory 2′ originating from the point $\bar{x}'$ and shown by the dashed line in Fig. 2.13. But in such a case a "better" trajectory than 1–2 could be constructed, and for the original problem only a control $\bar{u}$ would need to be chosen such that the trajectory 1, and then 2′, would be described. Meanwhile we started from the fact that the trajectory 1–2 is optimal. The contradiction proves the impossibility of a trajectory 2′ existing which provides a smaller value of Q than the trajectory 2. Thus trajectory 2 is optimal.

The optimality principle formulated above is a very general necessary condition for an optimal process which is valid for both continuous and discrete systems.

The optimality principle seems almost trivial and, at first glance, poor in the content of the statement. But as R. Bellman showed, methodical reasoning can derive necessary conditions for an optimal trajectory that are not at all trivial. In the main, the optimality principle is not really so trivial as may at first appear. This is seen if only from the fact that the statement concerning its generalization, "any section of an optimal trajectory is an optimal trajectory," is in general false. Thus, for example, the first section of the trajectory $\bar{x}^{(0)}\bar{x}'\bar{x}^{(T)}$ in Fig. 2.13 cannot be an optimal trajectory "by itself," i.e., it cannot give a minimum for the integral (2.169) for the time interval from t_0 to t', if the initial conditions $\bar{x}^{(0)}$ were only prescribed. We will clarify this assertion with an elementary illustration. How does a good runner distribute his strength for a race at a considerable distance? Does he act according to the principle "Run in each section as fast as you can"? Or "Seek a maximum for the distance traversed during each small time interval"?

Of course not. But the runner can be "spent" long before approaching the end. By reasonably distributing his resources in accordance with the final aim, the runner at first rations his strength in order to get out in front, or in any case not to be "spent" at the end of the distance. In an analogous fashion any control must not be "short-sighted." It must not be guided only by the achieving of the best instant, local effect. It must be "far-sighted"; it must be subordinate to the final aim, i.e., minimizing the value of Q [see (2.169)] on the "entire" interval from t_0 to T. Only in the case when the end point $\bar{x}'$ of the first section is "given" for $t = t'$, is the first section by itself also an optimal trajectory.

Another formulation of the optimality principle can be given:

The optimal strategy does not depend on the "previous history" of the system and is determined only by its state at the time instant under consideration.

The equivalence of this and the preceding formulation is evident if by the "previous history" of the system is meant the trajectory 1, along which the representative point arrived at the position $\bar{x}'$ (Fig. 2.13). In this case, "the state of the system at the time instant under consideration" means just "the state corresponding to the point $\bar{x}'$ at the time instant $t = t'$."

We will explain R. Bellman's method of reasoning first by the simple example of a controlled object, whose motion is characterized by the first-order equation

$$dx/dt = f_1(x, u), \tag{2.170}$$

where x is the sole coordinate of the system, and u is the only control action restricted to some region (2.168). Let the initial condition $x(0) = x^{(0)}$ be prescribed. We shall assume that it is required to find the control law $u(t)$ minimizing the integral

$$Q = \int_{t_0}^{T} G_1(x, u)\, dt + \varphi_1[x(T)], \tag{2.171}$$

where t_0 will usually be considered equal to zero, and for simplicity the value of T can be regarded as fixed. First of all we make the problem discrete, i.e., by approximately replacing the continuous system by a discrete-continuous one. The grounds for this are the following. First, making the problem discrete is an inevitable step in preparing it for solution on a computer. Second, it is simpler to explain the methods of reasoning by an example of a discrete-continuous system. Third, for the application of dynamic programming to continuous systems it is frequently necessary to introduce additional constraints on the class of functions under consideration, constraints that we can do without in the analysis of discrete-continuous systems. In general, as will be shown below, the basic range of application of the method of dynamic programming lies in the realm of discrete-continuous or purely discrete systems, or systems which approximately reduce to them.

Let us break up the interval $(0, T)$ into N equal parts of small length Δ, and consider only discrete values of $x = x(k)$ and $u = u(k)$ $(k = 0, 1, \ldots, N)$ at the time instants $t = 0, 1\Delta, 2\Delta, \ldots, (N-1)\Delta, N\Delta = T$. Then the differential equation (2.170) of the object can be approximately replaced by the finite difference equation

$$\frac{x(k+1) - x(k)}{\Delta} = f_1[x(k), u(k)], \tag{2.172}$$

or

$$x(k+1) = x(k) + f[x(k), u(k)], \tag{2.173}$$

where

$$f[x(k), u(k)] = \Delta f_1[x(k), u(k)]. \tag{2.174}$$

The initial condition remains the previous one:

$$x(0) = [x]_{t=0} = x^{(0)}. \tag{2.175}$$

The integral (2.171) is approximately replaced by the sum

$$Q = \sum_{n=0}^{N-1} G[x(k), u(k)] + \varphi[x(N)], \tag{2.176}$$

where

$$\begin{aligned} G[x(k), u(k)] &= G_1[x(k), u(k)]\Delta, \\ \varphi[x(N)] &= \varphi_1[x(N\Delta)] = \varphi_1[x(T)]. \end{aligned} \tag{2.177}$$

The problem now consists of determining the sequence of discrete values of the control action u, i.e., the quantities $u(0), u(1), \ldots, u(N-1)$ minimizing the sum (2.176) under the conditions (2.168), (2.173), and (2.175) imposed on the system. Thus it is required to find the minimum of a complicated function of many variables. But the method of dynamic programming gives the possibility of reducing this operation to a sequence of minimized functions of a "single" variable.

For solving the problem the device is used of the "retrograde" motion from the end of the process, i.e., from the instant $t = T$, to its beginning.

We assume that first the instant $t = (N - 1)\Delta$ is being considered. All the values of $u(i)$ $(i = 0, 1, \ldots, N - 2)$, besides the last one $u(N - 1)$, have already been realized in some manner, and in addition some value of $x(N - 1)$ has been obtained corresponding to the instant $t = (N-1)\Delta$. According to the optimality principle the action $u(N - 1)$ does not depend on the "previous history" of the system and is determined only by the state $x(N - 1)$ and the control purpose. Let us consider the last section of the trajectory, from $t = (N - 1)\Delta$ to $t = N\Delta$. The quantity $u(N - 1)$ only influences the terms of the sum (2.176) that relate to this section. Let us denote the sum of these terms by Q_{N-1}:

$$Q_{N-1} = G[x(N - 1), u(N - 1)] + \varphi[x(N)]. \tag{2.178}$$

From (2.173) we obtain

$$x(N) = x(N - 1) + f[x(N - 1), u(N - 1)]. \tag{2.179}$$

Hence $x(N)$ also depends on $u(N - 1)$. We will find the admissible value $u(N - 1)$ satisfying (2.178) and minimizing the quantity Q_{N-1}. Let us denote the minimal value of Q_{N-1} found by S_{N-1}. Evidently, this quantity depends on the state of the system at $t = (N - 1)\Delta$, i.e., on the value of $x(N - 1)$ entering into (2.178) and (2.179). Thus $S_{N-1} = S_{N-1}[x(N - 1)]$. We write out the expression for S_{N-1}:

$$\begin{aligned} S_{N-1}[x(N - 1)] &= \min_{u(N-1)\in\Omega(u)} Q_{N-1} \\ &= \min_{u(N-1)\in\Omega(u)} \{G[x(N - 1), u(N - 1)] + \varphi[x(N)]\} \\ &= \min_{u(N-1)\in\Omega(u)} \{G[x(N - 1), u(N - 1)] \\ &\qquad + \varphi[x(N - 1) + f[x(N - 1), u(N - 1)]]\}. \end{aligned} \tag{2.180}$$

Let us turn our attention to the fact that to determine S_{N-1} the minimization need only be carried out with respect to the single variable $u(N - 1)$. By accomplishing this process, we will obtain S_{N-1} in the form of a function of $x(N - 1)$; it is required to store this function (for example, in some storage device for calculation on a computer) before passage to the last stage of solution.

Let us now pass to the next-to-last time segment. Considering the two segments—the last and next-to-last—together, it can be noted that the choice of $u(N - 2)$ and $u(N - 1)$ appears only in the terms of the sum (2.176) that enter into the composition of the expression

$$\begin{aligned} Q_{N-2} = {} & G[x(N - 2), u(N - 2)] \\ & + \{G[x(N - 1), u(N - 1)] + \varphi[x(N)]\}. \end{aligned} \tag{2.181}$$

We will regard as given the quantity $x(N-2)$ at the initial instant of the next-to-last interval, obtained as a result of the "previous history" of the process. From the optimality principle it follows that only the value of $x(N-2)$ and the control purpose—the minimization of Q_{N-2}—determine the optimal control on the time segment under consideration. Let us find the quantity S_{N-2}—the minimum of Q_{N-2} with respect to $u(N-2)$ and $u(N-1)$. But the minimum with respect to $u(N-1)$ of the term contained in the braces of expression (2.181), was already found above for "every" value of $x(N-1)$, and the latter depends on $u(N-2)$. Besides, in the minimization of Q_{N-1} the corresponding optimal value of $u(N-1)$ was found in passing; we will denote this optimal value by $u^*(N-1)$. If it is also considered that the first term in (2.181) does not depend on $u(N-2)$, then we can write

$$\begin{aligned} S_{N-2}[x(N-2)] &= \min_{\substack{u(N-2)\in\Omega(u) \\ u(N-1)\in\Omega(u)}} Q_{N-2} \\ &= \min_{u(N-2)\in\Omega(u)} \{G[x(N-2), u(N-2)] + S_{N-1}[x(N-1)]\} \\ &= \min_{u(N-2)\in\Omega(u)} \{G[x(N-2), u(N-2)] \\ &\qquad + S_{N-1}[x(N-2) + f[x(N-2), u(N-2)]]\}, \end{aligned}$$

since from (2.173) it follows that

$$x(N-1) = x(N-2) + f[x(N-2), u(N-2)].$$

We note that here the minimization is also carried only with respect to the single variable $u(N-2)$. Here we find $u^*(N-2)$—the optimal value of $u(N-2)$—and the quantity S_{N-2}—the minimum of the function Q_{N-2}. Both $u^*(N-2)$ and S_{N-2} are functions of $x(N-2)$. Now the function S_{N-2} can be put in a cell of the memory block, and after obtaining S_{N-2} the henceforth unnecessary function $S_{N-1}[x(N-1)]$ can be "erased" from the memory, having been situated in the memory block earlier.

It is important to note that the optimal value $u^*(N-2)$ we have found minimizes the "entire" expression in the braces of the formula for S_{N-2}, and not just the one term $G[x(N-2), u(N-2)]$. Hence the strategy in which each value of $u(N-j)$ is chosen by a minimization of only "its own" term $G[x(N-j), u(N-j)]$ in the sum (2.176) is not optimal at all. It is too "short-sighted," which was already referred to above. The optimal strategy takes into account the final aim, i.e., the minimization of the entire expression in the curly brackets depending on $u(N-j)$.

The procedure of "retrograde" motion described above can be continued from the end to the beginning of the segment $(0, T)$. The calculation for the third section from the end requires examination of that part of the sum Q which depends on $u(N-3)$. Let us denote this part by Q_{N-3}:

$$Q_{N-3} = G[x(N-3), u(N-3)] + \{G[x(N-2), u(N-2)] + G[x(N-1), u(N-1)] + \varphi[x(N)]\}.$$

On the basis of expression (2.179) we can write

$$x(N-2) = x(N-3) + f[x(N-3), u(N-3)].$$

Further, the minimum of the expression in the braces in the form for Q_{N-3} is equal to $S_{N-2}[x(N-2)]$. Therefore the minimum S_{N-3} of the expression for Q_{N-3} is equal to

$$\begin{aligned} S_{N-3}[x(N-3)] &= \min_{u(N-3)\in\Omega(u)} \{G[x(N-3), u(N-3)] + S_{N-2}[x(N-2)]\} \\ &= \min_{u(N-3)\in\Omega(u)} \{G[x(N-3), u(N-3)] \\ &\quad + S_{N-2}[x(N-3) + f[x(N-3), u(N-3)]]\}. \end{aligned}$$

Passing in a completely analogous fashion to $S_{N-4}, \ldots, S_{N-k}$, we obtain the recurrence formula for determining $S_{N-k}[x(N-k)]$:

$$\begin{aligned} S_{N-k}[x(N-k)] &= \min_{u(N-k)\in\Omega(u)} \{G[x(N-k), u(N-k)] \\ &\quad + S_{N-k+1}[x(N-k) + f[x(N-k), u(N-k)]]\}. \end{aligned} \tag{2.182}$$

Parallel to the process of minimizing the right side of this formula, the optimal value u^* depending on $x(N-k)$ is determined:

$$u^*(N-k) = u^*[x(N-k)], \tag{2.183}$$

as well as the minimizing expression in the braces of (2.182).

By computing the S_{N-k} successively for $k = 1, 2, \ldots, N$ from formula (2.182), we finally arrive at the determination of the optimal value $u^*(0)$, i.e., the value of the control action required at the initial time instant. It is also necessary to find out just this value as the final result, since the given time instant taken as the current one can be regarded as coinciding with the initial one, while the subsequent instants now refer to the future. Simultaneously with the determination of the value $u^*(0)$, S_0, the minimal value of the criterion Q for optimal control, is also obtained.

In certain very simple cases we succeed in carrying out the whole procedure described analytically. But in the general case an analytical expression of the results of the minimization proves to be impossible; therefore the given procedure can be considered only as a computing program, performed by hand in simple cases, but in more complex ones on a discrete-operating calculator, for example on a universal digital computer.

The entire process of solution is carried over without difficulties to an object of arbitrary order n with Eq. (2.167) and any number of control actions u_l $(l = 1, ..., r)$. The scalars x, u, f in the formulas given above need only be replaced by the vectors $\bar{x}$, $\bar{u}$, and $\bar{f}$. Here the vectors for the kth time instant $t = k\Delta$ must be introduced:

$$\bar{x}(k) = \{x_1(k), ..., x_n(k)\} \quad , \quad \bar{u}(k) = \{u_1(k), ..., u_r(k)\}. \tag{2.184}$$

Here u_j $(N - k)$ is the jth control action and $x_j(N - k)$ the jth coordinate at the instant $t = (N - k)\Delta$.

We replace the differential equations (2.167) by finite difference equations, and the integral (2.169) by a sum. Then arguments completely analogous to those given above show that formula (2.182) is replaced by the expression

$$\begin{aligned} S_{N-k}[\bar{x}(N-k)] = \min_{\bar{u}(N-k)\in\Omega(\bar{u})} \{G[\bar{x}(N-k), \bar{u}(N-k)] \\ + S_{N-k+1}[\bar{x}(N-k) + \bar{f}[\bar{x}(N-k), \bar{u}(N-k)]]\}. \end{aligned} \tag{2.185}$$

The calculation procedure will not change if an explicit dependence on time enters into $\bar{f}$.

At each stage it is now required to find the minimum of a function of the r variables $u_1(N - k), ..., u_r(N - k)$. Further, the optimal variables—the scalar S_{N-k} and the vector $\bar{u}^*(N - k)$—are functions of the vector $\bar{x}(N - k)$, i.e., functions of the n variables $x_1(N - k), ...,$ $x_n(N - k)$.

What was presented above may disappoint readers who conceive of dynamic programming as some magic formula for obtaining the solutions of any problems. These solutions are sometimes thought of in the form of prepared general formulas. But to obtain a solution in such a form is usually impossible, and sometimes is unnecessary as well. Usually the solution is required in the form of graphs or tables.

Let the obtaining of the solution indicated above be the computing procedure for getting the required result. The simpler the computing procedure, the better the method. Dynamic programming differs just in the radical simplification of the computing procedure compared to a direct method of solving the problem. Actually, the problem of minimizing the sum (2.176) could in principle be regarded as the problem of minimizing a function of the N variables $u(0)$, $u(1)$, ..., $u(N-1)$. But in order to carry out this minimization in practice, it is first of all necessary to express "each" $x(k)$ in the form of a function of "all" the preceding control actions $u(0)$, ..., $u(k-1)$ (and the initial conditions) after using formula (2.173). That is, it is necessary to find a solution for $x(k)$ in general form. As a result of such a substitution, even if it is possible to carry it out, the expression (2.176) will be incredibly complicated, and only in the simplest cases does one succeed in reducing it to visible form. Then it will be necessary to seek the smallest of the minima (and there can be several of them!) of the obtained function of a large number of variables. In an overwhelming majority of cases such a procedure is really impractical.

Meanwhile dynamic programming permits the minimization of a complicated function of many variables to be replaced by a "sequence of minimizations." Here, in each of the minimization processes, as was emphasized above, the minimum of a far less complicated function of one or several variables (n variables for an object of nth order) is determined. Therefore with the aid of dynamic programming it is possible to solve a series of problems that have been unsolvable by the method of direct minimization.

Of course, from what was presented above it does not follow at all that the direct method is always implausible. In individual cases it has been successfully applied, for example, in the theory of impulse systems [3.31] (when the number of variables was not great).[†] But in general dynamic programming provides the essential rationale for the calculations as compared to the direct method.

It should be noted, however, that in general the solution of problems by the method of dynamic programming can nevertheless turn out to be exceedingly tedious. In fact, at each stage of the computations it is necessary to find and store the functions $S_{N-k}(\bar{x})$ and $S_{N-k+1}(\bar{x})$, i.e., in the general case two functions of n variables. The storage of such functions for large values of n requires an immense memory size, and in complicated cases is only practicable with the aid of some approximations. Certain methods of computation applicable to particular cases and references to the literature on this question are contained in [2.10, 2.35]. In the following chapters the questions of approximation are briefly examined for some special problems.

[†] In the paper [3.31] an approach to the solution of the problem by the method of dynamic programming has also been investigated.

The procedures described are carried over without essential changes to optimal systems with random processes as well. For illustration let us consider the example in which, aside from u, just a random disturbance z acts on a first-order object. Then Eq. (2.173) is replaced by the equality

$$x(k+1) = x(k) + f[x(k), u(k), z(k)], \tag{2.186}$$

where $z(k)$ is the discrete value of the disturbance. Now $x(k)$ and the criterion (2.176) become random variables. Therefore as the new criterion Q, whose value it is required to minimize, we will choose the mathematical expectation of expression (2.176), where in the arguments of G we will also introduce z for generality:

$$Q = M\left\{\sum_{n=0}^{N-1} G[x(k), u(k), z(k)] + \varphi[x(N)]\right\}. \tag{2.187}$$

Here M is the notation for mathematical expectation.

In this example we will regard the quantities $z(i)$ and $z(j)$ as independent for $i \neq j$, and assume that the probability densities $P[z(0)]$, $P[z(1)]$, ..., $P[z(N)]$ are known. Using the procedures presented above, we first find for each fixed $x(N-1)$ the function

$$\begin{aligned}
S_{N-1}[x(N-1)] &= \min_{u(N-1)\in\Omega(u)} Q_{N-1} \\
&= \min_{u(N-1)\in\Omega(u)} M\{G[x(N-1), u(N-1), z(N-1)] \\
&\quad + \varphi[x(N-1) + f[x(N-1), u(N-1), z(N-1)]]\} \\
&= \min_{u(N-1)\in\Omega(u)} \int_{-\infty}^{\infty} P[z(N-1)] \\
&\quad \times \{G[x(N-1), u(N-1), z(N-1)] + \varphi[x(N-1) \\
&\quad + f[x(N-1), u(N-1), z(N-1)]]\}\, dz\,(N-1).
\end{aligned} \tag{2.188}$$

In the minimization the optimal value $u^*[x(N-1)]$ is simultaneously determined as well. After storing $S_{N-1}[x(N-1)]$ we then find the function

$$\begin{aligned}
S_{N-2}[x(N-2)] &= \min_{u(N-2)\in\Omega(u)} M\{G[x(N-2), u(N-2), z(N-2)] \\
&\quad + S_{N-1}[x(N-1)]\} \\
&= \min_{u(N-2)\in\Omega(u)} \int_{-\infty}^{\infty} P[z(N-2)]\{G[x(N-2), u(N-2), z(N-2)] \\
&\quad + S_{N-1}[x(N-2) \\
&\quad + f[x(N-2), u(N-2), z(N-2)]]\}\, dz\,(N-2),
\end{aligned} \tag{2.189}$$

and so on. Thus the methods of solution prove to be essentially the same as for regular systems. Analogous procedures are applicable to an object of arbitrary order. More general problems can be considered in which the $P[z(i)]$ are not known in advance, and some optimal procedure for processing the observations permits information about the distribution densities to be accumulated [2.38, 5.32, 5.33].

Under certain additional assumptions the method of dynamic programming can be applied to the investigation of continuous systems. Let the motion of the object be characterized by the equation

$$d\bar{x}/dt = \bar{f}(\bar{x}, \bar{u}, t). \tag{2.190}$$

At the initial time instant t_0 the vector $\bar{x}$ equals $\bar{x}^{(0)}$, and the optimality criterion has the form

$$Q = \int_{t_0}^{T} G(\bar{x}, \bar{u}, t)\, dt, \tag{2.191}$$

where for simplicity we take $T = \text{const}$. (In the general case this condition is not compulsory; see, for example, Chapter IV.)

We will assume that the optimal trajectory has been found, leading from the initial point $\bar{x}^{(0)}$ to the end point $\bar{x}^{(T)}$ (Fig. 2.13). Let us denote by $S(\bar{x}^{(0)}, t_0)$ the minimal value of the criterion Q corresponding to the optimal trajectory. According to the optimality principle, the section of the trajectory from the point $\bar{x}$, corresponding to the instant $t > t_0$, to the end point $\bar{x}^{(T)}$ (Fig. 2.14) is also an optimal trajectory, and the

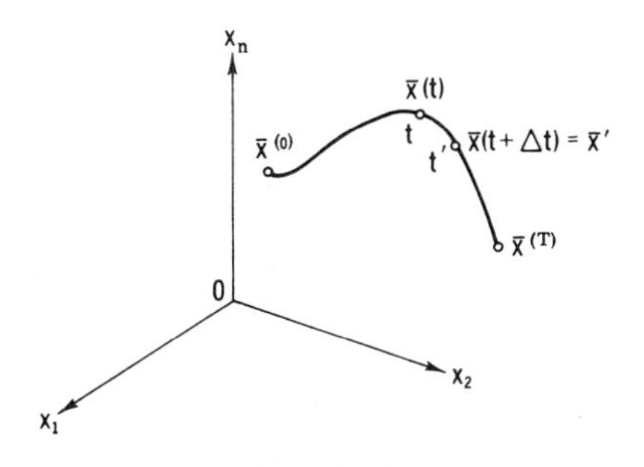

FIG. 2.14

part of the criterion Q that corresponds to this section and to the time interval from t to T has the minimal possible value. Let us denote this value by $S[\bar{x}(t), t]$.

Let Δt be a small time interval, and $S[\bar{x}(t + \Delta t), t + \Delta t] = S[\bar{x}', t']$ be the minimal value of the part of the integral for Q that corresponds to the section of the optimal trajectory from the point $\bar{x}(t + \Delta t) = \bar{x}'$ to the end point $\bar{x}^{(T)}$, and hence to the time interval from $t + \Delta t = t'$ to T. The relation between $S[\bar{x}', t']$ and $S[\bar{x}, t]$ is quite analogous to formula (2.185); we need only write $S[\bar{x}, t]$ instead of $S_{N-k}[\bar{x}(N - k)]$, $S[\bar{x}', t']$ instead of $S_{N-k+1}[\bar{x}(N - k + 1)]$, and finally, $G[\bar{x}(t), \bar{u}(t), t]\,\Delta t$ instead of $G[\bar{x}(N - k), \bar{u}(N - k)]$. The latter substitution would just be made in the first of Eqs. (2.177). Since Δt is a small but finite time interval and the replacement of the differential equation by the finite difference expression is inexact, it is still necessary to add an expression $o_1(\Delta t)$ to some of the parts of the equality, i.e., a quantity of small order higher than Δt. This means that

$$\lim_{\Delta t \to 0} \frac{o_1(\Delta t)}{\Delta t} = 0. \tag{2.192}$$

Thus instead of Eq. (2.185) we can now write

$$S[\bar{x}, t] = \min_{\bar{u}(t) \in \Omega(\bar{u})} \{G[\bar{x}, \bar{u}, t]\,\Delta t + S[\bar{x}', t']\} + o_1(\Delta t). \tag{2.193}$$

Equation (2.193) can be obtained irrespective of the discrete case considered above. In fact, according to the definition,

$$S[\bar{x}, t] = \min_{\bar{u}(\tau) \in \Omega(\bar{u})} \int_t^T G(\bar{x}, \bar{u}, \tau)\, d\tau \qquad (t \leqslant \tau \leqslant T). \tag{2.194}$$

Here S is the minimal value of the integral obtained over the set of all admissible controls $\bar{u}(\tau)$ in the interval from t to T.

The integral (2.194) can be represented in the form of a sum of two terms, corresponding to the intervals from t to $t + \Delta t$ and from $t + \Delta t$ to T. Since Δt is small,

$$S[\bar{x}, t] = \min_{\bar{u}(\tau) \in \Omega(\bar{u})} \left[G(\bar{x}, \bar{u}, t)\,\Delta t + \int_{t'=t+\Delta t}^T G(\bar{x}, \bar{u}, \nu)\, d\nu\right] + o_1(\Delta t), \tag{2.195}$$

where we regard Δt as small, and $o_1(\Delta t)$ is of small order higher than Δt. Since the first term in the brackets of (2.195) depends only on the value of $\bar{u}(t)$ at the instant t, and only the integral in the square brackets still depends on the values of $u(\nu)$ in the interval of variation of ν from $t' = t + \Delta t$ to T, we can write

$$\begin{aligned} S[\bar{x}, t] &= \min_{\bar{u}(t) \in \Omega(\bar{u})} \left[G(\bar{x}, \bar{u}, t)\,\Delta t + \min_{\bar{u}(\nu) \in \Omega(\bar{u})} \int_{t'}^T G(\bar{x}, \bar{u}, \nu)\, d\nu\right] + o_1(\Delta t) \\ &= \min_{\bar{u}(t) \in \Omega(\bar{u})} \{G(\bar{x}, \bar{u}, t)\,\Delta t + S[\bar{x}', t']\} + o_1(\Delta t). \end{aligned} \tag{2.196}$$

Here under the minimum sign before the bracket is the value of $\bar{u}(t)$ at the time instant t. Formulas (2.196) and (2.193) coincide.

In the same way as in formula (2.185), it should be taken into account that $\bar{x}' = \bar{x}(t + \Delta t)$ depends on $\bar{u}(t)$. From (2.190) we find for small Δt:

$$\begin{aligned}\bar{x}' = \bar{x}(t + \Delta t) &= \bar{x}(t) + (d\bar{x}/dt)\,\Delta t + o_2(\Delta t)\\ &= \bar{x}(t) + \bar{f}[\bar{x}(t), \bar{u}(t), t]\,\Delta t + o_2(\Delta t),\end{aligned} \tag{2.197}$$

where $o_2(\Delta t)$ is a quantity of higher small order compared to Δt. Formula (2.197) is analogous to expression (2.173).

Let us assume that S has partial derivatives with respect to the variables x_i $(i = 1, ..., n)$ and to t, i.e., that all the $\partial S/\partial x_i$ $(i = 1, ..., n)$ and $\partial S/\partial t$ exist. While this hypothesis is not based on anything, the correctness of the entire subsequent derivation depends on the validity of this assumption. If it is not justified, then the subsequent arguments have only a heuristic character. Meanwhile there exist cases where the assumption stated above is untrue, which is why the application of dynamic programming to continuous systems still requires a supplementary basis in the general case (see [2.31, 2.33]).

Let us substitute the expression for $\bar{x}'$ from (2.197) into formula (2.193) and expand $S[\bar{x}', t']$ in a Taylor series in the neighborhood of the point $(\bar{x}, t)$:

$$\begin{aligned}S[\bar{x}', t'] &= S[\bar{x}(t + \Delta t), t + \Delta t]\\ &= S[\bar{x}(t) + \bar{f}[\bar{x}(t), \bar{u}(t), t]\,\Delta t + o_2(\Delta t); t + \Delta t]\\ &= S[\bar{x}, t] + \sum_{i=1}^{n} \frac{\partial S[\bar{x}, t]}{\partial x_i} f_i[\bar{x}, \bar{u}, t]\,\Delta t + \frac{\partial S[\bar{x}, t]}{\partial t}\,\Delta t + o_3(\Delta t),\end{aligned} \tag{2.198}$$

where $o_3(\Delta t)$ is of higher small order compared to Δt.

This formula can be rewritten more compactly by introducing the gradient of the function $S[\bar{x}, t]$—the vector with coordinates $\partial S/\partial x_i$ $(i = 1, ..., n)$:

$$\operatorname{grad} S = (\partial S/\partial x_1, ..., \partial S/\partial x_n). \tag{2.199}$$

Then (2.198) takes the form

$$\begin{aligned}S[\bar{x}', t'] &= S[\bar{x}(t + \Delta t), t + \Delta t]\\ &= S[\bar{x}, t] + \langle \operatorname{grad} S[\bar{x}, t] \cdot \bar{f}[\bar{x}(t), \bar{u}(t), t]\rangle\,\Delta t\\ &\quad + \frac{\partial S[\bar{x}, t]}{\partial t}\,\Delta t + o_3(\Delta t).\end{aligned} \tag{2.200}$$

Here the scalar product of the vectors grad S and $\bar{f}$ is denoted by the brackets $< >$. Let us substitute (2.200) into (2.193) and introduce $S[\bar{x}, t]$ and $\partial S/\partial t$ into the braces on the right side, since they do not depend on $\bar{u}(t)$. Further, $S[\bar{x}, t]$ on the left and right sides can be canceled, and after dividing by Δt the formula takes the following form:

$$-\frac{\partial S[\bar{x}, t]}{\partial t} = \min_{\bar{u}(t)\in\Omega(\bar{u})} \{G[\bar{x}(t), \bar{u}(t), t] + \langle \text{grad } S[\bar{x}, t] \cdot \bar{f}[\bar{x}(t), \bar{u}(t), t]\rangle\} + \frac{o_4(\Delta t)}{\Delta t}, \tag{2.201}$$

where $o_4(\Delta t)$ is of higher small order compared to Δt. Now we let Δt tend to zero. Since $o_4(\Delta t)$ is subject to a condition of the type (2.192), the last term on the right side of (2.201) vanishes as $\Delta t \to 0$. Therefore in the limit we obtain

$$-\frac{\partial S[\bar{x}, t]}{\partial t} = \min_{\bar{u}(t)\in\Omega(\bar{u})} \{G[\bar{x}(t), \bar{u}(t), t] + \langle \text{grad } S[\bar{x}, t] \cdot \bar{f}[\bar{x}(t), \bar{u}(t), t]\rangle\}. \tag{2.202}$$

This expression is called Bellman's equation. It is a singular partial differential equation, since as a result of the minimization $\bar{u}$ vanishes from the right side for arbitrary time instants t. For illustration let us consider a simple example [2.17]. In the particular case let $r = 1$ and $n = 2$, where $G = G(x_1, x_2)$ and the sole control action is denoted by u. The equations of the object are

$$dx_1/dt = f_1 = ux_1 + x_2, \qquad dx_2/dt = f_2 = u^2. \tag{2.203}$$

Then equation (2.202) takes the form (for brevity we write S in place of $S[\bar{x}, t]$)

$$-\frac{\partial S}{\partial t} = \min_u \left\{G(x_1, x_2) + \frac{\partial S}{\partial x_1}(ux_1 + x_2) + \frac{\partial S}{\partial x_2} u^2\right\}. \tag{2.204}$$

Assuming that $\partial S/\partial x_2 > 0$, we find the minimum of the braces with respect to u by equating its derivative with respect to u to zero. The optimal value u^* minimizing the brackets is

$$u^* = -\frac{1}{2} x_1 \frac{\partial S}{\partial x_1} \cdot \frac{1}{\partial S/\partial x_2}. \tag{2.205}$$

Substituting this expression into equation (2.204), we obtain the partial differential equation written in the usual form:

$$-\frac{\partial S}{\partial t} = G(x_1, x_2) + \frac{\partial S}{\partial x_1} x_2 - x_1^2 \frac{(\partial S/\partial x_1)^2}{4\, \partial S/\partial x_2}. \tag{2.206}$$

The partial differential equation (2.206) can be solved, since the boundary conditions for it are known. Actually, $S[\bar{x}, T]$ is a known function. For example, for the criterion (2.171) it is equal to the known function $\varphi_1[x(T)]$, since for $t_0 = T$ the integral in (2.171) is equal to zero. For the criterion (2.191) the function $S[\bar{x}, T]$ is equal to zero. By knowing the boundary function $S[\bar{x}, T]$, equation (2.206) can be integrated by some known method. One of the usual methods of approximate integration consists of making the problem discrete and solving the recurrence relations of the type (2.185) obtained. In a number of cases an approximate solution can be found by other means (see, for example, [2.35]) or even an exact solution may be obtained in closed form. The value u^* obtained in passing is the optimal control.

4. The Maximum Principle

In 1956 Academician L. S. Pontryagin and his pupils V. G. Boltyanskii and R. V. Gamkrelidze published a note [2.11] stating a principle, in the form of a hypothesis, leading to the solution of the general problem of finding the transient process in continuous systems which is optimal in speed. The discovery of this principle emerged as a result of the work of L. S. Pontryagin and his coauthors on the solution of optimal control problems; a number of such problems had been posed by the author of this book in some lectures on optimal system theory in 1954 at a seminar directed by L. S. Pontryagin. These problems had also been posed in 1953 in the author's lecture at the second All-Union Conference on the Theory of Automatic Control [3.10]. In a subsequent series of papers by L. S. Pontryagin, V. G. Boltyanskii, and R. V. Gamkrelidze [2.12-2.14, 2.16, 2.18, 2.21], starting in 1956, the maximum principle was substantiated as a necessary and sufficient test for an optimal process in linear systems, and a necessary test for an optimal process in nonlinear systems. In addition, the maximum principle has been generalized to the case of minimizing an integral and to the case of bounded coordinates for the object. Other methods of proof were given later on by L. I. Rozonoer [2.15, 2.17, 2.19]. In the papers of L. I. Rozonoer the connection between the maximum principle and dynamic programming was first established, and also the proof of the validity of the maximum principle for linear discrete-continuous systems was given (see, in addition, [1.22, 2.20]). In the papers of A. G. Butkovskii [3.43, 3.44] the maximum principle has been generalized to definite classes of integral equations corresponding to distributed parameter systems.

The proofs of the validity of the maximum principle given in the papers of L. S. Pontryagin and his coauthors do not have any direct relation to R. Bellman's optimality principle and dynamic programming. But from methodological considerations it is appropriate first to derive the maximum principle from Bellman's equation [2.17, 2.20], in order to show its connection with this relation. Later on a derivation of the maximum principle which is independent of dynamic programming will be given.

Let us rewrite equation (2.202) in a more compact form. For this we will introduce the additional coordinate x_{n+1}, where $(x_{n+1})_{t=0} = 0$; let the equation for the coordinate x_{n+1} have the form

$$\frac{dx_{n+1}}{dt} = f_{n+1} = 1. \tag{2.207}$$

If at the initial instant $t = 0$, then $x_{n+1} = t$. Then instead of t, x_{n+1} can be written, and in place of $\partial S/\partial t$ we write $\partial S/\partial x_{n+1}$. In addition, let us introduce the coordinate x_0 [see (1.56)] with the equation

$$dx_0/dt = f_0 = G[\bar{x}, \bar{u}, t] = G[\bar{x}, \bar{u}, x_{n+1}], \tag{2.208}$$

where $(x_0)_{t=0} = 0$. Then the problem of minimizing the integral for Q will reduce to the problem of minimizing the quantity $(x_0)_{t=T} = x_0^{(T)}$ [see (1.57)].

We shall now introduce the generalized vectors in $(n + 2)$-dimensional space:

$$\tilde{x} = (x_0, x_1, ..., x_n, x_{n+1}), \qquad \tilde{f} = (f_0, f_1, ..., f_n, f_{n+1}) \tag{2.209}$$

and

$$\tilde{\psi} = (-1, -\partial S/\partial x_1, ..., -\partial S/\partial x_n, -\partial S/\partial x_{n+1}). \tag{2.210}$$

Let us transpose $-\partial S/\partial t$ in expression (2.202) to the right side, and then consider that the minimum of the expression on the right side means the maximum, with minus sign, of the expression opposite in sign to it. In fact, for any μ the relation

$$\max(-\mu) = -\min \mu \tag{2.211}$$

holds. By taking (2.207), (2.208), and (2.211) into account, expression (2.202) can be rewritten in the form

$$0 = \max_{\bar{u}(t)\in\Omega(\bar{u})} \Big\{ G[\bar{x}, \bar{u}, x_{n+1}] \cdot (-1) \\ -\langle \text{grad } S[\bar{x}, x_{n+1}] \cdot \bar{f}[\bar{x}, \bar{u}, x_{n+1}] \rangle - \frac{\partial S}{\partial x_{n+1}} (+1) \Big\}. \tag{2.212}$$

The comparison of this relation with expressions (2.209) and (2.210) for the vectors $\tilde{f}$ and $\tilde{\psi}$ shows that condition (2.212) can be given an extremely compact form:

$$0 = \max_{\bar{u}(t)\in\Omega(\bar{u})} \langle \tilde{\psi}\tilde{f} \rangle. \tag{2.213}$$

Now let us introduce the so-called Hamiltonian; this is the scalar

$$\tilde{H} = \langle \tilde{\psi}\tilde{f} \rangle = \sum_{i=0}^{n+1} \tilde{\psi}_i \tilde{f}_i, \tag{2.214}$$

where $\tilde{\psi}_i$ and $\tilde{f}_i$ are the ith coordinates of the vectors $\tilde{\psi}$ and $\tilde{f}$, respectively. Then equation (2.213) takes the form

$$0 = \max_{\bar{u}(t)\in\Omega(\bar{u})} \tilde{H}. \tag{2.215}$$

This is the maximum principle of L. S. Pontryagin.

From expression (2.215) two conclusions follow.

(a) If the process is optimal—and indeed we start from this with the derivation of (2.202)—then at an arbitrary time instant t the optimal control $\bar{u}^*(t)$ is a control that maximizes the quantity $\tilde{H}$, where

$$\tilde{H}_{\max} = \max_{\bar{u}(t)\in\Omega(\bar{u})} \tilde{H} = \max_{\bar{u}(t)\in\Omega(\bar{u})} \langle \tilde{\psi}\tilde{f} \rangle = \max_{\bar{u}(t)\in\Omega(\bar{u})} \sum_{i=0}^{n+1} \tilde{\psi}_i \tilde{f}_i. \tag{2.216}$$

In this formula the quantity $\tilde{H}$ depends on $\bar{u}$, since the vector $\tilde{f}$ depends on $\bar{u}$. At a given point $\tilde{x}$ of $(n + 2)$-dimensional space the quantity $\tilde{H}$ is completely determined as a function of $\bar{u}$, whenever the vector $\tilde{\psi}$ is known, and this vector is completely determined if the function $S[\bar{x}, x_{n+1}]$ is known, and hence its partial derivatives $\partial S/\partial x_i$ $(i = 1, ..., n + 1)$ are known.

Thus the formula for the choice of optimal control $\bar{u}$ turns out to be very simple in principle: at each time instant $\bar{u}$ must be selected such that the maximal possible value of the Hamiltonian $\tilde{H}$ will be ensured (with the constraints imposed on $\bar{u}$ taken into account).

(b) At any point of the optimal trajectory the maximal value of the quantity $\tilde{H}$ is the same: it is equal to zero.

The maximum principle has a definite geometric meaning. In order to explain this, let us introduce the auxiliary function $\tilde{S}$—a function of the point x in $(n + 2)$-dimensional space defined by the formula

$$\tilde{S} = x_0 + S(x_1, ..., x_n, x_{n+1}). \tag{2.217}$$

We will consider in the $(n + 2)$-dimensional space of $\tilde{x}$ the trajectory of a representative point moving from the initial position $\tilde{x}^{(0)}$ to the end position $\tilde{x}^{(T)}$.

Since $\tilde{S}$ is a continuous function of the point $\tilde{x}$ of the space, a certain surface can be found in this space—the geometric locus of points $\tilde{S} = \text{const}$. We will call such surfaces "isosurfaces."

Comparing (2.217) with (2.210), it is not difficult to be convinced that the coordinates of the vector $\tilde{\psi}$ are related to $\tilde{S}$ by the equations

$$\tilde{\psi}_i = -\partial\tilde{S}/\partial\tilde{x}_i. \tag{2.218}$$

Hence the vector $\tilde{\psi}$ is the gradient of the scalar $\tilde{S}$ in the $(n + 2)$-dimensional space of $\tilde{x}$, taken with a minus sign:

$$\tilde{\psi} = -\text{grad}\,\tilde{S}. \tag{2.219}$$

As is known from vector analysis, the gradient is orthogonal to the surface $\tilde{S} = \text{const}$. Let us consider at the point $\tilde{x}$ the vector

$$\tilde{f} = d\tilde{x}/dt. \tag{2.220}$$

The condition $\tilde{H} = \max$ coincides with the condition for maximizing the scalar product of the vectors $\tilde{\psi}$ and $\tilde{f}$ or, since at the given point $\tilde{x}$ the vector $\tilde{\psi}$ is prescribed and does not depend on $\bar{u}$, it coincides with the condition for the "maximum of the projection of the vector $\tilde{f}$ on the direction of $\tilde{\psi}$." Thus the geometric meaning of the maximum principle is the following: a control $\bar{u}$ must be selected such that the projection of the velocity vector $d\tilde{x}/dt$ of the representative point on the direction of the normal to the isosurface at the given point $\tilde{x}$ will be maximal. In general the projection turns out to be negative here and its maximal value is equal to zero according to (2.215).

Let us consider the special case when an explicit dependence on time t in the equations of motion and the function G is absent, and it is required to ensure a minimal time T for the transient process. In this case we must set $G = 1$ in Eq. (2.208); in addition, $\partial S/\partial t = 0$. Then from (2.212) we find

$$\max_{\bar{u}\in\Omega(\bar{u})} \langle -\text{grad } S[\bar{x}, t] \cdot \bar{f}[\bar{x}, \bar{u}]\rangle = 1. \tag{2.221}$$

It should be emphasized that here $\bar{x}$ and $\bar{f}$ are vectors in n-dimensional space.

Let us set

$$H = -\langle \text{grad } S \cdot \bar{f}\rangle = \langle \bar{\psi}\bar{f}\rangle, \tag{2.222}$$

where the n-dimensional vector $\bar{\psi}$ is defined by the expression

$$\bar{\psi} = -\text{grad } S. \tag{2.223}$$

Then condition (2.221) takes the form

$$\max_{\bar{u}\in\Omega(\bar{u})} H = 1. \tag{2.224}$$

This is the form that the maximum principle takes in this special case. Now the trajectory of the representative point in the n-dimensional space of $\bar{x}$ can be examined (Fig. 2.15). The optimal control $\bar{u}$ must be selected such that at each time instant the scalar H be maximized, and moreover the maximal value of H at any point of the trajectory is equal to 1. Since in this case

$$S[\bar{x}, t] = \int_t^T 1 \cdot dt = T - t, \tag{2.225}$$

where t is the time instant under consideration, then the quantity S—the time to attain the final point—decreases with an increase in t. Hence the vector $\bar{\psi}$ of (2.223), coinciding with the direction of the fastest decrease in S, is turned "toward the interior" of the isosurface $S =$ const (Fig. 2.15) containing the end point $\bar{x}^{(T)}$. In this case the isosurfaces $S =$ const become the surfaces of equal times $\tau_i = T - t_i$ of attaining the end point $\bar{x}^{(T)}$; such

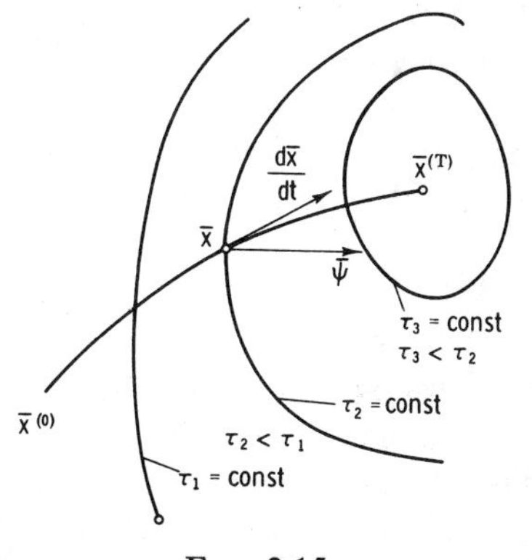

FIG. 2.15

surfaces have been called isochrone surfaces [3.6]. In this case the maximum principle requires $\bar{u}$ to be selected such that the projection of the velocity $d\bar{x}/dt$ of the representative point in phase space on the direction of $\bar{\psi}$ normal to the isosurface will be maximal. This is also evident from purely intuitive physical considerations. Actually, the motion "along" the isosurface, for example, $\tau_2 =$ const (Fig. 2.15), does not give any useful effect, because after displacement the representative point remains in positions from which, in the best case, the end point $\bar{x}^{(T)}$ can be attained after the same τ_2 seconds. Meanwhile, the faster the motion along the "normal" to the isosurface takes place, the more rapidly the representative point is transferred to the "following" isosurface, $\tau_2 - \Delta t =$ const, and hence later on less and less time for attaining the end point $\bar{x}^{(T)}$ becomes possible.

As was shown above, finding the function $S[\bar{x}, t]$ or $\tilde{S}$ is required for the application of the method of dynamic programming, which is associated with tedious operations such as the solution of a partial differential equation. Meanwhile for using the maximum principle we need only know the vector $\bar{\psi}$ examined on the optimal trajectory, and this vector, it turns out, can be found without constructing the surface $\tilde{S} =$ const. In order to find $\tilde{\psi}$, the so-called adjoint equation must be solved.

The derivation of adjoint equations by various methods is contained in [2.12, 2.21]. Below is given a derivation similar to one presented in the monograph [2.21].

We assume that the function $\tilde{S}(x)$ has second partial derivatives with respect to all the $\tilde{x}_i$ $(i = 0, 1, ..., n+1)$, i.e., that it is twice differentiable. Let us consider a variation in the vector $\tilde{\psi}$ with the motion of the representative point along the optimal trajectory. Since $\tilde{\psi} = \tilde{\psi}[\tilde{x}(t)]$, the vector $\tilde{\psi}$, depending on $\tilde{x}(t)$, is in the end a function of time. Therefore we can find the derivatives

$$\frac{d\tilde{\psi}_i}{dt} = -\frac{d}{dt}\left(\frac{\partial \tilde{S}}{\partial \tilde{x}_i}\right) = -\sum_{j=0}^{n+1} \frac{\partial}{\partial \tilde{x}_j}\left(\frac{\partial \tilde{S}}{\partial \tilde{x}_i}\right)\frac{d\tilde{x}_j}{dt}$$

$$= -\sum_{i=0}^{n+1} \frac{\partial^2 \tilde{S}}{\partial \tilde{x}_i\, \partial \tilde{x}_j} \tilde{f}_j \qquad (i = 1, ..., n+1). \tag{2.226}$$

Here the values $\tilde{f}_j$ have been substituted in place of $d\tilde{x}_j/dt$.

As is seen from (2.210), the coordinate $\tilde{\psi}_0$ is always equal to -1. Therefore

$$d\tilde{\psi}_0/dt = 0. \tag{2.227}$$

If the optimal control $\tilde{u}^*(t)$ along the optimal trajectory under consideration is substituted into expression (2.214) for $\tilde{H}$ instead of an arbitrary $\tilde{u}(t)$, then according to (2.213) we will obtain

$$\tilde{H} = \langle \tilde{\psi} \tilde{f} \rangle = \sum_{j=0}^{n+1} \tilde{\psi}_j \tilde{f}_j = - \sum_{j=0}^{n+1} \frac{\partial \tilde{S}}{\partial \tilde{x}_j} \tilde{f}_j = \tilde{H}_{\max} = 0. \qquad (2.228)$$

We will now consider a "fixed" time instant t. Then $\tilde{u}^*(t)$ will also be a fixed quantity. For points of $\tilde{x}$-space different from one that lies on the optimal trajectory, this control $\tilde{u}^*(t)$ will now not be optimal; hence for them the quantity $\tilde{H}$ will now not attain its maximum. From this argument it follows that for a fixed t and $u(t) = u^*(t)$, the quantity $\tilde{H} = \langle \tilde{\psi} \tilde{f} \rangle$ indeed attains its maximum (equal to zero) at a point of the optimal trajectory, and hence the derivatives of $\tilde{H}$ with respect to the $\tilde{x}_i$ vanish at this point. Differentiating the expression (2.228) for $\tilde{H}$, where $\tilde{\psi}$ must also be differentiated with respect to the $\tilde{x}_i$ in order to take into account the increment in $\tilde{\psi}$ at the displaced points, we arrive at the equations

$$\frac{\partial}{\partial \tilde{x}_i} \left(- \sum_{j=0}^{n+1} \frac{\partial \tilde{S}}{\partial \tilde{x}_j} \tilde{f}_j \right) = - \sum_{j=0}^{n+1} \frac{\partial^2 \tilde{S}}{\partial \tilde{x}_j \, \partial \tilde{x}_i} \tilde{f}_j - \sum_{j=0}^{n+1} \frac{\partial \tilde{S}}{\partial \tilde{x}_j} \cdot \frac{\partial \tilde{f}_j}{\partial \tilde{x}_i} = 0 \qquad (i = 1, \ldots, n+1). \qquad (2.229)$$

Hence it follows that

$$- \sum_{j=0}^{n+1} \frac{\partial^2 \tilde{S}}{\partial \tilde{x}_i \, \partial \tilde{x}_j} \tilde{f}_j = \sum_{j=0}^{n+1} \frac{\partial \tilde{S}}{\partial \tilde{x}_j} \cdot \frac{\partial \tilde{f}_j}{\partial \tilde{x}_i} \qquad (i = 1, \ldots, n+1). \qquad (2.230)$$

Noting that the left side of (2.230) is identical to the right side of (2.226), let us substitute the right side of expression (2.230) into the latter equation. As a result we find

$$\frac{d\tilde{\psi}_i}{dt} = \sum_{j=0}^{n+1} \frac{\partial \tilde{S}}{\partial \tilde{x}_j} \cdot \frac{\partial \tilde{f}_j}{\partial \tilde{x}_i} = - \sum_{j=0}^{n+1} \tilde{\psi}_j \frac{\partial \tilde{f}_j}{\partial \tilde{x}_i} \qquad (i = 1, \ldots, n+1). \qquad (2.231)$$

This is the set of adjoint equations, which combined with (2.227) determine the variation in the vector $\tilde{\psi}$ on the optimal trajectory. It should be noted that equations (2.231) are "linear" with respect to the coordinates $\tilde{\psi}_j$ of the vector $\tilde{\psi}$.

In the expression for $\tilde{H}$ being examined for the given point $\tilde{x}$ of the optimal trajectory, $\tilde{\psi}$ is prescribed, and only the vector $\tilde{f}$ depends on $\tilde{x}$

in explicit form. Therefore the partial derivative of $\tilde{H}$ with respect to $\tilde{x}_i$ takes the form

$$\frac{\partial \tilde{H}}{\partial \tilde{x}_i} = \sum_{j=0}^{n+1} \tilde{\psi}_j \frac{\partial \tilde{f}_j}{\partial \tilde{x}_i}. \tag{2.232}$$

A comparison of this expression with equation (2.231) shows that the latter can be rewritten in the compact form

$$d\tilde{\psi}_i/dt = -\partial \tilde{H}/\partial \tilde{x}_i \qquad (i = 1, ..., n+1). \tag{2.233}$$

We will note that from formula (2.228) for $\tilde{H}$ it also follows that

$$\partial \tilde{H}/\partial \tilde{\psi}_i = \tilde{f}_i, \tag{2.234}$$

since $\tilde{f}$ does not depend on $\tilde{\psi}_i$. Hence the equations of motion of the object can be rewritten thus:

$$d\tilde{x}_i/dt = \partial \tilde{H}/\partial \tilde{\psi}_i \qquad (i = 0, 1, ..., n+1). \tag{2.235}$$

Systems of equations of the type (2.233) and (2.235) are called canonical adjoints.

Thus the values of the vector $\tilde{\psi}$ are determined from the ordinary differential equations (2.231) and (2.233). But if the vector $\tilde{\psi}$ is known, then it is not necessary to calculate the function $\tilde{S}$. In applying the maximum principle two systems of equations are solved simultaneously—the fundamental and adjoint.

In a particular case let an explicit dependence on time be missing in the equations for the object, and in addition let it be required to ensure a minimum time for the transient process. Here $G = 1$ and the quantity H of (2.222) can be used instead of $\tilde{H}$, and likewise the n-dimensional phase space of $\bar{x}$ in place of the $(n+2)$-dimensional space of $\tilde{x}$. Then the equations of motion of the object can be written in the form

$$dx_i/dt = \partial H/\partial \psi_i \qquad (i = 1, ..., n), \tag{2.236}$$

and the adjoint equations (2.231) [see also (2.233)] take the form

$$d\psi_i/dt = -\partial H/\partial x_i \qquad (i = 1, ..., n). \tag{2.237}$$

Let us now derive the maximum principle as a necessary test for optimality by another method, basically following the course pointed out by L. S. Pontryagin and his coauthors but with certain simplifications, with the aim of facilitating an understanding of the material. This way of deriving it is useful in two respects. First, it permits the dynamic essence of the maximum principle to be understood from a point of view different from that presented above. Second, a method of derivation not related to dynamic programming permits the elimination of assumptions about the differentiability of the function $S[\bar{x}, t]$ which do not arise from the essence of the problem.

As a simplification let us consider the problem with a fixed time T and a free end of the trajectory. Let the equations of motion for the object have the form

$$d\tilde{x}/dt = \tilde{f}(\tilde{x}, \bar{u}) \tag{2.238}$$

and let it be required to minimize the quantity

$$Q = \tilde{x}_0^{(T)}. \tag{2.239}$$

We will regard the function $\tilde{f}_i$ as bounded and continuous with respect to all its arguments and differentiable with respect to the $\tilde{x}_j$ $(j = 0, 1, \ldots, n + 1)$. Let the control action $\bar{u}(t)$ belong to the class of piecewise-continuous functions, satisfying the condition

$$\bar{u}(t) \in \Omega(\bar{u}). \tag{2.240}$$

We take the initial value of the vector $\tilde{x}$ as given:

$$(\tilde{x})_{t=0} = \tilde{x}^{(0)}. \tag{2.241}$$

Under these conditions it is required to find the optimal trajectory $\tilde{x}^*(t)$ $(0 \leqslant t \leqslant T)$ and the optimal admissible control action $\bar{u}^*(t)$ ensuring a minimum for the criterion Q.

For simplicity the case $r = 1$ is considered below with one control action; however, analogous arguments and conclusions are also valid for $r > 1$.

We assume that $\tilde{x}^*(t)$ and $u^*(t)$ have been found.

Let us examine the curve of $u^*(t)$ (Fig. 2.16). As was shown above, this curve can contain a finite number of points with discontinuities of the first kind. We will concentrate our attention on the choice of optimal control in the infinitesimally small time interval

$$\tau - \epsilon < t < \tau, \tag{2.242}$$

where ϵ is an infinitesimally small quantity, $0 < \tau < T$. We will modify the control only on the one infinitesimally small interval (2.242) by varying the control from the optimal u^* to some other value $u \neq u^*$.

Here the control remains fixed on all the remaining intervals $(0, \tau - \epsilon)$ and (τ, T) and is equal to $u^*(t)$. Such a variation in the control u, in which only a "needle-shaped" change in u takes place on an infinitesimally small section, bears the name of "needle-shaped" variation.

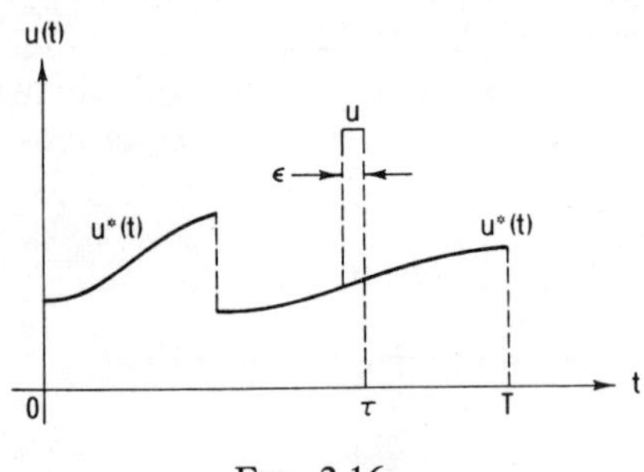

FIG. 2.16

We note that the magnitude of the increase $u - u^*$ for a needle-shaped variation must not be arbitrarily small at all. On the contrary, the quantities u and u^* can be arbitrary but they should not fall outside the allowed limits. For example, if the control is bounded by the condition

$$| u | \leqslant U = \text{const}, \qquad (2.243)$$

then both u and u^* must satisfy only the condition (2.243); the absolute value of the difference between them can be within the limits from 0 to $2U$ in this case.

The idea of a needle-shaped variation is of interest in the respect that in spite of a "finite" magnitude for the difference $u - u^*$, the influence of this variation on the subsequent motion of the object is "infinitesimally small". This property is evident from simple physical arguments. In fact, as is known, the influence of any short impulse on a system is estimated by the magnitude of its area. Since the area $(u - u^*)\epsilon$ of the increase is infinitesimally small, then the influence of this increase on the subsequent motion $\tilde{x}(t)$ $(t > \tau)$ is also infinitesimally small. The validity of this argument is proved below.

We will also note that the needle-shaped variation is a different construction from the variation used in the classical calculus of variations. In the latter the variation must be a function with a sufficient degree of smoothness within the time that the needle-shaped variation is a "jump-function." Only by including the class of piecewise-continuous functions $u(t)$ in the analysis can a needle-shaped variation be dealt with (Fig. 2.16). Meanwhile the needle-shaped variation is just the starting point for the derivation of the maximum principle. Thus the discontinuity which is the stumbling block in the classical calculus of variations becomes a useful tool in the theory of the maximum principle.[†]

As a result of modifying the control on the infinitesimally small interval $\tau - \epsilon < t < \tau$ the subsequent motion $\tilde{x}(t)$ for $t > \tau$ is now different from the optimal motion $\tilde{x}^*(t)$. The difference between these quantities at the time instant $t = \tau$, to within higher-order small terms, is equal to the difference of the velocity variations, i.e., the quantity $(d\tilde{x}/dt - d\tilde{x}^*/dt)_{t=\tau}$, multiplied by the time segment ϵ:

$$\begin{aligned}\tilde{x}(\tau) - \tilde{x}^*(\tau) &= \epsilon[(d\tilde{x}/dt)_{t=\tau} - (d\tilde{x}^*/dt)_{t=\tau}] \\ &= \epsilon\{\tilde{f}[\tilde{x}(\tau), u(\tau)] - \tilde{f}[\tilde{x}(t), u^*(\tau)]\}.\end{aligned} \tag{2.244}$$

This difference is infinitesimally small but different from zero. Therefore a divergence between the trajectories $\tilde{x}(t)$ and $\tilde{x}^*(t)$ will generally exist for $t > \tau$. Actually, although for $t > \tau$, $u^*(t)$ would be identical for both trajectories, yet as is seen from (2.244) at the time instant $t = \tau$, the values $x(\tau)$ and $x^*(\tau)$—which would be "initial" conditions for the interval $\tau < t < T$—are nonidentical. Hence, for $t > \tau$, $\tilde{x}(t)$ and $\tilde{x}^*(t)$ do not coincide. But in view of the fact that the difference $\tilde{x}(\tau) - \tilde{x}^*(\tau)$ is infinitesimally small, the entire subsequent motion $\tilde{x}(t)$ will differ from $\tilde{x}^*(t)$ in an "infinitesimally small way." This situation is illustrated in Fig. 2.17 by the closeness of the trajectories $\tilde{x}(t)$ and $\tilde{x}^*(t)$ for $t > \tau$.

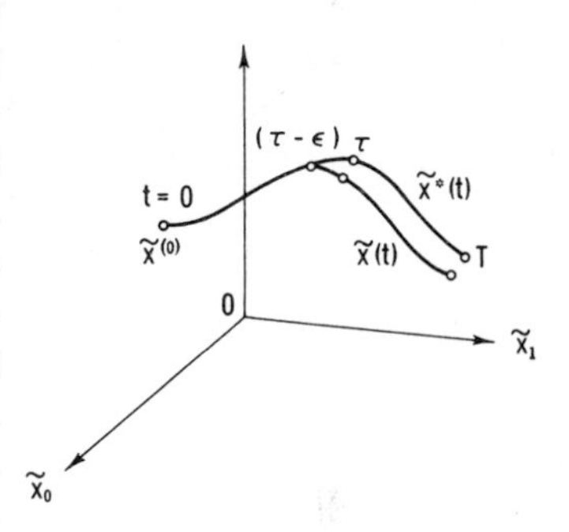

FIG. 2.17

Let us introduce the vector variation $\delta\tilde{x}(t)$ in the trajectory with coordinates $\delta\tilde{x}_j$ ($j = 0, \ldots, n + 1$), defined by the equation

$$\delta\tilde{x}(t) = \tilde{x}(t) - \tilde{x}^*(t) \qquad (\tau \leqslant t \leqslant T). \tag{2.245}$$

From (2.244) it follows that the "initial" value of the variation for $t = \tau$ is equal to

$$\delta\tilde{x}(\tau) = \epsilon\{\tilde{f}[\tilde{x}(\tau), u(\tau)] - \tilde{f}[\tilde{x}(\tau), u^*(\tau)]\}. \tag{2.246}$$

Since $\tilde{x}(t)$ differs in an arbitrarily small way from $\tilde{x}^*(t)$ for $t > \tau$, then the variation $\delta\tilde{x}(t)$ in the trajectory is infinitesimally small. Therefore

[†] It should be indicated that the needle-shaped variation has also been applied to other generalizations of classical variational methods.

its law of variation with time can be found from the linear equations for small changes in $\tilde{x}(t)$, which are called the "variational equations." The variational equations are obtained from the fundamental equations

$$d\tilde{x}_j/dt = \tilde{f}_j(\tilde{x}_0, \tilde{x}_1, ..., \tilde{x}_{n+1}, u) \qquad (j = 0, ..., n+1). \tag{2.247}$$

If all the $\tilde{x}_j$ are replaced by $\tilde{x}_j + \delta\tilde{x}_j$ ($j = 0, ..., n+1$), and then after expanding the $\tilde{f}_j$ in series in the $\delta\tilde{x}_i$, the terms of higher small orders are discarded,

$$\begin{aligned} \frac{d(\tilde{x}_j + \delta\tilde{x}_j)}{dt} &= \tilde{f}_j(\tilde{x}_0 + \delta\tilde{x}_0, ..., \tilde{x}_i + \delta\tilde{x}_i, ..., \tilde{x}_{n+1} + \delta\tilde{x}_{n+1}, u) \\ &= \tilde{f}_j(\tilde{x}_0, ..., \tilde{x}_i, ..., \tilde{x}_{n+1}, u) \\ &\quad + \sum_{i=0}^{n+1} \delta\tilde{x}_i \frac{\partial\tilde{f}_j}{\partial\tilde{x}_i}(\tilde{x}_0, ..., \tilde{x}_i, ..., \tilde{x}_{n+1}, u) + o(\delta\tilde{x}). \end{aligned} \tag{2.248}$$

Discarding the term $o(\delta\tilde{x})$ in which the factors of small order higher than the first are contained, and taking (2.247) into account, we arrive at the linear variational equations for the $\delta\tilde{x}_i$:

$$\frac{d(\delta\tilde{x}_j)}{dt} = \sum_{i=0}^{n+1} \delta\tilde{x}_i \frac{\partial\tilde{f}_j(\tilde{x}, u)}{\partial\tilde{x}_i} \qquad (j = 0, ..., n+1). \tag{2.249}$$

These equations can be integrated with the initial conditions (2.246). First, of interest to us is the quantity $(\delta\tilde{x})_{t=T}$, particularly the value of the coordinate $\delta\tilde{x}_0$ at $t = T$. In fact, according to (2.239) this quantity is the change δQ in the value of the criterion Q, arising because of the needle-shaped variation having occurred in the interval $\tau - \epsilon < t < \tau$. Since the optimal control $u^*(t)$ ensures the smallest value of Q, then for any other control $u(\tau)$ the value of Q can only increase. Hence

$$\delta Q = (\delta\tilde{x}_0)_{t=T} \geqslant 0. \tag{2.250}$$

This relation can be rewritten in the following manner:

$$-\delta Q = -(\delta\tilde{x}_0)_{t=T} = \langle\delta\tilde{x}(T), \tilde{\psi}(T)\rangle \leqslant 0, \tag{2.251}$$

where $\tilde{\psi}(T)$ is a vector chosen in a way such that the scalar product of $\delta\tilde{x}(T)$ and $\tilde{\psi}(T)$ equals $\delta\tilde{x}_0(T)$. Evidently, the coordinate of this vector $\tilde{\psi}_0(T) = -1$, and the remaining coordinates $\tilde{\psi}_j(T) = 0$ ($j = 1, ..., n+1$). Thus,

$$\tilde{\psi}(T) = (-1, 0, ..., 0). \tag{2.252}$$

The scalar product (2.251) represents the effect which the needle-shaped variation $u(\tau)$, appearing at the instant $t = \tau$, renders on the final purpose, i.e., on the value of the optimality criterion Q or $\tilde{x}_0$ at the instant $t = T$. The basic idea in deriving the maximum principle is that this effect can be estimated by the "linear" variational equations (2.249), because it is arbitrarily small. The linearity of the equations has as a consequence an extraordinary simplification of the analysis. Thus for example, the effects of two needle-shaped variations arising in different infinitesimally small time intervals can be considered independent of one another as a consequence of the additivity properties inherent in linear equations. Thus, the choice of an optimal value of $u(\tau)$ at some time instant can be based, "formally, independently of the entire remaining control process," only on making the magnitude of the corresponding increment—δQ as large as possible. For any nonoptimal controls this magnitude, equal to the scalar product

$$-\delta \underset{\sim}{Q} = \langle \delta\tilde{x}(T), \tilde{\psi}(T) \rangle \tag{2.253}$$

as seen from (2.251), will be negative. Only for $u(\tau)$ equal to $u^*(\tau)$ does it vanish, attaining its maximal value here. Essentially, condition (2.251) means only that any nonoptimal control is "worse" than the optimal one: it gives a smaller effect than the optimal control.

Expression (2.253) is not convenient enough, since for its calculation it is necessary to integrate equations (2.249) beforehand and find $\delta\tilde{x}(T)$ depending on the "initial" condition $\delta\tilde{x}(\tau)$. Meanwhile as is seen from (2.246), the value of $\delta\tilde{x}(\tau)$ is indeed directly related to the value of $u(\tau)$. We will pose the problem of obtaining a vector $\tilde{\psi}(t)$ which would satisfy the condition

$$\langle \delta\tilde{x}(t), \tilde{\psi}(t) \rangle = \langle \delta\tilde{x}(T), \tilde{\psi}(T) \rangle \qquad (\tau \leqslant t \leqslant T). \tag{2.254}$$

Then in the particular case for $t = \tau$ we can obtain the equality

$$\langle \delta\tilde{x}(\tau), \tilde{\psi}(\tau) \rangle = \langle \delta\tilde{x}(T), \tilde{\psi}(T) \rangle \tag{2.255}$$

and make the inference about $\delta \underset{\sim}{Q}$ according to the left side of this equation, which by the same token is directly related to the value of $u(\tau)$ we are required to determine.

It turns out that a differential equation which satisfies the vector $\tilde{\psi}(t)$ can be found. From (2.254) it follows that:

$$\langle \delta\tilde{x}(t), \tilde{\psi}(t) \rangle = \text{const} \qquad (\tau \leqslant t \leqslant T). \tag{2.256}$$

Hence we find

$$\frac{d}{dt}\langle \delta\tilde{x}(t), \tilde{\psi}(t)\rangle = 0 \qquad (\tau \leqslant t \leqslant T) \tag{2.257}$$

or

$$\left\langle \frac{d[\delta\tilde{x}(t)]}{dt}, \tilde{\psi}(t)\right\rangle + \left\langle \delta\tilde{x}(t), \frac{d\tilde{\psi}(t)}{dt}\right\rangle = 0 \qquad (\tau \leqslant t \leqslant T). \tag{2.258}$$

Let us rewrite this equation in expanded form:

$$\sum_{j=0}^{n+1} \frac{d[\delta\tilde{x}_j(t)]}{dt} \cdot \tilde{\psi}_j(t) + \sum_{i=0}^{n+1} \delta\tilde{x}_i(t) \frac{d\tilde{\psi}_i(t)}{dt} = 0. \tag{2.259}$$

Substituting the values of $d(\delta\tilde{x}_j)/dt$ from (2.249), we find

$$\sum_{j=0}^{n+1} \tilde{\psi}_j(t) \sum_{i=0}^{n+1} \delta\tilde{x}_i \frac{\partial \tilde{f}_j(\tilde{x}, u)}{\partial \tilde{x}_j} + \sum_{i=0}^{n+1} \delta\tilde{x}_i(t) \frac{d\tilde{\psi}_i(t)}{dt} = 0. \tag{2.260}$$

Interchanging the order of the summations with respect to i and j in the first term, we arrive at the expression

$$\sum_{i=0}^{n+1} \delta\tilde{x}_i \left[\sum_{j=0}^{n+1} \tilde{\psi}_j(t) \frac{\partial \tilde{f}_j(\tilde{x}, u)}{\partial \tilde{x}_i} + \frac{d\tilde{\psi}_i(t)}{dt}\right] = 0. \tag{2.261}$$

As is seen from (2.257), the left side of (2.261) is identically equal to zero for "arbitrary" $\delta\tilde{x}_j$. The equating to zero of the brackets in (2.261) is the necessary and sufficient condition for this, from which it follows that

$$\frac{d\tilde{\psi}_i(t)}{dt} = -\sum_{j=0}^{n+1} \tilde{\psi}_j(t) \frac{\partial \tilde{f}_j(\tilde{x}, u)}{\partial \tilde{x}_i} \qquad (i = 0, ..., n+1). \tag{2.262}$$

The equalities (2.262) obtained are a set of differential equations which are linear with respect to the $\tilde{\psi}_j$. It is not difficult to see that Eq. (2.262) and (2.231) are identical. Hence the equations that are adjoint to the fundamental system (2.249) are also obtained by the method presented. These equations must be solved with the boundary conditions (2.252).

Now, as is seen from (2.255) and (2.251), the quantity

$$-\delta Q = \langle \delta\tilde{x}(\tau), \tilde{\psi}(\tau)\rangle \leqslant 0 \tag{2.263}$$

can be considered. Substituting $\delta\tilde{x}(\tau)$ from (2.246), we will obtain after cancellation of ϵ:

$$\langle \tilde{f}[\tilde{x}(\tau), u(\tau)], \tilde{\psi}(\tau)\rangle - \langle \tilde{f}[\tilde{x}(\tau), u^*(\tau)], \tilde{\psi}(\tau)\rangle \leqslant 0. \tag{2.264}$$

Now let us introduce the quantity

$$\tilde{H} = \langle \tilde{f}[\tilde{x}(\tau), u(\tau)], \tilde{\psi}(\tau) \rangle. \tag{2.265}$$

From inequality (2.264) it is seen that the quantity $\tilde{H}$ attains a maximum for the optimal control $u^*(\tau)$. Hence the maximum principle follows: $u(\tau)$ must be chosen such that the quantity $\tilde{H}$ attain a maximum value. All these arguments are easily generalized to the case of any $r > 1$; u need only be replaced by $\bar{u}$.

We will now consider the techniques for application of the maximum principle. For determining the optimal trajectory $\tilde{x}^*(t)$ and the optimal control $\bar{u}^*(t)$, two systems of equations are solved simultaneously—the fundamental and adjoint. The process of solution can be presented in the following form. We assume that at $t = 0$ the system starts from some point $\tilde{x} = \tilde{x}^{(0)}$. We also assign some initial value $\tilde{\psi}^{(0)}$ of the vector $\tilde{\psi}$, since it is not known beforehand. The value of the vector $\bar{u}$ at the initial point $\tilde{x}^{(0)}$ is chosen from the condition $\tilde{H} = \max$ such that the scalar product

$$(\tilde{H})_{\tilde{x}=\tilde{x}^{(0)}} = \langle \tilde{f}^{(0)}\tilde{\psi}^{(0)} \rangle = \left\langle \frac{d\tilde{x}[\tilde{x}^{(0)}, \bar{u}]}{dt}, \tilde{\psi}^{(0)} \right\rangle \tag{2.266}$$

be maximal. By choosing the control $(\bar{u})_{t=0}$ in this way, the increments $\Delta\tilde{\psi}_i$ and $\Delta\tilde{x}_i$ for sufficiently small time intervals Δt can be found from the systems of equations (2.235) and (2.233)—the fundamental and adjoint. Hence the values of the vectors $\tilde{\psi}$ and $\tilde{x}$ can be found at a new point of the optimal trajectory, close to the initial and corresponding time instant $t = \Delta t$. At the new point the entire procedure described is repeated, the new optimal value $\bar{u}^*$ found, the new increments $\Delta\tilde{\psi}_i$ and $\Delta\tilde{x}_i$ determined, the shift of the representative point taking place to the subsequent position corresponding to the near time instant $t = 2\,\Delta t$, and so on. Thus, by operating step by step the entire optimal trajectory, or the extremal as it is often called, can be drawn. By the minimization of $\tilde{H}$ the optimal control $\bar{u}^*$ is simultaneously determined at each point.

In the procedure given for constructing the extremal one obscure point still exists. It is not known by what method the initial values $\tilde{\psi}_i^{(0)}$ of the coordinates of the vector $(\tilde{\psi})_{t=0} = \tilde{\psi}^{(0)}$ are to be chosen. The choice of these values turns out to be related to the boundary conditions of the problem.

Let us consider how this choice must be made for certain particular cases.

(1) *The problem with a free end of the trajectory and a fixed time* T. As was shown above, with the aid of the introduction of the variable

$\tilde{x}_0$ the problem of minimizing the functional Q reduces to the problem of minimizing the value of $x_0(T)$. From the preceding it is known that for this it is required to choose a control $\bar{u}$ such that the projection of the vector $d\tilde{x}/dt$ on the direction of $\tilde{\psi}$ be maximal. For the "last" infinitesimally small time interval $T - \Delta t < t < T$, the corresponding direction of $\tilde{\psi}$ is given by formula (2.252), whose meaning is easily explained. In order that the increment in $\tilde{x}_0(T)$ be as small as possible, it is necessary to direct the vector $\tilde{\psi}$ opposite to the direction of $\tilde{x}_0$ for $t = T - \Delta t$. But as is not difficult to see, this direction of $\tilde{\psi}$ coincides with the direction of $\tilde{\psi}(T)$ defined by formula (2.252).

Thus in this problem it is required to provide a solution $\tilde{\psi}(t)$ which satisfies the final conditions:

$$\tilde{\psi}_0^{(T)} = -1, \quad \tilde{\psi}_i^{(T)} = 0 \quad (i = 1, ..., n + 1). \tag{2.267}$$

It is necessary to choose the initial values $\tilde{\psi}_i^{(0)}$ such that the final values $\tilde{\psi}_i^{(T)}$ be equal to the prescribed values defined by (2.267).

Thus the solutions of the two systems of equations—fundamental and adjoint—must satisfy the prescribed "initial" conditions $\tilde{x}^{(0)}$ for the vector $\tilde{x}$ and the assigned "end" conditions (2.267) for the vector $\tilde{\psi}$. Therefore in general the solution of the problem of integrating the systems of equations for $\tilde{x}$ and $\tilde{\psi}$ under prescribed "boundary" conditions is required. Here the $n + 2$ initial conditions for $\tilde{x}^{(0)}$ and the $n + 2$ end conditions (2.267) for $\tilde{\psi}^{(T)}$ give all the boundary conditions required for solving the problem.

(2) *The problem of maximal speed of response with a fixed end of the trajectory $\bar{x}^{(T)}$ in the phase space of $\bar{x}$ and the time T not fixed beforehand.* It is required to select a control such that during a minimal time T the representative point $\bar{x}$ be transferred from the prescribed initial position $\bar{x}^{(0)}$ to the given final position $\bar{x}^{(T)}$ (Fig. 2.15).

For simultaneously solving the two systems of equations (2.236) and (2.237)—fundamental and adjoint—with a parallel choice of control $\bar{u}$ maximizing the quantity H at each point of the optimal trajectory, it is necessary to know all the $2n$ initial conditions. These are the initial values $x_1^{(0)}, ..., x_n^{(0)}$ of the coordinates of the point $\bar{x}$ in phase space and the initial values $\psi_1^{(0)}, ..., \psi_n^{(0)}$ for the vector $\bar{\psi}$. The first n values are given. But as far as the n initial coordinates of the vector $\bar{\psi}$ are concerned, it is required to choose them in such a way as to satisfy the prescribed boundary conditions at the end of the optimal trajectory, i.e., the assigned values $x_1^{(T)}, ..., x_n^{(T)}$ of the coordinates of the end point $\bar{x}^{(T)}$. Thus n conditions for the n unknowns $\psi_i^{(0)}$ $(i = 1, ..., n)$ are obtained.

There are no general rules for selecting the initial values $\psi_i^{(0)}$ of the coordinates of the vector $\bar{\psi}$. We will assume that we are given the vector $\bar{\psi}^{(0)}$ at random (Fig. 2.18). Let us construct the extremal M_0M_1 in the manner indicated above. However, in general it will not pass through the required point M_T for which $\bar{x} = \bar{x}^{(T)}$. Then we are given another value of the vector $\bar{\psi}^{(0)}$, construct another extremal M_0M_2, and so on, as long as the extremal does not pass through M_T. This procedure can be rationalized [3.25] by introducing a measure of the distance r in phase space from the extremal M_0M_1 to the required point M_T. For example, such a measure can be the minimal of the Euclidean distances of points of the curve M_0M_1 from M_T. Then by a search according to a known procedure, the coordinates $\psi_1^{(0)}, \ldots, \psi_n^{(0)}$ of the vector $\bar{\psi}^{(0)}$ should be chosen such that the quantity r depending on them becomes minimal:

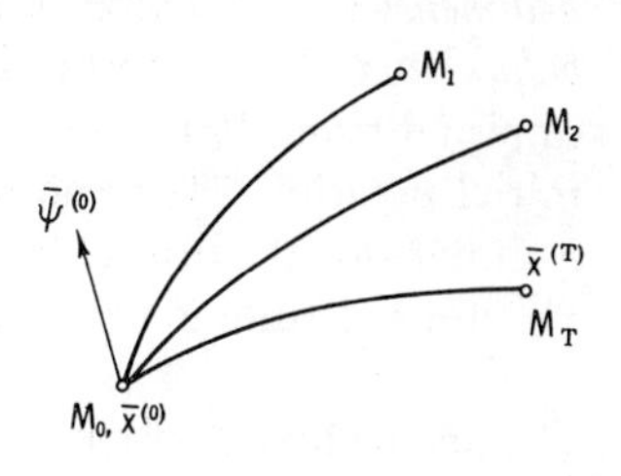

FIG. 2.18

$$r = r(\psi_1^{(0)}, \ldots, \psi_n^{(0)}) = \min. \tag{2.268}$$

Of course, this minimum must turn out to be equal to zero. The function r can have several minima. The desired minimum is the "minimum minimorum," i.e., the smallest of all the minima.

Thus, in the first place a series of minimizations of H with respect to $\bar{u}$ for each small time interval Δt enters into the procedure for solving the problem, as a result of which the extremal M_0M_j is "drawn" in phase space. For each extremal obtained in this manner its corresponding value of r is calculated. Then by the choice of the $\psi_i^{(0)}$ $(i = 1, \ldots, n)$ the function r (2.268) of them undergoes the minimization procedure, as a result of which it reduces to zero. Only then is the solution of the problem considered complete. At the present time a means for automatic synthesis of optimal systems has been developed [3.25, 3.71, 6.4, 6.6] in which the procedure indicated above has been automated. Automatic search underlies the operation of this equipment. In the process of solving the problem a "rapid" automatic search for the optimal control $\bar{u}$ is carried out, providing a maximum for the function $H(\bar{u})$ in each interval Δt. As a result of this operation the extremal M_0M_i is "drawn" and its distance r from the point $\bar{x}^{(T)}$ determined. In addition, a "slow" automatic search is carried out, i.e., a selection of the values $\psi_i^{(0)}$ minimizing the function r.

For systems with linear objects there are methods permitting the values $\psi_i^{(0)}$ to be found with the aid of iteration [2.39].

(3) *The problem of maximal speed of response with a fixed finite n-dimensional region P in the phase space of $\bar{x}$ and the time T not fixed beforehand.* It is required to control in such a manner that during a minimal time T the representative point $\bar{x}$ is transferred from the given initial position $\bar{x}^{(0)}$ to some point $\bar{x}^{(T)}$, belonging to a certain prescribed n-dimensional subset P of phase space (Fig. 1.9). Neither the point $\bar{x}^{(T)}$ nor the time T is fixed in advance.

The problem will be solved if the n initial values $\psi_i^{(0)}$ $(i = 1, ..., n)$ of the vector $\bar{\psi}$ turn out to be known, since of the $2n$ initial conditions for the two systems of equations—fundamental and adjoint—the n values of the coordinates $x_1^{(0)}, ..., x_n^{(0)}$ are given. Hence it is still necessary to add n boundary conditions. These conditions turn out to be the so-called "transversality conditions" imposed on the coordinates of the vector $\bar{\psi}$ at the end point $\bar{x}^{(T)}$ of the trajectory.

In order to obtain the transversality conditions, let us consider the isosurface $T =$ const in phase space (Fig. 2.19). We agree to so designate the locus of points of phase space which can be reached after a time T from the initial point $\bar{x}^{(0)}$ with the optimal control $\bar{u}^*$. For example, during the time T_1 it is possible to go from $\bar{x}^{(0)}$ as far as points of the isosurface T_1; during the time $T_2 > T_1$, as far as points of the isosurface T_2; and so on. It can be proved that these isosurfaces are convex. In Fig. 2.19 the isosurface T_2 does not intersect the surface separating the region P from the rest of phase space. This means that the time T_2 is insufficient for reaching the region P. By increasing T we obtain for the larger value of T a new isosurface in which the old isosurface with a smaller T is "imbedded." Continuing this process of forming new isosurfaces $T =$ const containing points of phase space still further removed from the initial one, we finally obtain (if the solution to the problem exists) the isosurface $T = T_3$ "tangent" to the surface separating the region P from the remaining space. Let $\bar{x}^{(T)}$ be the tangent point. Then this is the end point of the required optimal trajectory, shown by the solid curve in Fig. 2.19. In fact, the point $\bar{x}^{(T)}$ can be reached during the time $T = T_3$ under optimal control. Some other point M of the region P can be reached only during a time $T_4 > T_3$. Therefore the trajectory joining $\bar{x}^{(0)}$ with the point M, for example, will not be optimal.

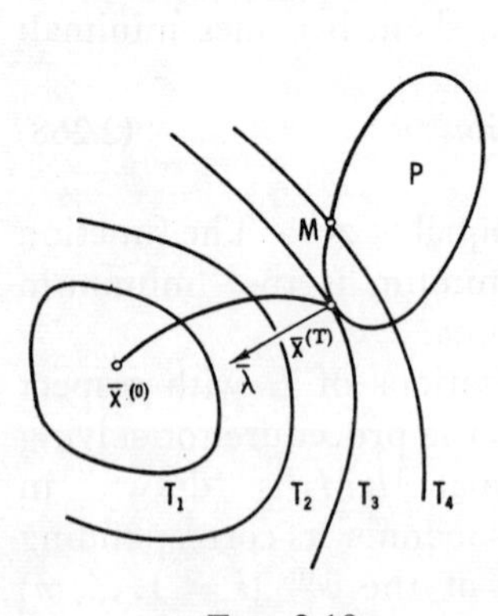

FIG. 2.19

Let the equation of the surface separating P from the rest of phase space have the form

$$\varphi(x_1, ..., x_n) = 0, \tag{2.269}$$

where φ is a function that is differentiable with respect to all the parameters x_i $(i = 1, ..., n)$.

From the construction in Fig. 2.19 it is seen that the surface bounding the region P, if it satisfies definite conditions, and the isosurface $T = T_3$ have a common normal vector $\bar{\lambda}$ at the point $\bar{x}^{(T)}$.

Then the gradient of φ can be taken as the normal $\bar{\lambda}$, i.e., the vector

$$\bar{\lambda} = \text{grad}\, \varphi = (\partial\varphi/\partial x_1, ..., \partial\varphi/\partial x_n). \tag{2.270}$$

We will stipulate that the surface $\varphi = 0$ does not have exceptional points at which all the $\partial\varphi/\partial x_i$ vanish simultaneously. Then the vector grad φ is defined for any point of the surface (2.269).

From Fig. 2.19 it is also seen that in the "last" infinitesimally small time interval $T - \Delta t < T$, where $\Delta t \to 0$, the optimal control consists in bringing about a possibly more rapid motion from the isosurface $T_3 - \Delta t = \text{const}$ to the isosurface $T_3 = \text{const}$. For this the maximal value of the projection of the vector $d\bar{x}/dt$ on the direction of the vector $\bar{\lambda}$ must be provided. Hence in this case the vector $\bar{\psi}(T)$ coincides with $\bar{\lambda}$ in direction, and we can take

$$\bar{\psi}(T) = \text{grad}\, \varphi = (\partial\varphi/\partial x_1, ..., \partial\varphi/\partial x_n), \tag{2.271}$$

from which it follows that

$$\psi_i(T) = \partial\varphi/\partial x_i \qquad (i = 1, ..., n). \tag{2.272}$$

Conditions (2.272) are called the transversality conditions. Since the time T is not fixed, the values $\psi_i^{(0)}, ..., \psi_n^{(0)}$ and the quantity T are not known in advance—there are $n + 1$ unknowns for whose determination $n + 1$ conditions are necessary. These conditions are given by the $n + 1$ relations (2.269) and (2.272). In fact, if we are given the values $\psi_i^{(0)}$ and T, then the coordinates $x_i(T)$ and the values $\psi_i(T)$ will be functions of these $n + 1$ unknowns. Substituting these functions into the $n + 1$ relations (2.269) and (2.272), we can expect that the $n + 1$ equations obtained with the $n + 1$ unknowns $\psi_i^{(0)}$ and T have a solution. Of course, in this case the difficulties in solving the problem, in view of the more complicated way of prescribing the boundary conditions, increase further compared to the preceding cases. If the region P shrinks to a point, then we revert to the preceding problem. In this case the transversality conditions lose meaning, but on the other hand the coordinates of the end point of the trajectory emerge. Thus the total number of equations for determining the unknown values $\psi_i^{(0)}$ proves to be sufficient.

The conditions of the problem can be generalized further, if instead of the fixed initial point $\bar{x}^{(0)}$ we consider some initial subset P_0 of points of phase space from which the representative point can start. In this case transversality conditions analogous to those presented above are imposed on the initial value of the vector $\bar{\psi}$. The reader who is interested in these questions in more detail can turn to the monograph of L. S. Pontryagin and others [2.21], where a rigorous and general derivation of the maximum principle with transversality conditions is given.

(4) *The problem of maximal speed of response with a fixed finite s-dimensional region* ($s < n$). In this more general case, the end point $\bar{x}^{(T)}$ of the extremal must be situated on a manifold M_s of dimension s, where $1 < s < n$, and moreover the coordinates of points $\bar{x}$ belonging to M_s are given by the system of equations

$$\varphi_i(x_1, ..., x_n) = \varphi_i(\bar{x}) = 0 \qquad (i = 1, ..., n - s). \tag{2.273}$$

For example, prescribing two equations of the type (2.273) in three-dimensional space defines a one-dimensional manifold M_s, i.e., a certain curve.

The vector $\bar{g}$ normal to the manifold M_s can be given in the following form:

$$\bar{g} = \sum_{\beta=1}^{\beta=n-s} \lambda_\beta \operatorname{grad} \varphi_\beta(\bar{x}), \tag{2.274}$$

where $\bar{x}$ satisfies Eqs. (2.273), and the λ_β are certain numbers. Here we consider that none of the vectors grad $\varphi_\beta(\bar{x})$ vanishes at a single point of the manifold M_s.

It is not difficult to verify that the vector $\bar{g}$ is orthogonal to any infinitesimally small vector $\delta\bar{x}$ belonging to the manifold M_s and emanating from the point $\bar{x}$. In fact, the vector $\delta\bar{x}$ lies in each of the surfaces $\varphi_\beta(\bar{x}) = 0$ of (2.273) and hence is orthogonal to each grad $\varphi_\beta(\bar{x})$. Therefore the scalar product is equal to zero:

$$\langle \bar{g}, \delta\bar{x} \rangle = \sum_{\beta=1}^{\beta=n-s} \lambda_\beta \langle \operatorname{grad} \varphi_\beta(\bar{x}), \delta\bar{x} \rangle = 0, \tag{2.275}$$

i.e., the vectors $\bar{g}$ and $\delta\bar{x}$ are orthogonal.

At first we will assume that the equations of the object do not depend explicitly on time and have the form

$$d\bar{x}/dt = \bar{f}(\bar{x}, \bar{u}), \tag{2.276}$$

where $\bar{x}$ is n-dimensional and $\bar{u}$ is an r-dimensional vector. Let it be required to choose a control $\bar{u}$ minimizing the integral

$$Q = \int_0^T G(\bar{x}, \bar{u})\, dt = \int_0^T f_0(\bar{x}, \bar{u})\, dt = x_0(T). \tag{2.277}$$

The time T may not be fixed. The isosurfaces $S_t =$ const can be represented in the n-dimensional space of the vector $\bar{x}$, where

$$S_t = \int_0^t G(\bar{x}, \bar{u}^*)\, dt \tag{2.278}$$

and $\bar{u}^*$ is the optimal control. In the special case when $G = 1$ and the minimum time T is required, these isosurfaces are shown in Fig. 2.19. But they can be constructed for a more general case as well, in which G is not equal to 1. The vector $\bar{\psi}$ is the gradient to the isosurface $S_t =$ const. This is seen from the fact that the vector $\bar{\psi}$ with coordinates $-\partial S/\partial x_i$ is the gradient of the isosurface $S =$ const, which will be easily discerned from (2.210). Meanwhile, by reversing the reading of time from end to beginning, the identity of the isosurfaces $S =$ const for such a new problem with the isosurfaces $S_t =$ const can be established [see (2.278)]. The only difference here is in sign; therefore

$$\bar{\psi} = \text{grad}\, S_t \tag{2.279}$$

(without the minus sign, which would be present in the formula for S).

As in the preceding case, we will conceive of a set of expanding isosurfaces S_t corresponding to a larger and larger value of t. If for a small value of t such an isosurface still does not have points in common with the manifold M_s, then with an increase in t such an instant t_1 occurs (if the solution to the problem exists) when the isosurface $S_t =$ const will have just a single point in common with the manifold M_s (assuming that for $t < t_1$ there were no such points in common). Since for the tangency of the isosurface and the manifold their normal vectors $\bar{\psi}$ and $\bar{g}$ must be collinear, we obtain the transversality condition in the form[†]

$$\bar{\psi}(t_1) = \bar{g} = \sum_{\beta=1}^{n-s} \lambda_\beta\, \text{grad}\, \varphi_\beta[\bar{x}(t_1)]. \tag{2.280}$$

[†] As a consequence of the linearity of the adjoint equations it suffices to determine the vector $\bar{\psi}$ to within a constant factor. Therefore multiplication of the final conditions by any constant factor will give a new value of $\bar{\psi}(t)$ that is also a solution.

In other words, for an end point $\bar{x}(t_1)$ satisfying conditions (2.273), numbers λ_β must exist, not simultaneously equal to zero, such that Eq. (2.280) hold.

The vector equation (2.280) is equivalent to n scalar equations. If further the $n - s$ conditions (2.273) are added to them, then in all we will obtain $2n - s$ equations in which are contained $2n - s$ unknowns, i.e., the $n - s$ constants λ_β and the n unknown values $\psi_i(t_1)$.

If Eqs. (2.276) contain t in explicit form, then by the substitution $x_{n+1} = t$ the problem reduces to the preceding case. The only difference here is that the construction will be carried out in $(n + 1)$-dimensional space. In addition, the manifold M_{s+1} in this new space, which is equivalent to the previous M_s in the n-dimensional space of $\bar{x}$, will be a cylinder whose base is the manifold M_s in the previous space, and the elements are straight lines parallel to the coordinate axis of $x_{n+1} \equiv t$. Here the transversality conditions have the previous form. If the time of passage from the initial point to the manifold M_s is fixed and equal to T, then yet another equation is added to the equations of the manifold:

$$\varphi_{n-s+1}(\tilde{x}) = x_{n+1} - T = 0, \tag{2.281}$$

which must be used in composing the transversality conditions.

As an illustration of what was presented above, we give a simple example of the application of the maximum principle to the problem with a free end of the trajectory and a fixed time T [2.17]. Let it be required to find the control minimizing the integral

$$Q = \tfrac{1}{2} \int_0^T (x^2 + u^2)\, dt, \tag{2.282}$$

where the object is described by the first-order equation

$$dx/dt = -ax + u. \tag{2.283}$$

The value $(x)_{t=0} = x^{(0)}$ is given. We also assume that no additional constraints have been imposed on the control action u.

Let us introduce the variables

$$\tilde{x}_1(t) = x(t), \qquad \tilde{x}_0(t) = \tfrac{1}{2} \int_0^t (\tilde{x}_1^2 + u^2)\, dt, \tag{2.284}$$

where $(\tilde{x}_0)_{t=0} = 0$. Then we obtain the system of equations in the form

$$d\tilde{x}_0/dt = \tfrac{1}{2}\tilde{x}_1^2 + \tfrac{1}{2}u^2 = \tilde{f}_0, \qquad d\tilde{x}_1/dt = -a\tilde{x}_1 + u = \tilde{f}_1. \tag{2.285}$$

Now we will constitute the function $\tilde{H}$ according to expression (2.228):

$$\tilde{H} = \tilde{\psi}_0(d\tilde{x}_0/dt) + \tilde{\psi}_1(d\tilde{x}_1/dt) = \tilde{\psi}_0\tfrac{1}{2}(\tilde{x}_1^2 + u^2) + \tilde{\psi}_1(-a\tilde{x}_1 + u). \tag{2.286}$$

Let us compose the adjoint equations for $\tilde{\psi}_0$ and $\tilde{\psi}_1$ according to (2.227) and (2.231):

$$\begin{aligned} \frac{d\tilde{\psi}_0}{dt} &= 0, \\ \frac{d\tilde{\psi}_1}{dt} &= -\sum_{j=0}^{1} \tilde{\psi}_j \frac{\partial \tilde{f}_j}{\partial \tilde{x}_1} = -\tilde{\psi}_0 \frac{\partial \tilde{f}_0}{\partial \tilde{x}_1} - \tilde{\psi}_1 \frac{\partial \tilde{f}_1}{\partial \tilde{x}_1} = -\tilde{\psi}_0 \tilde{x}_1 + \tilde{\psi}_1 a. \end{aligned} \tag{2.287}$$

According to (2.252) the final values of $\tilde{\psi}_i$ are

$$\tilde{\psi}_0(T) = -1, \qquad \tilde{\psi}_1(T) = 0. \tag{2.288}$$

The prescribed initial conditions for $\tilde{x}$ are

$$\tilde{x}_0^{(0)} = 0, \qquad \tilde{x}_1^{(0)} = x^{(0)}. \tag{2.289}$$

The control action u must be chosen at each time instant such that $\tilde{H}$ will be maximized. By virtue of the first of conditions (2.287) and (2.288)

$$\tilde{\psi}_0(t) = -1 = \text{const} \tag{2.290}$$

holds. Therefore from (2.286) we find

$$\tilde{H} = -\tfrac{1}{2}\tilde{x}_1^2 - \tfrac{1}{2}u^2 - a\tilde{\psi}_1\tilde{x}_1 + \tilde{\psi}_1 u. \tag{2.291}$$

Setting the derivative $\partial\tilde{H}/\partial u$ equal to zero, we find the optimal value

$$u^* = \tilde{\psi}_1 . \tag{2.292}$$

Substituting this value into the equations for $\tilde{x}_1$ and $\tilde{\psi}_1$ (the function $\tilde{x}_0$ does not enter into these equations and therefore is not of interest to us), we arrive at the system of equations

$$d\tilde{x}_1/dt = -a\tilde{x}_1 + \tilde{\psi}_1 , \qquad d\tilde{\psi}_1/dt = a\tilde{\psi}_1 + \tilde{x}_1 . \tag{2.293}$$

According to (2.288) and (2.289) the boundary conditions for this system have the form

$$\tilde{x}_1(0) = x^{(0)}, \qquad \tilde{\psi}_1(T) = 0. \tag{2.294}$$

It is not difficult to integrate the linear equations (2.293). As a result of the integration we obtain

$$\tilde{x}_1(t) = C_1 e^{pt} + C_2 e^{-pt}, \qquad \tilde{\psi}_1(t) = D_1 e^{pt} + D_2 e^{-pt}, \tag{2.295}$$

where

$$p = \sqrt{(a^2 + 1)} \tag{2.296}$$

is the root of the characteristic equation.

Conditions (2.294) take the form

$$x^{(0)} = C_1 + C_2, \qquad 0 = \tilde{\psi}_1(T) = D_1 e^{pT} + D_2 e^{-pT}. \tag{2.297}$$

From expression (2.293) for $t = 0$,

$$\begin{aligned} [d\tilde{x}_1/dt]_{t=0} &= C_1 p - C_2 p = -a\tilde{x}_1(0) + \tilde{\psi}_1(0) = -ax^{(0)} + D_1 + D_2, \\ [d\tilde{\psi}_1/dt]_{t=0} &= D_1 p - D_2 p = a\tilde{\psi}_1(0) + \tilde{x}_1(0) = a(D_1 + D_2) + x^{(0)}, \end{aligned} \tag{2.298}$$

together with (2.297), we determine the constants C_1, C_2, D_1, D_2. In particular we find

$$D_1 = \frac{x^{(0)}}{(p-a) + (p+a)e^{2pT}}, \qquad D_2 = -\frac{x^{(0)} e^{2pT}}{(p-a) + (p+a)e^{2pT}}. \tag{2.299}$$

Hence from (2.292) and (2.295) it follows that

$$u^*(t) = \tilde{\psi}_1(t) = \frac{x^{(0)}(e^{pt} - e^{2pT} \cdot e^{-pt})}{(p-a) + (p+a)e^{2pT}}. \tag{2.300}$$

Expression (2.300) gives the optimal control law as a function of time. By finding $\tilde{x}_1(t)$ and eliminating time t between $u^*(t)$ and $\tilde{x}_1(t)$, we will obtain the control law $u^* = u^*(\tilde{x}_1)$.

The maximum principle is extended in a natural way to discrete-continuous systems with "linear" objects [2.17]. In fact, for these objects the equation for the deviations—be they even large—is also linear. Therefore those arguments that underlay the derivation of the maximum principle for continuous systems can likewise be repeated with little modification for discrete-continuous systems with linear objects. But in the general case of nonlinear discrete-continuous systems such a proof cannot be carried out. Actually, the time intervals between the individual samples and also the total number of discrete values are finite quantities. Therefore the effect of a finite variation in one of the samples $u(k)$—the analog of the "needle-shaped" variation for this case—on the final purpose (i.e., on the value of Q) will also be a finite quantity, and not infinitesimally small at all as in continuous systems. But then the entire construction given above breaks down, since it is now impossible to talk about "small" deviations of the varied trajectory $\bar{x}$ from the optimal trajectory; hence, for determining the deviations "linear" variational equations cannot be used. Moreover, in this case we even succeed in constructing a "counterexample" in which for optimal control the maximum principle in the form given above proves to be invalid.

However, another considerably "weaker" formulation of the maximum principle turns out to be valid for a wide class of discrete-continuous systems. In order to obtain this "weak" formulation, a variation $\delta u(k)$ of the discrete value $u^*(k)$ must be considered, whose influence on the final value $Q = \bar{x}^{(T)}$ would be infinitesimally small. Evidently, this is possible in general only in the case when the quantity $\delta u(k)$ itself is infinitesimally small. In this case (see [1.22]) the arguments given above can be repeated and a similar result apparently obtained. But this result will have only a "local" character; it is valid only for sufficiently small variations $\delta u(k)$ and $\delta\tilde{H}$ in the quantities $u^*(k)$ and $\tilde{H}$. In this case it can only be proved that if the control u is optimal, then it gives a local maximum for the quantity $\tilde{H}$. Let us consider Fig. 2.20, in which

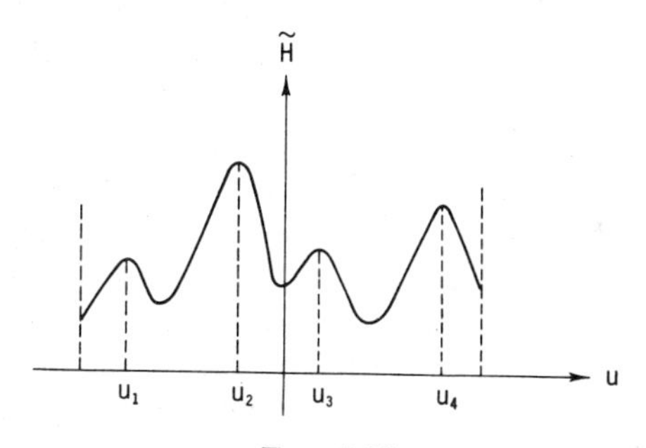

FIG. 2.20

the dependence of $\tilde{H}$ on u is depicted at some time instant t at some point of the trajectory. From the statement formulated above it follows that the values u_1, u_2, u_3, u_4 corresponding to the local maxima of $\tilde{H}$ are "suspect" in the sense that one of them is the optimal control. But just which one of them is the optimal control? This remains unknown. Therefore the value of the maximum principle in such a "weak" local formulation is diminished.

The methods presented above—dynamic programming and the maximum principle—in general cannot provide a solution to complex problems without the use of computers. Such is the situation dictating the need for solving problems sometimes in the form of a very long sequence of complicated calculations—minimizations of functions of several variables, integrations, and so on. For applying dynamic programming, functions of many variables have to be stored; for applying the maximum principle, it is necessary to solve a problem with boundary conditions, which is accompanied by the complex process of searching for missing initial conditions along the given boundaries.

The methods described in this chapter can in many cases give only a general prescription for the solution, while the step in obtaining a final solution is associated with calculations that are sometimes incredibly complicated. At this stage the techniques of approximate evaluations, including the ability to approximate rationally expressions close to them but simpler, frequently gain decisive value. Thus the theory of approximate evaluations proves to be a discipline of paramount importance for the synthesis of optimal systems. First, general principles are important here, based on functional analysis, which would be applicable to a wide class of problems arising in the theory of optimal systems. Some examples of approximate calculations will be given in subsequent chapters. But on the whole these extremely important problems have still not been developed quite enough.

This chapter has not presented all the mathematical methods that have been proposed and successfully applied to the solution of problems in optimal system theory. Thus for example, the interesting papers of N. N. Krasovskii [2.22, 2.23], have proposed a method based on the investigations of M. G. Krein [2.24] in functional analysis. The papers of the Polish theoretician R. Kulikowskii [2.25, 2.26], also related to functional analysis, deserve attention. The so-called "sliding conditions" have also not been considered in this book. In these conditions an infinitely large number of jumps in the control action $\bar{u}(t)$ is observed. In this case the questions of the existence of an optimal control have been worked out for certain classes of problems in the paper of L. S. Kirillova [2.27]. For these problems the maximum principle requires some generalization, which has been carried out by R. V. Gamkrelidze [2.28]. The theory developed by V. F. Krotov gives an entirely different approach to the same problems [2.29].

CHAPTER III

Optimal Systems with Complete Information about the Controlled Object

1. The Problem of Maximal Speed of Response; the Phase Space Method

The theory of systems with complete information in the controller about the controlled object was first developed as the theory of systems that are optimal in speed of response. Systems optimal in speed of response and those close to them have become the primary object of investigations in view of their practical importance. As far back as 1935 a patent [3.1] had been registered by D. I. Mar'yanovskii and D. V. Svecharnik on a system for moving the cylinders of a rolling mill, in which quadratic feedback was used to provide maximal speed of response. A similar principle had been applied somewhat later on in the automatic potentiometer "Speedomax" of the American firm Leeds and Northrup. Subsequently the publication of theoretical papers began. In 1949 in the paper [3.2], it was proved that the optimal process for a second-order linear object consisting of two successively joined integrating elements is made up of two intervals after the adjustment of the initial error. On the first of these the control action u, bounded by the condition

$$|u| \leqslant U, \tag{3.1}$$

is maintained at one of its limiting levels $\pm U$; on the second interval, at the other limiting level. In this paper optimal trajectories on the phase plane have been analyzed.

In 1951 in the paper [3.3], the optimal trajectories in the phase plane were investigated for a second-order linear object under various types of initial conditions. A comparison of the results of theory and simulation was also given.

In the paper [3.4] published in 1952, the statement of the problem was generalized to nth-order systems under one type of initial condition (the adjustment of the initial error), and the idea was expressed that in this case as well the absolute values of the restricted variables must be maintained at the maximum admissible level. Not all the processes proposed in this paper are strictly optimal; but they are close to optimal. But there the problem concerning certain constraints was posed and a hypothesis stated about the character of the optimal process in this case, which was justified for a number of problems.

Parallel to this there is also a series of papers that considers either various cases of optimal second-order systems or special principles whose application, while permitting the dynamic characteristics of the system to be improved, still does not make it optimal [3.25].

In 1953 the paper [3.5] introduced the general concept of the optimal process in n-dimensional phase space under arbitrary initial conditions and admissible external actions. In this paper the theorem concerning n intervals was formulated and proved. This theorem gave the possibility, in [3.11], of constructing a method of synthesizing a definite class of nth-order optimal systems. The account in this section has been based on the papers [3.2, 3.5, 3.11].

Theoretical results in other directions—for second-order systems with complex conjugate roots—were obtained by the American mathematician Bushaw in 1953 [3.15, 3.17, 3.23].

Starting in 1954, the stream of papers in the field of the theory of systems optimal in speed of response or close to it began to grow sharply [3.6–3.8, 3.12–3.14, 3.16, 3.18–3.24].

We will present in detail a statement of the problem of a system that is optimal in speed of response.

Let the continuously controlled object B in a feedback system be characterized in the general case by the vector equation of motion

$$d\bar{x}/dt = \bar{f}(\bar{x}, \bar{u}, t), \tag{3.2}$$

where $\bar{x}$ is an n-dimensional vector, and the vector $\bar{u}$ has r coordinates. The control action $\bar{u}$ is restricted by the condition

$$\bar{u} \in \Omega(\bar{u}), \tag{3.3}$$

where $\Omega(\bar{u})$ is a certain admissible closed region. The expression (3.1) serves as a specific example of condition (3.3). It is important to note that the end of the vector $\bar{u}$ can be situated not only in the interior of the region $\Omega(\bar{u})$, but also on its boundary. We will regard as the ideal process $\bar{x}(t)$ one for which the equalities

$$x_i(t) = x_i^*(t) \qquad (i = 1, 2, ..., n) \tag{3.4}$$

are ensured. Here the $x_i(t)$ are the coordinates of the object, and the $x_i^*(t)$ are given time functions which we will consider as coordinates of the vector $\bar{x}^*$:

$$\bar{x}^* = (x_1^*, x_2^*, \ldots, x_n^*). \tag{3.5}$$

The functions $x_i^*(t)$ must also satisfy definite constraints. In order to clarify their meaning, we will examine the n-dimensional phase space of the vector $\bar{x}$ (Fig. 3.1). If the state of the object B changed in exact correspondence with the ideal conditions (3.4), then the equality

$$\bar{x} = \bar{x}^* \tag{3.6}$$

would hold and the phase trajectory—the hodograph of the vector $\bar{x} = \bar{x}^*$—would be the trajectory $N_0^*N^*N_T$, for example. Let N^* be the position of the representative point on this trajectory at the current time instant $t > 0$, and N_0^* the position of this point at the initial time instant $t = 0$. However, the actual state of the object at the time instant $t = 0$ is characterized by the representative point N_0, not coinciding with N_0^*. Also let the current state of the object at an instant $t > 0$ be given by the point N, while the phase trajectory of the object has the form of the curve N_0NN_T. The automatic control system must join the actual state N with the required N^* in a minimal time. Let the points N and N^* be joined at the position N_T for $t = T$. We will assume that after this instant, i.e., for $t > T$, control actions $\bar{u}(t)$ can be chosen such that Eq. (3.6) will be ensured. But not every trajectory $\bar{x}^*(t)$ is realizable with the restricted resources of the control. We call the trajectories $\bar{x} = \bar{x}^*(t)$ that can be realized "admissible." Evidently, these trajectories are the solutions of Eq. (3.2) under condition (3.3). This condition is also the constraint imposed on the trajectories $\bar{x}^*(t)$. We restrict them to the possible motions of the system with the vector $\bar{u}$ lying in the interior† of the region $\Omega(\bar{u})$.

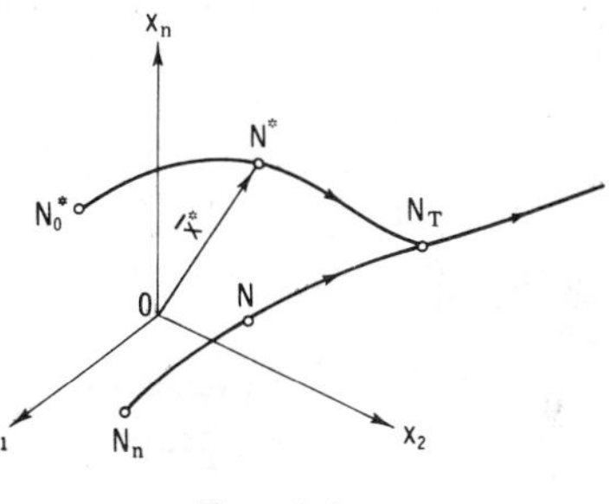

FIG. 3.1

The transient process is the process of passing from the initial state N_0 of the object to the required N_T. This process lasts an interval of

† In the interior but not on the boundary, since the point N must have the possibility of "overtaking" the point N^* for any initial conditions.

time equal to T. The system is called optimal with respect to the speed of transient processes, if the latter satisfy the condition

$$T = \min \tag{3.7}$$

for arbitrary initial conditions and any function $\bar{x}^*(t)$ belonging to the class of admissible functions, or to a precisely defined subclass of this class. In the latter case the system is called optimal in speed of response for the stated subclass of admissible functions $\bar{x}^*(t)$. The process $\bar{x}(t)$ in such a system is called the optimal process, and the corresponding control action $\bar{u}(t)$ is called the optimal control.

As is seen from Fig. 3.1, the problem of determining the optimal process $\bar{x}(t)$ and the optimal control $\bar{u}(t)$ can be interpreted as the problem of the quickest contacting of the two points N and N^* in phase space.

Instead of the phase space for $\bar{x}$ it is often convenient to use a phase space of the same order for the error

$$\bar{\epsilon} = \bar{x}^* - \bar{x} \tag{3.8}$$

with coordinates $\epsilon_i = x_i^* - x_i$ ($i = 1, 2, ..., n$). By the end of the transient process the error vector $\bar{\epsilon}$ vanishes. Hence in the new space the representative point of the system, starting from some initial position, must pass to the origin by the quickest admissible motion.

The new phase space is obtained from the previous one if the origin is carried over into the moving point N^*—the end of the vector $\bar{x}^*$ (Fig. 3.1).

In the new space we replace the notations ϵ_i ($i = 1, 2, ..., n$) by the letters x_i now signifying the error coordinates (Fig. 3.2). The problem of the quickest passage of the representative point of the system from the initial position N_0 to the origin O now does not differ at all from the problem of optimal speed of response considered in Chapter II.

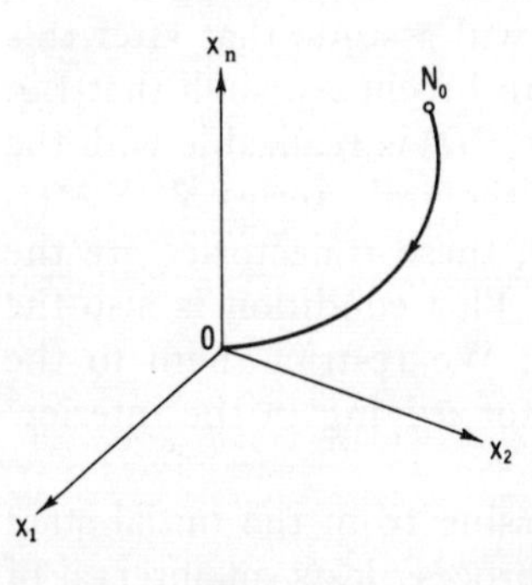

FIG. 3.2

Why is it that the representative points N and N^* in Fig. 3.1 cannot meet after an arbitrarily small time interval? Because of the limitations existing in any real system, for example, the constraints (3.3). These constraints do not permit an infinitely large rate of displacement of the representative point $\bar{x}$ in phase space to be developed.

In some cases the limitations are imposed not only on the control actions $u_j(t)$, but also on the coordinates $x_i(t)$ or on functions of these coordinates. Sometimes the latter type of constraint can be reduced to a restriction imposed on the control actions, though not in the actual configuration, in some other equivalent one. To illustrate this situation let us consider a simple example.

Let the equation of motion of the object have the form

$$b_0\, d^2x/dt^2 + b_1\, dx/dt = u(t), \tag{3.9}$$

where b_i = const. For example, the equation of motion of a constant-current servomotor has such a form, if the input variable is the voltage on the armature and the output variable is the rotation angle of the shaft. We assume that the constraint is imposed not on $u(t)$ but on the second derivative d^2x/dt^2:

$$| d^2x/dt^2 | \leqslant M. \tag{3.10}$$

For the example of the constant-current servomotor, the second derivative of x is proportional to the armature current if the load moment is negligibly small. In this case, instead of the system (3.9) another equivalent one, whose equation is determined not by equality (3.9) but only by the constraint condition (3.10), can be considered:

$$d^2x/dt^2 = v(t), \tag{3.11}$$

where v is the equivalent control action or control function satisfying the condition

$$| v(t) | \leqslant M. \tag{3.12}$$

Since we will find the optimal process $x_{\text{opt}}(t)$ for the equivalent system (3.11) with condition (3.12), the expression obtained can be substituted into (3.9) and hence the optimal (actual and not equivalent) control found:

$$u^*(t) = b_0 \frac{d^2x_{\text{opt}}}{dt^2} + b_1 \frac{dx_{\text{opt}}}{dt}. \tag{3.13}$$

It should be noted that in this case the optimal process is determined not by Eq. (3.9) of the object, but only by the constraint condition (3.10). But for determining the actual optimal control $u^*(t)$ the equation (3.9) of the object must be known.

A more general case of constraints imposed both on the coordinates of the object and on the control actions has been analyzed by R. V. Gamkrelidze in the paper [2.18] (see also [2.45, 2.46]).

We will now narrow down the statement of the problem. Let there be only a single control action $u(t)$ and let there be imposed on the object with output coordinate x a constraint of the type

$$\left| a_0 \frac{d^n x}{dt^n} + a_1 \frac{d^{n-1}x}{dt^{n-1}} + \cdots + a_n x \right| \leqslant M, \tag{3.14}$$

where $a_i = \text{const}$ and $a_0 > 0$. If in a particular case the object is characterized by the equation

$$a_0 \frac{d^n x}{dt^n} + a_1 \frac{d^{n-1}x}{dt^{n-1}} + \cdots + a_n x = u(t), \tag{3.15}$$

then the constraint (3.14) reduces to the condition

$$| u(t) | \leqslant M. \tag{3.16}$$

But as was shown above, in a more general case, the left side of the actual equation of the object cannot coincide with the expression under the absolute value sign in (3.14). In any of these cases only condition (3.14) is necessary for constructing the optimal process.

Further, we will restrict the analysis to only the case when the roots of the equation

$$a_0 p^n + a_1 p^{n-1} + \cdots + a_n = 0 \tag{3.17}$$

are real and nonpositive. In other words, the roots of (3.17) can be only negative and real, or equal to zero. We will call (3.17) the characteristic equation. Let us set

$$a_0 \frac{d^n x}{dt^n} + a_1 \frac{d^{n-1}x}{dt^{n-1}} + \cdots + a_n x = v. \tag{3.18}$$

Then the constraint (3.14) can be rewritten

$$| v | \leqslant M. \tag{3.19}$$

In a particular case, if the equation of the object is of type (3.15), then the expressions for $v(t)$ and $u(t)$ coincide. But as was shown earlier in a more general case, they may be nonidentical.

Under the above-mentioned conditions the theorem concerning n intervals, which is the following, holds:

The optimal process $x(t)$ consists of n intervals; in each of these intervals the process is described by the equation

$$v = a_0 \frac{d^n x}{dt^n} + a_1 \frac{d^{n-1}x}{dt^{n-1}} + \cdots + a_n x = \sigma M, \tag{3.20}$$

where the quantity σ is constant on each interval and

$$\sigma = \pm 1. \tag{3.21}$$

Here the signs of σ alternate on adjacent intervals.

If in a particular case $u = v$, then the formulation of the theorem concerning n intervals can be interpreted intuitively in the following way: the system receives an optimal control action $u(t)$ of the type "full speed forward," then "full speed backward," and so on, in all n times in succession. The proof of this theorem was obtained in 1953 in [3.5] by an elementary method and almost without the aid of calculation. In order to understand the idea of the proof better, we first investigate the simplest special case (already considered in 1949 in [3.2]).

In the special case let the equation of the object have the form

$$a_0 \frac{d^2x_1}{dt^2} = u; \tag{3.22}$$

the variable u is subject to the constraint (3.16). Further, let the initial conditions have the form

$$(x_1)_{t=0} = 0, \qquad (dx_1/dt)_{t=0} = 0. \tag{3.23}$$

Let it be required to transfer the object during the minimal possible time $T_{\min}$ to the state

$$x_1 = x_{1f} = \text{const}, \qquad dx_1/dt = 0. \tag{3.24}$$

It turns out that for this it is necessary to first realize a "momentum" with maximal acceleration for $u = M$. Here according to (3.22) the function dx_1/dt will vary in time according to a linear law:

$$\frac{dx_1}{dt} = \int_0^t \frac{d^2x_1}{dt^2}\,dt = \int_0^t \frac{M}{a_0}\,dt = \frac{Mt}{a_0}. \tag{3.25}$$

In Fig. 3.3a, curve 1 is depicted for the velocity dx_1/dt of the optimal process. At the middle of the path, for $t = T/2$, it is necessary to change the "momentum" by a maximal "braking," i.e., during the second interval to maintain the value $u = -M$. Thus the optimal curve for dx_1/dt has "triangular" form. Since the distance x_1 traversed is defined by the formula

$$x_1 = \int_0^t \frac{dx_1}{dt}\,dt, \tag{3.26}$$

then for $t < T_{\min}$ the optimal process has the form of two segments of a parabola (Fig. 3.3b), while for $t \geqslant T_{\min}$ it becomes a horizontal straight line $x_1 = x_{1f} = \text{const}$.

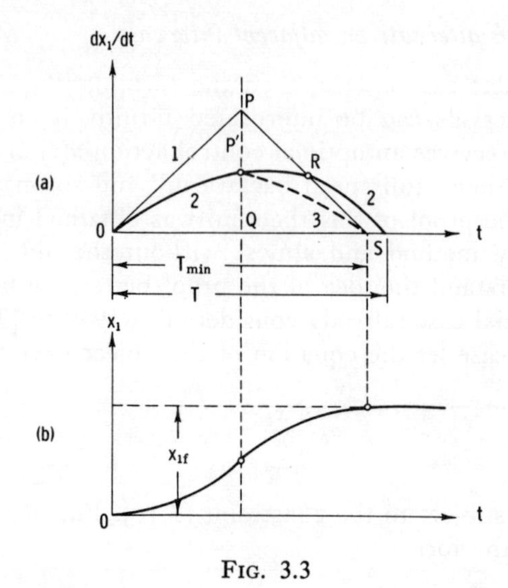

FIG. 3.3

We will prove that the process depicted in Fig. 3.3b is optimal. First let us note that the area bounded by the abscissa and any curve dx_1/dt of the transient process must be of constant magnitude, equal to x_{1f}, in this problem. In fact, if the time of the transient process is equal to T, and at its end the equality $x_1 = x_{1f}$ holds, then from (3.26) it follows that

$$\int_0^T \frac{dx_1}{dt}\, dt = x_{1f}\,. \tag{3.27}$$

In particular, this condition must be satisfied for the "triangular" curve 1 in Figure 3.3a.

Now we consider for dx_1/dt some curve 2 different from the "triangular" curve 1. By virtue of the constraint (3.16) and Eq. (3.22), the condition

$$| d^2x_1/dt^2 | = | u/a_0 | \leqslant M/a_0 = \text{const} \tag{3.28}$$

holds. Therefore on the first interval, for

$$0 \leqslant t \leqslant T_{\min}/2,$$

curve 2 can go either along curve 1 or below it, but it cannot be above curve 1. In fact, according to (3.28) the slope of curve 2 is less than or equal to the slope of curve 1, and the initial values of these curves are identical. Hence it follows that the ordinate of the point P' is less than the ordinate of the point P, and for any curve 2 the integral $\int_0^{T_{\min}/2}(dx_1/dt)\,dt$ is less than that for curve 1. But the integrals (3.27) for both curves must be identical, since x_{1f} is the same. Hence for $t > T_{\min}/2$ curve 2 cannot go as shown by the dashed line 3 in Fig. 3.3. It must necessarily intersect curve 1 at some point R. Further, curve 2 cannot intersect curve 1 for the second time, because the slope of curve 2 is, in absolute value, less than or equal to the slope of curve 1. Therefore curve 2 meets the abscissa at a $t = T$ which is larger than $T_{\min}$. Hence it follows that the time T of the transient process, for any admissible curve different from 1, is larger than $T_{\min}$. This means that curve 1 corresponds to the optimal process. The time $T_{\min}$ for this process is easily determined from condition (3.27). For curve 1 this condition means that the area of the triangle formed with the base OS and the height PQ is equal to x_{1f}

$$\tfrac{1}{2} PQ \cdot OS = \frac{1}{2}\left(\frac{M}{a_0}\,\frac{T_{\min}}{2}\right) \cdot T_{\min} = x_{1f}\,. \tag{3.29}$$

Hence it follows that

$$T_{\min} = 2\sqrt{\left(\frac{a_0 x_{1f}}{M}\right)}\,. \tag{3.30}$$

As is seen from this formula, the larger the admissible maximal value M of the control action, the smaller the time $T_{\min}$ of the optimal transient process. But for a finite value of M the quantity $T_{\min}$ is also finite.

In the general case of arbitrary initial conditions for x, any admissible driving action x^* and any order n of the equation (3.18) of the constraints [if $v = u$, then it reduces to Eq. (3.15) for the object], according to the theorem concerning n intervals the control function $v(t)$ has the form of the curve depicted in Fig. 3.4. For the example the case is shown when on the first and last intervals the quantity σ is equal to $+1$, i.e., $v = +M$. The entire process of change in $v(t)$ consists of n intervals which in general are different in length, where in each interval $\sigma =$ const and in adjacent intervals the signs of σ are different. This means that "full speed forward" when

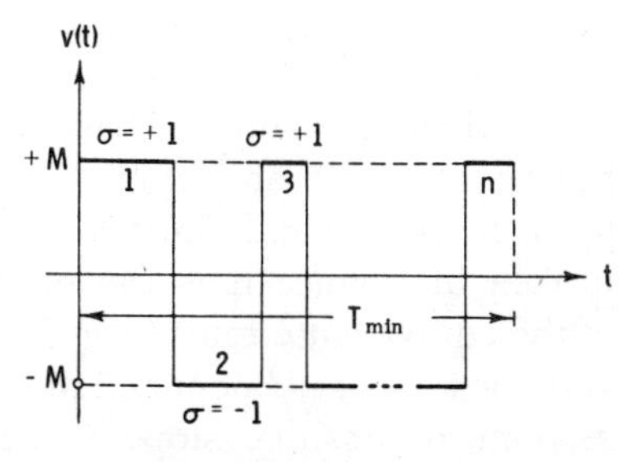

FIG. 3.4

$v = +M$ changes to "full speed backward" when $v = -M$, and so on. The signs of σ and the lengths of the intervals must be selected such that from a given initial state a certain prescribed final state occurs, for example, the origin (Fig. 3.2). The question about the manner in which all these parameters can be selected will be considered later on.

We will outline the proof of the theorem concerning n intervals by the example of a particular case when the constraint (3.14) has the form

$$| d^n x/dt^n | \leqslant M, \tag{3.31}$$

and Eq. (3.18) turns into the equality

$$d^n x/dt^n = v. \tag{3.32}$$

The relation of x to v has the same form as would the quantity v being fed to the input of a chain of n integrating elements (Fig. 3.5), while the quantity x would be the output variable of this system. It coincides with the actual block diagram of the object only in the case when the equation of the object coincides with (3.32), i.e., only when it has the form

$$d^n x/dt^n = u. \tag{3.33}$$

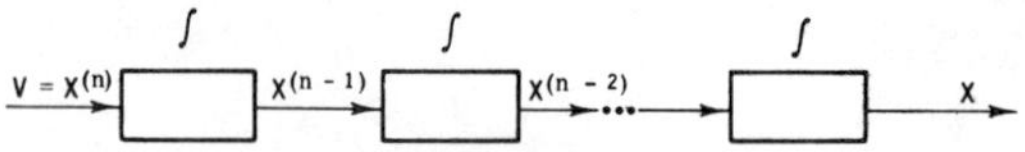

FIG. 3.5

In this case it is evident that $v = u$. But in the general case $v \neq u$. We will prove that for obtaining the optimal process $x(t)$ it is necessary to feed to the input of an equivalent circuit the quantity $v = \sigma M$, where $\sigma = \pm 1$, and the facts that the signs of σ alternate on adjacent intervals and the total number of intervals is equal to n. We will assume that such a process exists and that the time of the transient process is equal to T_0. At the instant $t = T_0$ the equalities

$$x = x^*, \quad x^{(1)} = x^{*(1)}, \ldots, x^{(n-1)} = x^{*(n-1)} \tag{3.34}$$

hold, where $x^{(k)}$ and $x^{*(k)}$ are kth derivatives. We will first consider the functions $x^{(n-1)}$ and $x^{*(n-1)}$ (Fig. 3.6). According to the preceding these curves coincide for $t \geqslant T_0$. The curve of $x^{(n-1)}(t)$ has the form of a broken line, since it is the output variable of the integrating element of the equivalent circuit (Fig. 3.5), at whose input the stepwise-changing variable $v = \pm M$ depicted in Fig. 3.4 acts. The broken line of $x^{(n-1)}$ has sections of positive slope $+M$ alternating with sections of negative slope $-M$ which are the same in absolute value.

We will suppose that there exists another curve $x_1(t)$ with the same initial conditions that $x(t)$ has, but with a smaller time T_0' of the transient process than $x(t)$ has. Thus $T_0' < T_0$. It is proved below that the existence of an admissible curve $x_1(t)$, i.e., satisfying the constraint (3.31), with the properties stated above is impossible.

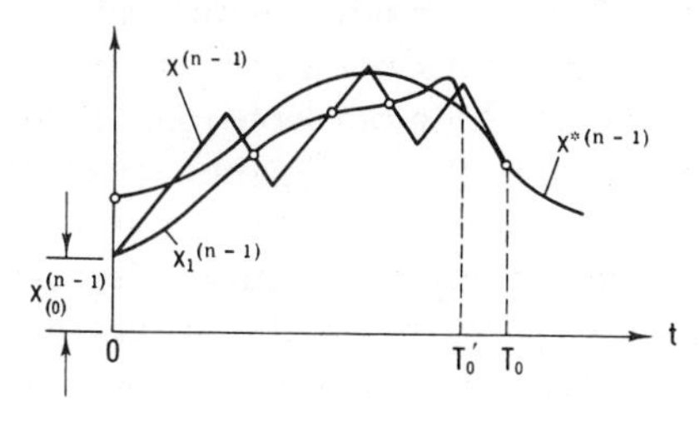

FIG. 3.6

We will add to the curve of $x_1(t)$ a segment of the curve of $x^*(t)$ on the interval $T_0' \leqslant t \leqslant T_0$. In what follows we will subtend under the curve of $x_1(t)$ just such an "extended" curve, prescribed on the interval $0 \leqslant t \leqslant T_0$. Hence, in particular it follows that

$$x_1^{(k)}(T_0) = x^{(k)}(T_0) \qquad (k = 0, 1, ..., n - 1). \tag{3.35}$$

The curve of $x_1^{(n-1)}$ cannot intersect more than once each of the segments of the broken line $x^{(n-1)}$. In fact, otherwise its derivative, i.e., the function $x_1^{(n)}$ will exceed the function $x^{(n)} = \pm M$ in absolute value; but then the curve of $x_1(t)$ will turn out to be inadmissible. Further, the curve of $x_1^{(n-1)}$ cannot intersect either the first or the last (on the interval $0 \leqslant t \leqslant T_0$) of the segments of the broken line $x^{(n-1)}$, since at the end points of these segments both curves coincide; otherwise the curve of $x_1(t)$ will again turn out to be inadmissible. Hence, "the number of intersection points of the curves of $x^{(n-1)}$ and $x_1^{(n-1)}$ cannot be greater than $n - 2$."

Let us now turn to the curves of $x^{(n-2)}(t)$ and $x_1^{(n-2)}(t)$. They coincide at the points $t = 0$ and $t = T_0$, and as was shown above their derivatives coincide at not more than $n - 2$ intermediate points. Hence it follows that the curves of $x^{(n-2)}$ and $x_1^{(n-2)}$ do not have more than $n - 3$ intersection points inside the interval $0 \leqslant t \leqslant T_0$ (here the coincidence points of the curves on the boundaries of this interval do not figure in the calculation). In fact, if two continuous and differentiable curves intersect at some two points for the two values t_1 and $t_2 (t_1 < t_2)$, then their derivatives must be equal to one another at some intermediate point for $t_1 < t < t_2$. Therefore if the curves of $x^{(n-2)}$ and $x_1^{(n-2)}$ had more than $n - 3$ intersection points, then by taking into account their coincidence at the boundaries of the interval, we would obtain in all more than $n - 1$ coincidence points for these curves. From this it would follow that their derivatives $x^{(n-1)}$ and $x_1^{(n-1)}$ coincide at more than $n - 2$ points inside the interval $0 \leqslant t \leqslant T_0$, and as was shown above this is impossible. Thus the curves of $x^{(n-2)}$ and $x_1^{(n-2)}$ intersect at not more than $n - 3$ points inside the interval $0 \leqslant t \leqslant T_0$.

By reasoning in an analogous manner it is not difficult to show that the curves of $x^{(n-3)}$ and $x_1^{(n-3)}$ do not have more than $n - 4$ intersection points inside the interval $0 \leqslant t \leqslant T_0$; the curves of $x^{(n-4)}$ and $x_1^{(n-4)}$, not more than $n - 5$ intersection points, and so on. Progressing in this manner from left to right along the chain of elements of the equivalent circuit of Fig. 3.5, we finally reach the curves of $d^2x/dt^2 = x^{(2)}$ and $d^2x_1/dt^2 = x_1^{(2)}$. Evidently, these curves do not have more than one intersection point. Further, from here it follows that the curves of $x^{(1)}$ and $x_1^{(1)}$ do not have any intersection points at all inside the interval $0 \leqslant t \leqslant T_0$. But this condition means that

$$x(T_0) = x_0 + \int_0^{T_0} x^{(1)}\, dt \neq x_0 + \int_0^{T_0} x_1^{(1)}\, dt = x_1(T_0), \tag{3.36}$$

i.e., the first equality of (3.35) is violated, which must hold if the time T_0' for the curve of $x_1(t)$ is to be smaller than the time T_0 for the curve of $x(t)$. The contradiction proves the impossibility of realizing an admissible curve of $x_1(t)$ with a time $T_0' < T_0$. Hence the curve of $x(t)$ is the optimal process, and the theorem concerning n intervals with the constraint (3.31) has been proved.

These arguments are easily generalized to the case of the constraint (3.14), when the elements of the equivalent circuit are either integrating or inertial ones [3.5, 3.25].†

By itself the theorem concerning n intervals still does not give a rule for selecting the sign of σ on the first interval or for choosing the lengths of the intervals. But after using this theorem we can realize the synthesis of the block diagram, or else the synthesis of the algorithm of the optimal controller A, which will automatically accomplish the required selection [3.11]. If the value of the required control function v at a given instant becomes known, then it is not difficult to find the corresponding value of the actual control action u. If a constraint of the type (3.16) is imposed on the object, then u and v coincide. But if the equation of the object and the constraint equation (3.18) do not coincide, then $v \neq u$; however, the relation between u and v is found by a simple

† Indeed, this case occurs with the constraint imposed on the roots of Eq. (3.17) (see p. 122).

method. For example, for Eq. (3.19) and the constraint (3.10) for $(dx/dt)_{t=0} = 0$, the relation

$$u^*(t) = b_0 v(t) + b_1 \int_0^t v\, dt \tag{3.37}$$

holds. Thus the basic problem consists in determining the control function v for any point of the phase space of $\bar{x}$ at the time instant t:

$$v = v(\bar{x}, t). \tag{3.38}$$

In order to find this function let us consider the phase space for the errors. By virtue of the theorem concerning n intervals, the value of v at any point of phase space can only be $+M$ or $-M$; i.e., if $v = \sigma M$, then $\sigma = +1$ or $\sigma = -1$. Thus at each time instant every ordinary point of phase space corresponds either to the value $\sigma = +1$ or the value $\sigma = -1$. Therefore at each time instant the entire phase space is divided into two regions, characterized by the values $\sigma = +1$ and $\sigma = -1$. In Fig. 3.7 a three-dimensional phase space with regions $\sigma = +1$ and $\sigma = -1$ is shown as an example. These regions are separated from one another by a boundary which is an $(n - 1)$-dimensional hypersurface S. The synthesis problem reduces just to determining this hypersurface at an arbitrary time instant. S is called the "switching hypersurface."

In the general case the hypersurface S has a different form at different time instants. Such hypersurfaces are called "nonstationary." This case occurs if the equations of the object depend explicitly on time. In a more particular class of cases the hypersurface S is fixed, but its form depends on the parameters of the driving function $x^*(t)$. We will call such a hypersurface "quasi-stationary."

In a still more particular class of cases the form of the hypersurface S does not depend on the parameters of $x^*(t)$ at all; then the hypersurface S is called "stationary." In the latter case, which is considered in detail below, the error equation does not depend on the parameters of $x^*(t)$.

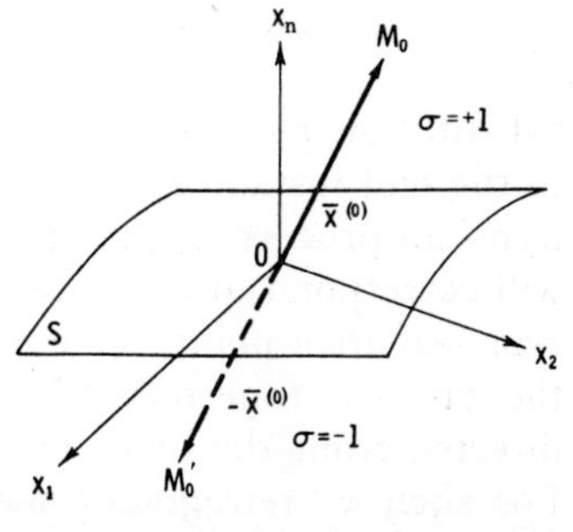

FIG. 3.7

Let us examine the two points M_0 and M_0' (Fig. 3.7) for the case of a stationary hypersurface S in the phase space of the errors. These points are situated symmetrically with respect to the origin and correspond to the values $\bar{x}^{(0)}$ and $-\bar{x}^{(0)}$ of the vector $\bar{x}$. We will assume that Eq. (3.18) is also the equation for the errors. For the point M_0 let the value of σ be equal to $+1$. Then for the point M_0' the value of σ is equal to -1. Actually, if the optimal process $\bar{x}(t)$ satisfying Eq. (3.18) starts from the point $\bar{x}^{(0)}$ and takes place under a determinate action $v(t)$, then the process $-\bar{x}(t)$ under the action $-v(t)$, starting from the point $-\bar{x}^{(0)}$, satisfies Eq. (3.18) and therefore is also optimal.

Since any points that are symmetrical with respect to the origin, and not lying on the hypersurface S, belong to different regions, then the hypersurface S passes through the origin. Further, from its very definition it follows that it does not contain "holes" through which a passage from one region of space to the other without intersecting the hypersurface S would be possible. Finally, it extends to infinitely far points of phase space.

The optimal trajectory in phase space is made up from n segments corresponding to the intervals of $\sigma = \text{const}$ in the general case. For example, in Fig. 3.8 the optimal phase trajectory M_0PQO is depicted for $n = 3$, going from the point M_0 (the vector $\bar{x}^{(0)}$) and consisting of three sections. For the first part M_0P the value of σ is equal to $+1$; for the second section PQ the value of σ is equal to -1; finally for the third segment QO the value of $\sigma = +1$. The trajectory $M_0'P'Q'O$ antisymmetrical to it, not shown in the figure, leads from the point M_0' (the vector $-\bar{x}^{(0)}$), and moreover on this trajectory the signs of σ alternate in the following order: -1, $+1$, -1. In particular, the last section $Q'O$ of this trajectory, shown by the dashed line in Fig. 3.8, corresponds to the value of $\sigma = -1$ and brings the representative point to the origin.

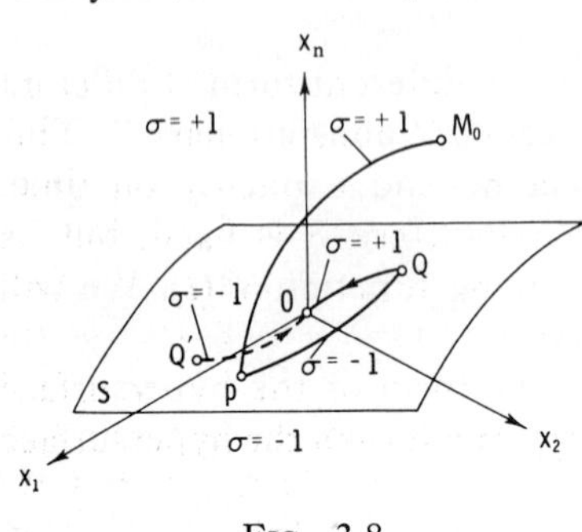

Fig. 3.8

Let the transient process start at $t = 0$. The trend of time can be "turned" the opposite way by the substitution $\tau = T - t$, where T is the value of time t corresponding to the end instant of the transient process. Now the end instant of the transient process, i.e., the representative point finding itself at the origin, will correspond to the value $\tau = 0$. With an increase in τ from zero the representative point M begins to perform a "retrograde" motion along the optimal trajectory—for example, along the trajectory $OQPM_0$—directed from the point O to Q, then from Q to P and from P to M_0. For such a "retrograde" motion the point M_0 will be reached at the instant $\tau = T$, corresponding to the value $t = 0$.

It is important to note that at the point O it is possible to find ourselves along only one of the two possible optimal trajectories: QO or $Q'O$. In fact, let us replace time t by $\tau = T - t$ and trace any possible "retrograde" motion. On the first section reckoned from the point O only two different values of the control function v are possible: $+M$ or $-M$, which correspond to the values $\sigma = +1$ and $\sigma = -1$. For $\sigma = +1$ the motion takes place along the trajectory OQ, while for $\sigma = -1$ it takes place along OQ'. We will denote these trajectories by L_1' and L_1'' respectively (Fig. 3.9).

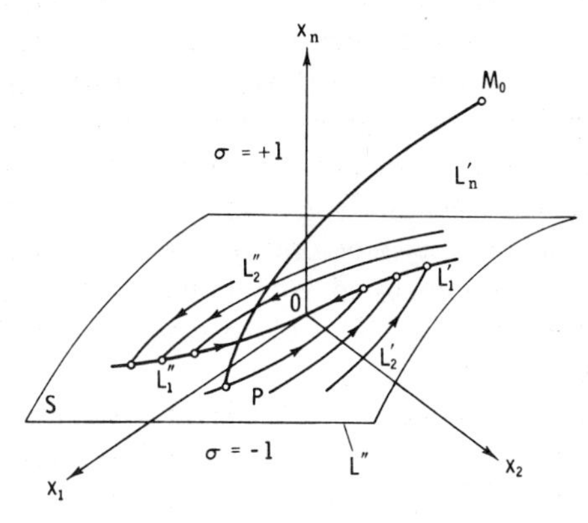

FIG. 3.9

It is possible to find ourselves on each of the trajectories L_1' and L_1'' by various paths. We can get to some "definite" point of the trajectory L_1' by moving along only one trajectory of the type PQ corresponding to the value $\sigma = -1$. We will denote by L_2' the set of points of the trajectories corresponding to $\sigma = -1$ and leading to the points of the trajectory L_1'. In turn one trajectory corresponding to the value $\sigma = +1$ (for example, PM_0 in Fig. 3.9) will lead to each definite point of the trajectory belonging to the set L_2'. We will denote by L_3' the set of points of the trajectories corresponding to $\sigma = +1$ and leading to L_2', and so on. In a like manner we construct the sets L_2', L_3', ..., L_{n-1}', and finally L_n'. The last set includes all ordinary points of phase space belonging to the region $\sigma = +1$ and not situated on the hypersurface S.

In a quite analogous fashion we will denote by L_2'' the set of points lying on trajectories leading to L_1''. Further, we will denote by L_3'' the set of points of the trajectories leading to L_2'', and so on. It is evident that L_n'' contains all ordinary points of phase space belonging to the region $\sigma = -1$ and not situated on the hypersurface S.

From Fig. 3.9 it is seen that n-dimensional "streams" of optimal phase trajectories become $(n - 1)$-dimensional after the first switching of the sign of σ, then $(n - 2)$-dimensional after the following switching, and so on; finally, in the nth interval a one-dimensional "stream" appears, flowing into the origin. There exist in all two such one-dimensional "streams": L_1' and L_1''. All this is easily shown by tracing the "retrograde" motion of the representative point from the origin.

At the point P of the optimal trajectory the first switching of the sign of σ takes place (if ordinary time t is considered). Since a switching of the sign of σ can take place only with an intersecting of the switching hypersurface S, then consequently the point P belongs to S. Since the collection of points P is the collection of sets L_{n-1}' and L_{n-1}'', which we will denote by L_{n-1}, the set L_{n-1} belongs to S. But it is possible to find ourselves at each point R belonging to S by motion along the trajectory of $\sigma = +1$ or $\sigma = -1$ from any ordinary point without switching, since switching takes place only on the hypersurface S. Hence all points R of the hypersurface S belong to the set L_{n-1}.

Thus, the $(n - 1)$-dimensional hypersurface S coincides with the $(n - 1)$-dimensional set L_{n-1}. Therefore finding S reduces to finding the set of points L_{n-1}.

Since the family of trajectories L_{n-1}'' is antisymmetrical to L_{n-1}' for a stationary hypersurface S and is obtained by replacing the signs of all the coordinates of the points of L_{n-1}' by their opposites, it suffices to consider only the construction of the family L_{n-1}'. It is convenient to carry it out by using the "retrograde" motion from the origin. Under this motion, with t replaced by $\tau = T - t$, the representative point will first move along the trajectory L_1' (Fig. 3.9), corresponding to the value $\sigma = +1$. We will assume that at τ_1 a switching of the sign of σ and a transition to the trajectory L_2' takes place, at $\tau_2 > \tau_1$ a new switching of the sign of σ and a transition to the trajectory L_3' takes place, and so on. Finally, for $\tau_{n-1} > \tau_{n-2} > \cdots > \tau_1$ the "last" (but in real time the first) switching of the sign of σ takes place. In the equation

$$a_0 \frac{d^n x}{dt^n} + a_1 \frac{d^{n-1}x}{dt^{n-1}} + \cdots + a_{n-1} \frac{dx}{dt} + a_n x = v, \tag{3.39}$$

going over to the new argument $\tau = T - t$, we obtain the new equation

$$a_0(-1)^n \frac{d^n x}{d\tau^n} + a_1(-1)^{n-1} \frac{d^{n-1}x}{d\tau^{n-1}} + \cdots + (-1)a_{n-1} \frac{dx}{d\tau} + a_n x = v. \tag{3.40}$$

We solve this equation for the "initial" conditions:

$$(x)_{\tau=0} = \left(\frac{dx}{d\tau}\right)_{\tau=0} = \cdots = \left(\frac{d^{n-1}x}{d\tau^{n-1}}\right)_{\tau=0} = 0 \tag{3.41}$$

and for $v = +M$, i.e., for $\sigma = +1$. Then we will obtain the equation of the trajectory L_1' in parametric form: $\bar{x} = \bar{x}(\tau)$. For $\tau = \tau_1$ the point of phase space corresponding to this trajectory has the coordinates $\bar{x}(\tau_1)$. If at this time instant $v = +M$ is replaced by $v = -M$, then the solution of Eq. (3.40) at the instant $\tau_2 > \tau_1$ will be a function both of τ_1 and τ_2: $\bar{x} = \bar{x}(\tau_1, \tau_2)$.

Performing the switching of the sign of σ at the instants $\tau_1, \tau_2, \ldots, \tau_{n-1}$, we will obtain the solution of Eq. (3.40) in the form

$$\bar{x} = \bar{x}(\tau_1, \tau_2, \ldots, \tau_{n-1}). \tag{3.42}$$

The equations for the coordinates $x_1, x_2, \ldots, x_n$ of the vector $\bar{x}$ have the form

$$x_i = x_i(\tau_1, \tau_2, \ldots, \tau_{n-1}) \qquad (i = 1, 2, \ldots, n). \tag{3.43}$$

In their aggregate these equations represent the equation of the hypersurface S in parametric form, and thus give the solution of the synthesis problem. By eliminating the parameters $\tau_1, \tau_2, \ldots, \tau_{n-1}$ from Eqs. (3.43), the equation of S can be obtained in explicit form in a number of cases, i.e., in the form of an equation relating the coordinates $x_1, x_2, \ldots, x_n$:

$$\psi(x_1, x_2, \ldots, x_n) = 0. \tag{3.44}$$

Let the function ψ be positive on one side of the hypersurface S and negative on the other side. For example, let $\psi > 0$ for points of the region $\sigma = +1$ and $\psi < 0$ for points of the region $\sigma = -1$. Then we can set

$$\sigma = \operatorname{sign} \psi \tag{3.45}$$

and

$$v = \sigma M = M \operatorname{sign} \psi = M \operatorname{sign} \psi(x_1, x_2, \ldots, x_n). \tag{3.46}$$

This is the required algorithm for operation of the optimal system, i.e., the equation $v = v(\bar{x})$. For each point $\bar{x}$ of phase space, Eq. (3.46) gives the value of v corresponding to the optimal control law. Analogous conclusions can be obtained for the case of a quasi-stationary surface S [3.11, 3.25], with the only difference in this case being that L'_{n-1} and L''_{n-1} are not antisymmetrical to one another.

By knowing the dependence (3.46) we can construct the block diagram of the optimal controller (Fig. 3.10). From the output of the object B the output variable $\bar{X}$ acts on the input to the block Σ belonging to the controller A, along the feedback path OO'. The quantity $\bar{X}$ can be a vector with several coordinates. Then the feedback circuit consists of several paths, along which the coordinates X_i of the vector $\bar{X}$ are transmitted, for example, X and dX/dt.

The vector $\bar{X}^*$ of the driving action is also fed to the block Σ. In the block Σ the coordinates $x_1, x_2, ..., x_n$ of the error vector are developed: $x_i = X_i^* - X_i$. If it is required, differentiators for determining the missing coordinates by the differentiation of certain input coordinates are included in the structure of the block Σ for this. The errors $x_1, x_2, ..., x_n$ are fed from the output of the block Σ to the input of the nonlinear transformer NT, generating the nonlinear function $\psi(x_1, x_2, ..., x_n)$ of them. The latter is fed to the input of the relay element RE; the output variable of this element, i.e., $M \operatorname{sign} \psi$, is also the quantity v. This variable acts on the input of the block A' transforming $v(t)$ into the control action $u(t)$. If $u(t) = v(t)$, then the block A' is not required.

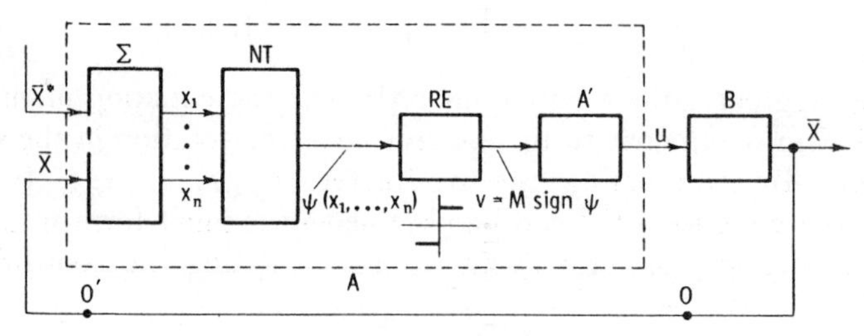

FIG. 3.10

It should be explained in what manner the process consisting of n intervals is obtained in the presence of only one $(n - 1)$-dimensional switching hypersurface S in n-dimensional phase space. In order to understand this, we should take into account that in any real system there exist fluctuations, which "knock" the representative point somewhat off the optimal phase trajectory. Therefore the real motion of the system depends not only on the character of the hypersurface S itself, but also on the structure of phase space in the neighborhood of this hypersurface. This structure is shown schematically in Fig. 3.11. For the example here the part L'_{n-1} of the hypersurface S corresponds to the value $\sigma = +1$, while L''_{n-1} corresponds to the value $\sigma = -1$.

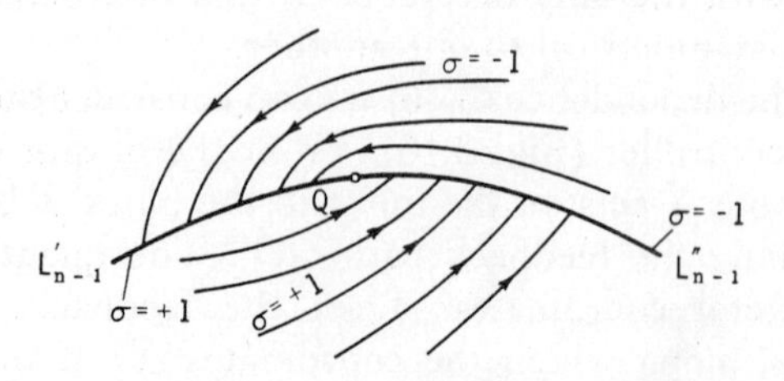

FIG. 3.11

On one side of L'_{n-1} (in Fig. 3.11, from below) the other trajectories with $\sigma = +1$, belonging to the region $\sigma = +1$ of phase space, go almost parallel to it. On the other side of L'_{n-1} (in Fig. 3.11, from above) go the trajectories of $\sigma = -1$, which will carry just to the set L'_{n-1}. As is seen from Fig. 3.11, an analogous situation exists in the neighborhood of the part L''_{n-1} of the hypersurface S, but with the difference that this time the trajectories of $\sigma = +1$ will now carry to L''_{n-1}.

If the representative point going along the section L'_{n-1} will be dislodged to a neighboring place of the region $\sigma = -1$ of phase space, then a change in sign of σ from $+1$ to -1 will occur. But after this the representative point will immediately (if the fluctuations are sufficiently small) return to the hypersurface. Therefore in this case the motion of the point will not change practically.

Another situation arises (inevitably, since in the system there exist very distinct fluctuations) if the representative point is dislodged to the region $\sigma = +1$ close to the section L'_{n-1} of the hypersurface S. In this case the representative point will not now return immediately to the hypersurface. The trajectory of the subsequent motion will pass to the region $\sigma = +1$ in close proximity to L'_{n-1}. Hence the motion will be "almost" the same as if the representative point would move exactly along the hypersurface. After the final time interval the representative point is found on the section L''_{n-1} of the hypersurface S in close proximity to the point Q, at which the "ideal" trajectory section L_{n-1} ends. If now a new fluctuation strikes the representative point, then it will go along the trajectory of $\sigma = -1$ which is close to one of the trajectories of L'_{n-2}, since a trajectory belonging to this set passes through Q. As a result of n such motions the representative point, moving in the regions $\sigma = -1$, $\sigma = +1$ in turn, is found in a small neighborhood of the origin, at which the transient process ends.

Hence the real trajectory passes "close" to the hypersurface S, where each switching is accompanied by a "piercing" of the hypersurface S by the representative point. In Fig. 3.12 a real trajectory in three-dimensional space ($n = 3$) is shown for illustration. The ideal trajectory has the form of the curve M_0PQO, where its sections PQ and QO lie on the switching surface S. If close to the position P a fluctuation knocks the representative point from the surface S, then it is found in the position P_1, and hence for $\sigma = -1$ it

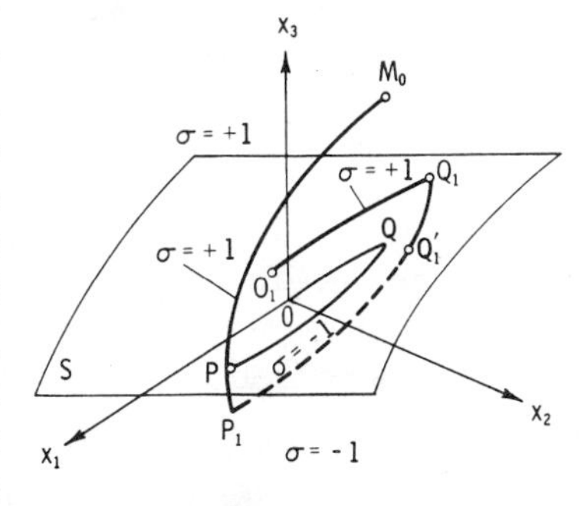

Fig. 3.12

will go along the trajectory P_1Q_1' close to the ideal trajectory PQ. At the point Q_1' the representative point reaches the surface S. If now the fluctuation knocks the point to the position Q_1, then for $\sigma = +1$ the subsequent motion will go along the trajectory Q_1O_1 close to the "ideal" trajectory QO. At the position O_1 the representative point reaches a small neighborhood of the origin, and the transient process can be regarded as completed.

Let us turn once again to the method of constructing the hypersurface S, and call to mind that it is the geometric locus of points of the "first" switching of the sign of σ. But all the remaining $n - 1$ switchings of the sign of σ are effected on this same hypersurface. In fact, from what was presented above it is not at all necessary to create some special surfaces or curves inside L_{n-1} for subsequent switchings of the sign of σ. It suffices to have only one "partition" in n-dimensional phase space corresponding to the first switching of the sign of σ. Because of the presence of fluctuations the real process is arbitrarily close but not identical to the ideal optimal process. In general the latter is impossible in a real system, since the probability of the representative point being exactly on the hypersurface S and moving along it is equal to zero.

We will show by the simplest example of a second-order object how the function ψ and surface S are determined. In this case the phase space of errors becomes a phase plane (Fig. 3.13), and the surface turns

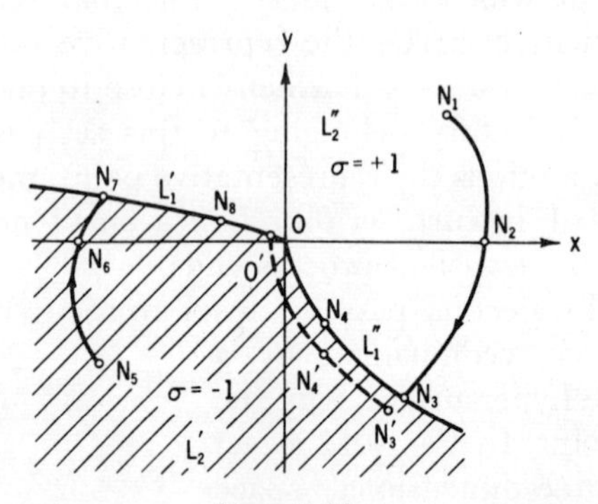

FIG. 3.13

out to be one-dimensional, i.e., it becomes a switching curve dividing the phase plane into two regions: $\sigma = +1$ and $\sigma = -1$. Let the object B be characterized by the equation

$$a_0 \frac{d^2x_1}{dt^2} = u, \tag{3.47}$$

and the constraint condition has the form

$$|u| \leqslant M. \tag{3.48}$$

We select a class of driving actions of the form

$$x_1^*(t) = A_0 + A_1 t + A_2 t^2, \tag{3.49}$$

where A_0, A_1, and A_2 are constants in each individual process. The admissible functions $x_1^*(t)$ are the solutions of Eq. (3.47) under the condition

$$|u| < M. \tag{3.50}$$

Here the symbol $\leqslant$ has been replaced by $<$ so that the point $\bar{x}$ could "reach" the point $\bar{x}^*$. Since

$$a_0 |d^2 x_1^*/dt^2| < M, \tag{3.51}$$

then by substituting the second derivative of expression (3.49) into condition (3.51) we find

$$|A_2| < M/2a_0. \tag{3.52}$$

Admissible functions of the type (3.49) are restricted by this condition.

The equation for the optimal process acquires the form

$$d^2 x_1/dt^2 = \sigma M/a_0, \tag{3.53}$$

where $\sigma = \pm 1$.

Let us denote the error by the letter x, i.e., the difference

$$x = x_1^* - x_1, \tag{3.54}$$

and its derivative by the letter y:

$$y = \frac{dx}{dt} = \frac{dx_1^*}{dt} - \frac{dx_1}{dt}. \tag{3.55}$$

From Eqs. (3.53), (3.54), and (3.49) it follows that

$$\frac{d^2x}{dt^2} = \frac{d^2x_1^*}{dt^2} - \frac{d^2x_1}{dt^2} = 2A_2 - \frac{\sigma M}{a_0} = \mu_0. \tag{3.56}$$

On each of the intervals the quantity μ_0 is constant, since $\sigma =$ const. We will now pass to "reversed" time $\tau = T - t$. Then with Eq. (3.56) taken into account we will obtain:

$$\frac{d^2x}{dt^2} = \frac{d}{dt}\left(\frac{dx}{dt}\right) = -\frac{d}{d\tau}\left(-\frac{dx}{d\tau}\right) = \frac{d^2x}{d\tau^2} = \mu_0. \tag{3.57}$$

Further,

$$y = \frac{dx}{dt} = -\frac{dx}{d\tau} = -\int \left(\frac{d^2x}{d\tau^2}\right) d\tau = -\mu_1 - \mu_0\tau \tag{3.58}$$

and

$$x = -\int y \, d\tau = \mu_2 + \mu_1\tau + \mu_0(\tau^2/2), \tag{3.59}$$

where μ_1 and μ_2 are constants. Let us find the curves L_1' and L_1'', in this case constituting the switching curve. Since for $\tau = 0$ the quantities y and x corresponding to these curves vanish, then from (3.58) and (3.59) we find $\mu_1 = \mu_2 = 0$. Hence,

$$x = \mu_0(\tau^2/2), \qquad y = -\mu_0\tau. \tag{3.60}$$

We will first find the curve L_1'' for which $\sigma = -1$. From formula (3.56) and condition (3.52) it follows that the sign of μ_0 is determined by the sign of σ:

$$\operatorname{sign} \mu_0 = -\operatorname{sign} \sigma. \tag{3.61}$$

But from the second equation of (3.60) it is seen that since $\tau > 0$, the condition

$$\operatorname{sign} y = -\operatorname{sign} \mu_0 \tag{3.62}$$

holds. Hence

$$\operatorname{sign} y = \operatorname{sign} \sigma = \sigma. \tag{3.63}$$

Thus for $\sigma = -1$ the quantity $\mu_0 > 0$, and $y < 0$. The motion along the curve L_1'' therefore takes place in the lower half-plane (Fig. 3.13). Eliminating the argument τ from the two equations of (3.60), we find the equation for the curve L_1'' in the form

$$x = y^2/2\mu_0 > 0, \qquad \sigma = \operatorname{sign} \sigma = \operatorname{sign} y = -1. \tag{3.64}$$

For the curve L_1' the quantity $\sigma = +1$. Hence as is seen from (3.56), $\operatorname{sign} \mu_0 = -1$. Therefore from (3.62) it follows that $y > 0$. In this case from Eq. (3.60) the expression

$$x = y^2/2\mu_0 < 0, \qquad \sigma = \operatorname{sign} \sigma = \operatorname{sign} y = +1 \tag{3.65}$$

is obtained. The equations for both curves L_1' and L_1'' can be combined into the form of a single equation for the curve $L_1 = S$:

$$x = \frac{y^2}{2\mu_0} = \frac{y^2}{2[2A_2 - (M_0/a_0) \operatorname{sign} y]}, \tag{3.66}$$

where instead of $\sigma = \text{sign } \sigma$ the quantity sign y equal to σ has been put. Expression (3.66) is the equation of the switching curve. This equation can also be written in a form analogous to (3.44), i.e., in the form $\psi = 0$, where

$$\psi(x, y) = x + \frac{y^2}{2[(M/a_0) \text{ sign } y - 2A_2]} = x + \frac{y^2 \text{ sign } y}{2[(M/a_0) - 2A_2 \text{ sign } y]}. \tag{3.67}$$

Figure 3.13 depicts the switching curve and optimal trajectories in the phase plane. The switching curve has the form of the curved line $N_3N_4ON_8N_7$. The representative point, starting from N_1, moves along the parabolic trajectory $N_1N_2N_3$, which is not difficult to obtain by solving Eq. (3.56) for $\sigma = +1$ and the initial conditions corresponding to the point N_1. At the position N_3 the representative point reaches the switching curve. After switching the representative point moves, in the ideal case, along the switching curve N_3N_4O itself to the origin. Actually because of the presence of fluctuations the representative point moves along the dashed curve $N_3'N_4'O'$ in the region $\sigma = -1$, i.e., along a trajectory close to the curve N_3N_4O, and at the point O' reaches a small neighborhood of the origin.

If at the initial time instant the representative point is in the position N_5, i.e., in the shaded region where $\sigma = -1$, then it will go along the parabolic trajectory $N_5N_6N_7$ [the solution of Eq. (3.56) for $\sigma = -1$] up to that part N_7N_8O of the switching curve which is in the second quadrant, and then along the switching curve to the origin.

Since $\psi(x, y)$ of (3.67) depends on the parameter A_2 of the driving action $x_1^*(t)$, then according to the definition given above the switching curve S is "quasi-stationary." This property could be expected, since the parameter A_2 enters into Eq. (3.56) for the error $x(t)$. In this case the parts L_1' and L_1'' of the switching curve S are now not antisymmetrical with respect to the origin. In fact, in formula (3.66) for $y < 0$ the absolute value of the denominator is equal to $|M/a_0 + 2A_2|$, while for $y > 0$ the absolute value of the denominator is equal to $|M/a_0 - 2A_2|$. Hence the semiparabolas L_1' and L_1'' are different and cannot be superimposed on one another by a rotation of one of them by 180^0. But if a more restricted class of driving actions is considered,

$$x_1^*(t) = A_0 + A_1 t, \tag{3.68}$$

then $A_2 = 0$ and the parameters of the driving action are missing in Eq. (3.56) for the error $x(\tau)$. In this case expression (3.67) takes the form

$$\psi(x, y) = x + \frac{y^2 \text{ sign } y}{2M/a_0} \tag{3.69}$$

and both semiparabolas L_1' and L_1'' are antisymmetrical with respect to the origin. The second term in formula (3.69) represents the quadratic feedback with respect to velocity, which has already appeared in [3.1].

By knowing Eq. (3.67), the synthesis of the optimal system can be carried out. After substituting a sufficiently small value of x into the expression for ψ, we obtain $\psi > 0$. Hence $\psi > 0$ above the switching curve. In an analogous manner it is easy to show that $\psi < 0$ below it. Therefore the quantity $w = \psi$ can be fed to the input of the relay element, whose characteristic is shown in Fig. 3.14a. The output variable of this element, equal to $\pm M$, can be fed to the input of the object. But in practice it is frequently found necessary to replace the relay element by a somewhat more complex element. In fact, the solution found above and the optimal controller corresponding to it makes sense only for use with large values of the error and its derivatives. When the representative point in the phase space of the errors is close to the origin or even to the

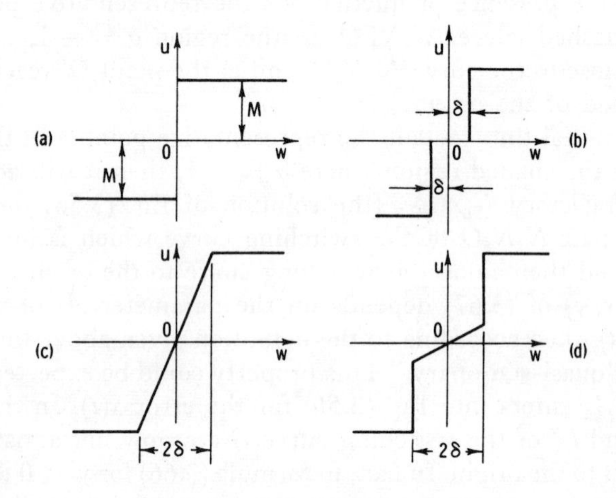

FIG. 3.14

hypersurface S, then the optimal law of motion can be replaced by some other one, for example, a linear one. As a result of this the total time of the transient process does not increase much. Meanwhile the oscillations close to the origin, arising from the frequent switchings, are usually considered undesirable, and they can be reduced by giving the dependence of u on w the form of one of the curves shown in Fig. 3.14b, c, d. For $|w| > \delta$, where δ is some small quantity, these dependences coincide with the characteristic of the relay element.

Instead of the control signal w, any other quantity w_1 having the same sign can be used, for example:

$$w_1 = 2w[(M/a_0) - 2A_2 \operatorname{sign} y]$$
$$= 2x[(M/a_0) - 2A_2 \operatorname{sign} y] + y^2 \operatorname{sign} y. \tag{3.70}$$

This expression is more convenient than (3.67), since the division operation is absent in it.

Figure 3.15 depicts the block diagram of a system in which the control law (3.70) is realized. Here the circuit of the object B is outlined by a double line. The output variable X of the object, being the controlled quantity, is brought to the summing device after a change of sign

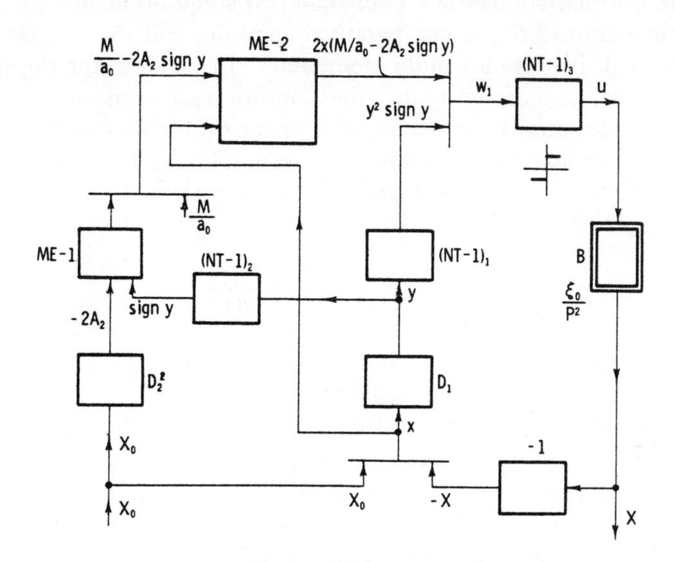

FIG. 3.15

produced by the inverter (-1). The driving action X_0 is also brought there. The difference $X_0 - X = x$ acts on the input to the differentiator D_1, and also on one of the inputs to the multiplier element $ME-2$. The quantity $M/a_0 - 2A_2 \operatorname{sign} y$ acts on the second input of this same element. Thus at the output of the element the product $2x(M/a_0 - 2A_2 \operatorname{sign} y)$ is formed, i.e., the first term of formula (3.70). The factor $M/a_0 - 2A_2 \operatorname{sign} y$ of this expression is obtained as a result of summing the quantity $M/a_0 = \text{const}$ and the output of the other multiplying element $ME-1$. The factors sign y and $-2A_2$ are fed to the input of the latter. The first of them is obtained at the output of the nonlinear transformer $(NT - 1)_2$ with a relay characteristic; the quantity y is fed to the input of the element $(NT - 1)_2$. The factor $- 2A_2$ is the acceleration $-d^2X_0/dt^2$, acting continuously through the double differentiator $D_2{}^2$. It is assumed that the driving action $X_0(t)$ has not been altered by noise and the differentiation takes place without distortions.

The sum appearing on the right side of Eq. (3.70) acts on the input of the nonlinear transformer $(NT - 1)_3$. The second term of this sum is obtained at the output of the nonlinear transformer $(NT - 1)_1$, at whose input the variable y acts. The characteristic of the transformer $(NT - 1)_3$ is one of the types depicted in Fig. 3.14.

Besides the blocks shown in Fig. 3.15, the diagram can still be provided with additional elements entering into operation only for small errors, when the optimal control law is changed to some other one [3.25].

The algorithm (3.67) is comparatively simple, and in the case (3.69) when $A_2 = 0$, it becomes quite elementary. In general, for the case of a second-order constraint under the condition of applicability of the theorem concerning n intervals, the algorithms of strictly optimal systems are comparatively simple. For example, let the object consist of successively joined inertial and integrating elements and have the transfer function

$$K(p) = \frac{\xi_0}{p(1 + pT)} = \frac{\bar{x}_1(p)}{\bar{u}(p)}, \tag{3.71}$$

where $\bar{x}_1(p)$ is the transform of the controlled variable $x_1(t)$ at the output of the object, and $\bar{u}(p)$ is the transform of $u(t)$.

We will assume that the control action $u(t)$ is bounded in absolute value:

$$|u(t)| \leqslant M_2. \tag{3.72}$$

Further, let the driving action belong to the class

$$x^*(t) = A_0 + A_1 t. \tag{3.73}$$

It is not difficult to show that in this case the admissible functions of this class must satisfy the condition

$$|A_1| < \xi_0 M_2 = M. \tag{3.74}$$

Then the formula for the quantity w, fed to the input of the relay element with the equation

$$u(t) = M_2 \operatorname{sign} w, \tag{3.75}$$

has the form

$$w = x + T\left[(A_1 + M \operatorname{sign} y) \ln\left(1 - \frac{y}{A_1 + M \operatorname{sign} y}\right) + y\right], \quad (3.76)$$

where $x = x^* - x_1$ is the error, and y its derivative [3.25].

For $A_1 = 0$ this formula takes a very simple form:

$$w = x + T\{M \operatorname{sign} y \ln[1 - (|y|/M)] + y\}. \quad (3.77)$$

The switching curve corresponding to formula (3.76) depends on the parameter A_1 of the driving action. Hence it is quasi-stationary. For the special case corresponding to expression (3.77), the switching curve does not depend on the parameters of the driving action and therefore is stationary.

As an example let us consider a problem with trajectories in three-dimensional phase space. Let the third derivative of the controlled variable x_1 be bounded in absolute value:

$$|d^3x_1/dt^3| \leqslant M. \quad (3.78)$$

If the driving action x^* belongs to the class of parabolas (3.49), then, as can be shown, in this case the switching surface S will be stationary. Therefore for convenience in the derivation of formulas we will set $x^* = A_0 = \text{const}$, which will not change the result since everything equal to S does not depend on the parameters of the function $x^*(t)$ [3.11, 3.25].

The constraint equation (3.39) is written in the form

$$d^3x_1/dt^3 = \sigma M = v. \quad (3.79)$$

After the reduction to "reversed" time $\tau = T - t$ this equation takes the form

$$d^3x_1/d\tau^3 = -\sigma M. \quad (3.80)$$

We will solve this equation for the zero initial conditions (3.41) and $\sigma = \text{const}$. Let us set

$$x = x^* - x_1, \quad y = \frac{dx}{dt} = -\frac{dx}{d\tau}, \quad z = \frac{d^2x}{dt^2} = \frac{d^2x}{d\tau^2} = -\frac{dy}{d\tau}. \quad (3.81)$$

Then from (3.80) for $\sigma = +1$ it follows that

$$dz/d\tau = -M, \quad (3.82)$$

from which for $\tau = \tau_1$ we obtain

$$z = -M\tau_1. \tag{3.83}$$

Further, we find for $\tau = \tau_1$

$$y = -\int_0^{\tau_1} z\, d\tau = M(\tau_1^2/2) \tag{3.84}$$

and

$$x = -\int_0^{\tau_1} y\, d\tau = -M(\tau_1^3/6). \tag{3.85}$$

Equations (3.83)—(3.85) represent the equation of the curve L_1' in parametric form. At the time instant $\tau = \tau_1$, the coordinates of the representative point N_1 leaving the origin are given by these equations (Fig. 3.16). Now at this instant let the value of σ change and become equal to $\sigma = -1$. Then the equation

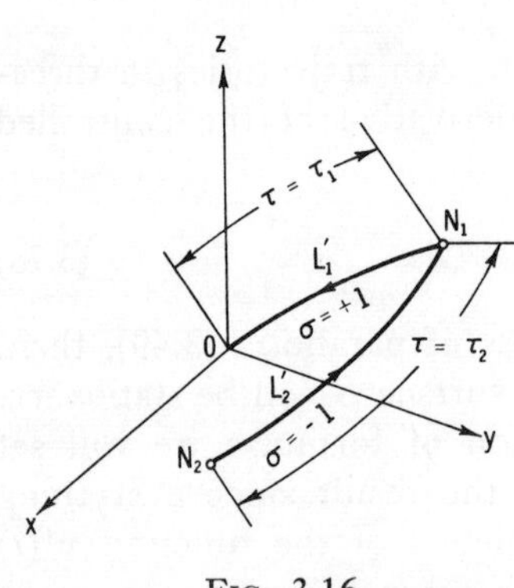

FIG. 3.16

$$dz/d\tau = +M \tag{3.86}$$

must be integrated with the initial conditions (3.83)–(3.85). If the beginning τ_2 of the reading of time is conducted from the instant of switching, then on the second section the formulas for z, y, x take the following form:

$$z = \int_0^{\tau_2} M\, d\tau_2 - M\tau_1 = M(\tau_2 - \tau_1), \tag{3.87}$$

$$y = -\int_0^{\tau_2} z\, d\tau_2 + M(\tau_1^2/2) = -\int_0^{\tau_2} M(\tau_2 - \tau_1)\, d\tau_2 + (M\tau_1^2/2)$$
$$= M\tau_1^2 - (M/2)(\tau_2 - \tau_1)^2 \tag{3.88}$$

and

$$x = -\int_0^{\tau_2} y\, d\tau_2 - M(\tau_1^3/6) = -\int_0^{\tau_2} [M\tau_1^2 - (M/2)(\tau_2 - \tau_1)^2]\, d\tau_2 - M(\tau_1^3/6)$$
$$= -M\tau_1^2\tau_2 + (M/6)(\tau_2 - \tau_1)^3. \tag{3.89}$$

Expressions (3.87), (3.88), and (3.89) represent the equations of the surface L_2' in parametric form. The equations for the surface L_2'' are easily obtained by replacing the sign of x, y, z by the opposite one:

$$\begin{aligned} x &= M\tau_1^2\tau_2 - (M/6)(\tau_2 - \tau_1)^3, \\ y &= -M\tau_1^2 + (M/2)(\tau_2 - \tau_1)^2, \\ z &= -M(\tau_2 - \tau_1). \end{aligned} \tag{3.90}$$

By changing the values of the parameters τ_1 and τ_2 in the preceding formulas in the interval

$$0 \leqslant \tau_1 < \infty, \qquad 0 \leqslant \tau_2 < \infty, \tag{3.91}$$

all points of the switching surface S can be obtained. In this example τ_1 and τ_2 can also be eliminated from both groups of equations, and after joining both the "half-surfaces" L_1' and L_2'' we can obtain the equation of the surface S in explicit form:

$$\psi(x, y, z) = x + \frac{z^3}{3M^2} + (\operatorname{sign} w)\left[\frac{yz}{M} + \sqrt{\left(\frac{1}{M}\right)}\left(\frac{z^2}{2M} + y \operatorname{sign} w\right)^{3/2}\right] = 0,$$

$$w = y + \frac{z^2}{2M}\operatorname{sign} z. \tag{3.92}$$

The control function is expressed by the formula

$$v = M \operatorname{sign} \psi. \tag{3.93}$$

Equation (3.92) is already comparatively complex. For other examples of third-order constraints, the algorithms of a strictly optimal system prove to be still more complicated. The algorithms are comparatively complex for $n = 2$ as well, if the characteristic equation has complex-conjugate roots [3.23]. But it is possible to find approximately optimal algorithms which are comparatively simple for arbitrary types of third-order systems (see, for example, [3.67]).

2. Application of Classical Variational Methods

Let us begin with the examination of simple examples.

Let the object be characterized by a linear equation with constant coefficients

$$b_m \frac{d^m X}{dt^m} + \cdots + b_0 X = u(t), \tag{3.94}$$

where $X(t)$ is the output variable, and $u(t)$ the control action. The transfer function of the object has the form

$$K_B(p) = \frac{\bar{X}(p)}{\bar{u}(p)} = \frac{1}{b_m p^m + \cdots + b_0} = \frac{1}{P_B(p)}\,. \tag{3.95}$$

Figure 3.17 depicts the block diagram of an automatic control system, in which the error x is the difference between the driving action X_0 and the controlled variable X:

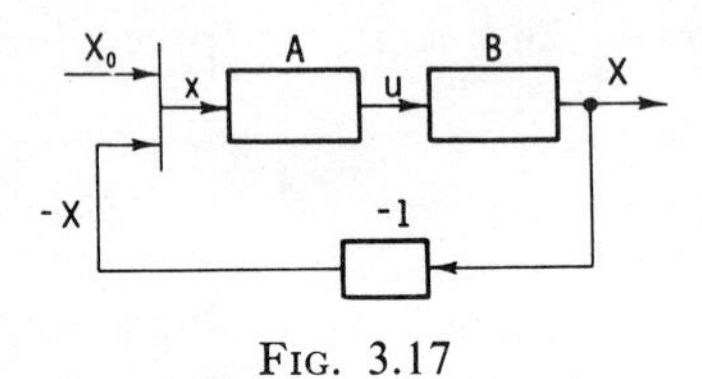

FIG. 3.17

$$x = X_0 - X.$$

For simplicity let

$$X_0 = \begin{cases} A_0 = \text{const}, & t \geqslant 0, \\ 0, & t < 0, \end{cases}$$

and thus let the transient process of the adjustment of the initial error be considered, where for $t < 0$ all the elements of the system are at rest, and hence both the controlled variable X and its derivatives are equal to zero. The error x is found to be the sum

$$x = x_s + x_d\,, \tag{3.96}$$

where x_s is the steady-state error (in this example a constant) and x_d the transient error, vanishing at the end of the transient process. It is required to choose a controller A for which the integral

$$I_V = \int_0^\infty V\,dt = \int_0^\infty \left[x_d^2 + \gamma_1 \left(\frac{dx_d}{dt}\right)^2 + \cdots + \gamma_{n-1}\left(\frac{d^{n-1}x_d}{dt^{n-1}}\right)^2\right] dt \tag{3.97}$$

reduces to a minimum.

Problems with linear objects and optimality criteria in the form of an integral of a quadratic form are the simplest type of problem in the synthesis of optimal systems, since in this case the optimal controller also turns out to be linear. In fact, in Chapter II it was shown that the extremal $x_d^*(t)$ giving the minimum for the integral (3.97) is the solution of the linear equation

$$\vartheta_n \frac{d^n x_d^*(t)}{dt^n} + \vartheta_{n-1} \frac{d^{n-1} x_d^*(t)}{dt^{n-1}} + \cdots + \vartheta_0 x_d^*(t) = 0. \tag{3.98}$$

The coefficients θ_i of this equation are related to the coefficients γ_i of the quadratic form V in (3.97) by functions that are easily obtained from Euler's equation and are given in Chapter II.

In a simple problem considered below the initial conditions for $x_d(t)$ do not vary. The problems with variations in the initial conditions can also be analyzed by classical variational methods. Instead of one general equation of the system here, it is often more convenient in engineering practice to consider the set of equations for its elements.

Since the equation of the entire system is linear and of nth order, and the equation of the object is also linear and of order $m < n$, then the equation of the controller will likewise be linear and of order $n - m$. Let $K_A(p)$ be the transfer function of this device. Then the transfer function of the closed-loop system is

$$K_E(p) = \frac{K_A(p)K_B(p)}{1 + K_A(p)K_B(p)} = \frac{\bar{X}(p)}{\bar{X}_0(p)}, \tag{3.99}$$

where $\bar{X}_0(p)$ is the transform of the driving action. By taking the equation prior to (3.96) into account we obtain:

$$\bar{X}(p) = A_0 \frac{K_A(p)K_B(p)}{1 + K_A(p)K_B(p)} = A_0 \frac{K_A(p)[P_B(p)]^{-1}}{1 + K_A(p)[P_B(p)]^{-1}} = \frac{A_0 K_A(p)}{P_B(p) + K_A(p)}.$$

In the general case let

$$K_A(p) = \frac{Q_A(p)}{P_A(p)}, \tag{3.100}$$

where the polynomial $P_A(p)$ has degree $n - m$, and the degree of the polynomial $Q_A(p)$ is less than the degree of $P_A(p)$. Then

$$\bar{X}(p) = \frac{A_0 Q_A(p)}{P_A(p)P_B(p) + Q_A(p)}. \tag{3.101}$$

The initial conditions for $x_d(t)$ will be fixed, i.e., independent of the parameters of the transfer function $K_A(p)$, only when $X(t)$ and its $n - 1$ derivatives will not depend on these parameters for $t = 0$ and turn out to be equal to zero. Since the initial value of the ith derivative is related to the transform† $\bar{X}(p)$ for a zero value of $(d^{i-1}X/dt^{i-1})_{t=0}$ by the equality

$$\left(\frac{d^i X}{dt^i}\right)_{t=0} = \lim_{p\to\infty} p^i \bar{X}(p) = \lim_{p\to\infty} \frac{A_0 p^i Q_A(p)}{P_A(p)P_B(p) + Q_A(p)}, \tag{3.102}$$

† Here transforms according to Carson have been taken.

then the condition

$$(d^iX/dt^i)_{t=0} = 0 \qquad (i = 1, 2, ..., n-1) \tag{3.103}$$

means that the degree of $Q_A(p)$ must be equal to zero, i.e., $Q_A(p)$ must be a constant. Thus, without loss of generality for this case we set $Q_A(p) = 1$ and

$$K_A(p) = \frac{1}{P_A(p)} = \frac{1}{a_{n-m}p^{n-m} + \cdots + a_1 p + a_0}\,. \tag{3.104}$$

Then as is seen from (3.101), the equation of the system in transforms will take the form

$$[P_A(p)P_B(p) + 1]\bar{X}(p) = A_0\,, \tag{3.105}$$

while the characteristic equation, identical for $X(t)$ and for the transient error $x_d(t)$, can be written in the form

$$P_A(p)P_B(p) + 1 = 0. \tag{3.106}$$

On the other hand, as is seen from (3.98) the characteristic equation of the optimal system has the form

$$H(p) = 0, \tag{3.107}$$

where

$$H(p) = \vartheta_n p^n + \vartheta_{n-1} p^{n-1} + \cdots + \vartheta_0\,. \tag{3.108}$$

By comparing (3.106) and (3.107), the equation

$$\alpha H(p) = P_A(p)P_B(p) + 1 \tag{3.109}$$

can be obtained, where α is any number different from zero. If the polynomial $P_A(p)$ is chosen such that Eq. (3.109) will come true, then the system as a whole will turn out to be optimal in the sense in which it was indicated above.

Equating coefficients for identical powers on the left and right sides of (3.109), we can obtain $n + 1$ equations. The unknowns in these equations are α, a_0, a_1, ..., a_{n-m}; i.e., the total number of unknowns is $n - m + 2$. The system will turn out to be simultaneous and determinate when $n + 1 = n - m + 2$, i.e., when $m = 1$. Thus in this case the given object can only be of the first order.

For example, let the object B be an integrating element with transfer function

$$K_B(p) = \frac{\xi_0}{p} = \frac{1}{(p/\xi_0)} = \frac{1}{P_B(p)}. \tag{3.110}$$

We will also assume that Eq. (3.107) has the form

$$H(p) = p^2 + 2d_0\omega_0 p + \omega_0^2 = 0, \tag{3.111}$$

i.e., it is required that the entire system be an oscillatory element and the integral (see Chapter II)

$$I_V = \int_0^\infty [x_d^2 + 2\sigma T_0^2(dx_d/dt)^2 + T_0^4(d^2x_d/dt^2)^2]\, dt \tag{3.112}$$

be minimized, where

$$T_0 = 1/\omega_0, \qquad d_0 = \sqrt{\left(\frac{1+\sigma}{2}\right)}. \tag{3.113}$$

Then the transfer function of the controller is only of the first order, i.e.,

$$P_A(p) = a_1 p + a_0. \tag{3.114}$$

Substituting into (3.109) the expressions for $P_A(p)$, $P_B(p)$, and $H(p)$ from (3.110), (3.111), and (3.114), we arrive at the relation

$$\alpha(p^2 + 2d_0\omega_0 p + \omega_0^2) = (a_1 p + a_0)(p/\xi_0) + 1, \tag{3.115}$$

from which we obtain three equations for the unknowns α, a_0, a_1:

$$\alpha = a_1/\xi_0, \qquad 2\alpha d_0\omega_0 = a_0/\xi_0, \qquad \alpha\omega_0^2 = 1.$$

Hence,

$$\alpha = 1/\omega_0^2, \qquad a_1 = \xi_0/\omega_0^2, \qquad a_0 = 2\xi_0 d_0/\omega_0,$$

and the optimal controller turns out to be an inertial element with transfer function

$$K_A(p) = \frac{1}{a_1 p + a_0} = \frac{1}{(\xi_0/\omega_0^2)p + (2\xi_0 d_0/\omega_0)} = \frac{K_A}{1 + pT_A}, \tag{3.116}$$

where

$$K_A = \omega_0/2\xi_0 d_0, \qquad T_A = 1/2\omega_0 d_0. \tag{3.117}$$

In this problem an arbitrarily large required value of ω_0 can be assigned, i.e., an arbitrarily small time of the transient process in the system. Then as is seen from (3.117), the quantity K_A must be chosen sufficiently large. But in this case the control action $u(t)$ will likewise turn out to be arbitrarily large. In order to bound this quantity if only in an integral sense, the constraint

$$I_u = \int_0^\infty u^2(t)\, dt \leqslant M$$

can be added. In this case the replacement of the inequality by an equality is permissible, if the most "sluggish" transient process is considered:

$$I_u = \int_0^\infty u^2(t)\, dt = M. \tag{3.118}$$

From the equation of the object

$$dX/dt = \xi_0 u \tag{3.119}$$

it follows that for this example, since $x_s = 0$ and $X_0 = \text{const}$,

$$\frac{dx_d}{dt} = \frac{d}{dt}(x - x_s) = \frac{dx}{dt} = \frac{d}{dt}(X_0 - X) = \frac{d}{dt}(A_0 - X) = -\frac{dX}{dt} = -\xi_0 u.$$

Therefore condition (3.118) turns into the equation

$$I_u = \int_0^\infty u^2(t)\, dt = \int_0^\infty \frac{1}{\xi_0^2}\left(\frac{dx_d}{dt}\right)^2 dt = M. \tag{3.120}$$

Let us introduce the Lagrange multiplier λ. If it is required to observe condition (3.120), then the integral

$$Q = I_V + \lambda I_u = \int_0^\infty \left[x_d^2 + \left(2\sigma T_0^2 + \frac{\lambda}{\xi_0^2}\right)\left(\frac{dx_d}{dt}\right)^2 + T_0^4\left(\frac{d^2 x_d}{dt^2}\right)^2\right] dt \tag{3.121}$$

is subject to minimization. Let us set

$$2\sigma' T_0^2 = 2\sigma T_0^2 + (\lambda/\xi_0^2).$$

Then the integral Q takes the same form as the integral I_V did earlier, but with the replacement of σ by σ'. Hence it follows that the solution also has the same form, but with the replacement of d_0 by d_0', where

$$d_0' = \sqrt{\left[\frac{\sigma' + 1}{2}\right]} = \sqrt{\left[\frac{1}{2}\left(\sigma + 1 + \frac{\lambda}{2T_0^2\xi_0^2}\right)\right]}$$
$$= \sqrt{\left[\frac{1}{2}\left(2d_0^2 + \frac{\lambda\omega_0^2}{2\xi_0^2}\right)\right]} = \sqrt{\left[d_0^2 + \frac{\lambda\omega_0^2}{4\xi_0^2}\right]}. \quad (3.122)$$

Moreover, the parameter λ must be selected such that condition (3.120) be satisfied. This condition (where the value A_0 of the driving action will also enter) gives just one relation between the parameters, and the two other relations are given by equations of the type (3.117):

$$k_A = \omega_0/2\xi_0 d_0', \qquad T_A = 1/2\omega_0 d_0'.$$

Now we will not succeed in making the transient process $x_d(t)$ one which minimizes the integral (3.112). The extremal of $x_d(t)$ is subject to the equation

$$d^2x_d/dt^2 + 2d_0'\omega_0\, dx_d/dt + \omega_0^2 x_d = 0. \quad (3.123)$$

For this curve let us calculate the integral

$$I = \int_0^\infty (dx_d/dt)^2\, dt = \int_0^\infty U\, dt \quad (3.124)$$

by the method presented in [3.25]. The initial conditions for x_d have the form (we consider that $x_s = 0$ and $x = x_d$):

$$(x_d)_{t=0} = (X_0 - X)_{t=0} = A_0,$$
$$\left(\frac{dx_d}{dt}\right)_{t=0} = \left[\frac{d}{dt}(X_0) - \frac{d}{dt}(X)\right]_{t=0} = 0.$$

Then

$$I = B_{11}A_0^2, \quad (3.125)$$

i.e., the quantity I is equal to the initial value of the quadratic form

$$W = B_{11}x_1^2 + 2B_{12}x_1x_2 + B_{22}x_2^2, \quad (3.126)$$

satisfying the condition

$$U = -dW/dt. \quad (3.127)$$

Here we set

$$x_1 = x_d, \qquad x_2 = dx_d/dt.$$

Hence it follows that

$$U = (dx_d/dt)^2 = x_2^2. \tag{3.128}$$

Instead of Eq. (3.123) two first-order equations can be written:

$$dx_1/dt = x_2, \qquad dx_2/dt = -\omega_0^2 x_1 - 2d_0' \omega_0 x_2. \tag{3.129}$$

We will write Eq. (3.127) in expanded form:

$$-U = \frac{\partial W}{\partial x_1}\frac{dx_1}{dt} + \frac{\partial W}{\partial x_2}\frac{dx_2}{dt}.$$

Here, after substituting expressions (3.128), (3.126), and (3.129), we arrive at the equation

$$\begin{aligned} -x_2^2 = 2B_{11}x_1x_2 + 2B_{12}x_2^2 - 2B_{12}\omega_0^2 x_1^2 - 2B_{22}\omega_0^2 x_1 x_2 \\ - 4B_{12}d_0'\omega_0 x_1 x_2 - 4d_0'\omega_0 B_{22}x_2^2. \end{aligned} \tag{3.130}$$

Equating coefficients for like terms on the left and right sides, we obtain the three equations:

$$\begin{gathered} -2B_{12}\omega_0^2 = 0, \qquad 2B_{11} - 2B_{22}\omega_0^2 - 4B_{12}d_0'\omega_0 = 0, \\ 2B_{12} - 4d_0'\omega_0 B_{22} = -1, \end{gathered} \tag{3.131}$$

from which we find:

$$B_{12} = 0, \qquad B_{22} = 1/4d_0'\omega_0, \qquad B_{11} = \omega_0/4d'_0. \tag{3.132}$$

From (3.125) it follows that

$$I = B_{11}A_0^2 = \omega_0 A_0^2/4d_0'. \tag{3.133}$$

By taking (3.133) into account, condition (3.120) gives

$$\omega_0 A_0^2/4d_0' = \xi_0^2 M, \tag{3.134}$$

from which

$$d_0' = A_0^2\omega_0/4\xi_0^2 M = \kappa\omega_0, \tag{3.135}$$

where

$$\kappa = A_0^2/4\xi_0^2 M. \tag{3.136}$$

By knowing d_0', we can also determine λ from (3.122), which we will not dwell upon.

The substitution of (3.135) into the expressions for k_A and T_A gives the parameters of the optimal controller A:

$$k_A = 2\xi_0 M/A_0^2, \qquad T_A = 2\xi_0^2 M/\omega_0^2 A_0^2. \tag{3.137}$$

Hence it follows that for a small admissible threshold M and large value of A_0, the magnitudes of the amplification coefficient k_A and time constant T_A must be small. It is not difficult to see that the solution of Eq. (3.123) for the extremal now will not be an arbitrarily rapid transient process, no matter how much the arbitrarily assigned quantity ω_0 is increased. Actually, the characteristic equation of the extremal has the form

$$p^2 + 2d_0'\omega_0 p + \omega_0^2 = 0. \tag{3.138}$$

Let us substitute the value of d_0' from (3.135). Then Eq. (3.138) can be rewritten in the form

$$p^2 + 2\kappa\omega_0^2 p + \omega_0^2 = 0. \tag{3.139}$$

The roots of this equation are

$$p_{1,2} = -\kappa\omega_0^2 \pm \sqrt{(\kappa^2\omega_0^4 - \omega_0^2)} = \omega_0^2\{-\kappa \pm \sqrt{[\kappa^2 - (1/\omega_0^2)]}\}. \tag{3.140}$$

As $\omega_0^2 \to \infty$ one of the roots

$$p_1 \cong \omega_0^2[-\kappa - \sqrt{(\kappa^2)}] = -2\kappa\omega_0^2 \tag{3.141}$$

tends to $-\infty$ and will stop affecting the transient process. As $\omega_0^2 \to \infty$ the second root

$$\begin{aligned} p_2 &= \omega_0^2\{-\kappa + \kappa\sqrt{[1 - (1/\kappa^2\omega_0^2)]}\} \\ &= \omega_0^2\kappa[-1 + 1 - (1/2\kappa^2\omega_0^2) + o(1/\omega_0^2)] \to -1/2\kappa \end{aligned} \tag{3.142}$$

now does not depend on ω_0^2. The motion of the roots p_1 and p_2 as $\omega_0 \to \infty$ is shown in Fig. 3.18. Thus the transient process becomes the exponential $A_0 \exp\{p_2 t\} = A_0 \exp\{-t/2\kappa\}$, not depending on ω_0^2; i.e., the optimal process cannot turn out to be arbitrarily rapid. Other analogous examples of the synthesis of optimal systems or the fitting of optimal parameters can be found in [3.15, 1.5, 1.9, 3.28].

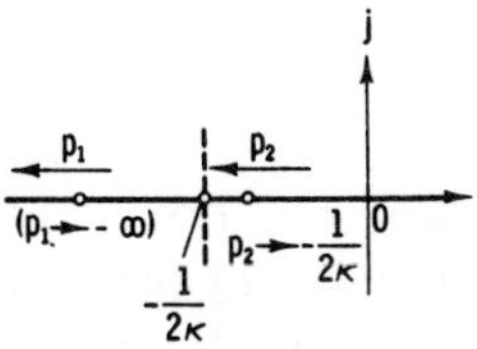

FIG. 3.18

Problems of this kind are solved easily by the methods of the classical calculus of variations for a higher-order equation of the object as well, if it is a linear equation with constant coefficients, and the optimality criterion and expressions restricted by equalities are integrals of quadratic forms.

The solution is hampered when constraints in the form of inequalities are added, and, as was indicated in Chapter II, the greatest difficulties arise in those cases where discontinuous functions have to emerge as the result of the solution. In order to avoid this and by the same token facilitate the process of solution by classical methods, we sometimes resort to the replacement of the original problem by some other one which is close to it. The new problem differs from the original one in that functions enter into its conditions, which are "smoothed" to the extent that a solution is obtained which is sufficiently smooth as well. As an example let us examine the problem cited in the paper [3.29]. Let the equation of the object of control have the form

$$dx_i/dt = f_i(x_1, x_2, ..., x_n, u_1, u_2, ..., u_r) \qquad (i = 1, 2, ..., n), \quad (3.143)$$

and let it be required to select the controls $u_1, u_2, ..., u_r$ such that the integral

$$Q = \int_0^T V(x_1, x_2, ..., x_n, u_1, u_2, ..., u_r)\, dt \quad (3.144)$$

be minimized, where T is not fixed, and the function V is assumed differentiable with respect to all its arguments, in like manner the functions f_i $(i = 1, 2, ..., n)$ as well. The initial and final values of the vector $\bar{x} = (x_1, x_2, ..., x_n)$ are also prescribed.

Let the additional constraint have the form

$$\psi(\bar{u}) = \psi(u_1, u_2, ..., u_r) \leqslant 0. \quad (3.145)$$

With the aim of an approximate calculation for this constraint a "penalty function" can be introduced, setting a very high "penalty" for nonfulfillment of condition (3.145). Let the magnitude of the "penalty" be

$$L(\bar{u}) = \begin{cases} 0, & \psi \leqslant 0, \\ k\psi^2, & \psi > 0, \end{cases} \quad (3.146)$$

where $k > 0$ is a sufficiently large quantity. Then a small violation of condition (3.145) will already give rise to a considerable "penalty" in magnitude. Now the problem can be posed in the following way. Ideally it is required to control in such a manner that the integral (3.144) becomes minimal under the simultaneous observance of the condition

$$\int_0^T L(\bar{u})\, dt = 0. \quad (3.147)$$

Condition (3.147) is exact, since if it is not violated, then ψ also does not become positive at a single time instant (with the exception perhaps of a set of points of measure zero, which is the case excluded from consideration). The method of replacing condition (3.145) by the equality (3.147), in which an integral appears, can simplify the analysis considerably (see also [3.27]). Now let us examine instead of the original integral (3.144) a new integral

$$Q' = Q + \int_0^T L(\bar{u})\, dt = \int_0^T V\, dt + \int_0^T L\, dt. \tag{3.148}$$

A more precise formulation requires the introduction of the Lagrange multiplier in front of the second term. But for a sufficiently large value of the coefficient k in formula (3.146), a small "penetration" of the value of $\bar{u}$ into the prohibited region $\psi > 0$ causes such a large increase in the second term, that Q' deviates strongly from the minimum. Therefore the point $\bar{u}$ corresponding to the minimum of Q' either is found inside the region $\psi \leqslant 0$, or is practically on the boundary of this region. Thus only the single criterion (3.148) may now be followed, not calling for (3.147).

In this case the Eqs. (3.143) of the object play the role of constraints in the form of the equalities

$$\dot{x}_i - f_i(x_1, x_2, ..., x_n, u_1, u_2, ..., u_r) = 0 \qquad (i = 1, 2, ..., n). \tag{3.149}$$

For their calculation we must introduce Lagrange multipliers λ_i [see Chapter II, Eqs. (2.153) and (2.154)] and instead of the quantity Q' minimize another quantity Q'', and moreover

$$\begin{aligned} Q'' &= \int_0^T V\, dt + \int_0^T L\, dt + \int_0^T \left\{ \sum_{i=1}^n \lambda_i[\dot{x}_i - f_i(x_1, ..., x_n, u_1, ..., u_r)] \right\} dt \\ &= \int_0^T F^*\, dt, \end{aligned} \tag{3.150}$$

where

$$\begin{aligned} F^* &= V(x_1, ..., x_n) + L(u_1, ..., u_r) \\ &\quad + \sum_{j=1}^n \lambda_j[\dot{x}_j - f_j(x_1, ..., x_n, u_1, ..., u_r)]. \end{aligned} \tag{3.151}$$

Regarding the x_i and u_j as ordinary variables, the Euler-Lagrange equations must now be constructed:

$$\frac{\partial F^*}{\partial x_i} - \frac{d}{dt}\frac{\partial F^*}{\partial \dot{x}_i} = 0 \qquad (i = 1, 2, ..., n) \tag{3.152}$$

and

$$\frac{\partial F^*}{\partial u_k} - \frac{d}{dt}\frac{\partial F^*}{\partial \dot{u}_k} = 0 \qquad (k = 1, 2, ..., r). \tag{3.153}$$

Moreover, the n equations (3.149) must still be satisfied. In all, $2n + r$ equations are obtained for the unknowns x_i ($i = 1, 2, ..., n$), λ_j ($j = 1, 2, ..., n$), and u_k ($k = 1, 2, ..., r$), whose total number is also equal to $2n + r$. Substituting expression (3.151) into (3.152) and (3.153), we arrive at the equations

$$\frac{\partial V}{\partial x_i} - \sum_{j=1}^{n} \lambda_j \frac{\partial f_j}{\partial x_i} - \frac{d\lambda_i}{dt} = 0 \qquad (i = 1, 2, ..., n) \tag{3.154}$$

and

$$\frac{\partial L}{\partial u_k} - \sum_{j=1}^{n} \lambda_j \frac{\partial f_j}{\partial u_k} = 0 \qquad (k = 1, 2, ..., r). \tag{3.155}$$

If the quantity u_k is expressed from (3.155) in terms of the remaining variables (the du_k/dt do not enter anywhere in the equations), then $2n$ equations remain with $2n$ unknowns $x_1, x_2, ..., x_n$ and $\lambda_1, \lambda_2, ..., \lambda_n$, where $2n$ boundary conditions are prescribed.

Let us consider the example of a linear object. Let its equation have the form

$$dx_i/dt = \sum_{j=1}^{n} a_{ij}x_j + \delta_{i1}u, \tag{3.156}$$

where

$$\delta_{i1} = \begin{cases} 0, & i \neq 1, \\ 1, & i = 1. \end{cases} \tag{3.157}$$

Thus u is introduced only in the first of Eqs. (3.156).

We will assume that it is required to ensure a minimum time of the transient process. Thus,

$$Q = \int_0^T 1 \cdot dt = T. \tag{3.158}$$

Hence $V = 1$. Further, let the sole control action $u(t)$ be bounded by the condition

$$|u| \leqslant M. \tag{3.159}$$

Then we can set

$$L(u) = L_1(u) + L_2(u), \tag{3.160}$$

where

$$\begin{aligned} L_1(u) &= \begin{cases} 0, & u + M > 0, \\ k(u + M)^2, & u + M < 0, \end{cases} \\ L_2(u) &= \begin{cases} 0, & M - u > 0, \\ k(M - u)^2, & M - u < 0. \end{cases} \end{aligned} \tag{3.161}$$

These functions are shown in Fig. 3.19a. According to (3.151), in this example the function F^* takes the form

$$F^* = 1 + [L_1(u) + L_2(u)] + \sum_{k=1}^{n} \lambda_k \left(\dot{x}_k - \sum_{j=1}^{n} a_{kj} - x_j - \delta_{k1} u \right). \quad (3.162)$$

In this case the system of Euler's equations (3.154) and (3.155) can be written in the form

$$d\lambda_i/dt = - \sum_{j=1}^{n} a_{ji}\lambda_j \qquad (i = 1, 2, ..., n) \quad (3.163)$$

and

$$\lambda_1 = \varphi(u) = \varphi_1(u) + \varphi_2(u), \quad (3.164)$$

where

$$\varphi(u) = dL/du$$

and

$$\varphi_1(u) = \begin{cases} 0, & u + M > 0, \\ 2k(u + M), & u + M < 0, \end{cases}$$
$$\varphi_2(u) = \begin{cases} 0, & M - u > 0, \\ -2k(M - u), & M - u < 0. \end{cases} \quad (3.165)$$

The curve of $\varphi(u)$ is shown in Fig. 3.19b. It consists of a section $\varphi(u) = 0$ for $|u| \leqslant M$ and inclined linear sections with slope k. The larger the value of k, the closer the sloping straight lines to the vertical half-lines (shown dashed).

The unknowns λ_i satisfy the system of equations (3.163) which are adjoint to the system of equations (3.156) of the object. If the roots of the characteristic equation corresponding to (3.156) are negative and real, then as can be shown the roots of the adjoint equation for the system of equations (3.163) are negative and real as well. Let these roots be equal to $\gamma_1, \gamma_2, ..., \gamma_n$. Then in the general case

$$\lambda_1 = C_1 e^{\gamma_1 t} + \cdots + C_n e^{\gamma_n t}. \quad (3.166)$$

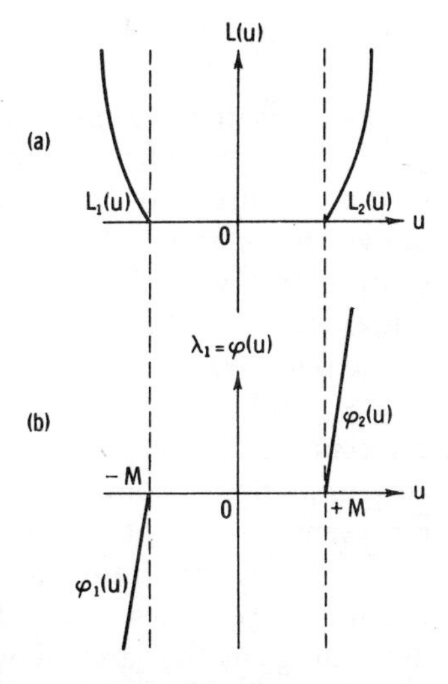

FIG. 3.19

It is known that this function can change sign not more than $n - 1$ times, and therefore in the general case has n intervals of constant sign. But from Eq. (3.164) the quantity $u(t)$ corresponding to the value of $\lambda_1(t)$ can be found at each time instant. From Fig. 3.19 it is seen that $\lambda_1 = \varphi(u)$ can change its sign only when u passes by a jump from the region $u < -M$ to the region $u > +M$, or vice versa. For a sufficiently large slope k of the linear sections, the corresponding values of u will be equal to $+M$ and $-M$. Thus the total number of intervals of constancy of $u = \pm M$ does not exceed n. Thus the theorem concerning n intervals is proved.

An interesting example of the construction of an optimal control system for a chemical reactor was given in the paper [3.32] (see also [3.33]). The problem consists of minimizing the flow time of the reactor, which corresponds to the maximal output of the assembly.

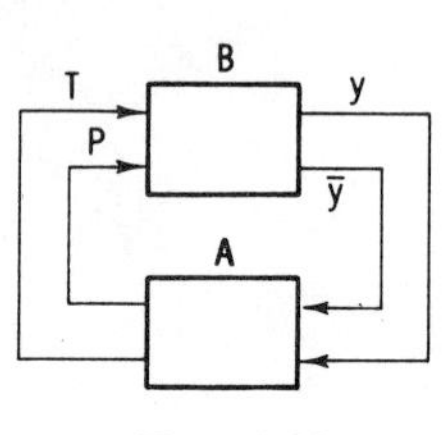

FIG. 3.20

In the reactor B (Fig. 3.20) let there be a mixture of three constituents whose relative concentrations are denoted by y, $\bar{y}$, and z. The sum of the relative concentrations is equal to unity; therefore

$$y + \bar{y} + z = 1. \tag{3.167}$$

As a result of the chemical reactions taking place in B, these concentrations change from the initial values y_0, $\bar{y}_0$, z_0, and at the end of the process pass to the required final values y_f, $\bar{y}_f$, z_f. These values can be attained more quickly or more slowly, depending on the laws by which the temperature T and pressure P in the reactor vary during the reaction. It is required to design a controller A realizing the optimal variation in the control parameters T and P, in order that by it the entire process will go through in the shortest time.

Since z can be computed for given y and $\bar{y}$ from Eq. (3.167), then only two independent components y and $\bar{y}$ of the mixture may be considered. Let us draw the phase plane with Cartesian coordinates y and $\bar{y}$ (Fig. 3.21). The initial state of the mixture corresponds to the point M_0 in this plane with coordinates y_0 and $\bar{y}_0$. The required final state corresponds to the point M_f with coordinates y_f and $\bar{y}_f$. From the position M_0 the representative point M can arrive at the position M_f by moving along various possible trajectories—for example, 1 or 2 depending on the laws of variation of the control parameters T and P. It is required to find a law (and by it the

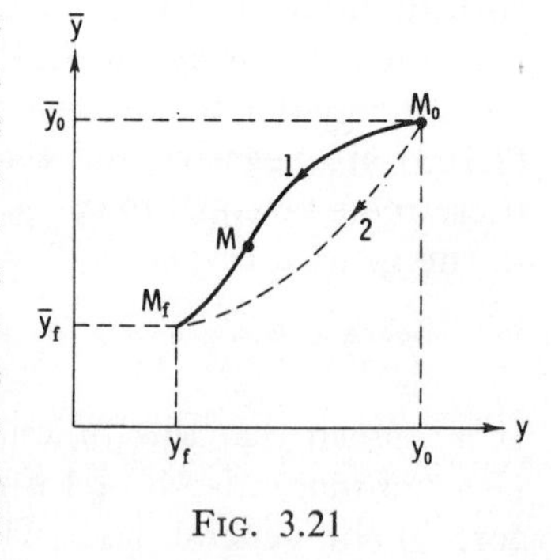

FIG. 3.21

trajectory of the point M in the phase plane as well) that corresponds to the minimum passage time of the representative point from the position M_0 to the position M_f.

In this case the time constants of the controller A are negligibly small compared to the reaction time. The equations of motion of the object are the equations of the kinetics of the reactor, composed on the basis of the law of effective masses. This law says that the speed of the reaction is proportional to the concentrations of the reacting substances. In this example the substances whose concentrations are denoted by y, $\bar{y}$, and z react with hydrogen fed into the reactor B. If the quantity of hydrogen entering in unit time is regarded as constant, then the speed of diminution of some component of the mixture is proportional to its amount. We will denote the coefficient of proportionality by the letter k with an associated index.

The diagram of the reactions is depicted in Fig. 3.22. The arrows denote the direction of reactions; under the arrow the corresponding coefficient of proportionality is indicated. From this diagram it is seen that the substance y turns into $\bar{y}$ with coefficient of proportionality k_3, and also into z with coefficient of proportionality k_1. In turn the substance $\bar{y}$ turns back into y with coefficient of proportionality k_4, and also into z with coefficient of proportionality k_2. On the basis of this diagram the following kinetic equations for the object B can be written down, in their accepted form:

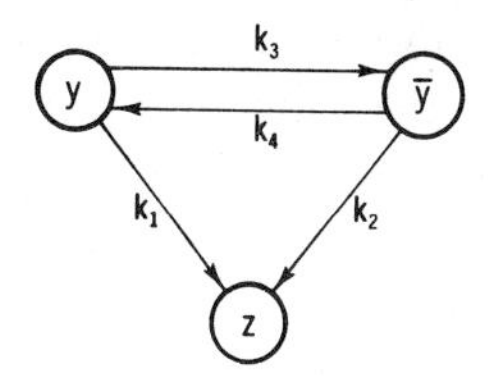

FIG. 3.22

$$dy/dt = -(k_1 + k_3)y + k_4\bar{y} \tag{3.168}$$

and

$$d\bar{y}/dt = -(k_2 + k_4)\bar{y} + k_3 y. \tag{3.169}$$

The coefficients k_i depend on the temperature T and pressure P. The dependence characteristic for chemical reactors has the following form:

$$k_i = a_i P^{n_i} \exp\{-b_i/T\}, \tag{3.170}$$

where a_i, n_i, b_i are constants. If for simplifying the investigation the temperature T is taken to be constant, which in general is not the best method of reactor control, then

$$k_i = A_i P^{n_i}, \tag{3.171}$$

where the A_i are constants ($i = 1, 2, 3, 4$). The problem consists of determining the optimal function $P(t)$, for which the passage from the point M_0 to M_f (see Fig. 3.21) would be accomplished in the shortest time.

To formulate the corresponding variational problem it is appropriate to pass to the new variables

$$w = \bar{y}/y \tag{3.172}$$

and

$$q = \ln(y_0/y). \tag{3.173}$$

At the initial time instant, $q = \ln 1 = 0$. With an increase of time t the quantity y decreases, since the concentration of the original substance decreases. Hence q increases monotonically with time. Therefore q can be taken for a new independent variable, which will simplify the equations.

For the transfer to the new variable we first transform Eq. (3.168) in the following way:

$$\frac{1}{y}\frac{dy}{dt} = -(k_1 + k_3) + k_4\frac{\bar{y}}{y}. \tag{3.174}$$

By taking (3.172) into account, we further find

$$\frac{d}{dt}(\ln y) = -(k_1 + k_3) + k_4 w. \tag{3.175}$$

Since $q = \ln y_0 - \ln y$, then the latter equation takes the form

$$\frac{dq}{dt} = (k_1 + k_3) - k_4 w. \tag{3.176}$$

Then let us expand the expression for dw/dt:

$$\frac{dw}{dt} = \frac{d}{dt}\left(\frac{\bar{y}}{y}\right) = \frac{y\, d\bar{y}/dt - \bar{y}\, dy/dt}{y^2} = \frac{1}{y}\frac{d\bar{y}}{dt} - \frac{\bar{y}}{y}\left[\frac{1}{y}\frac{dy}{dt}\right]$$

$$= \frac{1}{y}\frac{d\bar{y}}{dt} - w[-(k_1 + k_3) + k_4 w]. \tag{3.177}$$

Here the expression in the brackets has been substituted according to Eq. (3.174). From (3.169) we find

$$\frac{1}{y}\frac{d\bar{y}}{dt} = k_3 - (k_2 + k_4)w. \tag{3.178}$$

After substituting this expression into (3.177), we arrive at the equation

$$dw/dt = k_3 + (k_1 + k_3 - k_2 - k_4)w - k_4 w^2. \tag{3.179}$$

Now dividing Eqs. (3.179) and (3.176) termwise by one another, we will obtain

$$\frac{dw}{dq} = \frac{k_3 + (k_1 + k_3 - k_2 - k_4)w - k_4 w^2}{k_1 + k_3 - k_4 w}. \tag{3.180}$$

From Eq. (3.176) it follows that

$$dt = \frac{dq}{k_1 + k_3 - k_4 w}. \tag{3.181}$$

Hence the reaction time t_f can be found, since at the initial instant $q = 0$, and at the final instant $q = q_f = \ln(y_0/y_f)$:

$$t_f = \int_0^{q_f} \frac{dq}{k_1 + k_3 - k_4 w}. \tag{3.182}$$

Thus the problem reduces to determining a trajectory in the (q, w) plane and the corresponding law of variation of $P(q)$, which ensures the minimization of the integral (3.182).

The equations can be simplified, since for the reactor considered in the paper [3.32] the relations

$$k_2/k_1 \simeq k_4/k_3, \qquad n_1 = n_2, \qquad n_3 = n_4 \tag{3.183}$$

hold. In addition, $k_1/k_3 < 1$. Let us set

$$g = k_1/k_3 = (A_1/A_3)P^{(n_1-n_3)}, \qquad A = A_1^{m-1}/A_3^{m-1},$$
$$B = k_4/k_3 = A_4/A_3, \qquad C = k_2/k_3 = A_2/A_3, \qquad m = n_1/(n_1 - n_3). \tag{3.184}$$

Then

$$P = (g^m/AA_3)^{1/n_3}. \tag{3.185}$$

All the coefficients k_i can therefore be regarded as a function of the variable g, which in turn depends on q. Then the integral (3.182) can be written in the form

$$t_f = \int_0^{q_f} F(w, g)\, dq, \tag{3.186}$$

where the function F depends on w and g. The quantity w is related to q by Eq. (3.180), which we will rewrite in the form

$$\varphi(w, g, w') = f(w, g) - w' = 0. \tag{3.187}$$

Here $w' = dw/dq$. The differentiation with respect to q is denoted everywhere below by a prime.

Thus it is required to find the functions $g(q)$ and $w(q)$ minimizing the integral (3.186) under the auxiliary condition (3.187). In order to find Euler's equation, we construct the subsidiary function

$$F^* = F + \lambda(q)\varphi = F(w, g) + \lambda(q)[f(w, g) - w']. \tag{3.188}$$

Here $\lambda(q)$ is the Lagrange multiplier. To determine the three functions λ, w, and g the two Euler equations are used:

$$\frac{\partial F^*}{\partial w} - \frac{d}{dq}\left(\frac{\partial F^*}{\partial w'}\right) = 0, \tag{3.189}$$

$$\frac{\partial F^*}{\partial g} - \frac{d}{dq}\left(\frac{\partial F^*}{\partial g'}\right) = 0, \tag{3.190}$$

and also the connecting equation (3.187). Substituting the values of the derivatives of F^* from (3.188) into (3.189) and (3.190), we obtain

$$\lambda(q)\frac{\partial f(w, g)}{\partial w} + \frac{\partial F(w, g)}{\partial w} + \frac{\partial \lambda(q)}{\partial q} = 0 \tag{3.191}$$

and

$$\lambda(q)\frac{\partial f(w, g)}{\partial g} + \frac{\partial F(w, g)}{\partial g} = 0. \tag{3.192}$$

The function $\lambda(q)$ can be eliminated from the last two equations. From (3.191) we find

$$\lambda(q) = -\frac{\partial F(w, g)/\partial g}{\partial f(w, g)/\partial g} = \psi(w, g). \tag{3.193}$$

Therefore

$$\frac{d\lambda(q)}{dq} = \frac{\partial\psi}{\partial w}\frac{dw}{dq} + \frac{\partial\psi}{\partial g}\frac{dg}{dq} = \frac{\partial\psi}{\partial w}f(w, g) + \frac{\partial\psi}{\partial g}g'. \tag{3.194}$$

Substituting (3.193) and (3.194) into (3.191), we arrive at the equation

$$\psi(w, g)\frac{\partial f(w, g)}{\partial w} + \frac{\partial F(w, g)}{\partial w} + \frac{\partial\psi(w, g)}{\partial w}f(w, g) + \frac{\partial\psi(w, g)}{\partial g}g' = 0, \tag{3.195}$$

representing a first-order differential equation with respect to g. After the substitution of the values of the derivatives, this equation acquires the form

$$g' = (1 + w - w')\left[\frac{g}{(m-1)(1-Bw)} - \frac{1-Bw}{m}\right]. \tag{3.196}$$

By using the notations (3.184), Eq. (3.180) and (3.181) can be rewritten in the following form:

$$w' = w + 1 - \frac{g(1+Cw)}{g+1-Bw} \tag{3.197}$$

and

$$t' = \frac{Ag^{1-m}}{g+1-Bw}. \tag{3.198}$$

The simultaneous solution of Eqs. (3.196) and (3.197) gives the desired curves of $g(q)$ and $w(q)$. The initial value $w(0)$ is known for this time instant. The final value $w_f = w(q_f)$ is also known. But the value $g(0)$ is not known. It is not enough that just this quantity is sought as the final result, since by knowing g at the current time instant the required value of the pressure P can be computed from formula (3.185). In order to find $g(0)$, it is required to solve a boundary-value problem. This quantity must be chosen such that the representative point in the (q, w) plane passes from the initial position ($q_0 = 0$, $w = w_0 = \bar{y}_0/y_0$) to the final position ($q = q_f = \ln(y_0/y_f)$, $w = w_f = \bar{y}_f/y_f$). The motion must be subject to Eqs. (3.196) and (3.197). The quantity t_f, computed simultaneously from formula (3.186), gives the value of time that will be required for carrying out the reaction.

In an experimental model of a computer applied to the solution of this problem, the value of $g(0)$ has been selected automatically, and the current values y_0 and $\bar{y}_0$ taken for initial ones have been introduced by hand on the basis of measured and averaged data from measuring instruments. The choice of $g(0)$ has been made by a series of solutions performed at an accelerated rate.

This process consists of the following.

Equations (3.196) and (3.197) are integrated as a function of "time" q for an assigned initial value w_0 and some $g(0)$, where the value w_f' is measured at the "time" instant q_f. If $w_f - w_f' > 0$, i.e., the quantity w_f' is too small, then $g(0)$ must be increased. The increase in $g(0)$ was made proportional to the measured difference $w_f - w_f'$. After several turns the value of this difference decreases to a negligibly small quantity, and thus the value of $g(0)$ is automatically determined, which serves for computing the current value of the pressure P. It is required to maintain this pressure in the reactor at the given time instant.

The examples considered above show that in a definite domain, classical variational methods can find and do find fruitful application. But the application of new principles opens wider possibilities in the theory of optimal systems—dynamic programming and the maximum principle. The corresponding examples are given below.

3. Application of the Method of Dynamic Programming

In papers [3.34, 3.35] a linear object was analyzed, in the general case with variable parameters, for which the impulsive response was denoted by $g(t, \tau)$. This means that the output variable $x(t)$ observed at a time instant t is related to the input variables $u(\tau)$ $(-\infty < \tau \leqslant t)$ by the relation

$$x(t) = \int_{-\infty}^{t} g(t, \tau)u(\tau)\, d\tau. \tag{3.199}$$

For linear systems with constant coefficients $g(t, \tau) = g(t - \tau)$, and expression (3.199) turns into the ordinary Duhammel integral. Below, the state of the object at each time instant is characterized by the values of n coordinates $x_1, x_2, ..., x_n$ or the vector $\bar{x}$.

We will assume that as the optimality criterion a certain integral is chosen, taken with respect to "future" values of time on the interval from t to $t + T$:

$$Q = \int_{t}^{t+T} \{\lambda(\sigma) f_x[\bar{X}(\sigma) - \bar{x}(\sigma)] + f_u[\bar{U}(\sigma) - \bar{u}(\sigma)]\}\, d\sigma. \tag{3.200}$$

It is required to select a current value of the vector $\bar{u}(t)$ in such a manner that the quantity Q will be minimized. Under the integral in expression (3.200) there is a function f_x of the difference $\bar{X}(\sigma) - \bar{x}(\sigma)$, where $\bar{X}(\sigma)$ is the prescribed "ideal" function for the vector $\bar{x}$, and moreover $f_x(0) = 0$. The function f_u of the difference $\bar{U}(\sigma) - \bar{u}(\sigma)$, where $\bar{U}(\sigma)$ is the "ideal" action, also has the property $f_u(0) = 0$. Below it is assumed that

$$f_x(v) = f_u(v) = v^2.$$

In addition, $\lambda(\sigma)$ is some given "weighting" function, setting the relative "value" of the terms f_x and f_u in the general formula. The interval of integration is chosen finite, since in practical problems we can usually establish a finite value of T beyond whose limits a prediction of the future loses profit.[†]

For an optimal control the quantity Q assumes a minimal value S, depending only on the state of the object at the current time instant t taken as the initial one. Thus,

$$S = S(x_1, x_2, \dots, x_n, t) = S[\bar{x}(t), t] = \min_{\substack{u(\sigma)\in\Omega(u) \\ t\leqslant\sigma\leqslant t+T}} Q. \tag{3.201}$$

Substituting expression (3.200) for Q, we will obtain

$$S[\bar{x}(t), t] = \min_{\substack{u(\sigma)\in\Omega(u) \\ t\leqslant\sigma\leqslant t+T}} \int_t^{t+T} \{\lambda(\sigma) f_x[\bar{X}(\sigma) - \bar{x}(\sigma)] + f_u[\bar{U}(\sigma) - \bar{u}(\sigma)]\}\, d\sigma. \tag{3.202}$$

It should be noted that

$$S[\bar{x}(t+T), t+T] = 0. \tag{3.203}$$

In fact, as is seen from (3.200) the value of the integral Q is equal to zero when the lower limit of integration equals $t + T$. Therefore $S = \min Q$ is also equal to zero if the current value of time coincides with the final instant.

We will derive the necessary condition for optimality in the same way as in Chapter II. For this we will divide the whole time interval from t to $t + T$ into two intervals, $[t \leqslant \sigma \leqslant t + \epsilon]$ and $[t + \epsilon < \sigma \leqslant t + T]$, where ϵ is small. Then

$$S[\bar{x}(t), t] = \min_{\substack{u(\sigma)\in\Omega(u) \\ t\leqslant\sigma\leqslant t+T}} \epsilon\{\lambda(t) f_x[\bar{X}(t) - \bar{x}(t)] + f_u[\bar{U}(t) - \bar{u}(t)]\}$$

$$+ \int_{t+\epsilon}^{t+T} \{\lambda(\sigma) f_x[\bar{X}(\sigma) - \bar{x}(\sigma)] + f_u[\bar{U}(\sigma) - \bar{u}(\sigma)]\}\, d\sigma. \tag{3.204}$$

But the minimal value of the second term in the brackets is equal to $S[\bar{x}(t + \epsilon), t + \epsilon]$:

$$S[\bar{x}(t+\epsilon), t+\epsilon] = \min_{\substack{u(\sigma)\in\Omega(u) \\ t+\epsilon\leqslant\sigma\leqslant t+T}} \int_{t+\epsilon}^{t+T} \{\lambda(\sigma) f_x[\bar{X}(\sigma) - \bar{x}(\sigma)] + f_u[\bar{U}(\sigma) - \bar{u}(\sigma)]\}\, d\sigma. \tag{3.205}$$

[†] It should be indicated that the example under consideration can also be solved with the aid of the maximum principle, in the form in which it was generalized by A. G. Butkovskii to objects with integral equations (see [3.43, 3.44]).

Under the condition of providing the optimal control in the interval $t + \epsilon \leqslant \sigma \leqslant t + T$, the second term in the brackets of (3.204) can be replaced by $S[\bar{x}(t + \epsilon), t + \epsilon]$. Now it is required to find an optimal control $u(\sigma)$ on the small interval $t \leqslant \sigma \leqslant t + \epsilon$, such that the right side of the expression

$$S[\bar{x}(t), t] = \min_{u(t)\in\Omega(u)} [\epsilon\{\lambda(t) f_x[\bar{X}(t) - \bar{x}(t)] + f_u[\bar{U}(t) - \bar{u}(t)]\} + S[\bar{x}(t + \epsilon), t + \epsilon]] \tag{3.206}$$

be minimized.

It is not difficult to pass to the limit as $\epsilon \to 0$ in expression (3.206). Expanding $S[\bar{x}(t + \epsilon), t + \epsilon]$ in a Taylor series as in Chapter II, we find

$$S[x_1(t + \epsilon), \ldots, x_n(t + \epsilon), t + \epsilon] = S[\bar{x}(t), t] + \frac{\partial S[\bar{x}(t), t]}{\partial t}\epsilon + \sum_{k=1}^{n} \frac{\partial S[\bar{x}(t), t]}{\partial x_k}[x_k(t + \epsilon) - x_k(t)]. \tag{3.207}$$

Here we have neglected higher powers of ϵ. But again neglecting higher powers of ϵ, we further let

$$x_k(t + \epsilon) = x_k(t) + \epsilon x_k'(t), \tag{3.208}$$

where $x_k'(t)$ is the derivative of x_k with respect to time. Let us set

$$S_t = \partial S/\partial t, \qquad S_{x_k} = \partial S/\partial x_k ; \tag{3.209}$$

then from (3.207)–(3.209) it follows that

$$S[\bar{x}(t + \epsilon), t + \epsilon] = S[\bar{x}(t), t] + \epsilon S_t[\bar{x}(t), t] + \epsilon \sum_{k=1}^{n} x_k'(t) S_{x_k}[\bar{x}(t), t]. \tag{3.210}$$

Substituting (3.210) into (3.205), we find

$$S[\bar{x}(t), t] = \min_{u(t)\in\Omega(u)} \{\epsilon\{\lambda(t) f_x[\bar{X}(t) - \bar{x}(t)] + f_u[\bar{U}(t) - \bar{u}(t)]\} + S[\bar{x}(t), t] + \epsilon S_t[\bar{x}(t), t] + \epsilon \sum_{k=1}^{n} x_k'(t) S_{x_k}[\bar{x}(t), t]\}. \tag{3.211}$$

Canceling $S[\bar{x}(t), t]$ on both sides of the equality and then reducing the remaining equation by ϵ, we will arrive at the necessary condition for optimal control in the form of an equality [compare (2.198)]:

$$\min_{u(t)\in\Omega(u)} \Big(\lambda(t) f_x[\bar{X}(t) - \bar{x}(t)] + f_u[\bar{U}(t) - \bar{u}(t)] + S_t[\bar{x}(t), t] + \sum_{k=1}^{n} x_k'(t) S_{x_k}[\bar{x}(t), t]\Big) = 0. \tag{3.212}$$

From physical considerations it is clear that the function $\lambda(\sigma) \geqslant 0$, and the functions f_x and f_u must be strictly convex, for example $f_x(v) = f_u(v) = v^2$.

Below we consider the simplest example as an illustration. The object is characterized by only one coordinate x, and the impulsive response $g(t, \tau)$ has the form

$$g(t, \tau) = g_1 \exp[-(t - \tau)/T_1]. \tag{3.213}$$

Thus an inertial element appears as the object, with amplification coefficient $T_1 g_1$, time constant T_1, and equation of motion

$$x'(t) + (1/T_1)x(t) = g_1 u(t). \tag{3.214}$$

Further, let $f_x(v) = f_u(v) = v^2$. Then condition (3.212) takes the form

$$\min_{u(t)\in\Omega(u)} \{\lambda(t)[X(t) - x(t)]^2 + [U(t) - u(t)]^2 + S_t[x(t), t] + x'(t)S_x[x(t), t]\} = 0. \tag{3.215}$$

If the value of $x'(t)$ from (3.214) is substituted into (3.215), then we obtain the equation

$$\min_{u(t)\in\Omega(u)} \{\lambda(t)[X(t) - x(t)]^2 + [U(t) - u(t)]^2 + S_t[x(t), t] + [g_1 u(t) - (1/T_1)x(t)]S_x[x(t), t]\} = 0. \tag{3.216}$$

We will first assume that the constraints imposed on the control $u(t)$ are absent and a variation in $u(t)$ is permissible over the limits $-\infty < u(t) < \infty$. Then a minimum of the left side of expression (3.216) can be found by differentiating it with respect to u and equating the derivative to zero:

$$-2[U(t) - u(t)] + g_1 S_x[x(t), t] = 0. \tag{3.217}$$

Hence we find the optimal control

$$u^*(t) = U(t) - (g_1/2)S_x[x(t), t]. \tag{3.218}$$

Thus the optimal control $u^*(t)$ will be found if the function $S[x(t), t]$ is known. Substituting the value of (3.218) found into (3.216), we obtain the equation

$$\begin{aligned} &\lambda(t)[X(t) - x(t)]^2 + S_t[x(t), t] \\ &\quad + \{g_1 U(t) - (g_1^2/2)S_x[x(t), t] - (1/T_1)x(t)\}S_x[x(t), t] \\ &\quad - \{(g_1/2)S_x[x(t), t]\}^2 = 0, \end{aligned} \tag{3.219}$$

a partial differential equation. The solution of Eq. (3.219) must be found satisfying the boundary condition (3.203).

The highest power of $x(t)$ encountered in Eq. (3.219) is 2, since in the first term on the left side of the equation there is the factor $\lambda(t)x^2(t)$. This suggests that with the expansion of the function $S[x(t), t]$ in a series in x, it also suffices to restrict ourselves to the second power. Therefore we will set

$$S[x(t), t] = K(t) + K_1(t)x(t) + K_{11}(t)[x(t)]^2, \tag{3.220}$$

where $K(t)$, $K_1(t)$ and $K_{11}(t)$ are certain functions of time t which it is required to determine by the substitution of the assumed solution (3.220) into Eq. (3.219). As a preliminary we will find $S_t[x(t), t]$ and $S_x[x(t), t]$ by differentiating (3.220) with respect to t and x respectively. We find

$$S_t[x(t), t] = K'(t) + K_1'(t)x(t) + K_{11}'(t)[x(t)]^2. \tag{3.221}$$

Here the prime denotes the derivative of the corresponding function with respect to t. Further,

$$S_x[x(t), t] = K_1(t) + 2K_{11}(t)x(t). \tag{3.222}$$

Now we substitute (3.220), (3.221), and (3.222) into (3.219) and group terms with zero, first, and second powers of $x(t)$:

$$\begin{aligned} &\{K'(t) + g_1 K_1(t)U(t) - [(g_1/2)K_1(t)]^2 + \lambda(t)[X(t)]^2\} \\ &\quad + x(t)\{K_1'(t) + 2g_1 K_{11}(t)U(t) - (1/T_1)K_1(t) - (g_1^2/2)K_1(t)K_{11}(t) - 2\lambda(t)X(t)\} \\ &\quad + [x(t)]^2\{K_{11}'(t) - (2/T_1)K_{11}(t) - [g_1 K_{11}(t)]^2 + \lambda(t)\} = 0. \end{aligned} \tag{3.223}$$

Since Eq. (3.223) must be valid for all values of $x(t)$, the coefficients for zero, first, and second powers of $x(t)$ must be equal to zero. Hence three ordinary differential equations are obtained for the functions $K(t)$, $K_1(t)$, and $K_{11}(t)$:

$$K'(t) = [(g_1/2)K_1(t)]^2 - g_1K_1(t)U(t) - \lambda(t)[X(t)]^2, \qquad (3.224)$$

$$K_1'(t) = [(1/T_1) + (g_1^2/2)K_{11}(t)]K_1(t) - 2g_1K_{11}(t)U(t) + 2\lambda(t)X(t) \quad (3.225)$$

and

$$K_{11}'(t) = (2/T_1)K_{11}(t) + [g_1K_{11}(t)]^2 - \lambda(t). \qquad (3.226)$$

To solve these differential equations, the initial or boundary values of the functions $K(t)$, $K_1(t)$ and $K_{11}(t)$ must still be known. By comparing conditions (3.203) with Eq. (3.220), it is seen that (3.203) is satisfied for any finite $x(t + T)$ only if the coefficients $K_1(t + T)$ and $K_{11}(t + T)$ are set equal to zero:

$$K_1(t + T) = K_{11}(t + T) = 0. \qquad (3.227)$$

But then in view of the validity of (3.203) the equality

$$K(t + T) = 0 \qquad (3.228)$$

comes true. Thus the solutions of Eqs. (3.224)–(3.226) must satisfy the three boundary conditions (3.227) and (3.228).

When the functions $K(t)$, $K_1(t)$, and $K_{11}(t)$ have been found, then from (3.218) and (3.222) it is not difficult to find the optimal control law $u^*(t)$. Substituting S_x from (3.222) into (3.218), we obtain:

$$u^*(t) = [U(t) - (g_1/2)K_1(t)] - g_1K_{11}(t)x(t). \qquad (3.229)$$

Thus $u^*(t)$ depends on time t not only explicitly in terms of the functions $K_1(t)$, $K_{11}(t)$ and $U(t)$, but also implicitly in terms of $x(t)$.

In order to solve Eqs. (3.224)–(3.226) with the boundary conditions (3.227) and (3.228), the reading of time can be transformed by regarding the instant $t + T$ as a new beginning. Then conditions (3.227) and (3.228) will turn out to be "initial" ones, and the solution of the ordinary Cauchy problem with prescribed initial conditions will be required. The solution of these equations in closed form is impossible since they are nonlinear. But the solution can be obtained by the usual means with the aid of computers.

Figure 3.23 depicts a block diagram of the controller constructed according to Eq. (3.229). In this diagram the functions $K_1(t)$ and $K_{11}(t)$

are shown passing out of computers which solve Eqs. (3.224)–(3.226). The function $U(t)$ is also supplied from the outside. The function $K_1(t)$, passing through the amplifier element with amplification coefficient $-g_1/2$, gives the term $-g_1K_1(t)/2$. This term, when added in the summing element Σ to $U(t)$, gives the first term of the right side of (3.229) contained in the brackets. The second term of expression (3.229) enters from the output of the multiplier element ME. The output variable of the latter is equal to the product of input variables $K_{11}(t)$ and $-x(t)$ multiplied by the constant coefficient g_1. The factor $-x(t)$ is obtained at the output of the inverter -1 (an amplifier with an amplification coefficient equal to minus one); from the output of the controlled object B, $x(t)$ acts on the input to the inverter.

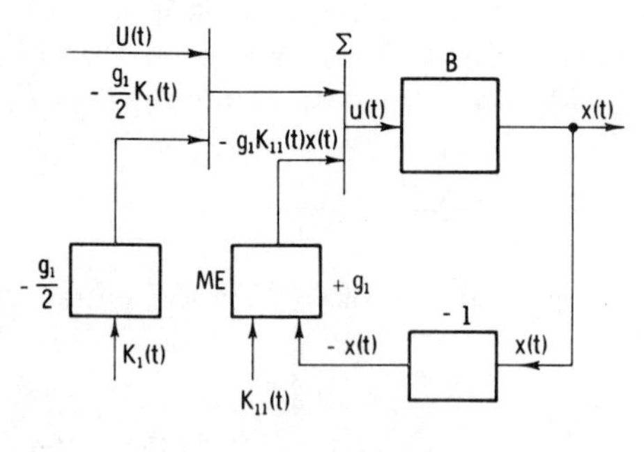

FIG. 3.23

Now let us consider the case when the constraint

$$L_m^-(t) \leqslant u(t) \leqslant L_m^+(t)$$

is imposed on $u(t)$. Now the value u giving a minimum of the left side of (3.216) cannot be found from Eq. (3.219), since the latter may give a value of u^* falling outside the limits of the permitted interval for $u(t)$. Clearly the following cases are possible.

(a) The solution $u_0(t)$ of Eq. (3.229), i.e., the expression

$$u_0(t) = [U(t) - (g_1/2)K_1(t)] - g_1K_{11}(t)x(t), \tag{3.230}$$

is within the required limits. Then nothing changes in the preceding expressions.

(b) The solution (3.230) is greater than the upper bound for $u(t)$:

$$u_0(t) > L_m^+(t). \tag{3.231}$$

In this case the value of $u^*(t)$ giving a minimum of the left side of (3.216) will be

$$u^*(t) = L_m{}^+(t). \tag{3.232}$$

(c) The solution (3.230) is less than the lower bound:

$$u_0(t) < L_m{}^-(t). \tag{3.233}$$

Then the left side of (3.216) attains a minimum for the admissible value

$$u^*(t) = L_m{}^-(t). \tag{3.234}$$

Combining all the cases, we obtain the control law:

$$u^*(t) = \begin{cases} L_m{}^+(t), & u_0(t) \geqslant L_m{}^+(t), \\ u_0(t), & L_m{}^-(t) \leqslant u_0(t) \leqslant L_m{}^+(t), \\ L_m{}^-(t), & u_0 \leqslant L_m{}^-(t). \end{cases} \tag{3.235}$$

However, it is important to note that in this case, the functions $K(t)$, $K_1(t)$, and $K_{11}(t)$ now cannot be found from the Eqs. (3.224)–(3.226). In fact, Eq. (3.219), from which these functions are found, does not hold for those time instants when $u_0(t)$ falls outside the limits of the permitted range. For example, at some time instant let condition (3.231) be satisfied, and hence Eq. (3.232) be realized. Then a partial differential equation for S may be obtained by substituting the expression $L_m{}^+(t)$ into the left side of (3.216) in place of $u(t)$. In this case we obtain the equation

$$\lambda(t)[X(t) - x(t)]^2 + [U(t) - L_m{}^+(t)]^2 + S_t[x(t), t] + [g_1 L_m{}^+(t) - (1/T_1)x(t)] \cdot S_x[x(t), t] = 0. \tag{3.236}$$

Substituting formulas (3.221) and (3.222) into this equation, we will find

$$\begin{aligned} &\{K'(t) + g_1 L_m{}^+(t) K_1(t) + [U(t) - L_m{}^+(t)]^2 + \lambda(t)[X(t)]^2\} \\ &\quad + x(t)[K_1'(t) + 2g_1 L_m{}^+(t) K_{11}(t) - (1/T_1)K_1(t) - 2\lambda(t)X(t)] \\ &\quad + [x(t)]^2[K_{11}'(t) + (2/T_1)K_{11}(t) + \lambda(t)] = 0. \end{aligned} \tag{3.237}$$

This equation will be satisfied for arbitrary $x(t)$, if the coefficients for the zero, first, and second powers of $x(t)$ will be zero. Hence the three equations for the functions $K(t)$, $K_1(t)$, and $K_{11}(t)$ follow:

$$\begin{aligned} K'(t) &= -g_1 L_m{}^+(t) K_1(t) - \{U(t) - L_m{}^+(t)\} - \lambda(t)[X(t)]^2, \\ K_1'(t) &= (1/T_1)K_1(t) - 2g_1 L_m{}^+(t) K_{11}(t) + 2\lambda(t)X(t), \\ K_{11}'(t) &= (2/T_1)K_{11}(t) - \lambda(t). \end{aligned} \tag{3.238}$$

Analogous equations, but with $L_m^+(t)$ replaced by $L_m^-(t)$, are obtained in the case when inequality (3.233) is satisfied and the optimal control is obtained from formula (3.234).

Thus the functions $K(t)$, $K_1(t)$ and $K_{11}(t)$ are obtained from a system of equations formed by a generalization of all three systems of equations, each of which holds in one of the three possible cases (a), (b), (c) enumerated above. Of course, instead of these systems of equations one common system can be written out, whose coefficients will depend on whether $u_0(t)$ is inside the admissible range or falls outside it on one side or the other. These equations are integrated under the same boundary conditions (3.227), (3.228). But now for integrating these equations at a rapid rate with time being read backward, an addition to the model by which Eq. (3.214) of the object is integrated is necessary, also with time being read backward, where the current value of $x(t)$ will now be "final." The integration of this equation is necessary, since as is seen from formula (3.230) the values of $u_0(\sigma)$ for $t < \sigma < t + T$ depend on $x(\sigma)$. Therefore the determination of the value of the optimal control $u^*(t)$ at the given current time instant requires a prior rapid calculation of the values of the functions $K(t)$, $K_1(t)$, $K_{11}(t)$ realized by the model, also "predicting" the future values of both these functions and $u^*(\sigma)$, $x(\sigma)$. When the values of these functions have been found, then the control is realized from formula (3.235), and the value of $u_0(t)$ is obtained from (3.230).

In general, for any linear object and quadratic optimality criterion, even for a control $u(t)$ restricted to an admissible interval, there exists a formula for an exact optimal control $u^*(t)$ consisting of a finite number of terms. Other optimality criteria will lead to optimal control laws which can only be roughly approximated by a finite number of terms.

The method examined above can be generalized to higher-order systems [3.34–3.36].

The method of dynamic programming is a very general method of formulating and preparing very diverse problems for an approximate solution. For example, for an object whose equation has the form

$$dx/dt = f(x, u), \tag{3.239}$$

with an initial condition

$$(x)_{t=0} = x^{(0)}, \tag{3.240}$$

the problem can be posed of finding the optimal control $u^* \in \Omega(u)$ minimizing the maximum deviation of $x(t)$ from some known function

$x^*(t)$ in the interval $0 \leqslant t \leqslant T$, where T is fixed. In this case the optimality criterion has the form

$$Q = \max_{0 \leqslant t \leqslant T} | x^*(t) - x(t) |. \tag{3.241}$$

It is required to choose $u(t)$ on this interval such that the condition

$$\min Q = \min_{u(t) \in \Omega(u)} \max_{0 \leqslant t \leqslant T} | x^*(t) - x(t) | = S(x^{(0)}) \tag{3.242}$$

is ensured. The minimal value of Q, depending on the initial condition $x^{(0)}$, is denoted by the letter S. The control $u(t)$ belongs to some admissible region $\Omega(u)$.

To solve the problem (see [2.9]) we first make it discrete by approximately replacing the differential equation (3.239) by the finite difference equation

$$x_{k+1} = x_k + f(x_k, u_k) \Delta, \tag{3.243}$$

where

$$x_k = x(k\Delta), \qquad u_k = u(k\Delta), \qquad \Delta = T/N, \tag{3.244}$$

and N is the total number of elementary discrete intervals into which the time segment T is broken up. Now instead of (3.241) we can write:

$$Q = \max_{k=0,1,\ldots,N} | x_k^* - x_k |, \tag{3.245}$$

and expression (3.242) is replaced by the relation

$$S(x^{(0)}) = \min Q = \min_{\substack{u_j \in \Omega(u), k=0,\ldots,N \\ j=0,1,\ldots,N-1}} \max | x_k^* - x_k |. \tag{3.246}$$

First let us examine only the single instant $t = N\Delta$. For this instant when $x = x^{(0)}$ the quantity

$$S_N(x_N) = \min Q_N = | x_N^* - x_N |, \tag{3.247}$$

and the variation in the control u generally does not affect the quantity Q_N. Moving backward from the instant $t = N\Delta$ to $t = (N-1)\Delta$, let us consider the function

$$\begin{aligned} Q_{N-1} &= \max_{k=N-1,N} | x_k^* - x_k | = \max\{| x_{N-1}^* - x_{N-1} |, | x_N^* - x_N |\} \\ &= \max\{|x_{N-1}^* - x_{N-1} |, | x_N^* - x_{N-1} - f(x_{N-1}, u_{N-1}) \Delta |\}. \end{aligned} \tag{3.248}$$

Here x_N is replaced according to (3.243). The notation $\max\{a, b\}$ means: the maximal of the two quantities a and b. For a given value of x_{N-1} the quantity Q_{N-1} depends on u_{N-1}. Let us choose a value $u_{N-1} \in \Omega(u)$ such that Q_{N-1} is minimized. Then we will obtain the optimal control u_{N-1}^* depending on x_{N-1}, x_{N-1}^*, x_N^*, and the minimal value of Q_{N-1}:

$$S_{N-1}(x_{N-1}) = \min_{u_{N-1} \in \Omega(u)} Q_{N-1} . \tag{3.249}$$

Here it can turn out that a variation in u_{N-1} over a certain range does not affect (3.248) at all, whenever the first of the quantities inside the braces of (3.248) prevails over the second. Hence it follows that the optimal control is determined ambiguously. We will always agree to determine the value of u minimizing the second term in the brackets of (3.248).

Let us now pass to the time instant $t = (N - 2)\,\Delta$. Considering the interval from $t = (N - 2)\,\Delta$ to $t = N\,\Delta$, we obtain

$$\begin{aligned} Q_{N-2} &= \max_{k=N-2, N-1, N} | x_k^* - x_k | \\ &= \max\{| x_{N-2}^* - x_{N-2} |, \max\{| x_{N-1}^* - x_{N-1} |, | x_N^* - x_N |\}\}. \end{aligned} \tag{3.250}$$

The control u_{N-1} affects only the second term inside the common braces in this expression. If it is always selected according to the law (3.249) found earlier, then

$$\begin{aligned} S_{N-2}(x_{N-2}) &= \min_{u_{N-1}, u_{N-2} \in \Omega(u)} Q_{N-2} \\ &= \min_{u_{N-2} \in \Omega(u)} \max \{| x_{N-2}^* - x_{N-2} |, S_{N-1}(x_{N-1})\} \\ &= \min_{u_{N-2} \in \Omega(u)} \max \{| x_{N-2}^* - x_{N-2} |, S_{N-1}(x_{N-2} + f(x_{N-2}, u_{N-2})\,\Delta)\} \\ &= \max\{| x_{N-2}^* - x_{N-2} |, \min_{u_{N-2} \in \Omega(u)} S_{N-1}(x_{N-2} + f(x_{N-2}, u_{N-2})\,\Delta)\}. \end{aligned} \tag{3.251}$$

Carrying out the minimization operation on the right side of (3.251), we will find the optimal control u_{N-2}, and so on. An iteration of this argument leads to a recursive sequence for calculating the functions $S_{N-k}(x_{N-k})$, where in passing the optimal controls u_{N-k}^* are also computed:

$$\begin{aligned} &S_{N-k}(x_{N-k}) \\ &\quad = \max\{| x_{N-k}^* - x_{N-k} |, \min_{u_{N-k} \in \Omega(u)} S_{N-k+1}(x_{N-k} + f(x_{N-k}, u_{N-k})\,\Delta)\}. \end{aligned} \tag{3.252}$$

Here $k = 1, 2, ..., N$. Since the function $S_N(x_N)$ is known [see (3.247)], then in principle the entire set of the S_{N-k} up to $S_0(x_0) = S_0(x^{(0)})$ can be found. The corresponding control u_0 is also sought.

Carrying out this procedure is only possible in numerical form; it does not require especially cumbersome calculations.

With the aid of dynamic programming the optimal control for a variable time T of the process can also be found, including the control which is optimal in speed of response. For example, let the equation of the object have the form

$$a_1 x + \sum_{i=1}^{n} a_{i+1} \frac{d^i x}{dt^i} = u(t). \tag{3.253}$$

Let us set

$$x = x_1, \qquad d^i x/dt^i = x_{i+1} \qquad (i = 1, 2, ..., n-1). \tag{3.254}$$

Then Eq. (3.253) can be replaced by a collection of n first-order equations:

$$\begin{aligned} \frac{dx_i}{dt} &= x_{i+1} \qquad (i = 1, 2, ..., n-1), \\ \frac{dx_n}{dt} &= -\sum_{i=1}^{n} \frac{a_i}{a_{n+1}} x_i + \frac{u}{a_{n+1}} = -\sum_{i=1}^{n} b_i x_i + v, \end{aligned} \tag{3.255}$$

where

$$b_i = a_i/a_{n+1}, \qquad u/a_{n+1} = v. \tag{3.256}$$

Let v be bounded in absolute value:

$$| v | \leqslant V. \tag{3.257}$$

We will find the optimal control, for which the time T is minimal for moving the representative point from the initial values $x_1^{(0)}, ..., x_n^{(0)}$, which correspond to a vector $\bar{x}^{(0)}$, to the origin in phase space.

Let us replace Eqs. (3.255) by finite difference equations, after setting $T = N\,\Delta t$, $x_i^{(k)} = x_i(k\,\Delta t)$, $v^{(k)} = v(k\,\Delta t)$:

$$\begin{aligned} x_i^{(k+1)} &= x_i^{(k)} + \Delta t \cdot x_{i+1}^{(k)} \qquad (i = 1, 2, ..., n-1), \\ x_n^{(k+1)} &= x_n^{(k)} - \Delta t \sum_{i=1}^{n} b_i x_i^{(k)} + \Delta t \cdot v^{(k)}. \end{aligned} \tag{3.258}$$

We will find the relation for the minimal time for being found inside a sphere of small radius δ for $k = N$:

$$\sum_{i=1}^{n} [x_i^{(N)}]^2 \leqslant \delta^2. \tag{3.259}$$

The minimal time T of occurring in this sphere depends only on the initial conditions, i.e., on the vector $\bar{x}^{(0)}$. Therefore $T = T(\bar{x}^{(0)})$. The transfer from the initial position in one step takes up a time Δt, after which the vector $\bar{x}^{(0)}$ will be replaced by the vector $\bar{x}^{(1)}$ depending on $v^{(0)}$. The minimal time of occurring in a small sphere from the point $\bar{x}^{(1)}$ equals $T(\bar{x}^{(1)})$, and the total time is $\Delta t + T(\bar{x}^{(1)})$. Depending on what the first step is, the total time may be different. It is evident that

$$T(\bar{x}^{(0)}) = \min_{|v^{(0)}| \leqslant V} [\Delta t + T(\bar{x}^{(1)})] = \Delta t + \min_{|v^{(0)}| \leqslant V} [T(\bar{x}^{(1)})]. \tag{3.260}$$

If the expression for $\bar{x}^{(1)}$ is expanded according to (3.258), then we will obtain:

$$T(x_1^{(0)}, \ldots, x_n^{(0)}) = \Delta t + \min_{|v^{(0)}| \leqslant V} T\left(x_1^{(0)} + \Delta t \cdot x_2^{(0)}, \ldots, x_{n-1}^{(0)} + \Delta t \cdot x_n^{(0)}, x_n^{(0)} - \Delta t \sum_{i=1}^{n} b_i x_i^{(0)} + \Delta t \cdot v^{(0)}\right). \tag{3.261}$$

We will assume that T is a differentiable function of the variables $x_i^{(0)}$. Then

$$T(x_1^{(0)}, \ldots, x_n^{(0)}) = \Delta t + \min_{|v^{(0)}| \leqslant V} \left\{ T(x_1^{(0)}, \ldots, x_n^{(0)}) + \sum_{i=1}^{n-1} \frac{\partial T}{\partial x_i^{(0)}} \Delta t \cdot x_{i+1}^{(0)} + \frac{\partial T}{\partial x_n^{(0)}} \left(-\Delta t \sum_{i=1}^{n} b_i x_i^{(0)} + \Delta t \cdot v^{(0)}\right) \right\} + o(\Delta t). \tag{3.262}$$

Only the last term in the braces depends on $v^{(0)}$. Hence the minimization with respect to $v^{(0)}$ only has a bearing on it. We will suppress the term $T(x_1^{(0)}, \ldots, x_n^{(0)})$ on both sides of Eq. (3.262) and then, after dividing by Δt and letting Δt tend to zero, we will arrive at the expression

$$0 = 1 + \sum_{i=1}^{n-1} \frac{\partial T}{\partial x_i^{(0)}} x_{i+1}^{(0)} + \min_{|v^{(0)}| \leqslant V} \left\{ \frac{\partial T}{\partial x_n^{(0)}} \left(- \sum_{i=1}^{n} b_i x_i^{(0)} + v^{(0)}\right) \right\}. \tag{3.263}$$

As $\Delta t \to 0$, the value of δ entering into (3.259) must also tend to zero. After replacing the values $x_i^{(0)}$ by the current values x_i which we can always take on the right for initial ones, we will obtain a partial differential equation for determining $T(x_1, x_2, ..., x_n)$:

$$0 = 1 + \sum_{i=1}^{n-1} \frac{\partial T}{\partial x_i} x_{i+1} - \frac{\partial T}{\partial x_n} \sum_{i=1}^{n} b_i x_i + \min_{|v| \leqslant V} \left\{ \frac{\partial T}{\partial x_n} \cdot v \right\}. \tag{3.264}$$

By associating (3.257) and (3.264), it is not difficult to conclude that the minimum of the last term on the right side of (3.264) is obtained from the observance of the condition

$$v^* = -V \operatorname{sign} \partial T / \partial x_n . \tag{3.265}$$

This is also the optimal control law. From this it follows that the values of $v(t)$ must always be selected on the boundaries of the admissible region:

$$v^* = \pm V. \tag{3.266}$$

This was already shown in Section 1 of this chapter.

Substituting (3.265) into (3.264), we will obtain the equation

$$0 = 1 + \sum_{i=1}^{n-1} \frac{\partial T}{\partial x_i} x_{i+1} - \frac{\partial T}{\partial x_n} \sum_{i=1}^{n} b_i x_i - V \left| \frac{\partial T}{\partial x_n} \right|. \tag{3.267}$$

The solution of Eq. (3.267) must satisfy the boundary condition

$$T(0, 0, ..., 0) = 0, \tag{3.268}$$

since for an initial value $\bar{x}^{(0)} = 0$, the time required for occurring at the origin is obviously equal to zero.

In order to obtain the optimal control v^* as a function of x, as is seen from (3.265), the function $T(\bar{x})$ must be found; i.e., the partial differential equation (3.267) must be solved.

A solution in explicit form can only be obtained for the simplest cases. As an example let us consider the equations

$$dx_1/dt = x_2 , \qquad dx_2/dt = u \tag{3.269}$$

for the object, with the constraint

$$| u | \leqslant 1. \tag{3.270}$$

It is required to pass in minimal time from the point $(x_1^{(0)}, x_2^{(0)})$ to the origin.

Then with the replacement of v by u, Eq. (3.264) takes the form

$$0 = 1 + \frac{\partial T}{\partial x_1} x_2 + \min_{|u| \leq 1} \left\{ \frac{\partial T}{\partial x_2} u \right\}, \tag{3.271}$$

and the optimal control law (3.265) can be written in the following way:

$$u^* = -\text{sign } \partial T/\partial x_2 . \tag{3.272}$$

According to (3.266) and (3.270) the optimal control assumes only the values ± 1. Therefore the (x_1, x_2)-phase plane can be decomposed into two regions (Fig. 3.13): the region L_2' in which $u = -1$, and the region L_2'' in which $u = +1$. For the first of these regions we obtain Eq. (3.271), in which -1 stands in place of u:

$$0 = 1 + \frac{\partial T}{\partial x_1} x_2 - \frac{\partial T}{\partial x_2} . \tag{3.273}$$

For the region L_2'', Eq. (3.271) takes the form

$$0 = 1 + \frac{\partial T}{\partial x_1} x_2 + \frac{\partial T}{\partial x_2} . \tag{3.274}$$

To solve these equations the boundary conditions on some curve must be known. For example, Eqs. (3.269) can be integrated for a variation in $u(t)$ along the "rectangular" curve $u = \pm 1$, and thus $T(x_1)$ can be obtained for $x_2 = 0$.

It can be shown [3.37] that the solution of Eq. (3.273) has the form

$$T = 2\sqrt{(\tfrac{1}{2}x_2^2 + x_1)} + x_2 , \tag{3.275}$$

while the solution for Eq. (3.274) is

$$T = 2\sqrt{(\tfrac{1}{2}x_2^2 - x_1)} - x_2 . \tag{3.276}$$

With the substitution of (3.272) into (3.271) we obtain the equation

$$0 = 1 + \frac{\partial T}{\partial x_1} x_2 - \left| \frac{\partial T}{\partial x_2} \right| . \tag{3.277}$$

Neither of formulas (3.275) and (3.276) separately is the solution of Eq. (3.277). This solution can be written in the form

$$T = \begin{cases} 2\sqrt{(\tfrac{1}{2}x_2^2 + x_1)} + x_2 , & \bar{x} \in L_2', \\ 2\sqrt{(\tfrac{1}{2}x_2^2 - x_1)} - x_2 , & \bar{x} \in L_2'' . \end{cases} \tag{3.278}$$

Let us find the region L_2'' within whose limits formula (3.276) holds. In order to do this we will substitute the partial derivatives from the right side of (3.276), i.e.,

$$\frac{\partial T}{\partial x_1} = \frac{-1}{\sqrt{(\frac{1}{2}x_2^2 - x_1)}}, \qquad \frac{\partial T}{\partial x_2} = \frac{x_2}{\sqrt{(\frac{1}{2}x_2^2 - x_1)}} - 1, \tag{3.279}$$

into formula (3.277); after that this expression will take the form

$$\left(1 - \frac{x_2}{\sqrt{(\frac{1}{2}x_2^2 - x_1)}}\right) - \left| \frac{x_2}{\sqrt{(\frac{1}{2}x_2^2 - x_1)}} - 1 \right| = 0. \tag{3.280}$$

This equality holds under the condition

$$1 - \frac{x_2}{\sqrt{(\frac{1}{2}x_2^2 - x_1)}} \geqslant 0 \tag{3.281}$$

or

$$\sqrt{(\tfrac{1}{2}x_2^2 - x_1)} \geqslant x_2 . \tag{3.282}$$

The latter condition holds for

$$x_2 > 0, \qquad x_1 + \tfrac{1}{2}x_2^2 \leqslant 0, \tag{3.283}$$

or for

$$x_2 < 0, \qquad \tfrac{1}{2}x_2^2 - x_1 \geqslant 0, \tag{3.284}$$

since the expression under the radical sign cannot be negative. From conditions (3.283) and (3.284) it follows that the boundary of the region L_2'' is characterized by the equations

$$\begin{aligned} x_2 > 0, &\qquad x_1 = -\tfrac{1}{2}x_2^2, \\ x_2 < 0, &\qquad x_1 = \tfrac{1}{2}x_2^2, \end{aligned} \tag{3.285}$$

or in more compact form, by the equality

$$x_1 = -\mathrm{sign}(x_2)\, \tfrac{1}{2}x_2^2. \tag{3.286}$$

This equation coincides with (3.69) since in this example $M/a_0 = 1$.

The same boundary can be found for L_2' as well if formula (3.275) is used.

From (3.279) and (3.272) it is seen that in the region L_2' the quantity $u = -1$. Since the switching curve is given by Eq. (3.286), by comparing

(3.283), (3.284), and (3.286) we see that the optimal switching law has the form

$$u^* = -\text{sign}[x_1 + \text{sign}(x_2) \tfrac{1}{2} x_2^2]. \tag{3.287}$$

Of course for this example the phase plane method is much simpler than the method of dynamic programming. But the latter is by far more general; as is shown in the following chapters, it is also applicable to the investigation of statistical problems.

The method of dynamic programming can be applied not only to systems with time quantization (discrete-continuous systems), in which arbitrary levels of variables are allowed, but also to purely discrete systems in which there is not only time quantization but level quantization as well [3.38]. In such systems the values of variables cannot be arbitrary; they must belong to a finite set of permitted levels.

Usually we assume that the system can be in the state s_i at the ith time instant. There exists a finite set of possible states of the system, denoted by $q_1, ..., q_m$ or simply 1, ..., m. The state s_i can be one of the numbers $q_1, ..., q_m$, which as was indicated above can be denoted by one of the numbers 1, ..., m. The external action u_i likewise can assume one of several possible values. For simplicity we assume that there are only two possible values of u_i, and just 1 and 0. The law of change in the states is given by the dependence

$$s_{i+1} = f(s_i, u_i). \tag{3.288}$$

Frequently we also assume that the output variable x_i is a discrete function of s_i and u_i:

$$x_i = \psi(s_i, u_i). \tag{3.289}$$

Below as a simplification the output variable x_i is identified with s_i, i.e., the dependence (3.289) has the form

$$x_i = s_i .$$

Then the discrete system is characterized by only one function of the two variables x_i and u_i:

$$x_{i+1} = f(x_i, u_i), \tag{3.290}$$

which we obtain by substituting $s_i = x_i$ into expression (3.288).

The dependence (3.290) is often depicted in the form of a table or diagram (see, for example, [3.25, Chapter XV]). For example, let the system be capable of being in the four possible states denoted by circles in Fig. 3.24 (the number of the states has been marked by figures inside the circles). The passage from one state to another is denoted in the figure by a curve with an arrow drawn from the state s_i to the state s_{i+1}. Since this passage depends on the value $u_i = 0, 1$, then next to the curve it must be marked for which of the two possible values of u_i this transition occurs. For example, from Fig. 3.24 it is seen that when $s_i = 2$ and $u_i = 1$, then the transition takes place to the state $s_{i+1} = 3$; but if $s_i = 2$ while $u_i = 0$, then the arrow will lead from $s_i = 2$ to the state $s_{i+1} = 1$.

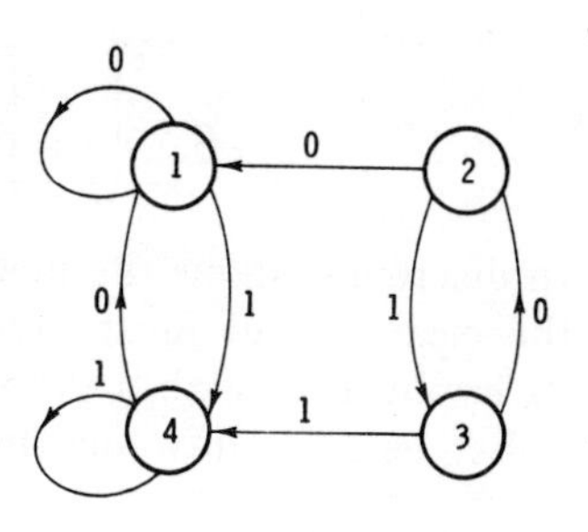

FIG. 3.24

Cases are possible when the state of the system does not change. For example, if $s_i = 4$ and $u_i = 1$, then the state remains the previous one, since the arrow leads to the same state $s_{i+1} = 4$. A like occurrence is observed for $s_i = 1$ and $u_i = 0$.

The same dependence can be represented in the form of a table with two entries u_i and s_i, where the corresponding value of s_{i+1} has been written in each square of the table.

The table corresponding to the diagram of Fig. 3.24 has the following form:

	s_i			
u_i	1	2	3	4
0	1	1	2	1
1	4	3	4	4

For example, for $s_i = 3$ and $u_i = 0$ we obtain $s_{i+1} = 2$.

At the initial time instant $i = 0$ let the system be in some initial state s_0. Further, we will assume that the process is examined at the discrete time instants $i = 0, 1, ..., N$, where N is fixed. It is required to find a sequence of control actions $u_0, u_1, ..., u_N$ such that a certain prescribed function $\varphi(s_N)$ of the final state s_N is maximized. This is a typical problem for optimal system theory, in which the optimality criterion is

$$Q = \varphi(s_N). \tag{3.291}$$

Evidently, it is required to prescribe the function $\varphi(s_N)$ in the form of a table. For example, let the function φ be characterized for the example under consideration by the following table:

s_N	1	2	3	4
$\varphi(s_N)$	-1	3	-2	2

In discrete systems the method of dynamic programming will appear in the clearest and most "revealing" form. It is convenient to associate its application with a graphical construction (Fig. 3.25).

Let us first draw for the instant N circles one under the other, representing the four possible states of s_N. They are depicted in the right-hand column in Fig. 3.25. The second column from the right represents the four possible states of s_{N-1} at the time instant $N - 1$; the next column situated farther left represents the states of s_{N-2} at the time instant $N - 2$ and so on. Only three columns are depicted in Fig. 3.25; in the general case $N + 1$ columns should be depicted corresponding to the states of $s_N, s_{N-1}, ..., s_0$.

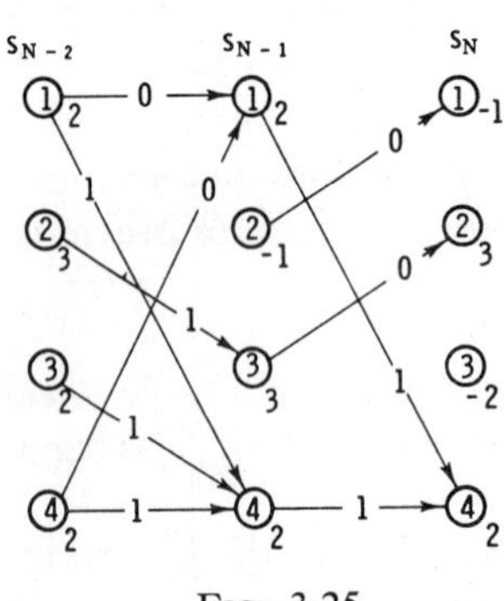

FIG. 3.25

We will start with the extreme right-hand column. Next to the states 1, 2, 3, 4 the corresponding values $\varphi(1)$, $\varphi(2)$, $\varphi(3)$, $\varphi(4)$ taken from the table for $\varphi(s_N)$ have been written. We will perform the "retrograde" motion from s_N to s_{N-1}. For example, let the state of the system be $s_{N-1} = 1$. Using the table or Fig. 3.24, we will find a value of u_{N-1} such that, starting from s_{N-1}, the largest possible value of s_N will be obtained. For $u_{N-1} = 1$ we obtain $s_N = 4$, which corresponds to the value of $\varphi(s_N)$ equal to 2. Meanwhile if we set $u_{N-1} = 0$, then we will obtain $s_N = 1$ and $\varphi(s_N) = -1$. Hence it follows that the optimal value of u_{N-1} turns out to be 1, where in this case the transition from $s_{N-1} = 1$ to $s_N = 4$ is accomplished. We will represent this transition accomplished with the optimal control in the form of a line, going from $s_{N-1} = 1$ to $s_N = 4$. The figure 1 has been plotted on the line corresponding to the optimal control $u_{N-1}^* = 1$.

In precisely the same way we draw the other lines between the states of s_{N-1} and s_N, corresponding to the optimal controls for various "initial" values of s_{N-1}, equal to 1, 2, 3, 4. These lines, with the values of the optimal control actions plotted on them, join each of the possible states of s_{N-1} with some of the states of s_N. By the same token the possibility is provided of obtaining the maximal value of the optimality criterion

$\varphi(s_N)$ corresponding to each of the "initial" values of s_{N-1}. For example, if $s_{N-1} = 4$, then with the aid of the optimal control $u_{N-1}^* = 1$ the transition to the same position $s_N = 4$ is accomplished, as a consequence of which the function $\varphi(s_N)$ assumes the value $\varphi(s_N) = 2$. This value of the criterion—maximal for the given "initial" condition $s_{N-1} = 4$—is written next to the state s_{N-1} in the form of the figure 2. The corresponding figures must be written next to each of the values $s_{N-1} = 1, 2, 3, 4$.

Then we will conduct a completely analogous operation for the retrograde motion from s_{N-1} to s_{N-2}, and obtain in the same figure lines joining each of the possible states $s_{N-2} = 1, 2, 3, 4$ to some of the states of s_{N-1} under the optimal control action u_{N-2}^*. Near the "initial" states of s_{N-2} the maximal attainable values of $\varphi(s_N)$ must be written. Repeating this procedure for s_{N-3}, s_{N-4}, and so on, we will finally come to s_0 and obtain the optimal strategies corresponding to arbitrary possible initial conditions.

For example, for $N = 2$ the value of $s_{N-2} = s_0$. From Fig. 3.25 it is seen that in this case for the initial position $s_0 = 3$ the optimal control $u_{N-2} = u_0 = 1$ must be chosen, which will lead to $s_{N-1} = s_1 = 3$, then $u_{N-1} = u_1 = 0$, from which the system will arrive in the position $s_N = s_2 = 2$ and the criterion $\varphi(s_N)$ will assume the maximal possible value $Q = 3$.

The optimal strategy is not necessarily unique. For example, from the position $s_{N-2} = 1$ it is possible to move in different ways by setting u_{N-2} equal to either 0 or 1. If in the first of these cases $u_{N-1} = 1$ is set, and in the second $u_{N-1} = 1$ as well, then we arrive at the state $s_N = 4$ and obtain the maximal possible value $\varphi(s_N) = 2$. The same case of ambiguity occurs for $s_{N-2} = 4$. Here there exist two optimal strategies with equal values of the criterion Q:

$$u_{N-2} = 1, \quad u_{N-1} = 1 \quad \text{and} \quad u_{N-2} = 0, \quad u_{N-1} = 1.$$

These examples can be generalized to more complicated cases, when, for example, the sequences of control actions $u_0, \ldots, u_N$ may not be arbitrary, but are restricted by additional conditions [3.38].

4. Application of the Maximum Principle

With the aid of the maximum principle it frequently turns out to be possible to clarify the typical features of optimal processes in continuous systems. For example, let us consider the problem of the minimum time of the transient process for an object whose equations have the form

$$dx_i/dt = f_i(x_1, x_2, \ldots, x_n) + b_i u \qquad (i = 1, 2, \ldots, n) \tag{3.292}$$

or in vector form

$$d\bar{x}/dt = \bar{f}(\bar{x}) + \bar{b} \cdot u. \tag{3.293}$$

Here u is the only control action, and $\bar{x}$, $\bar{f}$, and $\bar{b}$ are vectors, where

$$\bar{b} = (b_1, b_2, ..., b_n). \tag{3.294}$$

We regard the functions f_i as differentiable with respect to their arguments.

Let the constraint imposed on the control action have the form

$$|u| \leqslant 1. \tag{3.295}$$

We construct the adjoint equations

$$\frac{d\psi_i}{dt} = -\sum_{\alpha=1}^{n} \frac{\partial f_\alpha}{\partial x_i} \psi_\alpha . \tag{3.296}$$

Further, we form the Hamiltonian

$$H = \langle d\bar{x}/dt, \bar{\psi} \rangle = \langle \bar{f}(\bar{x}) + \bar{b}u, \bar{\psi} \rangle = \langle \bar{f}(\bar{x}), \bar{\psi} \rangle + u\langle \bar{b}, \bar{\psi} \rangle. \tag{3.297}$$

Since only the second term in this expression depends on u, it is evident† that the maximum of H will be obtained, with (3.295) taken into account, for

$$u = \operatorname{sign}\langle \bar{b}, \bar{\psi} \rangle = \operatorname{sign} \sum_{i=1}^{n} b_i \psi_i(t). \tag{3.298}$$

In order to obtain the control law in explicit form, the values of ψ_i must be found for each point $\bar{x}$ of phase space. But the important conclusion can already be made from formula (3.298), that for the class of problems being considered the control action must be on the boundary of the region of possible values (3.295). It is shown below that for other classes of problems this condition cannot be satisfied.

In the important special case of a linear object, Eqs. (3.292) take the form

$$dx_i/dt = \sum_{j=1}^{n} a_{ij} x_j + b_i u \qquad (i = 1, 2, ..., n). \tag{3.299}$$

† Here it is assumed that the case $\langle \bar{b}, \bar{\psi}(t) \rangle = 0$ does not occur, and hence the so-called singular equations are absent (see [2.17, 2.21]).

Let the characteristic equation for the free motion of the object (for $u = 0$) have only negative real roots. In this case the adjoint equations (3.296) can be rewritten in the form

$$d\psi_i/dt = -\sum_{\alpha=1}^{n} a_{\alpha i}\psi_\alpha \,. \tag{3.300}$$

These equations turn out to be adjoint to Eqs. (3.299) of the object. It is known that in the case when the roots of the fundamental equation are real and negative, this property is preserved for the roots of the characteristic equation of the adjoint system (3.300) as well. Hence the solutions $\psi_i(t)$ have the form

$$\psi_i(t) = \sum_{j=1}^{n} C_{ij} e^{p_j t}, \tag{3.301}$$

where the p_j are real and negative[†] and the constants C_{ij} are determined from the solution of the boundary-value problem. Therefore the right side of (3.298) is the sum of n exponentials

$$\begin{aligned} u &= \operatorname{sign} \sum_{i=1}^{n} b_i \sum_{j=1}^{n} C_{ij} e^{p_j t} \\ &= \operatorname{sign} \sum_{j=1}^{n} \left(\sum_{i=1}^{n} b_i C_{ij} \right) e^{p_j t} = \operatorname{sign} \sum_{j=1}^{n} D_j e^{p_j t}, \end{aligned} \tag{3.302}$$

where

$$D_j = \sum_{i=1}^{n} b_i C_{ij} \,. \tag{3.303}$$

But the expression $\sum_{j=1}^{n} D_j e^{p_j t}$ passes through zero not more than $n - 1$ times and hence does not have more than n intervals of constant sign. Therefore in the general case the control action u has n intervals of constant sign as well. Thus the theorem concerning n intervals is proved.

The control action does not, however, always have to be on the boundaries of the admissible region. For example, let us consider the problem for an object with Eqs. (3.299) and constraints (3.295) in the case when the optimality criterion has the form (see [3.39])

$$Q = \int_0^\infty \left(\sum_{j=1}^{n} a_j x_j^2 + c u^2 \right) dt. \tag{3.304}$$

[†] Not losing generality, the roots p_j can be considered distinct.

Here we regard the fact that in formula (3.304) all the a_i and also c are positive.

We will set

$$f_0(x_1, ..., x_n, u) = \sum_{j=1}^{n} a_j x_j^2 + cu^2,$$
$$f_i(x_1, ..., x_n, u) = \sum_{j=1}^{n} a_{ij} x_j + b_i u \qquad (i = 1, 2, ..., n). \tag{3.305}$$

Let us add to the coordinates $x_1, ..., x_n$ of the object yet another coordinate x_0 with equation

$$dx_0/dt = f_0(x_1, ..., x_n, u) \tag{3.306}$$

and initial condition $x_0(0) = 0$. Then according to (3.304) the quantity Q becomes equal to the limit of $x_0(t)$ as $t \to \infty$. We will construct the system of adjoint equations :

$$\frac{d\psi_0}{dt} = 0,$$
$$\frac{d\psi_i}{dt} = -\sum_{\alpha=0}^{n} \frac{\partial f_\alpha}{\partial x_i} \psi_\alpha = -2a_i\psi_0 x_i - \sum_{j=1}^{n} a_{ji}\psi_j \qquad (i = 1, 2, ..., n). \tag{3.307}$$

Now the expression for the Hamiltonian can be written

$$H = \sum_{\alpha=0}^{n} \psi_\alpha \frac{dx_\alpha}{dt} = \sum_{\alpha=0}^{n} \psi_\alpha f_\alpha(x_1, ..., x_n, u)$$
$$= \psi_0 \left(\sum_{i=1}^{n} a_i x_i^2 + cu^2\right) + \sum_{i=1}^{n} \psi_i \left(\sum_{j=1}^{n} a_{ij} x_j + b_i u\right). \tag{3.308}$$

On the right side of (3.308) the variable u will be contained in the expression

$$H' = c\psi_0(t)u^2(t) + u(t) \sum_{i=1}^{n} \psi_i b_i . \tag{3.309}$$

Therefore the condition for the maximum of H coincides with the condition

$$\max_{|u| \leqslant 1} H' = \max_{|u| \leqslant 1} \left[c\psi_0(t)u^2 + u \sum_{i=1}^{n} \psi_i b_i\right]$$
$$= \max_{|u| \leqslant 1} \left\{c\psi_0(t) \left[u + \frac{1}{2c\psi_0} \sum_{i=1}^{n} b_i\psi_i(t)\right]^2 - \frac{1}{4c\psi_0} \left[\sum_{i=1}^{n} b_i\psi_i(t)\right]^2\right\}. \tag{3.310}$$

In view of the validity of the first of conditions (3.307), the quantity ψ_0 is constant. Since its value can be selected as an arbitrary negative number (see Chapter II), we will set

$$\psi_0 = -1. \tag{3.311}$$

Substituting this expression into (3.310), it is not difficult to see that the maximum of the expression in the braces will be attained when the first negative term will vanish, if this is possible, or it will assume the smallest magnitude in absolute value. It is not difficult to see that the quantity

$$\left[u - \frac{1}{2c} \sum_{i=1}^{n} b_i \psi_i(t)\right]^2 \tag{3.312}$$

is equal to zero or is minimal in absolute value, if for the condition $|u| \leqslant 1$ the value of u is chosen equal to

$$u(t) = \begin{cases} \frac{1}{2c} \sum_{i=1}^{n} b_i \psi_i(t) & \text{if} \quad \left| \frac{1}{2c} \sum_{i=1}^{n} b_i \psi_i(t) \right| \leqslant 1, \\ 1 & \text{if} \quad \frac{1}{2c} \sum_{i=1}^{n} b_i \psi_i \geqslant 1, \\ -1 & \text{if} \quad \frac{1}{2c} \sum_{i=1}^{n} b_i \psi_i \leqslant 1. \end{cases} \tag{3.313}$$

The values of $\psi_i(t)$ will turn out to be known if the adjoint equations (3.307) have been solved. But for this the initial values $\psi_i(0)$ must be found beforehand.

We will first assume that $u(t)$ does not attain its boundary values. Then by substituting the upper of expressions (3.313) in place of $u(t)$ into Eqs. (3.299) and (3.307), we will obtain

$$\begin{aligned} \frac{dx_i}{dt} &= \sum_{j=1}^{n} a_{ij} x_j + \frac{b_i}{2c} \sum_{j=1}^{n} b_j \psi_j, \\ \frac{d\psi_i}{dt} &= 2a_i x_i - \sum_{j=1}^{n} a_{ji} \psi_j \qquad (i = 1, 2, ..., n). \end{aligned} \tag{3.314}$$

This system of equations must be solved with the initial conditions $x_1(0)$, $x_2(0)$, ..., $x_n(0)$, and also with the boundary (final) conditions

$$\lim_{t\to\infty} x_1(t) = \lim_{t\to\infty} x_2(t) = \cdots = \lim_{t\to\infty} x_n(t) = 0. \tag{3.315}$$

The $\psi_i(0)$ $(i = 1, 2, ..., n)$ must be selected in such a manner that both the initial and final conditions for the $x_i(t)$ will be satisfied.

In [3.39] it was shown that the desired values $\psi_i(0)$ are related to the $x_j(0)$ by the linear dependences

$$\psi_i(0) = \sum_{j=1}^{n} \gamma_{ij} x_j(0) \qquad (i = 1, 2, ..., n), \tag{3.316}$$

where the γ_{ij} are constants. Since any instant can be taken for the initial one, then generally in this problem for any time instant

$$u^* = \frac{1}{2c} \sum_{i=1}^{n} k_i x_i \, , \tag{3.317}$$

where

$$k_i = \sum_{j=1}^{n} b_j \gamma_{ji} \, . \tag{3.318}$$

It can also be assumed by analogy with (3.313) that in the general case the optimal control law has the form

$$u = \begin{cases} \dfrac{1}{2c} \sum\limits_{i=1}^{n} k_i x_i & \text{for} \quad \left| \dfrac{1}{2c} \sum\limits_{i=1}^{n} k_i x_i \right| \leqslant 1, \\ 1 & \text{for} \quad \dfrac{1}{2c} \sum\limits_{i=1}^{n} k_i x_i \geqslant 1, \\ -1 & \text{for} \quad \dfrac{1}{2c} \sum\limits_{i=1}^{n} k_i x_i \leqslant -1. \end{cases} \tag{3.319}$$

But in contrast to (3.313) a rigorous proof of the law (3.319) does not exist. From (3.319) it follows that in the phase plane there exist two hyperplanes with the equations

$$\frac{1}{2c} \sum_{i=1}^{n} k_i x_i = +1, \quad \frac{1}{2c} \sum_{i=1}^{n} k_i x_i = -1. \tag{3.320}$$

If the representative point is between these hyperplanes, then the optimal control law has the form of Eq. (3.317). Thus in this case u is not on the boundary of the admissible region $| u | \leqslant 1$. But if the representative point leaves the region between the two hyperplanes (3.320), then the optimal control proves to be on one or the other of the boundaries, depending on in which of the regions corresponding to (3.319) the representative point was found.

For a second-order system the validity of an optimal control law of the type (3.319) has been rigorously proved. For higher-order systems the question remains open.[†]

The papers [3.40, 3.41] describe the theory and give the results of the development of a tracking servo with two control parameters. The system is close to optimal in speed of response. The problem was solved with the aid of the maximum principle. Apparently, these papers were first in which the maximum principle was applied to the solution of an engineering problem. An original feature of the system is control by a constant-current motor with independent excitation. Two control actions enter this motor. One of them is the voltage U_1 supplied to the input of the electromagnetic amplifier (EMA), feeding the armature circuit of the motor. The other control action is the voltage U_2 supplied to the excitation winding of the motor. The control is performed simultaneously by both actions U_1 and U_2.

The diagram of the power portion of the tracking servo is shown in Fig. 3.26. Here I_y is the current of the excitation winding of *EMA* having the time constant T_1. The current of the excitation winding of the motor, having the time constant T_2, is denoted by I_M. We will denote the time constant of the transverse winding of *EMA* by T_3, and we will denote the emf of *EMA* by E_G. Further let R_A be the total resistance of the armature circuit of *EMA* and motor, Ω the shaft speed, $E_M = c\Omega I_M$ the counter-emf of the motor, I_A the armature current, $M_M = k_3 I_A I_M$ the turning moment, J the moment of inertia given to the motor shaft, X the angular position of the output shaft. We neglect the inductance in the armature circuit, the armature reaction and the drag moment on the motor shaft. Then the equations of the diagram of Fig. 3.26 can be written in the form

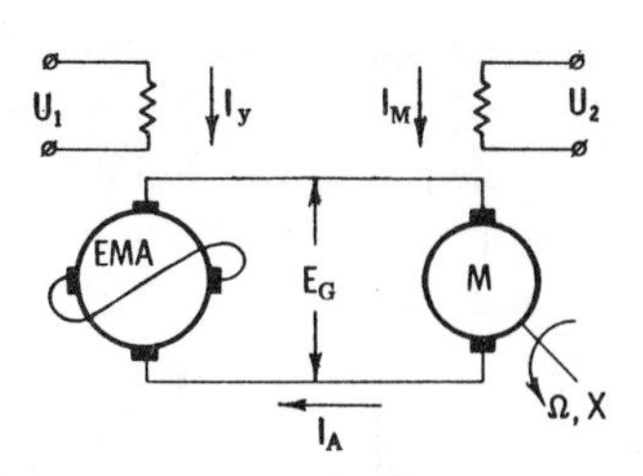

Fig. 3.26

$$dX/dt = \zeta\Omega, \qquad J\, d\Omega/dt = k_3 I_A I_M,$$
$$T_1\, dI_y/dt + I_y = k_1 U_1, \qquad T_2\, dI_M/dt + I_M = k_2 U_2, \qquad (3.321)$$
$$T_3\, dE_G/dt + E_G = k_4 I_y, \qquad E_G - c\Omega I_M = I_A R_A.$$

[†] In recent time this question has been investigated in the paper [3.53] by the method of dynamic programming, and some clarity has been brought into it.

Here ζ, k_1, k_2, k_3, k_4 are the constant amplification coefficients of the elements. If $T_1 \ll T_3$, then after transformations and reducing to relative time, the system of equations of the object depicted in Fig. 3.26 can be written in the following form:

$$\begin{aligned} dx/dt &= \zeta_0 \omega, & d\omega/dt &= (1-\mu)^{-1} i_M e_G - \omega i_M^2, \\ de_G/dt &= -\beta_1 e_G + \beta_1 u_1, & di_M/dt &= -\beta_2 i_M + \beta_2 u_2 . \end{aligned} \tag{3.322}$$

In these equations one denotes

$$\begin{gathered} x = \frac{X}{X_{\max}}, \quad e_G = \frac{E_G}{E_{G\max}}, \quad i_M = \frac{I_M}{I_{M\max}}, \\ \omega = \frac{\Omega}{\Omega_{\text{nom}}}, \quad u_1 = \frac{U_1}{U_{1\max}}, \quad u_2 = \frac{U_2}{U_{2\max}}, \\ \beta_1 = \frac{T_{EM}}{T_3}, \quad \beta_2 = \frac{T_{EM}}{T_2}, \\ T_{EM} = \frac{JR_A}{k_3 c I_M{}^2{}_{\max}}, \quad \eta = \frac{E_{G\max}}{E_{G\text{nom}}} > 1, \quad \mu = \frac{I_{A\text{nom}} R_A}{E_{G\text{nom}}} < 1, \end{gathered} \tag{3.323}$$

and the subscripts "max" and "nom" stand for the maximal and nominal values of the quantity, respectively.

As in [3.40] the notation t for dimensionless time has been preserved in equations (3.322); the electromagnetic time constant T_{EM} has been taken as the base quantity.

The voltage U_1 cannot exceed the maximal admissible magnitude in absolute value. Moreover, the voltage U_2 cannot be excessively reduced in order to avoid delivery for a small loading moment. Therefore the constraints

$$|u_1| \leqslant 1, \qquad 0 < \lambda \leqslant u_2 \leqslant 1 \tag{3.324}$$

are imposed on u_1 and u_2, where $\lambda = \text{const} > 0$.

Let us consider the transient process of adjusting the initial error, equivalent to a jump A_0 in the driving action, where A_0 is an arbitrary constant. Let us introduce new coordinates, related to the old coordinates by the equalities

$$x_1 = A_0 - x, \qquad x_2 = dx_1/dt = -\zeta_0 \omega, \qquad x_3 = -e_G, \; x_4 = i_M . \tag{3.325}$$

Substituting these expressions into (3.322), we arrive at the system of equations

$$\begin{aligned} dx_1/dt &= x_2, & dx_2/dt &= \alpha x_3 x_4 - x_2 x_4^2, \\ dx_3/dt &= -\beta_1 x_3 - \beta_1 u_1, & dx_4/dt &= -\beta_2 x_4 + \beta_2 u_2, \end{aligned} \tag{3.326}$$

where

$$\alpha = \eta/(1 - \mu). \tag{3.327}$$

Let the initial values $x_i(0)$ $(i = 1, 2, 3, 4)$ be assigned. It is required to find the optimal controls $u_1(t)$ and $u_2(t)$ transferring the representative point to the zero position in minimal time. The latter is the point with coordinates $(0, 0, 0, \gamma)$, where $\lambda < \gamma < 1$ is some fixed number.

Equations (3.326) in vector form can be written in the following way:

$$d\bar{x}/dt = \bar{f}(\bar{x}) + B\bar{u}, \tag{3.328}$$

where $\bar{x}$ and $\bar{f}$ are vectors whose components are shown in (3.326), the vector $\bar{u}$ has components u_1 and u_2, and B is the rectangular matrix

$$B = \begin{vmatrix} 0 & 0 \\ 0 & 0 \\ -\beta_1 & 0 \\ 0 & \beta_2 \end{vmatrix}. \tag{3.329}$$

Let us compose the system of adjoint equations

$$\begin{aligned} d\psi_1/dt &= 0, \qquad d\psi_2/dt = -\psi_1 + x_4{}^2\psi_2, \\ d\psi_3/dt &= -\alpha x_4\psi_2 + \beta_1\psi_3, \\ d\psi_4/dt &= (2x_2x_4 - \alpha x_3)\psi_2 + \beta_2\psi_4. \end{aligned} \tag{3.330}$$

We form the Hamiltonian

$$H = \langle\bar{\psi}, d\bar{x}/dt\rangle = \langle\bar{\psi}, \bar{f}(\bar{x}) + B\bar{u}\rangle = \langle\bar{\psi}, \bar{f}(\bar{x})\rangle + \langle\bar{\psi}, B\bar{u}\rangle. \tag{3.331}$$

Only the second term in this expression depends on $\bar{u}$. We will write it out in expanded form:

$$\langle\bar{\psi}, B\bar{u}\rangle = -\psi_3(t)\beta_1u_1 + \psi_4(t)\beta_2u_2. \tag{3.332}$$

Hence by taking (3.324) into account it is seen that the maximum of H occurs under the following conditions:

$$u_1 = -\operatorname{sign} \psi_3(t) \tag{3.333}$$

and

$$u_2 = \begin{cases} 1 & \text{for} \quad \psi_4(t) > 0, \\ \lambda & \text{for} \quad \psi_4(t) < 0. \end{cases} \tag{3.334}$$

Thus if the optimal control exists, then it consists of several intervals, in each of which the actions u_1 and u_2 are maintained at one of its limiting values.

In order to find the optimal control law in explicit form, the vector $\bar{\psi}$ must be known at each point $\bar{x}$ of the optimal trajectory. In the general case, as was noted in Chapter II, the solution of the corresponding boundary-value problem is only possible with the aid of computers. But in certain cases the problem of the synthesis of an optimal system can be solved analytically. Let us consider the simplest case, when the time constants T_2 of the transverse generator winding, T_3 of the excitation winding of the motor are negligibly small compared to T_{EM}—the electromagnetic time constant (in [3.40] and [3.41] more complicated cases have also been investigated). Then $\beta_1 = \infty$ and $\beta_2 = \infty$ and the equations of the object take the form

$$dx_1/dt = x_2, \qquad dx_2/dt = -u_2^2 x_2 - \alpha u_1 u_2. \tag{3.335}$$

The constraints imposed on u_1 and u_2 are expressed as before by conditions (3.324). The adjoint system is now written in the form

$$d\psi_1/dt = 0, \qquad d\psi_2/dt = -\psi_1 + u_2^2 \psi_2. \tag{3.336}$$

We form the Hamiltonian

$$H = \langle \bar{\psi}, \bar{f}(\bar{x}) \rangle = \psi_1 x_2 + \psi_2(-u_2^2 x_2 - \alpha u_1 u_2). \tag{3.337}$$

The solutions of Eqs. (3.336) have the form

$$\begin{aligned} \psi_1(t) &= \psi_{10} = \text{const}, \\ \psi_2(t) &= \exp\left\{\int_0^t u_2^2(\tau)d\tau\right\}\left[\psi_{20} - \int_0^t \psi_{10} \exp\left\{-\int_0^\tau u_2^2(s)\, ds\right\} d\tau\right]. \end{aligned} \tag{3.338}$$

From the latter expression it is seen that $\psi_2(t)$ can change sign no more than once.

In formula (3.337) for the Hamiltonian H, the quantities u_2 and α are always positive. Therefore the choice of u_1 maximizing H reduces to the expression

$$u_1 = -\text{sign}\, \psi_2(t). \tag{3.339}$$

In order to ascertain the control law for $u_2(t)$, let us rewrite formula (3.337) in the following form:

$$H = \psi_1 x_2 + \psi_2 \left[-x_2 \left(u_2 + \frac{\alpha u_1}{2x_2}\right)^2 + \frac{\alpha^2 u_1^2}{4x_2}\right]. \tag{3.340}$$

Two cases are possible.

(a) $\psi_2(t) < 0$, $u_1 = +1$. For $x_2 > 0$ the quantity u_2 must also assume its largest value, equal to unity. But if $x_2 < 0$, then as is seen from (3.340), we should choose for the maximization of H

$$u_2 = \begin{cases} 1, & \text{if} \quad |\alpha/2x_2| \geqslant 1, \\ |\alpha/2x_2|, & \text{if} \quad \lambda \leqslant |\alpha/2x_2| \leqslant 1, \\ \lambda, & \text{if} \quad |\alpha/2x_2| \leqslant \lambda. \end{cases} \tag{3.341}$$

(b) $\psi_2(t) > 0$, $u_1 = -1$. Then in order to maximize H, we should choose for $x_2 > 0$

$$u_2 = \begin{cases} 1, & \text{if} \quad \alpha/2x_2 \geqslant 1, \\ \alpha/2x_2, & \text{if} \quad \lambda \leqslant \alpha/2x_2 \leqslant 1, \\ \lambda, & \text{if} \quad \alpha/2x_2 \leqslant \lambda. \end{cases} \tag{3.342}$$

But if $x_2 < 0$, then the quantity $u_2 + \alpha u_1/2x_2$ must assume the largest value, and for this we must set $u_2 = 1$.

Since $\psi_2(t)$ changes sign no more than once, then in general $u_1(t)$ also consists of two intervals; in one of them $u_1 = +1$, while in the other $u_1 = -1$. The curve of $u_2(t)$ consists of four intervals. Three of them correspond to the first interval of u_1, while the fourth corresponds to the second interval of u_1. In the first interval u_2 is maximal, corresponding to the first line of (3.341). In the second interval u_2, varying continuously, runs through the segment $[\lambda, 1]$ [see the second lines of formulas (3.341) and (3.342)]. In the third interval $u_2 = \lambda$. Finally, in the last interval the action u_2 must again assume the maximal value, in order that at the end of it, for $x_1 = x_2 = 0$, the Hamiltonian H be maximal.

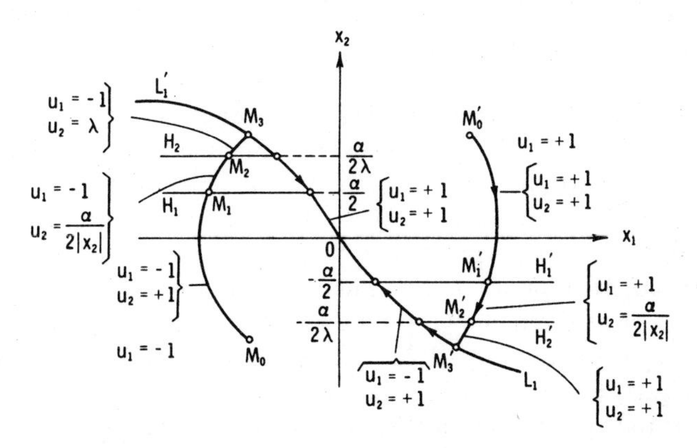

FIG. 3.27

The optimal trajectories on the phase plane are shown in Fig. 3.27; illustrative curves of $u_1(t)$, $u_2(t)$, $x_1(t)$, in Fig. 3.28. Only two optimal trajectories come to the origin O. One of them, denoted by L_1, corresponds to the values $u_1 = -1$, $u_2 = +1$. The other trajectory, denoted by L_1', corresponds to the values $u_1 = +1$, $u_2 = +1$. The trajectories of the families L_2 and L_2' come to these trajectories. For example, let us consider the trajectory $M_0M_1M_2M_3$ leading to L_1'. The entire region to the left of the switching curve L_1, L_1' for u_1 corresponds to the value $u_1 = -1$, and the entire region to the right of this curve to the value $u_1 = +1$. Thus the process u_1 takes place with one switching action, as is depicted in Fig. 3.28. As far as u_2 is concerned, the process of change in this action is more complicated. In the region $u_1 = -1$ let us draw the two horizontal straight lines H_1 (the line $x_2 = \alpha/2$) and H_2 (the line $x_2 = \alpha/2\lambda$). The analogous lines in the region $u_1 = +1$ are denoted by H_1' (the line $x_2 = -\alpha/2$) and H_2' (the line $x_2 = -\alpha/2\lambda$).

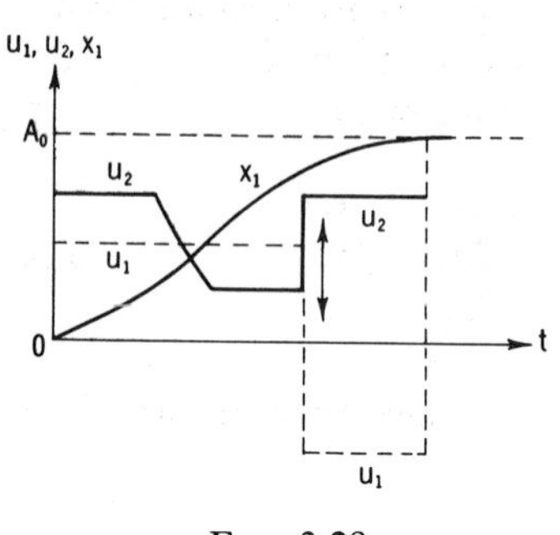

FIG. 3.28

If the initial point M_0 is in the third quadrant, then it moves along the optimal trajectory $M_0M_1M_2M_3O$. On the section $M_0M_1M_2M_3$ of this trajectory the value of u_1 is equal to -1, while on the section M_3O $u_1 = +1$. On the section M_0M_1 the quantity $u_2 = 1$; on the section M_1M_2 in the zone between the lines H_1 and H_2, the quantity u_2 varies continuously, obeying the law $u_2 = \alpha/2x_2$. At the point M_2 the value of u_2 becomes equal to λ and is maintained at this level on the section M_2M_3. At the point M_3 not only u_1 but also u_2 changes by a jump; the latter becomes equal to $+1$ and is maintained at this level on the entire section M_3O.

The variation in u_1 and u_2 takes place analogously on the optimal trajectory $M_0'M_1'M_2'M_3'O$, starting from the point M_0' which is situated in the first quadrant. The difference is only that on the section $M_0'M_1'M_2'M_3'$ the quantity $u_1 = +1$, while on the section $M_3'O$ the value of $u_1 = -1$. The quantity u_2 varies in the same manner as for the trajectory $M_0M_1M_2M_3O$. The illustrative graphs of $x_1(t)$, $u_2(t)$, and $u_1(t)$—the last of them shown by a dashed line on a scale different from u_2—are depicted in Fig. 3.28. They correspond to the trajectory $M_0'M_1'M_2'M_3'O$. From Fig. 3.28 it is seen that the curve of $u_2(t)$ consists of four intervals and the curve of $u_1(t)$ of two intervals.

CHAPTER IV

Optimal Systems with Maximal Partial Information about the Controlled Object

1. Continuous Systems with Maximal Information about the Object

In Chapter I the definition was given of systems with complete and partial information about the controlled object. We will assume that at least one of the components of information about the object is missing or is incomplete. This means that there is only partial information about the object in the controller. We will always regard as complete the information about the operator F of the object and the control purpose Q. In this chapter it is accepted that the information about the state $\bar{x}$ of the object is also complete. In other words, the knowledge of the state of the object B enters the controller A through a feedback circuit without distortions. Moreover, let the driving action $\bar{x}^*$ and noise $\bar{z}$ be measured without error and be supplied to the controller A. Further, the device A can measure without error and store its output values $\bar{u}$.

In Fig. 4.1 the driving action $\bar{x}^*$ being fed to the controller is shown; the noise $\bar{z}$ is also measured, and the result of the measurement enters the device A (the dashed line). We will now assume that one of the actions $\bar{x}^*$ and $\bar{z}$ or both of them are random processes. Then even a complete knowledge of the "previous history" of these processes for a time τ varying in the interval $-\infty < \tau \leqslant t$, where t is the current instant, does not permit their value in the future to be predicted exactly. Therefore the information about the controlled object, existing in the controller A, turns out to be incomplete.

It should be noted that the measurement of the noise $\bar{z}$ can be carried out in two ways.

(a) *The direct or immediate measurement.* In this case the dashed line in Fig. 4.1 must be replaced by a solid one. The noise $\bar{z}$ is measured by some measuring instrument, and the result of the measurement enters

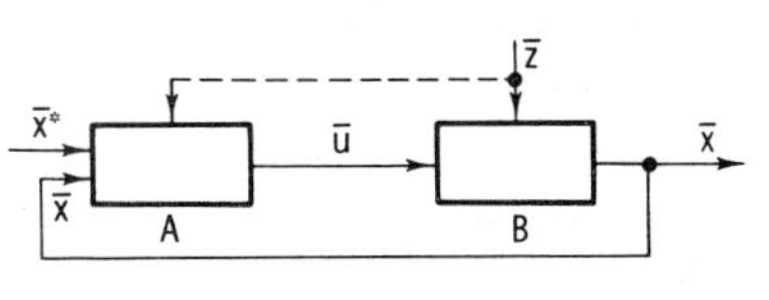

FIG. 4.1

the controller A. Here we consider that the error in measurement is negligibly small.

(b) *The indirect measurement.* In a number of cases the value of $\bar{z}$ can be obtained by indirect means, by measuring the values of $\bar{u}$ and $\bar{x}$ acting on the device A [4.1]. In fact, for example, let the equation of the object B have the form

$$dx_i/dt = f_i(\bar{x}, \bar{u}, z) \qquad (i = 1, ..., n), \tag{4.1}$$

where z is a scalar. We will consider one of Eqs. (4.1), say the first,

$$dx_1/dt = f_1(\bar{x}, \bar{u}, z). \tag{4.2}$$

Let f_1 be a single-valued and monotonic function of the scalar z for arbitrary values of $\bar{x}$ and $\bar{u}$. By measuring the values of x_i in an arbitrarily small time interval or by differentiating x_i with an ideal differentiator, the magnitude of the derivative dx_i/dt can in principle be found with an arbitrarily small error, and hence the value of $f_1(\bar{x}, \bar{u}, z)$ as well. Now if $\bar{x}$ and $\bar{u}$ are known, the value of z can be determined.

If an object without memory (inertialess object) is characterized by the equation

$$x = F(u, z), \tag{4.3}$$

where F depends monotonically on z, then from the current values of x and u the value of z is determined as well. For example, in the simplest case

$$x = \varphi(u) + z \tag{4.4}$$

the quantity z is equal to $x - \varphi(u)$. Evidently, in many cases the measurement of several components of the noise vector $\bar{z}$ is also possible. Thus we regard that $\bar{u}$ and $\bar{x}$ are measured without error, and the operator F is also known exactly and has a form such that $\bar{z}$ can also be determined from $\bar{u}$ and $\bar{x}$. In this case the result of indirect measurement does not differ at all from the result of direct measurement.

In an optimal system any possibility of measuring the noise $\bar{z}$ must necessarily be used, if it arises. Therefore in the case when the indirect method is possible, it should be considered that information about the entire "previous history" of the noise $\bar{z}$ enters the controller A. From now on we will assume that for the optimal device the "principle of complete utilization of information" is valid, which can be formulated in the following form:

The optimal controller completely uses all the information arriving at it for the purpose of control. Of course, there is in view only useful information for control.

The statement of problems in optimal control must be verified in the light of the principle of complete utilization of information. Here it is sometimes found that a statement of the problem which is correct in form will indeed contain inaccuracies. These errors can be of two types.

(a) *Redundancy of information or redundancy of system structure.* For example, in Fig. 4.2a the noise $\bar{z}$ is measured and the result of the measurement sent to the controller A. But if the object is of a type such that there is the possibility of indirect measurement of the quantity $\bar{z}$, then the measurement circuit going from $\bar{z}$ to A is not necessary. Redundant, superfluous information passes along it, and the measurement device itself is a redundant element of the general structure.

(b) *Partial utilization of information.* We will assume that the noise $\bar{z}$ is measured with an error, or random noise $\bar{e}$ is added to the result of measurement in the time of its passage through the channel E (Fig. 4.2b). Let the probabilistic characteristics of $\bar{e}$ and $\bar{z}$ be prescribed. We will also assume that there is the possibility of exact indirect measurement of the noise $\bar{z}$. In such a case the measurement circuit in the channel E is superfluous. The information about the probabilistic characteristics of the noise $\bar{e}$ is also unnecessary. But the inaccuracy of the problem statement is aggravated if it is guided only by the data about $\bar{z}$ entering through the channel E, and it is considered that the noise $\bar{z}$ is known inexactly to the controller A. Such a problem statement contradicts the principle of complete utilization of information. The same contradiction will be obtained if the circuit for direct measurement of $\bar{z}$ is missing, and it is assumed that only *a priori* information about the noise $\bar{z}$ is known to the controller (for example, the *a priori* probability distribution of the random variable $\bar{z}$), and meanwhile there is the possibility of exact indirect measurement of $\bar{z}$. If $\bar{z}$ is a random variable, then in general its value may now prove to be exactly known in such objects after an infinitesimally small time interval after the system starts functioning. In this case the replacement of the exact value by an *a priori* probability distribution can substantially worsen the properties of the system and move it significantly away from optimal.

In real systems no one quantity can be measured absolutely precisely. In any system for the transmission and processing of information there is noise, for example, the noise $\bar{g}$ in the transmission channel G from A to B (Fig. 4.2c) or the noise $\bar{h}$ in the transmission channel H of the feedback circuit (Fig. 4.2d). In this case an exact indirect measurement of the noise $\bar{z}$ is impossible, and the channel E of direct measurement can give supplementary information about the noise. The question of the quantity of this supplementary information and of the appropriateness of including the channel of direct measurement of $\bar{z}$ in the system can be resolved only by a specific design. However, we study the problems in idealized systems. Those inaccuracies which were noted above make the problem statement in such systems "improper", if known mathematical terms are applied. Below it is assumed that we will have to do with problems in a proper (in the sense indicated above) formulation.

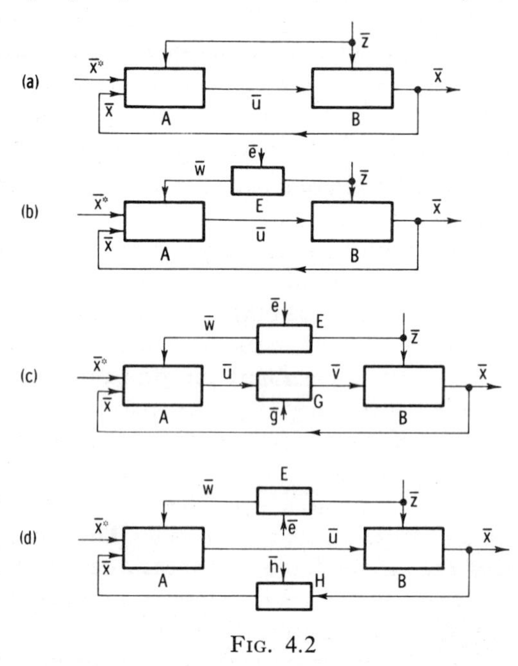

FIG. 4.2

The question arises, is complete information about the object necessary in all cases for optimal control? In particular, is the result of optimal control made worse with the disappearance of information about the noise z? This question is very important in practice, since it is frequently more difficult just to extract information about the noise z, or, what is the same, about the unexpectedly varying characteristics of the controlled object. It turns out that in some cases the information about the noise z is useless for determining the optimal control. For example, let the object be described by the equations

$$dv/dt = u, \qquad x = F(v, z), \tag{4.5}$$

where the noise z is constant during the transient process, and the function F is monotonic with respect to v; in addition, $\partial F/\partial v > 0$. We will take the constraint in the form $|u| \leqslant U$. We will assume that it is required to realize the system which is optimal in speed of response, in which the error $x^* - x$ is eliminated in minimal time, where $x^* = \text{const.}$ In a system with an object described by Eqs. (4.5), the optimal control has the following form:

$$u^* = U \operatorname{sign}(x^* - x), \tag{4.6}$$

and does not depend on the value of z. Therefore information about the noise z is redundant here.

Examples of such a kind are very rare, however, and in a known sense degenerate. In the overwhelming majority of cases the use of information about the noise improves the result of control. For example, if the first of Eqs. (4.5)—the equation of an integrating element—is replaced by an equation of a circuit consisting of two successively joined integrating elements, then as was shown in Chapter III the process of optimal control consists of two intervals, in which $u = \pm U$. The instant of switching the sign of u now depends on the quantity z. Hence in this case information about the noise z permits the time of the transient process to be reduced.

If the driving action $\bar{x}^*$ and noise $\bar{z}$ are measured exactly and it is known that they are regular time functions, then an infinitesimally small time interval is sufficient in order to find out precisely the future values of these functions. In this case we return to systems with complete information about the controlled object. But if $\bar{x}^*$ and $\bar{z}$ are random processes, then as was indicated above the controller now does not have complete information about the object. Here two cases are possible.

(a) In the general case, by observing the "previous history" of the random process, supplementary information about it can be accumulated, which will permit its future values to be estimated more precisely and the *a priori* probabilistic characteristics to be replaced by *a posteriori* ones. Such systems relate to the class of systems with storage of information about the object. They are considered in subsequent chapters.

(b) In special cases, when $\bar{x}^*$ and $\bar{z}$ are purely random or Markov processes, then under the condition of exact measurement of their current values the storage of information will not occur. Actually, a purely random process differs in that its future values do not depend on the past. Therefore the observation of its "previous history" will not give any additional information about values possible in the future, if only the *a priori* probabilistic characteristic is known. Here we do not consider the case when this characteristic is known, and the observation of the "previous history" permits it to be estimated, where the more precise it is, the longer the observation is made. Further, if $\bar{x}^*$ and $\bar{z}$ are Markov processes, then the probabilistic characteristics of their future values depend only on the current observation, and a knowledge of the "previous history" of the processes will not give any additional information. Therefore in this case as well the storage of information is not required. In the cases indicated above the controller obtains the maximal possible information about the object, whenever the current values of $\bar{x}^*$ and $\bar{z}$, measured exactly, act on its input. It is natural to call such systems ones with "maximal" (but partial) information about the object.

It should be emphasized that in the systems being studied in this chapter with maximal information about the object, this information is essentially partial since the exact values of $\bar{x}^*$ and $\bar{z}$ in the future are unknown. Thus the class of systems with maximal information about the object differs from the class of systems with complete information. But it is appropriate to study systems with maximal information immediately after systems with complete information, since much in these two classes of systems is similar. For both classes of systems it is distinctive that the storage of information about the object does not take place in them. In this chapter noise added to useful signals is not considered. Here it is regarded as negligibly small. But if its influence cannot be neglected, then storage of information in the optimal system is required (see Chapters V and VI).

It should be noted, however, that depending on the formulation of the problem the same system can be formed as a system with or without storage of information. Thus for example, if the process $\bar{x}^*$ or $\bar{z}$ acting on the system is purely random or Markov, but its probabilistic characteristics are completely or partially known, then they can be estimated or refined by accumulating information about the action. Then such a system will be a system with information storage. Besides, the very concept of a purely random or Markov process is an idealization; in actuality, by observing the real process during a long time, additional information about it can in principle be obtained. Nevertheless, the idealization that permits us to obtain a problem statement about a system with maximal information is completely justifiable in definite terms; recently this class of systems has undergone investigation in several papers.

Certain types of random process acting on the input of an automatic control system have been described in [4.2]. Only purely random and

Markov random processes are considered below. In general, the Markov random process can also be regarded as the result of passing a purely random process through a certain dynamical system. Let the equation of such a system with an output $\bar{z}$ and input $\bar{\xi}$ have the form

$$d\bar{z}/dt = \bar{\varphi}(\bar{z}, \bar{\xi}, t). \tag{4.7}$$

Here the coordinates φ_i of the vector $\bar{\varphi}$ ($i = 1, \ldots, m$) are certain generally nonlinear functions of their arguments (they can be considered differentiable).

We will regard the vector $\bar{\xi}$ as a purely random vector process acting on the input of the system described by Eq. (4.7). A conclusion regarding the future values of $\bar{\xi}$ at a given time instant can only be drawn from a known *a priori* probability density $P(\bar{\xi})$. The observation of a purely random process does not give any new information. But the value of $\bar{z}$ at some future time instant now does not depend only on the future values of $\bar{\xi}$. It also depends on the position the representative point $\bar{z}$ in the m-dimensional phase space of the system (4.7) has at the current time instant. Hence the density distribution of the variable $\bar{z}$ at the future time instant $t + \tau$ depends on the value of $\bar{z}$ at the current time instant t. Here a knowledge of the "previous history", i.e., of the values of $\bar{z}$ at past time instants, does not add anything to the information about the future course of the process $\bar{z}$. Hence it is seen that $\bar{z}$ is a Markov process.

If Eq. (4.7) is added to the equations of the object, then we obtain an "equivalent" object on which the purely random process $\bar{\xi}(t)$ acts.

The essential distinction of the problems being considered in this chapter from the problems of the previous chapter, still lies in the character of the optimality criterion as well. The input variables $\bar{x}^*$ and $\bar{z}$ acting on the system are random processes. Therefore the output variable $\bar{x}$ of the object B and also the control action $\bar{u}$ are random. Hence if some primary optimality criterion $Q_1 = Q_1(\bar{x}^*, \bar{x}, \bar{u}, \bar{z})$ is chosen, then Q_1 is also a random variable, changing from trial to trial in a manner not expected in advance. But the best possible optimality criterion Q must not be a random measure of the quality of system operation. Therefore as a measure of Q we usually take either the mathematical expectation of the primary criterion Q_1 — the quantity $Q = M\{Q_1(\bar{x}^*, \bar{x}, \bar{u}, \bar{z})\}$, or the probability that the criterion Q_1 will be found sufficiently small (smaller than a given threshold), or a threshold Q such that the event $Q_1 < Q$ has a prescribed probability.

In the general case Q_1 is a functional depending on the processes $\bar{x}^*$, $\bar{x}$, $\bar{u}$, $\bar{z}$ in a finite or infinitely large time interval. A more restricted problem is considered below when the time interval is finite. Let $t_0 \leqslant t \leqslant T$, where the values t_0 and T are fixed.

We will examine the simplest equation of a first-order object

$$dx/dt = f(x, u, t) + \xi, \tag{4.8}$$

where ξ is a purely random process. Let us first represent the process $\xi(t)$ in the form of a discrete sequence of Gaussian independent random variables with mean value $m(x, t)$ and dispersion $\sigma^2(x, t)/\Delta t$. Let the interval between samples be equal to Δt.

As $\Delta t \to 0$ in the limit a purely random process is obtained with infinitely large dispersion. It should be noted that white noise—a stationary random process with a constant spectral density $S(\omega) = S_0$ (see Chapter II)—also has an infinitely large dispersion. In fact, substituting S_0 in place of $S_x(\omega)$ into formula (2.94), we obtain the dispersion of white noise in the form

$$K_x(0) = \frac{1}{2\pi} \int_{-\infty}^{\infty} S_0 \, d\omega \to \infty. \tag{4.9}$$

The process $\xi(t)$ described above with mean value $m(x, t)$ and dispersion $\sigma^2(x, t)/\Delta t$ is nonstationary, since its exponents depend on time t. Thus it can be regarded as a generalization of normal white noise.

From (4.8) for small Δt it follows that

$$\Delta x = f\,\Delta t + \xi\,\Delta t. \tag{4.10}$$

Evidently, the quantity $\xi\,\Delta t$ is also normally distributed and has a mean $m\,\Delta t$ and dispersion $\sigma^2(\Delta t)^2/\Delta t = \sigma^2\,\Delta t$.

Thus

$$P(\xi\,\Delta t) = \frac{1}{\sqrt{(2\pi\sigma^2\,\Delta t)}} \exp\left\{-\frac{(\xi\,\Delta t - m\,\Delta t)^2}{2\sigma^2\,\Delta t}\right\}. \tag{4.11}$$

Hence from (4.10) it follows that the conditional probability density for the displacement Δx for fixed x and t is obtained, if its expression from (4.10) is substituted into (4.11) in place of $\xi\,\Delta t$:

$$P(\Delta x \mid x, t) = \frac{1}{\sqrt{(2\pi\sigma^2\,\Delta t)}} \exp\left\{-\frac{(\Delta x - f\,\Delta t - m\,\Delta t)^2}{2\sigma^2\,\Delta t}\right\}. \tag{4.12}$$

Thus the probability density for the displacement Δx depends on x and t, since f, m, and σ^2 depend on x and t. Hence the process $x(t)$ is Markov.

Now let the primary criterion have the form

$$Q_1 = \int_{t_0}^{T} G(x, u, t)\, dt. \tag{4.13}$$

Then the statistical criterion Q takes the form

$$Q = M\{Q_1\} = M\left\{\int_{t_0}^{T} G(x, u, t)\, dt\right\}. \tag{4.14}$$

Actually $x(t)$ is a random process. Hence the integral Q_1 is also a random variable, and in order to obtain Q the mathematical expectation of the quantity Q_1 must be found.

The physical meaning of the quantity Q can be explained thus: If the control $u(t)$ selected is not random, nevertheless the process $x(t)$ is random. Therefore the trajectories in phase space during the time $t_0 < t < T$ for various trials will be different. The "fan" of possible trajectories is shown in Fig. 4.3a. If $\int_{t_0}^{T} G\, dt$ is measured on each of the possible trajectories and afterward the arithmetic mean is taken, then we will obtain $Q = M\{\int_{t_0}^{T} G\, dt\}$.

In order to determine the optimal control $u^*(t)$, let us consider the integral (4.13) and break it up into two integrals. If Δt is small, then the representation of this integral in the following form is valid:

$$\int_{t_0}^{T} G(x, u, t)\, dt = \int_{t_0}^{t_0+\Delta t} G(x, u, t)\, dt + \int_{t_0+\Delta t}^{T} G(x, u, t)\, dt$$

$$\cong G(x_0, u_0, t_0)\, \Delta t + \int_{t_0+\Delta t}^{T} G(x, u, t)\, dt. \tag{4.15}$$

Here the first term has been transformed to within second-order small factors. For the present we assume that a certain "determinate" displacement Δx is made for some fixed value u_0 during the time Δt (Fig. 4.3b). We will consider that later on for $t > t_0 + \Delta t$ the optimal control occurs. Then the conditional mathematical expectation M' of the integral (4.15), under the condition that Δx is determinate, has the form

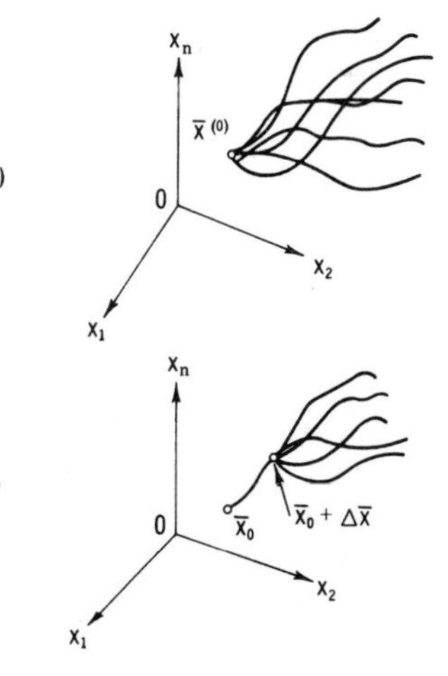

FIG. 4.3

$$M'\left\{\int_{t_0}^{T} G(x, u, t)\, dt\right\} = G(x_0, u_0, t_0)\, \Delta t$$

$$+ \min_{\substack{u\in\Omega(u) \\ t_0+\Delta t\leqslant t<T}} M\left\{\int_{t_0+\Delta t}^{T} G(x, u, t)\, dt\right\}. \tag{4.16}$$

The indices under the symbol min mean that the minimum is found from arbitrary admissible controls $u(t)$ on the interval from $t_0 + \Delta t$ to T.

Let the minimal value of Q be denoted by Q^*. It depends on x_0, the value of x at the instant $t = t_0$:

$$Q^*(x_0, t_0) = \min_{\substack{u \in \Omega(u) \\ t_0 \leqslant t < T}} M \left\{ \int_{t_0}^{T} G(x, u, t)\, dt \right\}. \tag{4.17}$$

With such a notation (4.16) takes the form

$$M' \left\{ \int_{t_0}^{T} G(x, u, t)\, dt \right\} = G(x_0, u_0, t_0)\, \Delta t + Q^*(x_0 + \Delta x, t_0 + \Delta t). \tag{4.18}$$

But in reality the displacement Δx is "random." Therefore the true quantity $Q(x_0, t_0)$ must be the mathematical expectation of expression (4.18), where the averaging takes place over all possible values of Δx. Let $M_{\Delta x}$ denote the averaging operation over all possible Δx. Then

$$\begin{aligned} Q(x_0, t_0) &= M_{\Delta x}\{G(x_0, u_0, t_0)\Delta t + Q^*(x_0 + \Delta x, t_0 + \Delta t)\} \\ &= G(x_0, u_0, t_0)\, \Delta t + M_{\Delta x}\{Q^*(x_0 + \Delta x, t_0 + \Delta t)\}. \end{aligned} \tag{4.19}$$

We obtain the minimal value Q^* of the quantity Q by dealing with the value $u_0 = (u)_{t=t_0}$:

$$Q^*(x_0 ; t_0) = \min_{u_0 \in \Omega(u)} \{G(x_0, u_0, t_0)\, \Delta t + M_{\Delta x}\{Q^*(x_0 + \Delta x, t_0 + \Delta t)\}\} \tag{4.20}$$

or in expanded form,

$$\begin{aligned} Q^*(x_0, t_0) = \min_{u_0 \in \Omega(u)} \Big\{ & G(x_0, u_0, t_0)\, \Delta t + \int_{\Omega(\Delta x)} Q^*(x_0 + \Delta x, t_0 + \Delta t) \\ & \times P(\Delta x \mid x_0, t_0)\, d\Omega(\Delta x) \Big\}, \end{aligned} \tag{4.21}$$

where $P(\Delta x \mid x_0, t_0)$ is the conditional probability density determined from (4.12). Not only the first but also the second term in the brackets on the right side of (4.21) depends on u_0, since $P(\Delta x \mid x_0, t_0)$ depends on u_0 through f, as follows from (4.8) and (4.12). Here $\Omega(\Delta x)$ is the region of all possible values of Δx, and $d\Omega(\Delta x)$ its infinitesimally small element.

In general x, t not x_0, t_0 can be written in Eq. (4.21), since any current point can be regarded as the initial one. Then $u = u(t)$ will enter into the equation in place of u_0, i.e., the current value. Formula (4.21) takes the form

$$\begin{aligned} Q^*(x, t) = \min_{u(t) \in \Omega(u)} \Big\{ & G(x, u, t)\, \Delta t + \int_{\Omega(\Delta x)} Q^*(x + \Delta x, t + \Delta t) \\ & \times P(\Delta x \mid x, t)\, d\Omega(\Delta x) \Big\}. \end{aligned} \tag{4.22}$$

We note that P is a probability density, so therefore

$$\int_{\Omega(\Delta x)} P(\Delta x \mid x, t)\, d\Omega(\Delta x) = 1. \tag{4.23}$$

From the integral relation (4.22) a partial differential equation can be arrived at in the same way as in physics (in diffusion theory) or in mathematics in the theory of Markov processes [4.23, 4.24, 2.1].

Let us expand Q^* under the integral in a series, restricting ourselves to no higher than second-order terms:

$$Q^*(x + \Delta x, t + \Delta t) = Q^*(x, t) + \frac{\partial Q^*}{\partial x} \Delta x + \frac{\partial Q^*}{\partial t} \Delta t + \frac{(\Delta x)^2}{2!} \frac{\partial^2 Q^*}{\partial x^2} + \frac{(\Delta t)^2}{2!} \frac{\partial^2 Q^*}{\partial t^2} + \Delta x \cdot \Delta t \cdot \frac{\partial^2 Q^*}{\partial x\, \partial t} + \cdots. \tag{4.24}$$

We will substitute (4.24) into (4.22). Then by taking (4.23) into account we find

$$\begin{aligned} Q^*(x, t) = \min_{u(t) \in \Omega(u)} \Big\{ & G(x, u, t)\, \Delta t + Q^*(x, t) \\ & + \frac{\partial Q^*}{\partial x} \int_{\Omega(\Delta x)} \Delta x\, P(\Delta x \mid x, t)\, d\Omega(\Delta x) + \frac{\partial Q^*}{\partial t} \Delta t \\ & + \frac{1}{2} \frac{\partial^2 Q^*}{\partial x^2} \int_{\Omega(\Delta x)} (\Delta x)^2 P(\Delta x \mid x, t)\, d\Omega(\Delta x) + \frac{(\Delta t)^2}{2} \frac{\partial^2 Q^*}{\partial t^2} \\ & + \Delta t \frac{\partial^2 Q^*}{\partial x\, \partial t} \int_{\Omega(\Delta x)} \Delta x\, P(\Delta x \mid x, t)\, d\Omega(\Delta x) + \cdots \Big\}. \end{aligned} \tag{4.25}$$

The quantity $Q^*(x, t)$ can be taken out from the brackets on the right side since it does not depend on $u(t)$. It is canceled on the left and right sides of (4.25). In an analogous manner the quantity $\partial Q^*/\partial t$ can also be taken out of the brackets by dividing both sides of the equation by Δt. We will now regard the fact that the integrals in the braces are the mathematical expectations of Δx and Δx^2 respectively. As is seen from (4.12) these mathematical expectations are equal to

$$\begin{aligned} \int_{\Omega(\Delta x)} \Delta x\, P(\Delta x \mid x, t)\, d\Omega(\Delta x) &= f\, \Delta t + m\, \Delta t, \\ \int_{\Omega(\Delta x)} (\Delta x)^2 P(\Delta x \mid x, t)\, d\Omega(\Delta x) &= \sigma^2\, \Delta t + (m\, \Delta t)^2 \cong \sigma^2\, \Delta t. \end{aligned} \tag{4.26}$$

The first of these formulas follows in an obvious manner from (4.12). The second formula is obtained from the expression known in probability theory[†]

$$y_{\text{rms}}^2 = D_y + (m_y)^2, \tag{4.27}$$

where y_{rms}^2 is the square of the root mean square value, D_y the dispersion, and m_y the mathematical expectation of the random variable y. The term $(m\,\Delta t)^2$ can be neglected in the second formula of (4.26) since it is of small second order.

As Δt tends to zero, we obtain from (4.25) in the limit with (4.26) taken into account,

$$-\frac{\partial Q^*}{\partial t} = \min_{u(t)\in\Omega(u)} \left\{ G(x, u, t) + (f + m)\frac{\partial Q^*}{\partial x} + \frac{\sigma^2}{2}\frac{\partial^2 Q^*}{\partial x^2} \right\}. \tag{4.28}$$

The remaining terms vanish since they are of order $(\Delta t)^2$.

By solving Eq. (4.28), we can find Q^*, and, in a parallel way, the optimal control $u^*(t)$. This method is generalized without any changes (see [4.27]) to higher-order systems with object equations

$$dx_i/dt = f_i(x_1, ..., x_n) + \xi \cdot \delta_{ij} \qquad (i = 1, ..., j, ..., n), \tag{4.29}$$

where the constant number j is in the interval $1 \leqslant j \leqslant n$, and δ_{ij} is the Kroneker symbol:

$$\delta_{ij} = \begin{cases} 1, & i = j, \\ 0, & i \neq j. \end{cases} \tag{4.30}$$

Let ξ be a purely random process with a mean value m and dispersion $\sigma^2/\Delta t$. In this case the partial differential equation, derived in the same way as above, takes the form

$$-\frac{\partial Q^*}{\partial t} = \min_{u\in\Omega(u)} \left\{ G(\bar{x}, u, t) + \sum_{i=1}^{n} (m \cdot \delta_{ij} + f_i)\frac{\partial Q^*}{\partial x_i} + \frac{\sigma^2}{2}\frac{\partial^2 Q^*}{\partial x_j^2} \right\}. \tag{4.31}$$

An equation of the type (4.7) describing the obtaining of a Markov process from a purely random one can be included in the set of equations (4.29). In such a case the set of equations (4.29) describes an "equivalent" object.

[†] It is easily obtained from (2.25) and (2.27).

Let us consider an elementary example when the equation of the object has the form

$$dx_1/dt = u. \tag{4.32}$$

The driving action—we will call it x_2—is a Markov process obtained in the form of the output variable of an inertial element, at whose input the purely random process ξ acts with a mean value $m = 0$:

$$dx_2/dt = -x_2 + \xi. \tag{4.33}$$

Equations (4.32) and (4.33) can be regarded as the equations of some "equivalent" object, at whose input the action ξ enters. Let the optimality criterion have the form

$$Q_1 = \int_{t_0}^{T} [(x_1 - x_2)^2 + u^2]\, dt. \tag{4.34}$$

From expression (4.34) it is seen that in the ideal case necessarily $x_1 = x_2$, i.e., the output variable x_1 of the object must be equal to the driving action x_2. Further let $m = 0$. Then Eq. (4.31) takes the form

$$-\frac{\partial Q^*}{\partial t} = \min_u \left\{(x_1 - x_2)^2 + u^2 + u\frac{\partial Q^*}{\partial x_1} - x_2\frac{\partial Q^*}{\partial x_2} + \frac{\sigma^2}{2}\frac{\partial^2 Q^*}{\partial x_2^{\,2}}\right\}. \tag{4.35}$$

If no constraints are imposed on the variable u, then it is easy to find the minimum of the right side by differentiating it with respect to u and equating the derivative to zero:

$$2u + \partial Q^*/\partial x_1 = 0. \tag{4.36}$$

Hence we determine the optimal value u^*:

$$u^* = -\frac{1}{2}\frac{\partial Q^*}{\partial x_1}. \tag{4.37}$$

Substituting (4.37) into (4.35) we obtain

$$-\frac{\partial Q^*}{\partial t} = (x_1 - x_2)^2 - \frac{1}{4}\left(\frac{\partial Q^*}{\partial x_1}\right)^2 - x_2\frac{\partial Q^*}{\partial x_2} + \frac{\sigma^2}{2}\frac{\partial^2 Q^*}{\partial x_2^{\,2}}. \tag{4.38}$$

It is convenient to go over to "reversed" time τ in this equation by setting $\tau = T - t$. Then instead of (4.38) we will obtain:

$$\frac{\partial Q^*}{\partial \tau} = (x_1 - x_2)^2 - \frac{1}{4}\left(\frac{\partial Q^*}{\partial x_1}\right)^2 - x_2\frac{\partial Q^*}{\partial x_2} + \frac{\sigma^2}{2}\frac{\partial^2 Q^*}{\partial x_2^{\,2}}. \tag{4.39}$$

The boundary conditions for solving Eq. (4.39) are obtained from natural considerations:

$$\begin{aligned} &Q^*(\bar{x}, \tau = 0) = 0 \text{ for all } \bar{x}, \\ &Q^*(\bar{x}, \tau) \to \infty \quad \text{as} \quad |\bar{x}| \to \infty. \end{aligned} \tag{4.40}$$

Here $|\bar{x}|$ is the absolute value of the vector $\bar{x}$ in the phase plane. Actually, for $\tau = 0$ the value $t_0 = t = T$, and the integral (4.34) vanishes.

As in Chapter III, a solution of Eq. (4.39) can be sought in the form of the series

$$Q^*(\bar{x}, \tau) = k_0(\tau) + \sum k_i(\tau)x_i + \sum\sum k_{ij}(\tau)x_i x_j + \sum\sum\sum k_{ijm}(\tau)x_i x_j x_m . \tag{4.41}$$

Here the functions k_0, k_i, k_{ij}, k_{ijm} depend only on time. Substituting (4.41) into (4.39) and equating coefficients for corresponding terms on the left and right sides of the equality, it can be found that only $k_0(\tau)$ and $k_{ij}(\tau)$ differ from zero, where we can choose $k_{ij} = k_{ji}$. We obtain ordinary differential equations for the functions $k_0(\tau)$ and $k_{ij}(\tau)$ as in Chapter III. These equations must be solved with the initial conditions $k_0(0) = k_{ij}(0) = 0$ arising from (4.40). In [4.3] graphs were given for the functions k_0, k_{11}, k_{12}, k_{22} of the problem under consideration where, as is seen from (4.41),

$$Q^*(x_1, x_2, \tau) = k_0(\tau) + k_{11}(\tau)x_1^2 + 2k_{12}(\tau)x_1 x_2 + k_{22}(\tau)x_2^2. \tag{4.42}$$

These graphs are shown in Fig. 4.4. From (4.37) and (4.42) we find the optimal control law:

$$u^* = -k_{11}(\tau)x_1 - k_{12}(\tau)x_2 . \tag{4.43}$$

FIG. 4.4

A block diagram of the optimal controller A constructed according to Eq. (4.43) is shown in Fig. 4.5. The block D has been formed according to Eq. (4.33). The purely random input ξ is applied to its input, and the Markov process x_2 appears at the output of block D, which acts as the driving action on the input of the controller A. Of course, block D is absent in a real system; it is depicted only in order to show the structure of the Markov process x_2. An integrating element has been indicated by the symbol $\int$ in this block and in block B. Block D can be regarded as part of an "equivalent" object incorporating blocks B and D.

The object B is an integrating element with output variable x_1 [see (4.32)]. The control action u in the controller is developed according to expression (4.43) in the form of the sum of outputs of two multiplying elements ME_1 and ME_2. The factors $-k_{11}(\tau)$ and $-k_{12}(\tau)$ act at the inputs of these elements from the computer C, where they are generated according to the curves depicted in Fig. 4.4. It should be taken into account that $\tau = T - t$, where t is real time varying from t_0 to T.

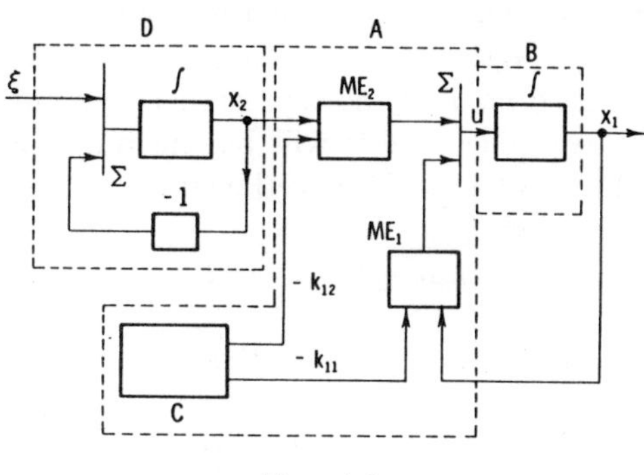

FIG. 4.5

The control that is optimal in speed of response can be determined in a like manner. Let the object equation be of order n; we will write the equations of the "equivalent" object in the form

$$dx_i/dt = f_i(\bar{x}, \bar{u}, \bar{\xi}) \qquad (i = 1, ..., n + m), \tag{4.44}$$

where $\bar{\xi}$ is a purely random vector with a given probability density. We will note that in composing the coordinates x_i of the "equivalent" object, the coordinates z_j of the noise generated by a block of dimension m are included as well, and the purely random process $\bar{\xi}$ enters into Eq. (4.44). We will find the control $\bar{u}$ that is optimal in speed of response, transferring the representative point $\bar{x}$ in the $(n + m)$-dimensional phase space of the "equivalent" system from the initial position $\bar{x}^{(0)}$ to the origin for the ordinary phase space, i.e., in the subregion $x_1 = x_2 = \cdots = x_n$ for the $(n + m)$-dimensional phase space. Since $\bar{\xi}$ is a random process, then the transition time T will also be a random process, and the mathematical expectation can be taken as the optimality criterion

$$Q = M\{T\}. \tag{4.45}$$

Further, instead of hitting the origin it is proper to consider occurring in some small neighborhood of the origin describable by an inequality, for example,

$$\sum_{i=1}^{n} x_i^2 \leqslant \delta^2, \tag{4.46}$$

where δ is sufficiently small.

We will consider two adjacent positions of the representative point: $\bar{x}$ at the time instant t and $\bar{x} + \Delta\bar{x}$ at the time instant $t + \Delta t$, where Δt is small. Let $Q^*(\bar{x})$ be the minimal value of the mathematical expectation of the transition time T, if the representative point is starting from the position $\bar{x}$. Then $Q^*(\bar{x} + \Delta\bar{x})$ will be the minimal transition time for a start from the point $\bar{x} + \Delta\bar{x}$, and the sum

$$\Delta t + Q^*(\bar{x} + \Delta\bar{x}) \tag{4.47}$$

gives the mathematical expectation of the time T, if a "determinate" first step $\Delta\bar{x}$ is made from the point $\bar{x}$ in some direction, and all subsequent steps now take place according to the optimal strategy. Since due to the presence of the random noise $\bar{\xi}$ the quantity $\Delta\bar{x}$ is in reality random, then the variable defined by expression (4.47) is also random, and the mathematical expectation $Q(\bar{x})$ of this quantity is obtained by averaging over all possible values of $\Delta\bar{x}$, which is denoted by the operation $M_{\Delta\bar{x}}$:

$$Q(\bar{x}) = M_{\Delta\bar{x}}\{\Delta t + Q^*(\bar{x} + \Delta\bar{x})\}. \tag{4.48}$$

The optimal control $\bar{u}^*(t)$ at the time instant t will be obtained if the minimum of $Q(\bar{x})$ with respect to the control $\bar{u}$ is found at the instant t. This minimal value is equal to $Q^*(\bar{x})$. Thus,

$$\begin{aligned} Q^*(\bar{x}) = \min_{\bar{u}\in\Omega(\bar{u})} Q(\bar{x}) &= \min_{\bar{u}\in\Omega(\bar{u})} M_{\Delta\bar{x}}\{\Delta t + Q^*(\bar{x} + \Delta\bar{x})\} \\ &= \Delta t + \min_{\bar{u}\in\Omega(\bar{u})} M_{\Delta\bar{x}}\{Q^*(\bar{x} + \Delta\bar{x})\}. \end{aligned} \tag{4.49}$$

If the conditional probability density for $\Delta\bar{x}$ with fixed $\bar{u}$ and $\bar{x}$ is denoted by $P(\Delta\bar{x} \mid \bar{u}, \bar{x})$, then Eq. (4.49) can be rewritten in expanded form as:

$$Q^*(\bar{x}) = \Delta t + \min_{\bar{u}\in\Omega(\bar{u})} \int_{\Omega(\Delta\bar{x})} Q^*(\bar{x} + \Delta\bar{x}) P(\Delta\bar{x} \mid \bar{u}, \bar{x})\, d\Omega(\Delta\bar{x}). \tag{4.50}$$

From this equality it is not difficult to arrive at a partial differential equation for $Q^*(\bar{x})$. Let us expand $Q^*(\bar{x} + \Delta\bar{x})$ in a series, neglecting small terms higher than second order:

$$Q^*(\bar{x} + \Delta\bar{x}) = Q^*(\bar{x}) + \sum_{i=1}^{n+m} \frac{\partial Q^*}{\partial x_i} \Delta x_i + \sum_{i,j=1}^{n+m} \frac{\partial^2 Q^*}{\partial x_i\, \partial x_j} \Delta x_i\, \Delta x_j + \cdots. \qquad (4.51)$$

Substituting (4.51) into (4.50) and taking into account that

$$\int_{\Omega(\Delta\bar{x})} P(\Delta\bar{x} \mid \bar{u}, \bar{x})\, d\Omega(\Delta\bar{x}) = 1, \qquad (4.52)$$

we obtain the following equation:

$$Q^*(\bar{x}) = \Delta t + \min_{\bar{u}\in\Omega(\bar{u})} \Big\{ Q^*(\bar{x}) + \sum_{i=1}^{n+m} \frac{\partial Q^*}{\partial x_i} \int_{\Omega(\Delta\bar{x})} \Delta x_i\, P(\Delta\bar{x} \mid \bar{u}, \bar{x})\, d\Omega(\Delta\bar{x})$$

$$+ \sum_{i,j=1}^{n+m} \frac{\partial^2 Q^*}{\partial x_i\, \partial x_j} \int_{\Omega(\Delta\bar{x})} \Delta x_i\, \Delta x_j\, P(\Delta\bar{x} \mid \bar{u}, \bar{x})\, d\Omega(\Delta\bar{x}) \Big\}. \qquad (4.53)$$

We will cancel $Q^*(\bar{x})$ on both sides of the equality, divide both sides by Δt, and introduce the notations

$$\lim_{\Delta t\to 0} \frac{1}{\Delta t} \int_{\Omega(\Delta\bar{x})} \Delta x_i\, P(\Delta\bar{x} \mid \bar{u}, \bar{x})\, d\Omega(\Delta\bar{x}) = a_i(\bar{u}, \bar{x}),$$

$$\lim_{\Delta t\to 0} \frac{1}{\Delta t} \int_{\Omega(\Delta\bar{x})} \Delta x_i\, \Delta x_j\, P(\Delta\bar{x} \mid \bar{u}, \bar{x})\, d\Omega(\Delta\bar{x}) = b_{ij}(\bar{u}, \bar{x}). \qquad (4.54)$$

Evidently, the a_i are the mean values of the rates of change in the Δx_i, i.e., the mean values of $\Delta x_i/\Delta t$, and the b_{ij} are the mean rates of change in the products of the quantities Δx_i and Δx_j. For $i = j$ the quantity b_{ij} is the mean rate of change in the square of Δx_i, i.e., the mean value of $(\Delta x_i)^2/\Delta t$. With the notations (4.54), Eq. (4.53) takes the form

$$0 = 1 + \min_{\bar{u}\in\Omega(\bar{u})} \Big\{ \sum_{i=1}^{n+m} \frac{\partial Q^*}{\partial x_i} a_i(\bar{u}, \bar{x}) + \sum_{i,j=1}^{n+m} \frac{\partial^2 Q^*}{\partial x_i\, \partial x_j} b_{ij}(\bar{u}, \bar{x}) \Big\}. \qquad (4.55)$$

This is the desired partial differential equation whose solution must satisfy the boundary conditions [see (4.46)]

$$Q^*(\bar{x}) = 0 \quad \text{for} \quad \sum_{i=1}^{n} x_i^2 = \delta^2 \quad \text{and any} \quad x_{n+j} \quad (j = 1, \ldots, m). \qquad (4.56)$$

Let us consider the particular case when there is a single action u and Eqs. (4.44) of the "equivalent" object have the form

$$dx_i/dt = f_i(\bar{x}, \xi) + \beta_i(\xi)u \qquad (i = 1, \ldots, n + m). \tag{4.57}$$

Here the coordinates of the noise are included in the set of coordinates x_i, represented in the form of the output of a dynamical system, to whose input the purely random process ξ is fed. Then to within first-order small terms the displacement Δx_i can be written in the form

$$\Delta x_i = f_i(\bar{x}, \xi)\,\Delta t + \beta_i(\xi)u\,\Delta t \qquad (i = 1, \ldots, n + m). \tag{4.58}$$

Hence it follows that the mean value of Δx_i is equal to

$$M\{\Delta x_i\} = a_i(\bar{u}, \bar{x})\,\Delta t = \Delta t[M\{f_i(\bar{x}, \xi)\} + uM\{\beta_i(\xi)\}], \tag{4.59}$$

and the mean value of the product of displacements is

$$\begin{aligned} M\{\Delta x_i\,\Delta x_j\} &= b_{ij}(\bar{u}, \bar{x})\,\Delta t \\ &= (\Delta t)^2 M\{[f_i(\bar{x}, \xi) + u\beta_i(\xi)][f_j(\bar{x}, \xi) + u\beta_j(\xi)]\} \\ &= (\Delta t)^2 (M[f_i(\bar{x}, \xi) f_j(\bar{x}, \xi)] + u\{M[\beta_i(\xi) f_j(\bar{x}, \xi)] \\ &\quad + M[\beta_j(\xi) f_i(\bar{x}, \xi)]\} + u^2 M[\beta_i(\xi)\beta_j(\xi)]). \end{aligned} \tag{4.60}$$

If the correlation functions on the right side of the equality are finite, which as was shown above cannot occur, then the b_{ij} are infinitesimally small and hence vanish from Eq. (4.55). Confining ourselves to this case and substituting the a_i into (4.55) from (4.59), we arrive at the equation

$$0 = 1 + \min_{u \in \Omega(u)} \sum_{i=1}^{n+m} \frac{\partial Q^*}{\partial x_i} [M\{f_i(\bar{x}, \xi)\} + uM\{\beta_i(\xi)\}]. \tag{4.61}$$

If u is bounded by the condition

$$|\, u \,| \leqslant 1, \tag{4.62}$$

then the minimum of the right side of (4.61) with respect to u will be ensured for

$$u = u^* = -\text{sign} \sum_{i=1}^{n+m} \frac{\partial Q^*}{\partial x_i} M\{\beta_i(\xi)\}. \tag{4.63}$$

This is the condition of optimal control. From this, it follows that in this type of system the optimal value u^* is always on the boundary of the admissible region and thus is equal to ± 1.

The substitution of (4.63) into (4.61) gives the equation

$$0 = 1 + \sum_{i=1}^{n+m} \frac{\partial Q^*}{\partial x_i} M\{f_i(\bar{x}, \bar{\xi})\} - \left| \sum_{i=1}^{n+m} \frac{\partial Q^*}{\partial x_i} M\{\beta_i(\bar{\xi})\} \right|. \tag{4.64}$$

By solving this equation with the boundary conditions (4.56) and substituting the solution into (4.63), the optimal control u^* can be found in explicit form, where it is found to be a function of the current values x_i $(i = 1, ..., n + m)$, into whose set the noises z_j also enter, as was indicated above, if they are outputs of the elements of the "equivalent" object.

Equation (4.64) can be solved approximately, but only in the simplest cases by computers. Such an attempt for a second-order system was carried out in [4.4]. The numerical result given in this article apparently requires correction (see [4.19]).

The optimal control algorithm for a continuous object in a system with maximal information has also been determined in [4.5], where the connection between the method of dynamic programming and Lyapunov's second method was indicated.

The solution of all the problems presented above represents the dependence of the optimal control vector $\bar{u}^*$ on the current state $\bar{x}$ of the object. Thus $\bar{u}^* = \bar{u}^*(\bar{x})$. The control at a given time instant depends only on the state of the object at the very same time instant. Such a form for the optimal control algorithm has been characterized only for systems with complete or maximal information about the object. Subsequent chapters will analyze fundamentally more complex algorithms in which the optimal control $\bar{u}^*$ at a given time instant will be calculated on the basis of the "previous history" of the system motion, i.e., the past values of those variables which act on the input of the controller A. A similar form of optimal algorithm has been characterized for systems with storage of information about the object.

2. Discrete-Continuous and Purely Discrete Systems with Maximal Information about the Object

The methods of solving problems for discrete-continuous systems, i.e., systems with quantization in time, coincide in an essential part with the one presented in the preceding section. The difference is only that differential equations for the object are replaced by finite difference equations, and, in addition, after the derivation of the integral relation for Q^* the passage to partial differential equations is not made. We will illustrate these methods by an example of the simplest first-order object with the equation

$$x_{k+1} = \alpha x_k + u_k + \xi_k , \tag{4.65}$$

where $\alpha =$ const and $x_k = x(k\,\Delta t)$, $u_k = u(k\,\Delta t)$, $\xi_k = \xi(k\,\Delta t)$. The process lasts during N times ($k = 0, 1, ..., N$), where N is given, and the time length is equal to Δt. Let ξ_k be a discrete purely random Gaussian process with zero mean value and dispersion σ^2. Since from (4.65) we obtain

$$\xi_k = x_{k+1} - \alpha x_k - u_k , \tag{4.66}$$

then the conditional probability density for x_{k+1} is defined by the expression

$$P(x_{k+1} \mid x_k , u_k) = \frac{1}{\sigma \sqrt{(2\pi)}} \exp \left\{ - \frac{(x_{k+1} - \alpha x_k - u_k)^2}{2\sigma^2} \right\}. \tag{4.67}$$

We will assume that the optimality criterion has the form

$$Q = M \left\{ \sum_{i=0}^{N} (\lambda^2 x_i^2 + u_i^2) \right\}, \tag{4.68}$$

where $\lambda^2 =$ const. It is required to choose an optimal sequence u_0, u_1, ..., u_N such that the condition $Q =$ min is ensured. No constraints are imposed on the values u_k.

The integral relation for this problem is obtained in the same form as Eq. (4.22), and by analogous arguments. The difference is only that $G\,\Delta t$ is replaced by $\lambda^2 x_k^2 + u_k^2$, and $Q^*(x, t)$ is denoted by $Q^*(x_k)$, since there is no explicit dependence on time t in (4.65) and (4.68). Moreover, $Q^*(x + \Delta x, t + \Delta t)$ is replaced by $Q^*(x_{k+1})$, and $P(\Delta x)$ by $P(x_{k+1})$. Thus an expression of the type (4.22) for this problem takes the form

$$Q^*(x_k) = \min_{u_k} \left[(\lambda^2 x_k^2 + u_k)^2 + \int_{\Omega(x_k)} Q^*(x_{k+1}) P(x_{k+1} \mid x_k , u_k)\, d\Omega(x_k) \right]; \tag{4.69}$$

here $Q(x_k)$ is the part of the entire sum (4.68) from $i = k$ to $i = N$, and Q^* is the minimum of Q with respect to u.

If it is taken into account that $-\infty < x_k < \infty$, and the expression for $P(x_{k+1})$ in the form of (4.67) is substituted into (4.69), then Eq. (4.69) assumes the form

$$Q^*(x_k) = \min_{u_k} \Big[(\lambda^2 x_k^2 + u_k^2) + \int_{-\infty}^{\infty} \frac{Q^*(x_{k+1})}{\sigma \sqrt{(2\pi)}} \exp \left\{ - \frac{(x_{k+1} - \alpha x_k - u_k)^2}{2\sigma^2} \right\} dx_{k+1} \Big]. \tag{4.70}$$

The recursive sequence of functions $Q^*(x_k)$ can be determined from (4.70) as soon as $Q^*(x_N)$ is known. Since from (4.68) it follows that

$$Q^*(x_N) = \min_{u_N} \{\lambda^2 x_N{}^2 + u_N{}^2\}, \tag{4.71}$$

then by finding the optimal value $u_N{}^* = 0$ and substituting it into (4.71), we obtain

$$Q^*(x_N) = \lambda^2 x_N{}^2. \tag{4.72}$$

In general, let

$$Q^*(x_{k+1}) = A_{k+1} + B_{k+1} x_{k+1}^2 \,. \tag{4.73}$$

Then from (4.70) we obtain

$$Q^*(x_k) = \min_{u_k} \left[(\lambda^2 x_k{}^2 + u_k{}^2) + \int_{-\infty}^{\infty} \frac{A_{k+1} + B_{k+1} x_{k+1}^2}{\sigma \sqrt{(2\pi)}} \times \exp\left\{ -\frac{(x_{k+1} - \alpha x_k - u_k)^2}{2\sigma^2} \right\} dx_{k+1} \right]. \tag{4.74}$$

We will set

$$x_{k+1} - \alpha x_k - u_k = w, \qquad dx_{k+1} = dw. \tag{4.75}$$

Then the integral in (4.74) is transformed in the following manner:

$$\begin{aligned} &\int_{-\infty}^{\infty} \frac{A_{k+1} + B_{k+1}(w + \alpha x_k + u_k)^2}{\sigma \sqrt{(2\pi)}} \exp\left\{ -\frac{w^2}{2\sigma^2} \right\} dw \\ &= \frac{A_{k+1} + B_{k+1}(\alpha x_k + u_k)^2}{\sigma \sqrt{(2\pi)}} \int_{-\infty}^{\infty} \exp\left\{ -\frac{w^2}{2\sigma^2} \right\} dw \\ &+ \frac{B_{k+1}}{\sigma \sqrt{(2\pi)}} 2(\alpha x_k + u_k) \int_{-\infty}^{\infty} w \exp\left\{ -\frac{w^2}{2\sigma^2} \right\} dw \\ &+ \frac{B_{k+1}}{\sigma \sqrt{(2\pi)}} \int_{-\infty}^{\infty} w^2 \exp\left\{ -\frac{w^2}{2\sigma^2} \right\} dw. \end{aligned} \tag{4.76}$$

We substitute the values of the integrals

$$\begin{aligned} &\int_{-\infty}^{\infty} \exp\left\{ -\frac{w^2}{2\sigma^2} \right\} dw = \sigma \sqrt{(2\pi)}, \\ &\int_{-\infty}^{\infty} w \exp\left\{ -\frac{w^2}{2\sigma^2} \right\} dw = 0, \\ &\int_{-\infty}^{\infty} w^2 \exp\left\{ -\frac{w^2}{2\sigma^2} \right\} dw = \sigma^3 \sqrt{(2\pi)}. \end{aligned} \tag{4.77}$$

Then the integral (4.76) turns out to be equal to

$$A_{k+1} + B_{k+1}(\alpha x_k + u_k)^2 + \sigma^2 B_{k+1}, \tag{4.78}$$

and the whole expression (4.74) can be written thus:

$$\begin{aligned} Q^*(x_k) &= \min_{u_k} [\lambda^2 x_k^2 + u_k^2 + A_{k+1} + B_{k+1}(\alpha x_k + u_k)^2 + \sigma^2 B_{k+1}] \\ &= \min_{u_k} [(\lambda^2 + B_{k+1}\alpha^2)x_k^2 + A_{k+1} + \sigma^2 B_{k+1} + u_k^2(1 + B_{k+1}) \\ &\qquad + 2B_{k+1}\alpha x_k u_k]. \end{aligned} \tag{4.79}$$

If $1 + B_{k+1} > 0$, then expression (4.79) has a minimum, which can be found by equating the derivative $dQ^*(x_k)/du_k$ to zero:

$$2u_k(1 + B_{k+1}) + 2B_{k+1}\alpha x_k = 0. \tag{4.80}$$

Thus the optimal control is

$$u_k^* = -\frac{\alpha B_{k+1} x_k}{1 + B_{k+1}}. \tag{4.81}$$

Substituting (4.81) into (4.79) we find

$$Q^*(x_k) = \left(\lambda^2 + \alpha^2 B_{k+1} - \frac{\alpha^2 B_{k+1}^2}{1 + B_{k+1}}\right) x_k^2 + A_{k+1} + \sigma^2 B_{k+1} = A_k + B_k x_k^2, \tag{4.82}$$

where

$$A_k = A_{k+1} + \sigma^2 B_{k+1}, \qquad B_k = \lambda^2 + \frac{\alpha^2 B_{k+1}}{1 + B_{k+1}}. \tag{4.83}$$

From (4.72) we determine

$$A_N = 0, \qquad B_N = \lambda^2 > 0. \tag{4.84}$$

Therefore all $B_k > 0$, and hence it follows that $A_k > 0$. Thus the optimal control is determined from (4.81) and the B_k from (4.83) and (4.84). It is interesting to note that in view of the linearity of the system and the quadratic form of the criterion, the noise ξ does not affect the choice of the optimal control at all. In fact, the quantities B_k starting from $B_N = \lambda^2$, and with them the values u_k^*, do not depend on σ. The presence of the noise manifests itself only in the quantity $Q^*(x_k)$, which depends on σ through A_k.

We will examine the class of purely discrete objects and the questions of their optimal control. These questions have been investigated in the papers [4.6–4.8, 4.12]. The optimal control by Markov discrete objects represents the basic interest. The theory and block diagrams of such objects have been considered in both the mathematical and the theoretical engineering literature. From the papers in the latter field, we will indicate [4.9–4.11]. In the same way as regular discrete objects described in Chapter III, Markov discrete objects are characterized by a set of possible states $q_1, q_2, \ldots, q_{N+1}$. We will denote the current state at the instant t by the symbol s_t. It can be any of the possible states q_j ($j = 1, \ldots, N+1$). With a change in time $t = 0, 1, 2, \ldots$, the state s_t also changes in general. Somehow the system "jumps" from one possible state to another. In [4.7] the analogy to a beetle sitting on a flower has been applied to describing these motions. The possible positions of the beetle can be identified with the number j of the flower q_j; the current position s_t of the beetle, when it flies from one flower to another, varies discretely. The passage from s_t to s_{t+1} is the flight of the beetle from one flower to another.

In contrast to the systems considered in Chapter III, the process of the transition of the object from one state to another is random. This means that by knowing the state s_t of the object at the time instant t and the control action u_t at the same time instant, we can only give a prediction of the probabilistic character of what the following state s_{t+1} of the object will be. Since all quantities are discrete, we will assume that u_t can assume only one of several possible values α_k ($k = 1, 2, \ldots, m$). By knowing $s_t = q_j$ and $u_t = \alpha_k$, only the transition probability of the object to some state q_j at the following time instant can be determined, i.e., the probability that s_{t+1} will be found equal to q_j. We will denote this transition probability depending on q_j and α_k by

$$p_{ij}(k) = p\{s_{t+1} = q_j \mid s_t = q_i, u_t = \alpha_k\}. \tag{4.85}$$

Thus $p_{ij}(k)$ is the transition probability from the ith state q_i to the jth state q_j, if the control α_k is applied.

The following remarks should be made concerning expression (4.85). First, the transition probability p_{ij} is a conditional probability and depends on what the state s_t and the control u_t are at the time instant t. The dependence of p_{ij} on the value of s_t defines the random process of the state change $s_0, s_1, \ldots, s_t, s_{t+1}, \ldots$ as a Markov process. Further, formula (4.85) is not the most general one even for Markov processes. In the general case the quantity p_{ij} can also depend on time t. For the present we restrict ourselves only to the case when p_{ij} does not depend explicitly on t. Finally, expressions (4.85) for different i and j cannot be arbitrary. In the first place, p_{ij} must be between zero and unity:

$0 \leqslant p_{ij} \leqslant 1$. In addition, the object inevitably transfers from the state s_t to some state s_{t+1}. In a special case this new state can coincide with the old one. But the object necessarily transfers to some state. Therefore the sum of the probabilities p_{ij} for a given i over all j must be equal to unity:

$$\sum_{j=1}^{N+1} p_{ij} = 1. \tag{4.86}$$

The maximal information about a Markov discrete object consists in that, at the time instant t, the state s_t of the object and the input control action u_t are known. This is partial information since it does not uniquely determine the future behavior of the object but only permits its probabilistic characteristics to be found. But more detailed information about this object cannot, in principle, be obtained.

A Markov discrete object is completely described by prescribing the several matrices (whose total number equals m)

$$\bar{\bar{P}}_k = \| p_{ij}(k) \| \tag{4.87}$$

of the transition probabilities for the various control actions α_k ($k = 1, \ldots, m$). Each of the matrices $\bar{\bar{P}}_k$ can be replaced by a descriptive diagram, on which the possible states q_j ($j = 1, \ldots, N+1$) are marked by circles with the indices j. For example, in Fig. 4.6, two possible states 1 and 2 are

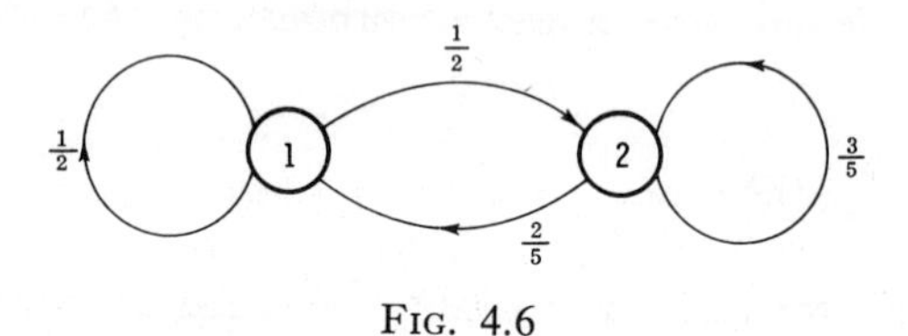

FIG. 4.6

shown. The possible transitions from one state to another are indicated by arrows, and the probabilities of these transitions are given near the arrows. For example, we can transfer from state 2 to state 1 with probability 2/5, and transfer to the same state 2 with probability 3/5; i.e., the system remains in the same state with probability 3/5.

Besides the transition probabilities, the probabilities of the states are also of interest in the Markov object. Let us denote by $p_i(t)$ the probability that the state s_t at the time t is found to be q_i, i.e., the ith of possible states (the probability that the beetle is found to be on the

ith flower at the instant t). Since at the instant t there will be some state, then the sum of the probabilities $p_i(t)$ is equal to unity:

$$\sum_{i=1}^{N+1} p_i(t) = 1. \tag{4.88}$$

To determine the Markov process the initial probabilities of the states must still be assigned. In particular, if a definite initial state is given, for example q_j, then this means that $p_j(0) = 1$, and all the remaining $p_i(0)$ are equal to zero ($i \neq j$). But in general a transition from the state q_j to some other one is now possible at the following time instant. Therefore for $t > 0$ the probability of the system staying in some state will turn out to be "spread" over various possible states.

It is not difficult to derive a formula that permits all the $p_i(t + 1)$ to be found for a given $u = \alpha_k$ at the instant t, whenever the $p_i(t)$ are prescribed ($i = 1, ..., N + 1$). We will consider the state q_j at the time instant $t + 1$. Let all the $p_i(t)$ be known. Further, we will consider the probability of the compound event, consisting of the transition from the state q_i at the instant t to the state q_j at the time instant $t + 1$, if the probability $p_i(t)$ is known. The stated probability of the compound event is equal to the product of the absolute probability of being in the state q_i at the instant t, i.e., $p_i(t)$, times the conditional probability $p_{ij}(k)$ of transition from the ith state to the jth. Thus the probability of this compound event is equal to $p_i(t) \cdot p_{ij}(k)$. Since the probability $p_j(t + 1)$ of being in the state q_j at the time instant $t + 1$ is the sum of the transition probabilities to the state q_j from arbitrary states q_i ($i = 1, ..., N + 1$), then the $N + 1$ equalities

$$p_j(t + 1) = \sum_{i=1}^{N+1} p_i(t)p_{ij}(k) \qquad (j = 1, ..., N + 1, \quad t = 0, 1, ...) \tag{4.89}$$

hold. These expressions permit the probabilities $p_i(1)$ of the states for $t = 1$ to be found from the initial probabilities $p_i(0)$ of the states. Then the quantities $p_i(2)$ can be found by the same formulas (but now in general for another control α_k), and so on.

Let us introduce the probability state vector (this is a row vector, or else a lower-case vector):

$$\bar{p}(t) = (p_1(t), p_2(t), ..., p_{N+1}(t)). \tag{4.90}$$

Then equalities (4.89) can be replaced by a single vector equality:

$$\bar{p}(t + 1) = \bar{p}(t)\bar{\bar{P}}_k, \tag{4.91}$$

where $\bar{P}_k$ is the matrix defined by expression (4.87). For an example let us consider the case when $u = \alpha_k$ is fixed and does not depend on t. Then

$$\bar{p}(1) = \bar{p}(0)\bar{P}_k, \qquad \bar{p}(2) = \bar{p}(1)\bar{P}_k = \bar{p}(0)\bar{P}_k^2 \tag{4.92}$$

and in general,

$$\bar{p}(t) = \bar{p}(0)\bar{P}_k^t, \tag{4.93}$$

where the matrix $\bar{P}_k^t$ is the tth power of the matrix $\bar{P}_k$. In principle Eq. (4.93) permits $\bar{p}(t)$ to be found for any t. For example, for the diagram depicted in Fig. 4.6 let

$$p_1(0) = 1, \qquad p_2(0) = 0 \tag{4.94}$$

be given. This means that at the initial instant $t = 0$ the object is in the state 1. At the initial instant the probability state vector is

$$\bar{p}(0) = (1, 0). \tag{4.95}$$

The matrix $\bar{P}_k$ has the form (see Fig. 4.6)

$$\bar{P}_k = \begin{vmatrix} p_{11} & p_{12} \\ p_{21} & p_{22} \end{vmatrix} = \begin{vmatrix} \frac{1}{2} & \frac{1}{2} \\ \frac{2}{5} & \frac{3}{5} \end{vmatrix}. \tag{4.96}$$

Therefore according to the known rules for matrix multiplication we find

$$\bar{p}(1) = \bar{p}(0)\bar{P}_k = |\,1, 0\,| \begin{vmatrix} \frac{1}{2} & \frac{1}{2} \\ \frac{2}{5} & \frac{3}{5} \end{vmatrix} = |\,\tfrac{1}{2}, \tfrac{1}{2}\,|, \tag{4.97}$$

$$\bar{p}(2) = \bar{p}(1)\bar{P}_k = |\,\tfrac{1}{2}, \tfrac{1}{2}\,| \begin{vmatrix} \frac{1}{2} & \frac{1}{2} \\ \frac{2}{5} & \frac{3}{5} \end{vmatrix} = |\,\tfrac{9}{20}, \tfrac{11}{20}\,|, \tag{4.98}$$

and so on. Here it is necessary to consider the vector of the lower-case matrix. The values of $p_1(t)$ and $p_2(t)$ are given in the following table:

t	0	1	2	3	4	5
$p_1(t)$	1	0.5	0.45	0.445	0.4445	0.44445
$p_2(t)$	0	0.5	0.55	0.555	0.5555	0.55555

From the table it is seen that with an increase in t the quantities $p_1(t)$ and $p_2(t)$ tend to constant limits: $p_1(t) \to 0.4444 \ldots$, and $p_2(t) \to 0.5555 \ldots$ These values are the so-called limiting or final probabilities. It can be shown that the values of the final probabilities obtained in this system are the same no matter what the initial conditions $p_1(0)$ and $p_2(0)$. From a physical point of view the final probabilities are the probabilities in steady-state behavior. A Markov process for which the limiting probabilities do not depend on the initial conditions is called completely ergodic [4.7]. Thus the process corresponding to the diagram of Fig. 4.6 is completely ergodic.

In order to find the limiting probabilities of a completely ergodic process, the sequence $p_j(t)$ $(t = 0, 1, 2, \ldots)$ need not be computed. For example, let us consider Eqs. (4.91) for the diagram of Fig. 4.6. We will write them in the form of two equations of the type (4.89):

$$p_1(t+1) = \tfrac{1}{2}p_1(t) + \tfrac{2}{5}p_2(t), \qquad p_2(t+1) = \tfrac{1}{2}p_1(t) + \tfrac{3}{5}p_2(t). \quad (4.99)$$

When the transient process ends, then the probabilities p_1 and p_2 assume the steady-state values $p_{1,f}$ and $p_{2,f}$. Substituting $p_{1,f}$ in place of $p_1(t+1)$ and $p_1(t)$, and $p_{2,f}$ in place of $p_2(t+1)$ and $p_2(t)$ into (4.99), we arrive at the equations

$$p_{1,f} = \tfrac{1}{2}p_{1,f} + \tfrac{2}{5}p_{2,f}, \qquad p_{2,f} = \tfrac{1}{2}p_{1,f} + \tfrac{3}{5}p_{2,f}. \quad (4.100)$$

The solution of these equations gives:

$$p_{1,f} = 4/9 = 0.4444 \ldots, \qquad p_{2,f} = 5/9 = 0.5555 \ldots, \quad (4.101)$$

which coincides with the limits to which the values of $p_j(t)$ tend as $t \to \infty$, as found in the table.

In this example both the final probabilities $p_{1,f}$ and $p_{2,f}$ differ from zero. But in some cases one of these probabilities is equal to unity, and the other one to zero. This means that in steady-state behavior the object is in a definite state. Such is the object whose diagram is depicted in Fig. 4.7. In this case it is clear without any calculations that having hit

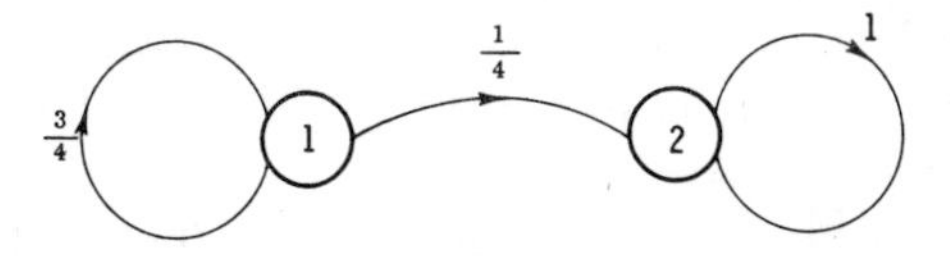

FIG. 4.7

state 2 (and the probability of this occurrence is greater than zero), the system will now no longer leave it.

Cases are possible where the distribution of limiting probabilities depends on the initial conditions. For example, such is the object with the diagram shown in Fig. 4.8. The object remains in position 1 if it is put into state 1 at the initial instant; but it remains in position 2 if it is found to be in it at the initial instant.

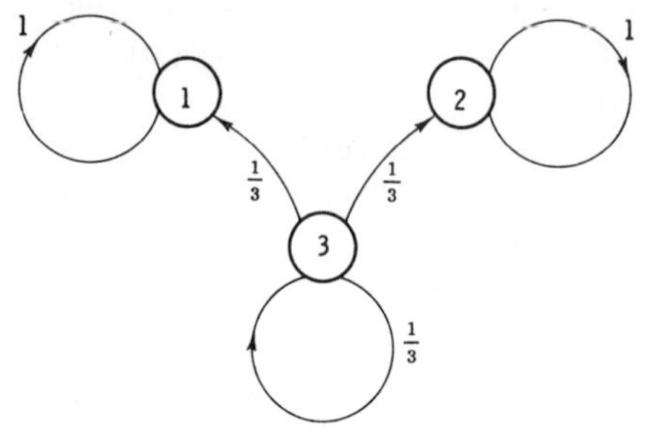

FIG. 4.8

In steady-state behavior the existence of limit cycles is possible. Let us consider the diagram depicted in Fig. 4.9. No matter what the initial conditions, it will finally "close" the system in a limit cycle consisting of the alternation of the states 2 and 3.

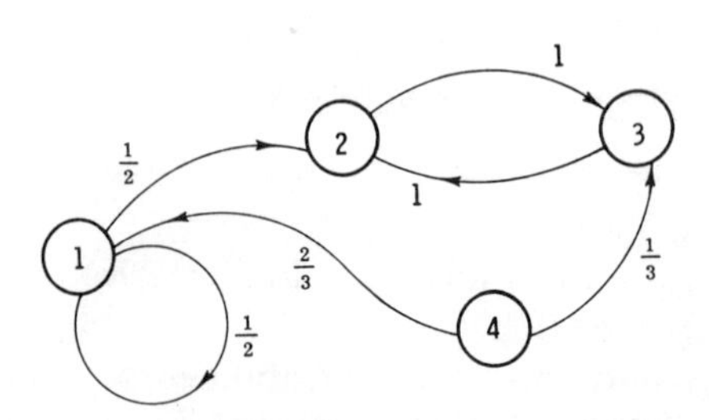

FIG. 4.9

From what was presented above it is seen that the determination of the $p_j(t)$ reduces to the investigation of finite difference equations of the type (4.89). These equations have been studied in the mathematical literature on Markov processes [4.13, 4.14], in works on the theory of finite difference equations [4.15], in the theory of impulse systems [4.16], in the theory of extremal systems [4.11], and also in learning theory in physiology [4.17, 4.18]. Although the object is purely discrete, nevertheless by finding the quantities $p_j(t)$ it is possible to forget about their origin and consider them as values at the time instant t of certain signals at the nodes of diagrams of the type in Fig. 4.6–4.8. The nodes correspond to the states of the object. In Eqs. (4.85) the quantities $p_{ij}(k)$ can in this case be regarded as amplification coefficients for the sections of the diagrams, corresponding to the paths leading from one state to another. Since in Eqs. (4.89) the value of $p_i(t+1)$ depends on the previous values $p_j(t)$, then each section joining two states must include a single-time delay element. Thus an equivalent diagram can be constructed corresponding to the discrete Markov process. The design of this equivalent diagram can be carried out by any method used for the investigation of impulse systems of the discrete-continuous type. For example, frequently the z-transform method [4.7] or the discrete Laplace transform [4.16] is applied. We will not consider equivalent diagrams. Those interested can become familiar with them from the literature cited above.

Let us consider in general form how to find the optimal control for a discrete Markov object [4.12]. Let the system have the $N+1$ states $q_1, q_2, \ldots, q_{N+1}$, of which q_{N+1} is the required final state, or else the purposeful state.

With any transition from q_i to q_j for which the probability $p_{ij}(k)$ is positive, we will associate a certain positive "cost" $c_{ij}(k) > 0$. Here the cost associated with transitions from the purposeful state to itself is regarded as equal to zero; in addition, we consider that transitions from the purposeful state to another state are missing. Thus,

$$p_{N+1,N+1}(k) = 1 \qquad (k = 1, 2, \ldots, m) \tag{4.102}$$

for arbitrary commands $u_k = \alpha_k$.

In the particular case of all i, j, k for which $p_{ij}(k) \neq 0$, $c_{ij}(k)$ is equal to 1. In this case the total cost of the entire process will be larger the larger the number of transitions, i.e., the greater the length of the process.

At the instant $t = 0$ the system is in some initial state s_0 which can be measured, as well as all the $s_t(t \geqslant 0)$. Precise information about the current state s_t acts from the controlled object to the controller, which generates the control action u_t. Since all the $p_{ij}(k)$ are known, then by knowing the current state the transition probabilities to all other states can be calculated. Thus for the current instant the information about the object is obtained as the maximal possible.

During l steps a certain sequence $(s_0, s_1, \ldots, s_l)$ of states of the object occurs, and also a sequence $(u_0, u_1, \ldots, u_{l-1})$ of command-control actions. We will call the collection $(\alpha_{k0}, \alpha_{k1}, \ldots, \alpha_{k,l-1}, q_{j0}, q_{j1}, \ldots, q_{jl})$

of command and state sequences possible if it can be realized, i.e., if the conditional probability of the appearance of this sequence ($\alpha_{k0}, \alpha_{k1}, \ldots, \alpha_{k,l-1}$) of commands and the sequence ($q_{j0}, q_{j1}, \ldots, q_{jl}$) of states with the initial state q_{j0} is greater than zero.

We associate with each possible sequence of length l the cost c in the following manner:

$$c = \sum_{\nu=0}^{l} c_{i_\nu i_{\nu+1}}(k). \tag{4.103}$$

Here $c_{i_\nu i_{\nu+1}}$ is the cost of a single transition from $s_\nu = q_{i_\nu}$ to $s_{\nu+1} = q_{i_{\nu+1}}$, when the command $u_\nu = \alpha_{k_\nu}$ appears at the input to the object. When the object hits the purposeful state, then further gain in the cost c ceases. In fact, the costs of subsequent transitions from the state q_{N+1} to itself are equal to zero. The value of c is also determined from formula (4.103) for such processes when the purposeful state is not attained at all. Here the total cost of the process turns out to be infinitely large, if the number of steps l is infinite.

In [4.12] the control algorithm, i.e., the law

$$u_t = \pi(s_t) \tag{4.104}$$

which associates with each state, besides the purposeful one, a certain command, has been called a strategy or policy. We must give this command u_t if the object is in the state s_t. The law (4.104) represents a perfect analog of the algorithm

$$\bar{u} = \bar{u}(\bar{x}) \tag{4.105}$$

for a continuous system. Actually, the algorithm (4.105) associates with each point $\bar{x}$ of phase space a definite control $\bar{u}$. But for the discrete object described above the phase space is also discrete and turns into a finite set of states 1, 2, ... in the diagram (see Fig. 4.6–4.9). Therefore the analog of (4.105) in discrete systems is the law (4.104). Both of these expressions are characteristic for systems with complete or maximal information about the object, when the memory of the "previous history," i.e., the sequence $s_0, \ldots, s_{t-1}$ of states, does not add anything to the information about the object. We will not call laws of the type (4.104) or (4.105) strategies.

It is convenient to regard the set of expressions $\pi_1, \pi_2, \ldots, \pi_N$ of the type (4.104) for any possible states $q_1, q_2, \ldots, q_N$ as a vector

$$\bar{\pi} = (\pi_1, \pi_2, \ldots, \pi_N). \tag{4.106}$$

Here it is accepted that the command π_i is applied when the object is in the state q_i. Moreover, the notations $p_{ij}(\pi_i)$ and $c_{ij}(\pi_i)$ are used below in place of $p_{ij}(k)$ and $c_{ij}(k)$. Here it is accepted that the command π_i assumes one of its possible values $\alpha_1, ..., \alpha_m$.

Since the transitions from one state to another are random, the total cost c of the process is also a random variable, whose distribution depends on the form of the algorithm $\bar{\pi}$. We will use as the optimality criterion the mathematical expectation of the cost of reaching the position $X_i(\bar{\pi})$. This quantity depends on what the initial state q_i and the adopted algorithm $\bar{\pi}$ are. We will denote by $\bar{X}(\bar{\pi})$ the vector

$$\bar{X}(\bar{\pi}) = (X_1(\bar{\pi}), X_2(\bar{\pi}), ..., X_N(\bar{\pi})), \tag{4.107}$$

whose coordinates are the mathematical expectations of the cost for starting from the states $q_1, q_2, ..., q_N$. We will call the vector $\bar{X}(\bar{\pi})$ the cost vector in abbreviated fashion. Let $\bar{X}$ and $\bar{Y}$ be the two cost vectors corresponding to two different algorithms. Then the expression $\bar{X} \leqslant \bar{Y}$ means that $X_i \leqslant Y_i$ for all $i = 1, 2, ..., N$. Thus the algorithm corresponding to the vector $\bar{X}$ is uniformly better than the algorithm corresponding to the vector $\bar{Y}$. If the cost corresponding to a process of fixed length l is considered, then it can be denoted by $X_i(\pi, l)$, or more simply $X_i(l)$.

The algorithm $\bar{\pi}^0$ is called optimal if the condition

$$\bar{X}(\bar{\pi}^0) \leqslant \bar{X}(\bar{\pi}) \tag{4.108}$$

holds for arbitrary $\bar{\pi}$. For the optimal algorithm $\bar{\pi}^0$ a relation of the same type as Bellman's equation can be derived. Let us consider beforehand some algorithm $\bar{\pi}$ and the expected cost $X_i(\bar{\pi})$ for starting from the initial state q_i. The transition probability for the first step from q_i to some state q_j is equal to $p_{ij}(\pi_i)$, and the cost of this transition equals $c_{ij}(\pi_i)$. We will first assume that the first step has brought the object to the state q_j. Then the total cost of this process is equal to the sum

$$c_{ij}(\pi_i) + X_j(\bar{\pi}), \tag{4.109}$$

where $X_j(\bar{\pi})$ is the mathematical expectation of the cost of the process starting from the state q_j. The quantity defined by Eq. (4.109) is random, since it is possible for different states q_j to occur after the first step. By averaging the quantity (4.109), we obtain the following equation for the mathematical expectation $X_i(\bar{\pi})$:

$$X_i(\bar{\pi}) = M\{c_{ij}(\pi_i) + X_j(\bar{\pi})\} \tag{4.110}$$

or in expanded form,

$$X_i(\bar{\pi}) = \sum_{j=1}^{N+1} p_{ij}(\pi_i)[c_{ij}(\pi_i) + X_j(\bar{\pi})] \qquad (i = 1, ..., N). \tag{4.111}$$

Since $X_{N+1}(\bar{\pi}) = 0$,

$$X_i(\bar{\pi}) = \sum_{j=1}^{N} p_{ij}(\pi_i)X_j(\bar{\pi}) + \sum_{j=1}^{N+1} p_{ij}(\pi_i)c_{ij}(\pi_i) \qquad (i = 1, ..., N). \tag{4.112}$$

Instead of N scalar equalities a single vector equation

$$\bar{X}(\bar{\pi}) = \bar{\bar{P}}(\bar{\pi})\bar{X}(\bar{\pi}) + \bar{c}(\bar{\pi}) \tag{4.113}$$

can be written, where $\bar{\bar{P}}(\bar{\pi})$ is a square matrix with N rows and with elements $p_{ij}(\pi_i)$, and $\bar{c}(\bar{\pi})$ is a vector whose ith coordinate is equal to

$$\sum_{j=1}^{N+1} p_{ij}(\pi_i) \cdot c_{ij}(\pi_i) = c_i(\bar{\pi}). \tag{4.114}$$

The meaning of $c_i(\bar{\pi})$ is that it is the mathematical expectation of the cost of the first step for a start from q_i and with the use of the algorithm $\bar{\pi}$. Therefore $c_i(\bar{\pi})$ can be called the cost vector of the first step.

Equation (4.113) holds for any algorithm. We will denote the cost vector for the optimal algorithm $\bar{\pi}^0$ by $\bar{X}^0(\bar{\pi}^0)$.

In order to find the relation for $\bar{X}^0(\bar{\pi}^0) = \bar{X}^0$, we make the following argument. We will assume that the first step from q_i is made by a certain stipulated method, by the application of the command π_i; then the system finds itself in the position q_j and the following steps are made in the optimal way. Then analogous to (4.109) the mathematical expectation of the cost of the process will be equal to

$$c_{ij}(\pi_i) + X_j^0(\bar{\pi}^0). \tag{4.115}$$

The mathematical expectation of this quantity, equal to X_i, is obtained in the form

$$X_i = M\{c_{ij}(\pi_i) + X_j^0\} = \sum_{j=1}^{N+1} p_{ij}[c_{ij}(\pi_i) + X_j^0]. \tag{4.116}$$

To obtain the optimal algorithm X_i^0 the minimum of the right side of the equality must be taken over all possible values of π_i. Thus,

$$X_i^0 = \min_{\pi_i} \left\{ \sum_{j=1}^{N+1} p_{ij}[c_{ij}(\pi_i) + X_j^0] \right\}. \tag{4.117}$$

Of course, the minimum can be taken with respect to $\bar{\pi}$ and not over the π_i, from which Eq. (4.117) will not change. In vector form it will be written as:

$$\bar{X}^0 = \min_{\bar{\pi}} [\bar{\bar{P}}(\bar{\pi})\bar{X}^0 + \bar{c}(\bar{\pi})]. \tag{4.118}$$

By solving this equation $\bar{X}^0$ can be found; in passing the optimal algorithm $\bar{\pi}^0$ is determined as well. For the solution the fact can be used that, as seen from (4.118), $\bar{X}^0$ is a fixed point of the transformation

$$T(\bar{X}) = \min_{\bar{\pi}} [\bar{\bar{P}}(\bar{\pi})\bar{X} + \bar{c}(\bar{\pi})]. \tag{4.119}$$

After prescribing some zero approximation to $\bar{X}$ and carrying out the operation indicated on the right side of this equality, we will obtain the first approximation $T(\bar{X})$ to the vector $\bar{X}^0$. Introducing this first approximation into the right side of (4.119) in place of $\bar{X}$ and carrying out the operation indicated in (4.119), we will obtain the second approximation $T^2(\bar{X})$, and so on. Thus it can be assumed that by iterations one will succeed in approaching arbitrarily close to the required value $\bar{X}^0$. Simultaneously with the minimization of the right side of (4.119), the algorithm $\bar{\pi}$ is also determined closer and closer to the optimal. Thus,

$$\bar{X}^0 = \lim_{r\to\infty} T^r(\bar{X}). \tag{4.120}$$

Of course, the assumptions we have stated cannot replace a proof at all. Does there exist in general a uniformly optimal algorithm for this problem? If it exists, then is it unique? Does the iteration process presented above converge to it for any initial value of $\bar{X}$?

In [4.12] an exhaustive answer to these questions has been given. It turns out that under certain conditions a uniformly optimal algorithm exists and is unique. The iteration process stated converges for any initial value of $\bar{X}$. The fundamental result obtained in [4.12] (whose proof we omit) can be formulated in the following way.

Let an algorithm be called a proper algorithm under which the purposeful state is attainable from an arbitrary initial state of the object with a positive probability.

If there exists at least one proper algorithm, then Eq. (4.118) has a unique solution defined by formula (4.120), where $\bar{X}$ is the initial vector and T is given by expression (4.119).

The latter expression is valuable in the respect that it gives an effective means for the calculation of $\bar{X}^0$ and the optimal algorithm $\bar{\pi}^0$ related to it.

The arguments given above can be generalized to the case when the purposeful state varies as a function of time. Then we obtain the problem of hitting a moving target. We will assume that the driving action entering the controller at the instant t is a discrete variable ρ_t belonging to the set of values $q_1, \ldots, q_N$. (There is no state q_{N+1} now. Instead of q_{N+1} one of the states q_i appears as the purposeful state at each time instant.) This means that ρ_t is the current value of the driving action. It can vary in a random manner and be a Markov process with a transition probability $p(\rho_{t+1} \mid \rho_t)$. This is the conditional probability that the purposeful state will be ρ_{t+1} at the time instant $t + 1$, if it was equal to ρ_t at the time instant t. In order to generalize to the random method presented above, we can define a certain composite "vector" state $\bar{s}_t = (s_t, \rho_t)$, i.e., the vector with coordinates s_t and ρ_t.

The conditional probability for such a vector is given by the expression following from the theorem on multiplication of probabilities:

$$\begin{aligned} p(\bar{s}_{t+1} \mid \bar{s}_t, u_t) &= p(s_{t+1}, \rho_{t+1} \mid s_t, \rho_t, u_t) \\ &= p(s_{t+1} \mid s_t, \rho_t, u_t) \cdot p(\rho_{t+1} \mid s_{t+1}, s_t, \rho_t, u_t) \\ &= p(s_{t+1} \mid s_t, u_t) \cdot p(\rho_{t+1} \mid \rho_t), \end{aligned} \tag{4.121}$$

since the probability for s_{t+1} depends only on s_t and u_t, and the probability for ρ_{t+1} depends only on ρ_t, and thus s_{t+1} and ρ_{t+1} are independent.

The vector $\bar{s}_t$ can have one of the possible values $\bar{q}_\lambda = (q_i, q_j)$. In all, there are N^2 such states. Hence the vector $\bar{s}_t$ is a Markov process for which the N^2 states and the transition probability defined by formula (4.121) are considered known.

When the state s_t of the original system coincides with the driving action ρ_t, then one of the states $\bar{s}_t = (q_i, q_i)$ is realized in the Markov process $\bar{s}_t$. There are N such states; each of them corresponds to "hitting the target." The set of all such states (q_i, q_i) $(i = 1, \ldots, N)$ represents a region which we will denote by Q. Hence the problem of optimally hitting the target has been reduced to the problem of the optimal transfer of the system $\bar{s}_t$ from a given initial state to any of the states of the region Q. Of course, for the new system the notation s_t can be used once again in place of $\bar{s}_t$. Then the problem remains essentially the same, the only difference being that occurring in one purposeful state is replaced by occurring in the prescribed region Q consisting of several states.

Let the states belonging to the region Q be denoted by $q_{L+1}, \ldots, q_N$. Since the process ends when the state q_t gets to belong to Q the first time, then $q_{L+1}, \ldots, q_N$ can be called "absorption points." In fact, now the process does not leave the region Q any more. The cost of occurring in the region Q itself from the region Q is evidently equal to zero:

$$X_{L+1} = X_{L+2} = \cdots = X_N = 0. \tag{4.122}$$

Therefore the equation for the costs $X_i(\bar{\pi})$ $(i = 1, ..., L)$ of occurring in the region Q, which is derived in a fashion analogous to relation (4.112), can be written as:

$$X_i(\bar{\pi}) = \sum_{j=1}^{L} p_{ij}(\pi_i) X_j(\bar{\pi}) + \sum_{j=1}^{L} p_{ij}(\pi_i) c_{ij} \qquad (i = 1, 2, ..., L). \tag{4.123}$$

These equations are obtained from (4.112) if (4.122) is taken into account and we set $p_{ij}(\pi_i) = 0$ for $i = L + 1, ..., N$.

In vector form relations (4.123) become

$$\bar{X}(\bar{\pi}) = \bar{\bar{P}}^*(\bar{\pi}) \bar{X}(\bar{\pi}) + \bar{c}^*(\bar{\pi}), \tag{4.124}$$

where

$$\begin{aligned} \bar{\pi} &= (\pi_1, ..., \pi_L), \\ \bar{X}(\bar{\pi}) &= (X_1(\bar{\pi}), ..., X_L(\bar{\pi})), \\ \bar{c}^*(\bar{\pi}) &= (c_1^*(\bar{\pi}), ..., c_L^*(\bar{\pi})), \end{aligned} \tag{4.125}$$

and $\bar{\bar{P}}^*(\bar{\pi})$ is a matrix with elements $p_{ij}(\pi_i)$ $(i, j = 1, ..., L)$. Finally,

$$c_i^*(\pi_i) = \sum_{j=1}^{L} p_{ij}(\pi_i) c_{ij}(\pi_i) \qquad (i = 1, 2, ..., L). \tag{4.126}$$

In this case the minimal cost vector is defined by an equation analogous to (4.118):

$$\bar{X}^0 = \min_{\bar{\pi}} [\bar{\bar{P}}^*(\bar{\pi}) \bar{X}^0 + \bar{c}^*(\bar{\pi})]. \tag{4.127}$$

Thus in this case the replacement of the original system by a more complicated one is first performed, and then the solution of the analogous problem for the system obtained is found.

In the case considered above the transient process in a discrete Markov object has been investigated, and the optimal control determined minimizing some criterion (for example, time) related to the transient process. But another kind of problem can be solved, related not only to the transient but also to the steady-state process in a discrete Markov object [4.7]. For example, instead of the loss c_{ij} let the payoff r_{ij} be obtained for each transition from the state q_i to the state q_j (the payoff can be represented as a negative loss or the negative cost of the transition). Instead of the mathematical expectation of the total cost X_i of the process for a start from the state q_i, the mathematical expectation of the total payoff $v_i(n)$ after n steps can be introduced.

Let us examine n steps (or transition times) with the initial state q_i. We will assume from the beginning that the first step consists of the transition from q_i to the fixed state q_j with payoff r_{ij}. Then, since the mathematical expectation of the total payoff during the remaining $n-1$ steps is equal to $v_j(n-1)$, the mathematical expectation of the total payoff under the conditions stated above can be obtained in the form

$$r_{ij} + v_j(n-1). \tag{4.128}$$

But in fact the transition from q_i to q_j is random. Therefore the quantity defined by (4.128) is also random. By averaging it over the various states q_j with probabilities p_{ij}, we obtain the mathematical expectation of the total payoff for a start from the state q_i:

$$\begin{aligned} v_i(n) &= M\{r_{ij} + v_j(n-1)\} \\ &= \sum_{j=1}^{N} p_{ij}[r_{ij} + v_j(n-1)] = h_i + \sum_{j=1}^{N} p_{ij} v_j(n-1) \\ &\qquad (i = 1, 2, \ldots, N, \quad n = 1, 2, 3, \ldots), \end{aligned} \tag{4.129}$$

where

$$h_i = \sum_{j=1}^{N} p_{ij} r_{ij} \qquad (i = 1, \ldots, N). \tag{4.130}$$

Equation (4.129) is analogous to (4.112), and (4.130) is analogous to (4.114). We will introduce the column vectors $\bar{v}(n)$ and $\bar{h}$:

$$\bar{v}(n) = \begin{vmatrix} v_1(n) \\ \vdots \\ v_N(n) \end{vmatrix}, \qquad \bar{h} = \begin{vmatrix} h_1 \\ \vdots \\ h_N \end{vmatrix}, \tag{4.131}$$

and also the matrix $\bar{\bar{P}}$ with elements p_{ij}, having N columns and N rows. Then the vector equation

$$\bar{v}(n) = \bar{h} + \bar{\bar{P}}\bar{v}(n-1) \tag{4.132}$$

holds. This equality is analogous to (4.113). If $\bar{h}$ and $\bar{\bar{P}}$ are known, then by knowing $\bar{v}(1) = \bar{h}$, $\bar{v}(2)$ can be found from Eq. (4.132), then successively $\bar{v}(3)$, and so on. This problem also has meaning for an unbounded increase in n. Then the steady-state process is obtained for large n.

Now we will consider that p_{ij} and r_{ij} depend on the control $u_n = \alpha_k$ ($k = 1, ..., m$). We will find the optimal control for which the total payoff $v(n)$ is maximal. We will write the quantities p_{ij} and r_{ij} in the form $p_{ij}(k)$ and $r_{ij}(k)$, since they depend on α_k. Then we reason in the same way as for the derivation of Eq. (4.118), but with minimum replaced by maximum. Let the first step from q_i to q_j be fixed, and likewise the control $u_0 = \alpha_k$, where at all the subsequent instants the control is optimal. Then the mathematical expectation $v(n)$ of the payoff after n steps is written analogous to (4.128) in the following form:

$$r_{ij}(k) + v_j^0(n-1), \tag{4.133}$$

where $v_j^0(n-1)$ is the mathematical expectation of the payoff after $n-1$ steps for a start from q_i and under the optimal control. By choosing the control $u = \alpha_k$ at the "first" step such that the mathematical expectation of the payoff (4.133) be maximized, we arrive at the equation

$$v_i^0(n) = \max_k \sum_{j=1}^{N} p_{ij}(k)[r_{ij}(k) + v_j^0(n-1)] \qquad (i = 1, ..., N) \tag{4.134}$$

or in vector form,

$$\bar{v}^0(n) = \max_k \{\bar{\bar{P}}_k \bar{v}^0(n-1) + \bar{h}\}. \tag{4.135}$$

From Eqs. (4.134) or (4.135) the optimal mathematical expectation $\bar{v}^0(n)$ of the payoff can be found by successive calculations. In fact, as soon as $\bar{v}^0(0) = 0$ is known, then by using these equations we can find successively $\bar{v}^0(1)$, $\bar{v}^0(2)$, $\bar{v}^0(3)$, and so on. The simultaneously carried out maximization with respect to the u_k gives the optimal control.

The quantity

$$h_i = h_i(k) = \sum_{j=1}^{N} p_{ij}(k) r_{ij}(k) \tag{4.136}$$

depends on the commands $u_n = \alpha_k$ ($k = 1, ..., m$). In turn, the number of commands k depends on n. Introducing the notation $h_i(k)$, we can rewrite (4.134) in the form

$$v_i^0(n) = \max_k \left[h_i(k) + \sum_{j=1}^{N} p_{ij}(k) v_j^0(n-1)\right]. \tag{4.137}$$

Since $p_{ij}(k)$ depends on k, then both terms in the square brackets of expression (4.137) depend on k.

Let us consider the example of [4.7]. Let there be only two possible commands $u = \alpha_k$, where $k = 1, 2$. Further, let there be two states q_1 and q_2. In the case $k = 1$ let the matrices $p_{ij}(1)$ and $r_{ij}(1)$ have the form

$$\bar{P}_1 = \| p_{ij}(1) \| = \begin{vmatrix} 0.5 & 0.5 \\ 0.4 & 0.6 \end{vmatrix}, \quad \bar{R}_1 = \| r_{ij}(1) \| = \begin{vmatrix} 9 & 3 \\ 3 & -7 \end{vmatrix}. \tag{4.138}$$

But if $k = 2$, then the matrices have the form

$$\bar{P}_2 = \| p_{ij}(2) \| = \begin{vmatrix} 0.8 & 0.2 \\ 0.7 & 0.3 \end{vmatrix}, \quad \bar{R}_2 = \| r_{ij}(2) \| = \begin{vmatrix} 4 & 4 \\ 1 & -19 \end{vmatrix}. \tag{4.139}$$

We calculate the values of $h_i(k)$:

$$\begin{aligned} h_1(1) &= p_{11}(1)r_{11}(1) + p_{12}(1)r_{12}(1) = 0.5 \cdot 9 + 0.5 \cdot 3 = 6, \\ h_2(1) &= p_{21}(1)r_{21}(1) + p_{22}(1)r_{22}(1) = 0.4 \cdot 3 + 0.6(-7) = -3. \end{aligned} \tag{4.140}$$

In an analogous fashion

$$\begin{aligned} h_1(2) &= p_{11}(2)r_{11}(2) + p_{12}(2)r_{12}(2) = 0.8 \cdot 4 + 0.2 \cdot 4 = 4, \\ h_2(2) &= p_{21}(2)r_{21}(2) + p_{22}(2)r_{22}(2) = 0.7 \cdot 1 + 0.3(-19) = -5. \end{aligned} \tag{4.141}$$

From formula (4.137) for $n = 1$, after setting $v_j{}^0(0) = 0$, we find

$$v_1{}^0(1) = \max_{k(1)} \{h_1(k)\} = \max_{k(1)} \begin{cases} h_1(1) = 6, & k(1) = 1, \\ h_1(2) = 4, & k(1) = 2. \end{cases} \tag{4.142}$$

Evidently we must choose $k(1) = 1$, since then $v_1(1)$ gains the maximal value $v_1{}^0(1) = 6$. If the start is made from the state q_2, then

$$v_2{}^0(1) = \max_{k(1)} \{h_2(k)\} = \max_{k(1)} \begin{cases} h_2(1) = -3, & k(1) = 1, \\ h_2(2) = -5, & k(1) = 2. \end{cases} \tag{4.143}$$

Here also we must choose $k(1) = 1$; then $v_2(1)$ gains the maximal value $v_2{}^0(1) = -3$.

In order to find $v_i(2)$, we use formula (4.137), which we will write for $n = 2$:

$$v_i{}^0(2) = \max_{k(2)} \left[h_i(k) + \sum_{j=1}^{2} p_{ij}(k) v_j{}^0(1) \right]. \tag{4.144}$$

Here the $v_j{}^0(1)$ are the values

$$v_1{}^0(1) = 6, \quad v_2{}^0(1) = -3 \tag{4.145}$$

found earlier. We will first find $v_1^0(2)$. For this we write the two possible values of $v_i(2)$ for different values of k:

$$\begin{aligned} [v_1(2)]_{k=1} &= h_1(1) + p_{11}(1)v_1^0(1) + p_{12}(1)v_2^0(1) \\ &= 6 + 0.5 \cdot 6 + 0.5(-3) = 7.5, \\ [v_1(2)]_{k=2} &= h_1(2) + p_{11}(2)v_1^0(1) + p_{12}(2)v_2^0(1) \\ &= 4 + 0.8 \cdot 6 + 0.2(-3) = 8.2. \end{aligned} \tag{4.146}$$

Since $v_i(2)$ assumes the greatest value for $k = 2$, we set $k = 2$. Then $v_1^0(2) = 8.2$.

In an analogous fashion we find $v_2^0(2) = -1.7$, where the optimal control in this case is $k(2) = 2$. Continuing this process further, we find the values of $k_1(n)$—the optimal control for a start from q_1 and $k_2(n)$—for a start from q_2 (see accompanying table). As is seen from the table, as the transient process ends and the steady-state process appears in the system, the increment $\Delta v_i^0(n)$ in the quantity $v_i^0(n)$ approaches some constant value. As is shown below, it is not random.

	n				
$n =$	0	1	2	3	4...
$k_1(n)$	—	1	2	2	2...
$v_1^0(n)$	0	6	8.2	10.22	12.222...
$\Delta v_1^0(n) = v_1^0(n) - v_1^0(n-1)$	—	6	2.2	2.02	2.002...
$k_2(n)$	—	1	2	2	2...
$v_2^0(n)$	0	−3	−1.7	0.23	2.223...
$\Delta v_2^0(n) = v_2^0(n) - v_2^0(n-1)$	—	−3	−1.3	1.93	1.993...

For example, if three times remain until the end of the process, and thus $n = 3$, then the expected payoff for a start from the state q_1 equals 10.22. Here the optimal control $k_1(3) = 2$ must be used.

The method presented above for determining the $v_i^0(n)$ becomes too cumbersome for large values of n. To determine the optimal algorithm in the steady-state process as $n \to \infty$, it is convenient to apply another method.

As $n \to \infty$ the state probabilities tend to the same limiting (final) values p_{if} independent of the initial state, if the process is completely ergodic. Here the mathematical expectation of the payoff after one step becomes a constant, which we will denote by g. But the total payoff

$v_i(n)$ after n steps receives a constant increment g as $n \to \infty$ with each new step. Hence for large n the asymptotic expression

$$v_i(n) \simeq ng + v_i \tag{4.147}$$

for $v_i(n)$ holds, where the v_i are constants which can be different for different i. And $v_1{}^0(n)$ and $v_2{}^0(n)$ acquire just such an asymptotic form for large n in the table given above.

Substituting expression (4.147) into formula (4.129), we obtain:

$$ng + v_i = h_i + \sum_{j=1}^{N} p_{ij}[(n-1)g + v_j] \qquad (i = 1, ..., N). \tag{4.148}$$

Since $\sum_{j=1}^{N} p_{ij} = 1$, from here it follows that

$$g + v_i = h_i + \sum_{j=1}^{N} p_{ij} v_j \qquad (i = 1, ..., N). \tag{4.149}$$

The N equations (4.149) must be used to determine the $N + 1$ unknowns $g, v_1, ..., v_N$. We will note that adding an arbitrary constant a to all the v_i will not change Eqs. (4.149). Hence it follows that in general the values of v_i by themselves cannot be determined from these equations. Only the differences $v_i - v_j$ can be determined from them. However, the values of v_i by themselves are not required in order to determine the optimal algorithm in the steady-state condition as an algorithm for which the average payoff g necessary at one step is maximal. In this case Eqs. (4.149) give all that is required.

However, the quantity g can be determined from other considerations. We will assume that we have found the final state probabilities p_{if}. Here as is seen from (4.129) the average payoff at one step is equal to h_i. But since arbitrary q_i with probabilities p_{if} are possible in the steady-state condition, the mathematical expectation of the payoff at one step is defined by the expression

$$g = \sum_{i=1}^{N} p_{if} \cdot h_i . \tag{4.150}$$

To use this formula the final probabilities p_{if} of the states must be determined as a preliminary. How this is done was shown above. It is simpler to make the determination of g and the v_i from Eqs. (4.149) by setting one of the v_i, for example v_N, equal to zero. Then the differences $v_i - v_N$ will actually be determined; but the value of g can be obtained from these equations. Meanwhile Eqs. (4.137) give the possibility of finding by iterations the optimal algorithm $k^0(i)$ (the collection of optimal commands k^0 for all states q_i in the steady-state condition), for which the expected payoff at the step g is maximal.

From (4.137) it follows that the optimal algorithm k^0 is obtained by a maximization over k of the following expression:

$$h_i(k) + \sum_{j=1}^{N} p_{ij}(k) v_j(n) \tag{4.151}$$

[in (4.137), $v_i(n+1)$ instead of $v_i(n)$ and $v_j(n)$ in place of $v_j(n-1)$ can be substituted].

For large n this expression, in accordance with (4.147), can be replaced by

$$h_i(k) + \sum_{j=1}^{N} p_{ij}(k)(ng + v_j) = h_i(k) + \sum_{j=1}^{N} p_{ij}(k) v_j + ng. \tag{4.152}$$

Only part of this expression depends on k:

$$h_i(k) + \sum_{j=1}^{N} p_{ij}(k) v_j\,. \tag{4.153}$$

In order to start the iteration process, we will be given some initial values of the v_j. For example, we can set $v_j = 0$ for all j. Then the v_j are fixed and for any i the commands k can be found which maximize expression (4.153). Now let us fix the algorithm k found and solve Eqs. (4.149) with the aim of determining g and the v_i (here we always set $v_N = 0$). By finding new values of the v_i, we again revert to (4.153) and maximize this expression with respect to k by defining a new iteration of the commands here. Finding k for all i, we again fix them, determine g and the v_i from (4.149) for a new iteration, and so on. It has been proved [4.7] that this process converges and in the limit gives the solution of the equation

$$g + v_i^0 = \max_k \left[h_i(k) + \sum_{j=1}^{N} p_{ij}(k) v_j^0 \right], \tag{4.154}$$

which is quite analogous to (4.134) for the case of steady-state behavior. It should not be forgotten that here v_i^0 and v_j^0 are certain numbers which do not depend on n.

Let us consider the example given in [4.7], for the case of the matrices (4.138) and (4.139). We will choose first $k = 1$ for both $i = 1$ and $i = 2$.

In this case Eqs. (4.149) take the form

$$\begin{aligned} g + v_1 &= 6 + 0.5v_1 + 0.5v_2\,, \\ g + v_2 &= -3 + 0.4v_1 + 0.6v_2\,. \end{aligned} \tag{4.155}$$

By setting $v_2 = 0$ and solving these equations, we find

$$g = 1, \qquad v_1 = 10, \qquad v_2 = 0. \tag{4.156}$$

Now the values of v_i obtained must be introduced into expression (4.153) and the maximization with respect to k carried out for each i. The result of the calculations is given in the accompanying table.

i	k	$h_i(k) + \sum_{j=1}^{N} p_{ij}(k)v_j$
1	1	$6 + 0.5 \cdot 10 + 0.5 \cdot 0 = 11$
	2	$4 + 0.8 \cdot 10 + 0.2 \cdot 0 = 12$
2	1	$-3 + 0.4 \cdot 10 + 0.6 \cdot 0 = 1$
	2	$-5 + 0.7 \cdot 10 + 0.3 \cdot 0 = 2$

Hence it follows that for $i = 1$ the optimal value is $k = 2$; for $i = 2$ the optimal value is likewise $k = 2$.

Let us continue the iteration process. Fixing $k = 2$ for any i, we obtain Eqs. (4.149) in the form

$$\begin{aligned} g + v_1 &= 4 + 0.8v_1 + 0.2v_2\,, \\ g + v_2 &= -5 + 0.7v_1 + 0.3v_2\,. \end{aligned} \tag{4.157}$$

After setting $v_2 = 0$ and solving these equations, we find

$$g = 2, \qquad v_1 = 10, \qquad v_2 = 0. \tag{4.158}$$

Comparing (4.158) with (4.156), one will be convinced that as a result of the new iteration twice as large an average payoff g is obtained than before. The iterations can be continued further as well. But the values of the v_i in (4.158) turned out to be the same as in (4.150). Hence their substitution into (4.153) and the maximization of this expression with respect to k will give the same result as the table quoted above. Thus the values of $k(i)$ obtained for the following iteration coincide with the values of the preceding iteration. This means that further iterations are not necessary and the algorithm found is optimal, satisfying condition (4.154). Thus the optimal algorithm in the example under consideration is $k^0(1) = k^0(2) = 2$, and the optimal average payoff at the step $g = 2$.

The method of iterating the algorithm presented above is only applicable for steady-state processes. In this case it permits the optimal algorithm to be found comparatively simply.

In [4.7] an interesting domain of problems with discrete objects was also considered, in which the transition instant from one state to another is a random (Poisson) process. Similar problems are not examined here. The problems dealing with the optimal process in a continuous system with Poisson input actions have been studied in [4.21].

CHAPTER V

Optimal Systems with Independent (Passive) Storage of Information about the Object

1. Fundamental Problems in the Theory of Optimal Systems with Independent Information Storage

As before we will consider that there is complete information in the controller A about the operator F of the object and about the control purpose, i.e., about the form of the optimality criterion Q. But information about the driving action $\bar{x}^*$, about the perturbation $\bar{z}$ acting on the object B, and about the output variable $\bar{x}$ of the object may be incomplete. Further, we will assume that the store of information about the quantities $\bar{x}^*$, $\bar{x}$, and $\bar{z}$ can be increased, accumulated in the course of time, where this storage process does not depend on the actions of the controller A. If the latter controls the object in an optimal manner, then we will call such systems optimal systems with independent or passive information storage (see also Chapter I).

Information storage can arise in two cases.

(a) Let the variable $\bar{x}^*$ (or $\bar{z}$), measured without error, be a more complicated random process than Markov. The probabilistic characteristics of this process can be refined by observing it during some time interval. In this case the observation permits information to be accumulated making the behavior of the process in the future more precise.

(b) The variable $\bar{x}^*$ (or $\bar{z}$) is measured with some error, or the result of the measurement passes through a channel with noise adding to the useful signal. In this case observation is necessary for refining the values of the useful signal. The larger the observation time is, the more exactly the behavior of $\bar{x}^*$ in the future can in general be determined.

The second case is most important.

A typical example of a system with independent information storage is shown in Fig. 5.1. In this case information about the value of the controlled variable $\bar{x}$ acts on the controller A through a feedback circuit from the output of the object B. Inside the closed loop of the system noise and fluctuations are absent. But the driving action $\bar{x}^*$ acts on the input of A through a channel H^* with noise $\bar{h}^*$. In this channel the noise is mixed with the useful signal. At the output of the channel H^* the variable $\bar{y}^*$ differing from $\bar{x}^*$ appears. It is also fed to the input of the controller A instead of $\bar{x}^*$. The separation of the useful signal from the noise is the task of the controller; it can be accomplished

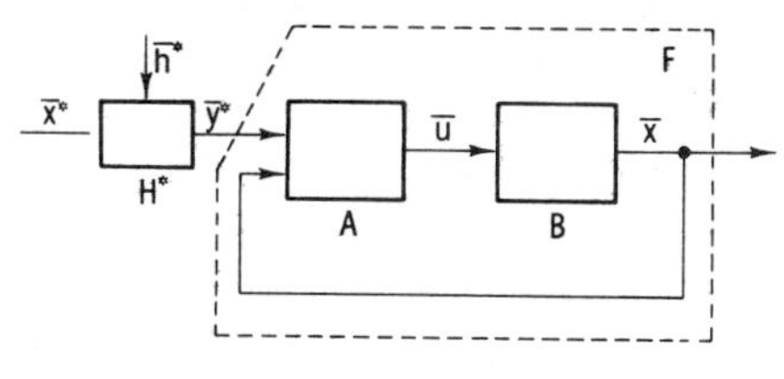

FIG. 5.1

with a definite degree of reliability if the values of $\bar{y}^*$ are observed during some time interval. An estimate of the value of $\bar{x}^*$ made at the expiration of this time interval will generally depend on the observed values of $\bar{y}^*$. Thus the estimate of $\bar{x}^*$, and hence the control action as well, undertaken by the device A at a given current time instant t, depends on the "previous history" of the input variable $\bar{y}^*(\tau)$ for $\tau < t$. In other words, the control action $\bar{u}(t)$ at the time instant t is a functional of the values of $\bar{y}^*(\tau)$ for $\tau < t$. But from here it follows that in contrast to the systems considered in Chapters III and IV, the optimal controller A in this case is now not inertialess. It must be a "dynamical system" whose output variable at a given time instant depends not only on the current values of the input variables, but also on its values in the past. In this and the following chapters the optimal controller is a dynamical system.

A block diagram of a closed-loop control system is shown in Fig. 5.1. The problem of the synthesis of the optimal controller A is posed—more precisely, the problem of finding its optimal algorithm. Since A is now a dynamical system, in order to differentiate this more complex case from the simpler ones considered earlier we will speak of the "optimal strategy" of the controller A.

Sometimes the problem of determining the optimal strategy of the device A in the diagram of Fig. 5.1 is solved by dividing the whole problem into two stages. In the first stage the entire system inside the dashed contour is regarded as one arrangement F, and the optimal algorithm of this system as a whole is determined. In the second stage F is analyzed by finding for a given operator of the object the operating law of the device A. Of course, difficulties can arise here related to the realizability or coarseness of the device A obtained. We will call a device realizable when its output variable depends perhaps on current and past values, but does not depend in any case on future values of the input variables, if the latter are not given in advance. The property of coarseness means that under sufficiently small changes in the parameters of the controller algorithm, the variation in any output variables or characteristics of this device or the system as a whole is arbitrarily small. Only as the controller becomes a dynamical system are we forced to impose constraints on possible types of such devices in the form of conditions of realizability and coarseness. Frequently the concept of physical realizability ordinarily used includes both the notions mentioned above as subordinate to it.

An open-loop system with independent information storage is shown in Fig. 5.2. As in Fig. 5.1, the driving action $\bar{x}^*$ passes through the

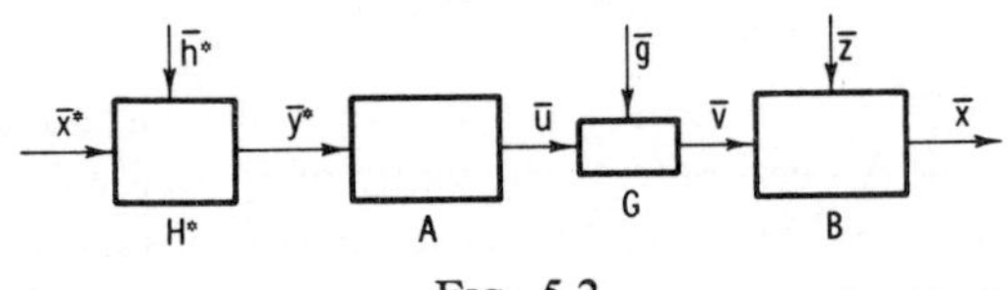

FIG. 5.2

channel H^* with noise $\bar{h}^*$. From this channel the action $\bar{y}^*$—a mixture of useful signal and noise—acts on the input of the controller A. We will assume that the latter operates on the object B through the channel G with noise $\bar{g}$. Therefore the true action $\bar{v}$ entering the input to the object B can differ from the action $\bar{u}$ at the output of the controller A. In this chapter we will mainly consider open-loop systems whose output variable $\bar{x}$ is not fed to the input of A.

The random noise $\bar{z}$ acting on the object B is not measured in Fig. 5.2. Therefore in this scheme only the *a priori* probabilistic characteristics of the noise $\bar{z}$ are known, which we will assume can be found by the statistical processing of trials in the past and put into the device A. In the scheme of Fig. 5.2 the device A does not receive any information about the specific behavior of the noise $\bar{z}$ in a given trial.

However, the case is possible when in the course of a trial the quantity $\bar{z}$ is measured, and the result of the measurement is sent to the input of the device A. Such a case is depicted in Fig. 5.3. Since any measurement is made with some error, this process can be represented in the form of the transmission of the result of an exact measurement of the noise $\bar{z}$ through the channel E with noise $\bar{e}$, which is added to the useful signal. In general the value of $\bar{w}$ at the output of the channel E, fed to the controller A, differs from the true value of $\bar{z}$.

Even if the noise $\bar{z}$ is measured approximately, the possibility does occur of its exact or approximate compensation. This means that the action $\bar{u}$ can be calculated in such a manner as to neutralize the effect of the action of the noise $\bar{z}$, and obtain exactly or approximately the required law relating $\bar{x}$ and $\bar{x}^*$, which in the ideal case must not depend on the value of $\bar{z}$.

Often the problem of constructing an automatic control system is formulated as a dual problem. On the one hand $\bar{x}$ must be subject to a certain required law (usually depending on $\bar{x}^*$); on the other hand the variable $\bar{x}$ must not depend on the noise $\bar{z}$. Sometimes the latter requirement is selected—the condition of $\bar{x}$ and $\bar{z}$ being independent—and it is

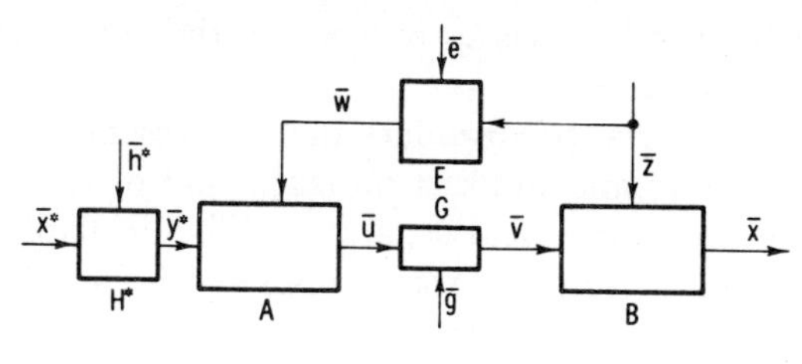

FIG. 5.3

formulated in the form of an invariance condition: $\bar{x}$ must be invariant with respect to $\bar{z}$. But in the general case it is more convenient to formulate the control purpose not in the form of two conditions, but in the form of a unique condition, as is done everywhere in this book. Let the control purpose be formulated in the form of the condition

$$Q(\bar{x}, \bar{x}^*) = \min. \tag{5.1}$$

For example, in the specific case of an nth-order system some criterion can be used, say,

$$Q = \int_0^T \sum_{i=1}^n (x_i - x_i^*)^2 \, dt. \tag{5.2}$$

Then in the optimal system the criterion Q assumes the minimal possible value, and by the same token a minimal degree of dependence of $\bar{x}$ on the noise $\bar{z}$ is guaranteed.

If the attainment of the absolute minimum of Q in the system is possible, then such a system can be called "ideal." For example, in an ideal system with the criterion (5.2) the values of the x_i must always be equal to the x_i^*; therefore in the ideal case Q assumes the minimal possible value, equal to zero. Just because of the fact that $\bar{x}$ depends only on $\bar{x}^*$, the complete independence of $\bar{x}$ from the noise $\bar{z}$ is realized in the ideal system. Of course, a non-ideal system is possible in which $\bar{x}$ is not exactly equal to the required value $\bar{x}^*$, but nevertheless does not depend on $\bar{z}$.

According to [5.1, 5.2], in an ideal system "complete invariance" holds. The error $\bar{\epsilon} = \bar{x} - \bar{x}^*$ is identically equal to zero and does not depend on $\bar{x}^*$ and $\bar{z}$.

The equating of $\bar{\epsilon}$ identically to zero, i.e., the fulfillment of the idealness condition for the system, represents the best conceivable solution. If we always succeeded in obtaining such a solution, then the theory of optimal systems would reduce to the theory of invariance. In order to find the interrelation between these two distinct theories, it is necessary to establish for what classes of systems the condition of complete invariance is attainable.

The theory of invariance considers the two essentially distinct cases of systems with direct and indirect measurement of the noise $\bar{z}$ respectively, sometimes without a clear separation. The principle of compensation for the first of these cases was proposed by Ponsel as far back as 1829. The condition of compensation with direct measurement of the noise $\bar{z}$ takes a very simple form, if the motions of all the elements of the system are described by linear equations with constant coefficients [5.2].

The investigations have also been extended to nonlinear systems (see, e.g., [3.25], pp. 501–502). In the paper [5.42] the conditions for linear systems and a very general class of nonlinear systems were obtained with the aid of variational methods.

With the indirect measurement of the noise $\bar{z}$ the circuit for direct measurement is missing. For illustration let us consider an elementary example ([5.3], pp. 123–124). Let the noise z act on the output of the object with transfer function $G(p)$ (Fig. 5.4a). The noise z can also be brought to the output of the object as noise acting at other places in the control system. We consider that the denominator of the transfer function $G(p)$ depends on p and its degree is higher than the degree of the numerator. Thus this element has inertia; its amplitude-frequency characteristic falls off at sufficiently large frequencies. Therefore by itself the object shown in Fig. 5.4a is not ideal, and with the variable x^* supplied to its input, another variable is in general obtained.

In order to make the system ideal (inertialess and not reacting to the presence of the noise z), an element with transfer function G^{-1} must be inserted in succession with G (Fig. 5.4b). Further, the quantity $-z$ must be inserted at the input to the system, compensating for the noise z.

The operation of this scheme can be treated as indirect measurement and neutralization of the noise z with forcing, i.e., by the introduction of the element G^{-1} neutralizing the inertial properties of the object G. By knowing the output variable u of the object and also x and $G(p)$, the quantity z can be determined (Fig. 5.4a). Using the same notations u, z, and x for transforms, we obtain

$$z = x - uG(p). \tag{5.3}$$

By feeding the quantity $-z$ to the input of the system, we obtain the circuit depicted in Fig. 5.4c.

As is shown in Fig. 5.4c, the right side of Eq. (5.3) is formed by summing the quantity u passing through the unit $G(p)$ and amplifier element with amplification coefficient k_0 equal to unity, with the variable $-x$. To obtain the latter term an inverting element with amplification coefficient -1 must be introduced.

But the diagram of Fig. 5.4c is equivalent to the scheme depicted in Fig. 5.4d. The transfer function of the element enclosed by the dashed line is defined by the expression

$$\frac{1/G}{1 - k_0 G(1/G)} = \frac{1}{1 - k_0}\frac{1}{G} = k_1 \frac{1}{G}. \tag{5.4}$$

As $k_0 \to 1$ the amplification coefficient k_1 of this element tends to infinity. As a result the diagram shown in Fig. 5.4e is obtained.

A sufficiently large amplification coefficient can theoretically be obtained either with the aid of a multistage amplifier with a sufficiently large number of stages, or by the inclusion of an amplifier with a small number of stages of positive feedback. Thus there exists the fundamental possibility of obtaining complete invariance.

From the diagram of Fig. 5.4e the unit G^{-1} can be deduced. Then another scheme is generated, shown in Fig. 5.5. In this circuit a sufficiently large amplification coefficient k_1 for the amplifying element permits an arbitrarily small error ϵ to be achieved [5.4]. In fact, the transform of the output variable can be written in the form

$$x = \frac{z}{1 + k_1 G} + \frac{k_1 G}{1 + k_1 G} x^*, \tag{5.5}$$

and the transform of the error in the form

$$\epsilon = x^* - x = \frac{x^*}{1 + k_1 G} - \frac{z}{1 + k_1 G}. \tag{5.6}$$

With k_1 tending to infinity, in the limit an error ϵ equal to zero can be obtained.

Of course free oscillations appear in this system apart from forced vibrations. They can also grow; the system may turn out to be unstable. However, we will not be concerned with stability questions. In [5.4] it has been shown in what way a feedback circuit can be made stable even for an amplification coefficient k which is large without limit.

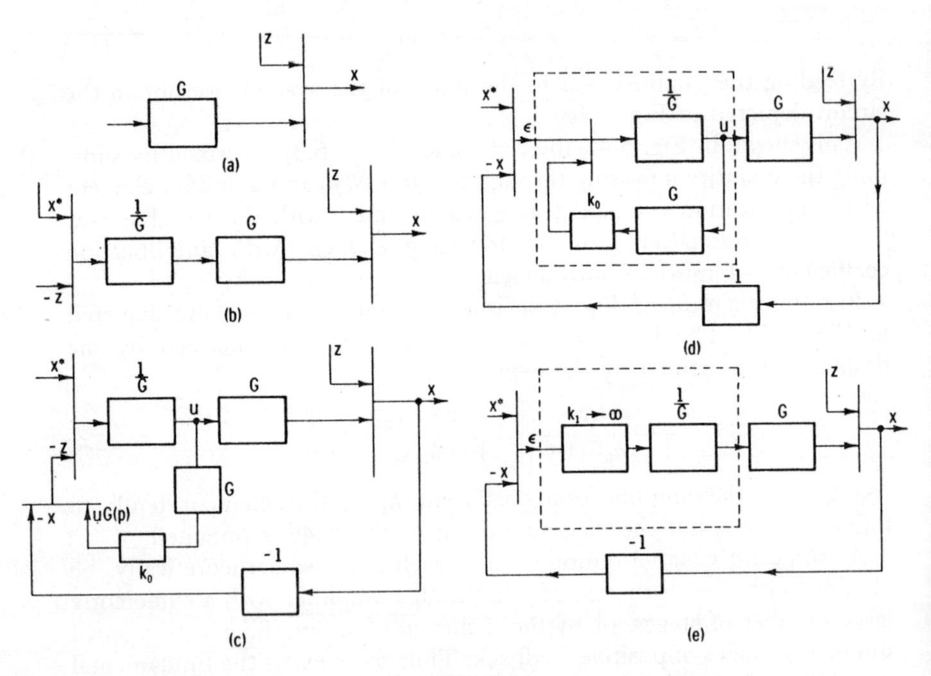

Fig. 5.4

The physical meaning of the phenomena occurring in the circuit of Fig. 5.4e or Fig. 5.5 is simple enough. The difference $\epsilon = x^* - x$ is fed to the input of the amplifier element with an infinitely large amplification coefficient k_0. This is the quantity which will deviate by an

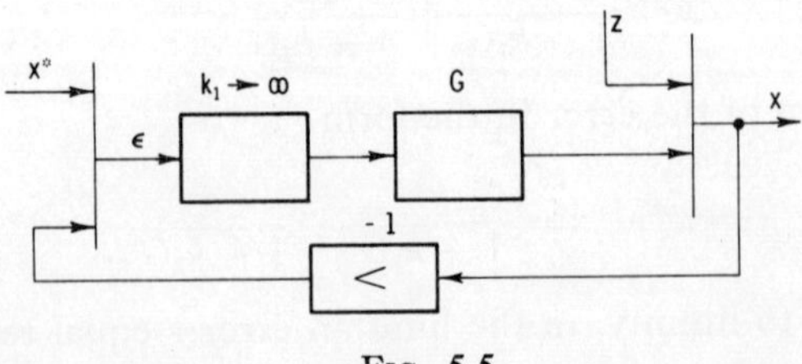

Fig. 5.5

appreciable amount from zero in some direction, when a sufficiently large value appears instantly at the output of the amplifier element, acting on the object and changing its output variable x in such a manner as to return the difference ϵ to zero.

If the obtaining of an error arbitrarily small or equal to zero could be achieved in the ideal case, then such a solution to the problem would be the best, and we would only have to apply the theory of invariance everywhere, instead of the theory of optimal systems. But unfortunately the realization of an ideal system is only possible for such a far-fetched idealization of real problems that its practical applicability is restricted to a comparatively narrow scope. Invariance can be obtained with:

(a) the observance of the conditions of realizability and coarseness;

(b) the confining of the action of random noise to a definite range (see below for more details about this);

(c) the absence of constraints imposed directly or indirectly on the control action and on the coordinates of the system.

Not being concerned with point (a) (but its observance is by no means possible in all cases), we will first consider the effect of the second of these factors. It turns out that even in the circuit with an amplifier element having a large amplification coefficient, the neutralization of an "arbitrary" noise z is impossible. Such a system is shown in Fig. 5.6a; however, it must be taken into account that any real amplifier element has a complex transfer coefficient $K(j\omega)$, whose absolute value decreases sharply for sufficiently large frequencies. Therefore if the fluctuation z contains noise of sufficiently high frequency, then its complete neutralization is impossible. A physically similar cause operates in the purely discrete circuit shown in Fig. 5.6b. Here all the variables are examined only at the discrete time instants $t = 0, 1, ..., s, ...$. Let the variable at the instant $t = s$ have the index s. If g_s is a sequence of independent random variables with an assigned distribution law, and for simplicity G is taken equal to 1, then an exact measurement of the random variable z_s is impossible. In fact, from Fig. 5.6b it is seen that

$$x_s = z_s + u_s + g_s , \tag{5.7}$$

from which

$$z_s = (x_s - u_s) - g_s . \tag{5.8}$$

The first term of this expression gives an estimate of the desired variable z_s, while the second term represents an error or mistake in measurement. In view of the fact that z_s is measured in an indirect way with error in this scheme, exact compensation cannot be achieved in it. This example is illuminated in more detail in Chapter VI, where closed-loop systems are analyzed.

If the noise is applied at some point which is between the output of the amplifier k_1 and the output variable x, then as is seen from the

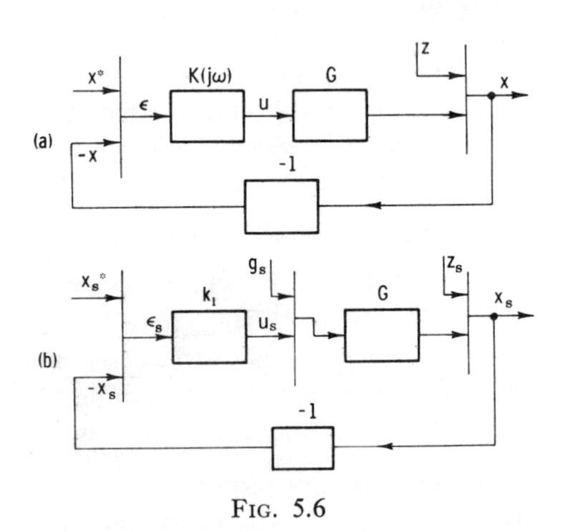

FIG. 5.6

foregoing its compensation to some extent is possible. But the situation is made considerably worse in other not more important cases when the noise is added to the driving action x^* (Fig 5.7a) or is found in the feedback circuit added to the output variable (Fig. 5.7b). These cases are fully workable, since the measurement of the variables x^* and x always takes place with some error; moreover, the measured quantities may be transmitted over a communication channel in which noise is added to them. In the circuits depicted in Fig. 5.7 invariance in general cannot be achieved, and in desiring to obtain the best results, problems are posed and solved naturally as in the theory of optimal systems. The error $\bar{e}$ likewise plays an analogous role in the compensation system of Fig. 5.3.

In the theory of invariance the absence of constraints is usually assumed, imposed directly or indirectly on the control action and in general on the coordinates of the system. This is a very severe condition, whose observance frequently deprives the problem of practical value. For example, the selection of an optimality criterion in the form

$$Q = \int_0^\infty [(x - x^*)^2 + \lambda^2 u^2]\, dt = \int_0^\infty (\epsilon^2 + \lambda^2 u^2)\, dt \tag{5.9}$$

is an indirect constraint. If $\lambda = 0$, then an ideal system is obtained with $\epsilon = 0$ and $Q = 0$. But adding the term $\lambda^2 u^2$ in the integrand in practice limits the integral effect of the action u and inhibits much larger values of it. In this case the minimum of Q now will not be equal to zero (see Chapter III) and the quantity ϵ is not identically equal to zero for an optimal system. The same result is also obtained with the application of the generalized integral criterion (see Chapter I).

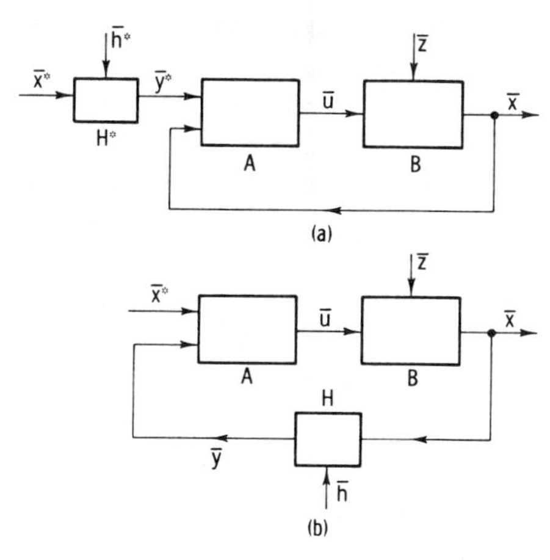

FIG. 5.7

However, the most important role is played by direct constraints, whose meaning is that the control action "cannot" or "must not" fall outside the admissible limits. Hence the amplifier output variable in Figs. 5.4, 5.5, 5.6 also cannot exceed the admissible value. Hence it immediately follows that achieving the value $\epsilon = 0$ is impossible. Meanwhile in the vast majority of problems the only idealization that is reasonable is one for which the requirement of a restriction on the variables by admissible values is not discarded.

It should be noted that the methods based on compensation with direct measurement (see Fig. 5.3) are a valuable auxiliary means for the development of systems. These methods based on Ponsel's principle do not represent the fundamental content of the theory of invariance, and form a separate branch developed over many years.

Further, the realization of partial invariance is possible, for example, the lack of error only in the steady-state process. The corresponding methods (for example, the introduction of integrating elements or other means) are also well known and do not have a direct relation to the theory of invariance. Furthermore, the accounting for random noise reduces questions of calculating steady-state behavior to problems which are typical for optimal system theory, when the error is not equal to zero and it is required to minimize some measure of it (for example, the mean-square value).

The reduction of error by the introduction of amplifiers with a large amplification coefficient and heavy negative feedback, as in the circuit in Fig. 5.5, permits a small steady-state error in combination with stability to be ensured [5.4]. For example, such schemes have found considerable dissemination in electronic simulation [3.25, 5.5]. This direction is likewise independent.

But the fundamental branch of the theory of invariance (first presented in [5.6]), being formally proper, has to do with hyper-idealized systems. In this theory the indirect measurement of noise and the elimination of inertia are achieved by forcing actions and the application of positive feedback, with the aim of achieving an infinite amplification coefficient. Here the effect of random noise and constraints is not taken into account. In addition, such systems do not satisfy the coarseness condition [5.6]. Often nonlinearities of various types play the role of constraints; therefore in the theory of invariance similar systems cannot be considered as well. Hence this theory can have only very narrow limits of applicability. These limits are not widened in perspective, but rather are narrowed, since in engineering definite progress is observed in increasing accuracy and a more rational use of system resources, which stipulates the operation at the limiting values, the accounting for constraints, and also the accounting for random noise.

There are papers in which the so-called "invariance to within ϵ" is considered. That is, in view of the impossibility of achieving complete invariance, these papers examine the questions of attaining the least error —but always differing from zero and capable of being arbitrary, even arbitrarily large. Such a statement of the problem now does not differ at all from the formulation of the problem in the theory of optimal systems, in view of which the name "theory of invariance" turns out to be superfluous here. In this field the problems represent a particular sphere and have a fundamental interest, where the conditions of weak sensitivity of the system to a change in the perturbing actions are determined. The investigation of sensitivity can be performed with a systems analysis. But if this concept is introduced for synthesis, then it can appear in the form of an optimality criterion or in the form of constraints, and thus it enters fully into the range of problems of optimal system theory.

The open-loop systems being considered in this chapter are of interest in communication engineering. Indeed at first the theory of optimal systems with information storage had been developed in connection with communication systems. But in perspective this theory also has great value for automatic control systems (see Chapter VI).

In spite of the immense diversity in the problems in communication theory during recent years, a comparatively small number of directions have crystallized, in which both formulations of problems and methods of solving them are evolved. There are three basic groups of theories.

(1) *Correlation theories.* The formulation of problems in this group of theories differs in that exhaustive probabilistic characteristics are not prescribed, but only the correlation functions of the random processes entering the system. It turns out that these data are quite sufficient for us to find, for example, the optimal filter, if we are restricted to linear systems and an optimality criterion in the form of minimum mean square error. This group of theories is not considered in this book. The literature relating to this field is given in Chapter I.

A correlation device can be used for designing nonlinear systems, but in this case its possibilities are considerably limited.

(2) *Information theory.* This theory investigates questions of the transmission of messages from a very general point of view. Its fundamental problem consists in finding the optimal code for the given properties of the communication channel and the prescribed statistical nature of the message source. This problem has fruitful solutions; but in the field of information processing, in contrast to questions related to its transmission and storage, the methods of information theory have for the present been developed to a highly inadequate degree.

(3) *Statistical decision theory.* This theory is apparently the most general and is applicable to any process in communication and control systems, both open-loop and closed-loop. Below an account is composed in connection with the concepts of this theory.

Mathematical statistics—an important branch of probability theory—has been one of the sources of statistical decision theory. In this field the theory of parameter estimation was worked out at the end of the nineteenth and in the first half of the twentieth century (the works of R. Fisher and H. Cramer, and very recently by the Soviet scientist Yu. V. Linnik [5.7–5.9]), and also the theory of hypothesis testing (the papers of J. Neyman, K. Pearson, and A. N. Kolmogorov).

The theory of games began to be developed independently of mathematical statistics. This discipline, whose foundation was laid by the French mathematician E. Borel, was developed by J. von Neumann and others (1928). In the course of its development it was linked in various forms to theories of operators, linear programming (whose foundations were laid by the Soviet mathematician L. V. Kantorovich in 1938), and dynamic programming. The basic object in the study of game theory is a game with the participation of two or more where the rules of the game are given, and in particular the possible payoffs to the players are specified. The problem is to design an optimal strategy for each of the players in the game, i.e., a method of making the "best" decisions. Such a strategy is also necessary in the worst situation, which can arise for a given player in the process of the game, to permit the most profitable course to be found.

The theory of games is not presented here. Those interested can turn to the literature [5.10–5.14].

The unification and development of a number of the ideas in the disciplines mentioned above brought the American mathematician A. Wald in 1948–1950 to the construction of a general mathematical-statistical discipline, which he called the theory of decision functions [5.15]. Apparently a number of R. Bellman's ideas (Bellman having developed dynamic programming) are related to some extent to the ideas of the theory of decision functions.

Independently of the development of the mathematical disciplines mentioned above, very general problems arose in communication theory. The first paper devoted to the problem of constructing an optimal receiver and an explanation of the properties of such a receiver was the investigation of the Soviet scientist V. A. Kotel'nikov, published in 1946 [5.16]. The optimal receiver provides for the reception of signals with minimal probability of error. Kotel'nikov called his theory the theory of potential noise stability. Later on both in the USSR and beyond the border it became interesting to develop papers on optimal methods of radio reception.

The mathematical and engineering branches of statistical decision theory were already closed after the first half of the twentieth century. In the papers of the American scientist D. Middleton and others, it was explained that the methods of the theory of decision functions can be used successfully for solving problems of the best means of receiving signals in a noise background [1.12, 1.13, 5.17–5.22]. As is shown in this and the following chapter, the problems of automatic control theory can also be solved by the methods of statistical decision theory.

Let us first consider the fundamental problems of statistical decision theory in connection with communication systems. The block diagram of a communication system is shown in Fig. 5.8. The transmitter is denoted by. the letters *XMTR*, the communication channel by *CC*, the receiver by *REC*. We will denote the signal sent by the transmitter by $s(t)$, the signal accepted by the receiver by $x(t)$. In the communication channel noise $n(t)$ is added to the signal $s(t)$. Therefore

$$x(t) = x[s(t), n(t)]. \tag{5.10}$$

For example, in a special case,

$$x(t) = s(t) + n(t). \tag{5.11}$$

Let certain parameters of the transmitted signal be known (if they were all known, then there would be no sense in transmitting this signal). But their *a priori* probability distribution may be known; for example, this distribution can be obtained by the processing of statistical data during a long past time period. Further, we will assume that the method of combination of signal and noise in the communication channel, i.e., the form of formula (5.10), is known. In addition, we will regard the probabilistic characteristics of the noise $n(t)$ as known.

The problem is thus posed: let the signal $x(t)$ act on the input of the receiver *REC* during a finite time interval from $t = 0$ to $t = T$, i.e., for $0 \leqslant t \leqslant T$. On the basis of the assumed model of the signal $x(t)$, it is required to decide, in a manner which is optimal in some sense, what the unknown parameters of the transmitted signal $s(t)$ are. It is assumed that this decision is made automatically in *REC* and is delivered to the receiver output in the form of a certain signal d (Fig. 5.8). Thus it is required to find out the algorithm of the optimal receiver.

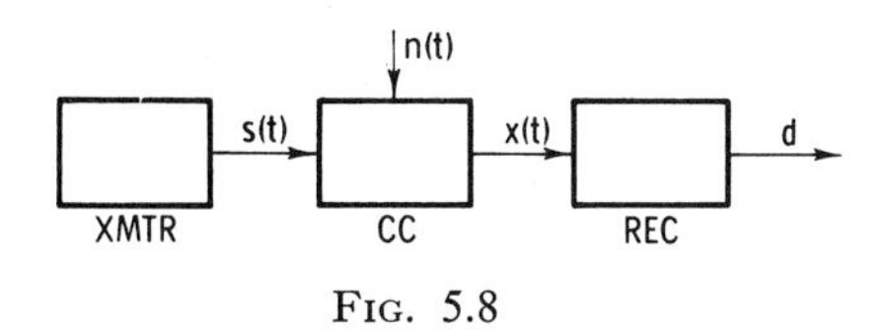

FIG. 5.8

In order to refine the problem, it is necessary to indicate the optimality criterion and the parameters of the signal $s(t)$ to be determined. Depending on alternative versions of the formulation various types of problems and different types of theories are possible.

(1) *Theory of two-alternative decisions.* Let there be only one unknown parameter A. For example, the signal $s(t)$ can have the form

$$s(t) = A \sin(\omega_0 t + \varphi_0), \tag{5.12}$$

where ω_0 and φ_0 are given. In the theory of two-alternative decisions we consider the case when the unknown parameter can assume only two possible values A_1 and A_2, with *a priori* probabilities $p(A_1)$ and $p(A_2)$ respectively. In a special case $A_1 \neq 0$, $A_2 = 0$ can occur. This problem bears the name of the signal detection problem.

(2) *Theory of multi-alternative decisions.* In this theory it is assumed that the unknown parameter A can take on r different possible values $A_1, A_2, ..., A_r$ with *a priori* probabilities $p(A_1), p(A_2), ..., p(A_r)$, respectively.

(3) *Theory of parameter estimation.* Let the unknown parameter A be capable of assuming any value in a certain region $\Omega(A)$ with *a priori* probability density $P(A)$. The theory of parameter estimation gives the possibility of constructing the optimal receiver, giving at the output the best estimate in a definite sense of the value of the parameter A of the transmitted signal $s(t)$, on the basis of the assumed model of $x(t)$.

(4) *Theory of estimation of processes.* It was assumed above that the parameter A of the transmitted signal is constant during one process of transmission. But it can turn out that A is a function of time: $A = A(t)$. The problem consists of determining the form and parameters of this function in an optimal manner.

These problems are not the most general. In the general case the transmitted signal $s(t)$ can contain several unknown parameters, which we will denote by $a_1, a_2, ..., a_m$:

$$s(t) = s(a_1, a_2, ..., a_m, t). \tag{5.13}$$

Each of these parameters can take on several possible values or even an infinite set of values in some region. The problems of estimating several parameters can be formulated in the same form as problems for one parameter, if the parameter vector A is introduced with coordinates $a_1, a_2, ..., a_m$:

$$A = (a_1, a_2, ..., a_m). \tag{5.14}$$

Here the problem formulations remain the previous; it is only necessary to consider that A is not a scalar but a vector.

We will note certain distinctive features of the problem formulations stated above.

(1) First, all realizations of x on the interval $0 \leqslant t \leqslant T$ are taken, and then the decision d about the parameter A is made. Of course, another formulation is possible under which an estimate of the values of A is generated, now starting from the time instant $t = 0$, and is refined gradually on the basis of information arriving at the receiver.

(2) In the majority of cases we will regard the *a priori* probabilities of the possible values of the parameter A as given. Problems of this type are called "Bayes' " problems. But such an approach is not always possible. Very often we lack the results of preliminary statistical analyses from which the *a priori* probabilities can be found. For example, let

us assume that the detection problem is solved as well as the estimate of the coordinates and velocities of hostile aircraft attacking some object. Where can information about the *a priori* probabilities of these quantities be obtained? They could only be obtained as a result of the storage of a long trial of the repulse of the attacks on the object, where all these attacks must occur under the same conditions (the same number and identical types of attacking aircraft, the same time of day, the same weather, and so on). It is evident that obtaining such data in the case under consideration is impossible. Therefore in many cases an "*a priori* obstacle" has to be encountered, including the difficulty or even impossibility of obtaining the *a priori* probabilities.

For the solution of problems considered in this chapter we can use geometric constructions which are instrumental for the visualization of the conception—the noise space, signal space, and observation space. Let us first introduce the concept of a noise space. We will consider that measurements are made not continuously, but only at the discrete time instants t_1, t_2, ..., t_k. Therefore we are interested in the values n_1, n_2, ..., n_k of the noise only at these time instants (Fig. 5.9). We will introduce the noise vector N with Cartesian coordinates n_1, n_2, ..., n_k:

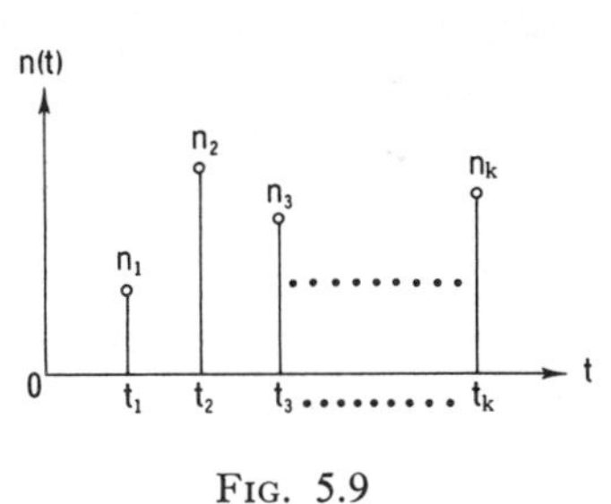

FIG. 5.9

$$N = (n_1, n_2, ..., n_k). \tag{5.15}$$

This radius vector in k-dimensional space, called the noise space, is depicted in Fig. 5.10. For the example a three-dimensional space is shown. We will denote by $P(N)$ the probability density of the vector N, i.e., the joint probability density of n_1, n_2, ..., n_k:

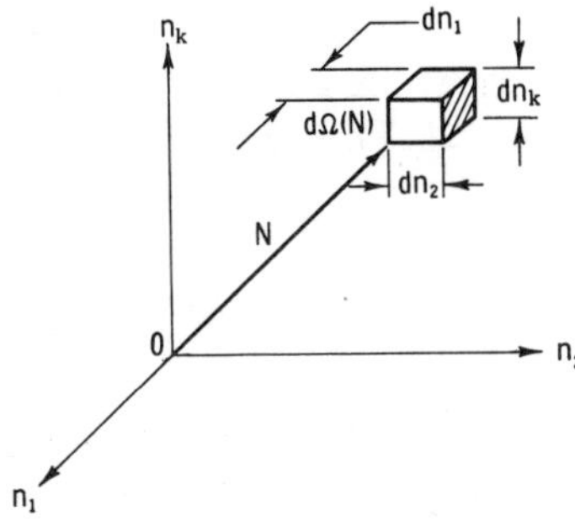

$$P(N) = P(n_1, n_2, ..., n_k). \tag{5.16}$$

FIG. 5.10

Let us determine the probability of the end of the vector N occurring in an infinitesimally small volume $d\Omega(N)$ of noise space, for example, in the infinitesimally small parallelopiped with volume

$$d\Omega(N) = dn_1\, dn_2 \cdots dn_k, \tag{5.17}$$

depicted in Fig. 5.10. Evidently this probability is equal to[†]

$$P(N)\, d\Omega(N) = P(n_1, ..., n_k)\, dn_1\, dn_2 \cdots dn_k . \tag{5.18}$$

Since the total probability of the end of the vector N occurring at some point of noise space $\Omega(N)$ is equal to unity, then the integral over all infinitesimally small volumes $d\Omega(N)$ of this space is equal to unity:

$$\int_{\Omega(N)} P(N)\, d\Omega(N) = 1. \tag{5.19}$$

The subscript under the integral means integration over the whole region $\Omega(N)$ of possible values of N, i.e., over the entire noise space.

The parameter space is related to the parameter vector A defined by (5.14), in the same way as the noise space is related to the vector N. This space is an m-dimensional space with Cartesian coordinates a_1, a_2, ..., a_m. If

$$P(A) = P(a_1, a_2, ..., a_m)$$

is the probability density for the vector A, then the product $P(A)\, d\Omega(A)$ gives the probability of the end of the vector A occurring in the infinitesimally small volume $d\Omega(A)$ of signal space.

Finally, let us introduce the "observation space," or else the space of the received signal. By considering its discrete values x_1, x_2, ..., x_k, the vector

$$X = (x_1, x_2, ..., x_k) \tag{5.20}$$

can be introduced [see also (2.97)].

The k-dimensional space of this vector is thus the observation space. If the probability density $P(X) = P(x_1, x_2, ..., x_k)$ is known, then the probability of the end of the vector X occurring in an infinitesimally small volume $d\Omega(X)$ of the observation space $\Omega(X)$ is defined by the product $P(X)\, d\Omega(X)$. It is also evident that

$$\int_{\Omega(X)} P(X)\, d\Omega(X) = 1. \tag{5.21}$$

It should be noted that $P(N)$ and $P(A)$ are given with the formulation of the problem, while $P(X)$ is not prescribed but can be determined in the process of solving the problem.

[†] Compare with Fig. 2.6 and the arguments related to it.

Thus let there be known:

(a) The form of the functions characterizing the transmitted signal:

$$s(t) = s(t, A);$$

(b) The *a priori* probability density $P(A)$ of the parameter vector A, or the probabilities $p(A_i)$ of the individual values A_i, if A can only assume a finite (or countable) set of values A_i;

(c) The probability density $P(N)$ of the noise vector N;

(d) The method of combination of the signal s and the noise n in the communication channel:

$$x = x[s, n].$$

We will assume that discrete values $x_1, x_2, ..., x_k$ of the signal x are taken and a specific realization of the vector X is defined. What can be said about the parameter vector A after this observation? First of all, it should be stressed that as a result of observing the vector X, in general the value of A cannot be found out exactly. Actually, as a matter of fact in the observation process the receiver absorbs the signal $s(t)$ mixed with random noise $n(t)$ inseparable from it. Therefore after the trial including the observation of the vector X, the probability distribution of different values of A is not concentrated solely at some one value, but is "spread" over various possible values of the vector A. But the probability density after the trial, i.e., the *a posteriori* probability density, is in general not equal to the *a priori* probability density possessed up to the trial. The *a posteriori* probability density shows the tendency to shrink to the true value of A in the transmitted signal.

Essentially all the information obtained as a result of the trial is concentrated in the *a posteriori* probability density for A. This probability density represents the conditional probability density $P(A \mid X)$, or else the probability density for A, provided that the observed vector X is given, stipulated. In expanded form

$$P(A \mid X) = P(a_1, a_2, ..., a_m \mid x_1, x_2, ..., x_k). \tag{5.22}$$

To determine the *a posteriori* probability density Bayes' formula (2.13) can be used. We will first consider the case when the vector A can assume a finite number r of possible values $A_1, A_2, ..., A_r$ with *a priori* probabilities $p(A_1), p(A_2), ..., p(A_r)$ respectively. Let $P(X \mid A)\, d\Omega(X)$ be the conditional probability that the end of the vector X is in the volume $d\Omega(X)$ of observation space, provided that the vector A is given.

Let us find the probability of the compound event that the parameter vector has the value A, and in addition the end of the vector X falls in the volume $d\Omega(X)$. According to the formula for multiplication of probabilities, this probability can be expressed in two ways. It is equal to

$$p(A)[P(X \mid A)\, d\Omega(X)] = p(A \mid X)[P(X)\, d\Omega(X)], \tag{5.23}$$

where $P(X)$ is the absolute probability density of the vector X. Hence after canceling the $d\Omega(X)$ Bayes' formula follows for the *a posteriori* probability $p(A \mid X)$:

$$p(A \mid X) = \frac{p(A)P(X \mid A)}{P(X)}. \tag{5.24}$$

The function $P(X \mid A)$ is called the likelihood function. By knowing the *a priori* probability $p(A_i)$ and recognizing the incoming signal X, the *a posteriori* probability can be found from formula (5.24):

$$p(A_i \mid X) = \frac{p(A_i)P(X \mid A_i)}{P(X)}. \tag{5.25}$$

The denominator of this expression can be determined in the following way: Summing the left and right sides of the expressions

$$p(A_j)P(X \mid A_j) = P(X)p(A_j \mid X) \tag{5.26}$$

for $j = 1, \ldots, r$, we obtain

$$\sum_{j=1}^{r} p(A_j)P(X \mid A_j) = P(X) \sum_{j=1}^{r} p(A_j \mid X), \tag{5.27}$$

from which, by taking into account that

$$\sum_{j=1}^{r} p(A_j \mid X) = 1, \tag{5.28}$$

it follows that

$$P(X) = \sum_{j=1}^{r} p(A_j)P(X \mid A_j). \tag{5.29}$$

Substituting this expression into (5.25), we finally find:

$$p(A_i \mid X) = \frac{p(A_i)P(X \mid A_i)}{\sum_{j=1}^{r} p(A_j)P(X \mid A_j)} \qquad (i = 1, \ldots, r). \tag{5.30}$$

An analogous formula can also be derived for the case of a continuous distribution of the vector A with *a priori* density $P(A)$. Let $P(A \mid X)$ be the probability density for A with a given X, i.e., the *a posteriori* probability density. The absolute probability of finding the end of the vector A in the volume $d\Omega(A)$ of parameter space is equal to $P(A)\, d\Omega(A)$, and the conditional probability for a fixed X is equal to $P(A \mid X)\, d\Omega(A)$. The probability of the compound event of the ends of the vectors A and X occurring in the volumes $d\Omega(A)$ and $d\Omega(X)$ of the respective spaces is equal to $P(A, X)\, d\Omega(A)\, d\Omega(X)$, where $P(A, X)$ is the corresponding joint probability density. Then according to the theorem on multiplication of probabilities,

$$\begin{aligned} P(A, X)\, d\Omega(A)\, d\Omega(X) &= [P(A)\, d\Omega(A)][P(X \mid A)\, d\Omega(X)] \\ &= [P(X)\, d\Omega(X)][P(A \mid X)\, d\Omega(A)]. \end{aligned} \tag{5.31}$$

Hence we obtain that

$$P(A)P(X \mid A) = P(X)P(A \mid X), \tag{5.32}$$

from which Bayes' formula follows for the *a posteriori* probability density

$$P(A \mid X) = \frac{P(A)P(X \mid A)}{P(X)}. \tag{5.33}$$

In order to find the denominator of this formula, we will multiply both sides of (5.32) by $d\Omega(A)$ and integrate over the entire region $\Omega(A)$ of possible values of the vector A. We find

$$\int_{\Omega(A)} P(A)P(X \mid A)\, d\Omega(A) = P(X) \int_{\Omega(A)} P(A \mid X)\, d\Omega(A). \tag{5.34}$$

The integral on the right side of this expression is equal to unity; therefore

$$P(X) = \int_{\Omega(A)} P(A)P(X \mid A)\, d\Omega(A). \tag{5.35}$$

This expression can be substituted into (5.33). Then we will obtain Bayes' formula in final form:

$$P(A \mid X) = \frac{P(A)P(X \mid A)}{\int_{\Omega(A)} P(A)P(X \mid A)\, d\Omega(A)}. \tag{5.36}$$

Formulas (5.30) and (5.36) give the possibility of determining the *a posteriori* probabilities of the values of A, if the *a priori* probabilities and the vector X are known; the latter is determined as a result of experiment.

The "method of *a posteriori* probabilities" is connected with these formulas, consisting in the selection of a value of A for which the *a posteriori* probability $p(A \mid X)$ is maximal. Thus with the use of this method the "most probable" value of A appears at the output of the optimal receiver as the decision d.

If the parameter A has a continuous distribution, then this method reduces to the selection of a value of A for which the *a posteriori* probability density $P(A \mid X)$ is maximal.

It should be noted, however, that the application of this method is only possible in the case when the *a priori* probabilities of different values of A are known. But if they are unknown, then another method described below can be used.

Let us consider the likelihood function $P(X \mid A)$. If the vector X is given, then this function depends only on A:

$$P(X \mid A) = L(A). \tag{5.37}$$

The simplest of the methods of estimating values of A is related to the likelihood function and is called the "maximum likelihood method." This method, proposed by the English mathematician R. Fisher, can be formulated in the form of the following rule:

The most likely value of the parameter A is the one for which the likelihood function $L(A)$ is maximal.

This rule is taken as a postulate; it results solely from concepts with sensible meaning. We give a simple example justifying the value of this rule. A student returning a test can receive either a good grade (pass) or a bad one (fail). The grade depends basically on his knowledge. But attendant circumstances, for example the state of his health or the disturbance of the corresponding random success or failure with the choice of paper, likewise can affect his answer, and hence the grade, on one side or the other. For simplicity we will consider that the knowledge A of the student has only two gradations. It is either good ($A = 1$) or bad ($A = 0$). We will also assume that the student's answer can be either good ($X = 1$) or bad ($X = 0$). On the basis of trial the instructor can make up a table of probabilities $p(X \mid A)$. In it will be four numbers: the probability $p(1 \mid 1)$ of a good answer with good knowledge; the probability $p(0 \mid 1)$ of a bad answer with good knowledge; the probability $p(1 \mid 0)$ of a good answer with bad knowledge, and finally the probability $p(0 \mid 0)$ of a bad answer with bad knowledge. It is natural that $p(1 \mid 1) > p(1 \mid 0)$ and $p(0 \mid 0) > p(0 \mid 1)$.

We will now assume that the examiner has received the signal (which has entered in the form of an answer from the student). It turned out that $X = 0$, i.e., the answer is bad. What decision must the examiner make about the student's knowledge? Evidently, it must be based on the probabilities $p(0 \mid 0)$ and $p(0 \mid 1)$ of a bad answer in the presence of bad or good knowledge. A reasonable argument is the following. The instructor reckons that for the given value $X = 0$ the more likely value of A is the one for which the probability

$$p(X = 0 \mid A) = p(0 \mid A)$$

(and this is just the likelihood function) is greater. Since

$$p(0 \mid 0) = p(X = 0 \mid A = 0) > p(X = 0 \mid A = 1) = p(0 \mid 1), \qquad (5.38)$$

then the examiner must give the student's knowledge a bad grade ($A = 0$).

In everyday life it is frequently necessary for each of us to use intuitively estimates of the maximum likelihood type. We regard as more likely a cause for which the consequence X actually observed is more probable.

The estimate of the value of the parameter A by the maximum likelihood method is made thus: as the result of observation let a signal—the vector X—be received. In the case of a discrete distribution of A, we write all possible values of the likelihood function

$$L(A_1), L(A_2), \ldots, L(A_r).$$

Comparing these values, we choose an A_j for which the value of the likelihood function is greater (or not less) than the other values of the function:

$$L(A_j) \geqslant L(A_k) \qquad (k = 1, \ldots, r). \qquad (5.39)$$

The value A_j selected by such a method is called "most likely."

This method is extended in a natural way to problems with a continuous distribution of the parameter A. A value of A is chosen for which the value of $P(X \mid A) = L(A)$ is maximal.

2. Theory of Two-Alternative Decisions

In the simplest case let the transmitted signal $s = s(a, t)$ depend on a single parameter a, where the latter can have only two possible values $a = a_1$ and $a = a_0$ with *a priori* probabilities p_1 and p_0, respectively:

$$a = \begin{cases} a_1 \to p_1 , \\ a_0 \to p_0 . \end{cases} \tag{5.40}$$

Since the parameter a takes on one of these two values for certain, then

$$p_1 + p_0 = 1. \tag{5.41}$$

We will consider a still more particular case, when the transmitted signal has the form

$$s(t) = af(t), \tag{5.42}$$

where $f(t)$ is a known function. For example,

$$s(t) = a \cos(\omega t + \varphi), \tag{5.43}$$

where ω and φ are considered known. If the detection problem is examined, then we set $a_1 \neq 0$, $a_0 = 0$. Further, in a typical problem which will be analyzed below, let the signal and noise in the communication channel be combined additively:

$$x(t) = s(t) + n(t). \tag{5.44}$$

We suppose that reception of discrete values $x_1, x_2, ..., x_k$ takes place at the time instants $t_1, t_2, ..., t_k$ respectively. We will set

$$s(t_i) = s_i , \qquad n(t_i) = n_i . \tag{5.45}$$

Then

$$x_i = s_i + n_i = af(t_i) + n_i . \tag{5.46}$$

We will assume that the values n_i of the noise are a sequence of independent random variables with the same normal distribution density, where the mean value is equal to zero and the dispersion equals σ^2:

$$P(n_i) = \frac{1}{\sigma \sqrt{(2\pi)}} \exp \left\{ - \frac{n_i^2}{2\sigma^2} \right\}. \tag{5.47}$$

The problem is to estimate the value of the parameter a in the transmitted signal $s(t)$ after receiving the signal $X = (x_1, x_2, ..., x_k)$. For detecting the signal ($a_1 \neq 0$, $a_0 = 0$) it is required to decide whether the transmitted signal $a_1 f(t)$ different from zero is present in the received signal $x(t)$, or the received signal only contains the noise $n(t)$.

By no matter what method we solve the problem formulated above, in the process of solution the likelihood function $P(X \mid a)$ will have to be found. For example, let us find it for $a = a_0$. Since in this case corresponding to (5.46)

$$x_i = a_0 f(t_i) + n_i , \tag{5.48}$$

then

$$n_i = x_i - a_0 f(t_i) \tag{5.49}$$

and the distribution density for x_i will be obtained if expression (5.49) is substituted into (5.47). Actually, according to (5.49) the probability of the received signal being between the values x_i and $x_i + dx_i$ is equal to the probability of the noise n_i being between $x_i - a_0 f(t_i)$ and $x_i - a_0 f(t_i) + dx_i$. Therefore the conditional probability density $P(x_i \mid a_0)$ is given by the expression

$$P(x_i \mid a_0) = \frac{1}{\sigma \sqrt{(2\pi)}} \exp \left\{ - \frac{[x_i - a_0 f(t_i)]^2}{2\sigma^2} \right\}. \tag{5.50}$$

Since the individual values n_i for different i are statistically independent, the x_i for different i are also independent. Hence the probability density for the set of variables $x_1, x_2, \ldots, x_k$ is equal to the product of the probability densities for all the x_i:

$$P(X \mid a_0) = P(x_1, x_2, \ldots, x_k \mid a_0) = \prod_{i=1}^{k} P(x_i \mid a_0)$$

$$= \prod_{i=1}^{k} \frac{1}{\sigma \sqrt{(2\pi)}} \exp \left\{ - \frac{[x_i - a_0 f(t_i)]^2}{2\sigma^2} \right\}$$

$$= \frac{1}{\sigma^k (2\pi)^{k/2}} \exp \left\{ - \frac{1}{2\sigma^2} \sum_{i=1}^{k} [x_i - a_0 f(t_i)]^2 \right\}. \tag{5.51}$$

Carrying out the same calculation for $a = a_1$, we find an analogous formula for the conditional probability density of the vector X for $a = a_1$:

$$P(X \mid a_1) = \frac{1}{\sigma^k (2\pi)^{k/2}} \exp \left\{ - \frac{1}{2\sigma^2} \sum_{i=1}^{k} [x_i - a_1 f(t_i)]^2 \right\}. \tag{5.52}$$

Expressions (5.51) and (5.52) are the values of the likelihood function for $a = a_0$ and $a = a_1$. If the x_i are known, then the values of the likelihood function are also known. According to (5.30) and (5.40) after receiving the signal X, the *a posteriori* probabilities for a_1 and a_0 are determined from the expressions

$$p(a_1 \mid X) = \frac{p_1 P(X \mid a_1)}{p_1 P(X \mid a_1) + p_0 P(X \mid a_0)} \tag{5.53}$$

and

$$p(a_0 \mid X) = \frac{p_0 P(X \mid a_0)}{p_1 P(X \mid a_1) + p_0 P(X \mid a_0)}. \tag{5.54}$$

Dividing both sides of these expressions by one another, we obtain:

$$\frac{p(a_1 \mid X)}{p(a_0 \mid X)} = \frac{p_1 P(X \mid a_1)}{p_0 P(X \mid a_0)} = \frac{p_1}{p_0} \frac{L(a_1)}{L(a_0)}, \tag{5.55}$$

where the values of the likelihood function are denoted by $L(a_1)$ and $L(a_0)$:

$$L(a_1) = P(X \mid a_1), \qquad L(a_0) = P(X \mid a_0). \tag{5.56}$$

Let us introduce the notation

$$\Lambda(X) = \frac{L(a_1)}{L(a_0)} = \frac{P(X \mid a_1)}{P(X \mid a_0)} \tag{5.57}$$

and call the quantity $\Lambda(X)$ the "likelihood ratio."† This quantity depends on what the received signal X is. We will assume that the method of maximum *a posteriori* probability is used. Then for

$$\frac{p(a_1 \mid X)}{p(a_0 \mid X)} = \frac{p_1}{p_0} \Lambda(X) > 1 \tag{5.58}$$

the decision $a = a_1$ is made, since the *a posteriori* probability density for a_1 is greater than for a_0. The inequality (5.58) can be rewritten as

$$\Lambda(X) > p_0/p_1 \rightarrow a = a_1. \tag{5.59}$$

† Sometimes this quantity is called the "likelihood coefficient."

The arrow indicates that from this inequality the decision $a = a_1$ follows. In the opposite case the decision $a = a_0$ is made, which can be written as

$$\Lambda(X) < p_0/p_1 \rightarrow a = a_0 . \tag{5.60}$$

Formulas (5.59) and (5.60) can be combined into the form of a general rule:

$$\Lambda(X) \begin{cases} > p_0/p_1 \rightarrow a = a_1 , \\ < p_0/p_1 \rightarrow a = a_0 . \end{cases} \tag{5.61}$$

If the decision is made by the maximum likelihood method, then the decision is made in using the value of a for which the likelihood function is greater. Therefore for $\Lambda(X) > 1$ the decision $a = a_1$ is made while for $\Lambda(X) < 1$ the decision $a = a_0$. Thus in this case the decision method can be written out in the form of the rule

$$\Lambda(X) \begin{cases} > 1 \rightarrow a = a_1 , \\ < 1 \rightarrow a = a_0 . \end{cases} \tag{5.62}$$

A comparison of (5.61) and (5.62) shows that in each of these cases the decision is made depending on whether the likelihood ratio $\Lambda(X)$ will turn out to be larger or smaller than some threshold. The methods of (5.61) and (5.62) differ from one another only in the value of this threshold: they coincide with one another only in the case of equal *a priori* probabilities, when $p_1 = p_0 = \frac{1}{2}$.

Substituting the values of the likelihood functions from (5.51) and (5.52) into (5.57), we find for the typical problem under consideration:

$$\Lambda(X) = \frac{P(X \mid a_1)}{P(X \mid a_0)} = \exp \left\{ \frac{- \sum_{i=1}^{k} [x_i - a_1 f(t_i)]^2 + \sum_{i=1}^{k} [x_i - a_0 f(t_i)]^2}{2\sigma^2} \right\}. \tag{5.63}$$

Instead of this expression it is more convenient to operate with the function

$$F(X) = 2\sigma^2 \ln \Lambda(X) = - \sum_{i=1}^{k} [x_i - a_1 f(t_i)]^2 + \sum_{i=1}^{k} [x_i - a_0 f(t_i)]^2. \tag{5.64}$$

Let us apply the maximum likelihood method and the rule (5.62).

Since for $F(X) > 0$ the value of $\Lambda(X) > 1$ and for $F(X) < 0$ the value of $\Lambda(X) < 1$, then the rule (5.62) can be replaced by the following rule:

$$F(X) \begin{cases} > 0 \rightarrow a = a_1 , \\ < 0 \rightarrow a = a_0 , \end{cases} \tag{5.65}$$

where $F(X)$ is defined by expression (5.64).

This decision method has a simple geometric interpretation. We will examine the observation space (Fig. 5.11). Let OM_0 and OM_1 be vectors

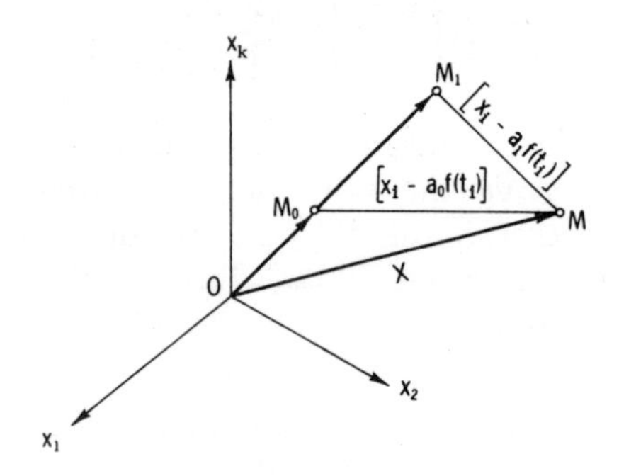

FIG. 5.11

with coordinates $a_0f(t_i)$ and $a_1f(t_i)$, respectively ($i = 1, ..., k$). Since their coordinates differ only by a constant factor, then the vectors OM_0 and OM_1 are collinear. Further, let the point M correspond to the received signal vector X.

If the noise were absent ($n_i = 0$) and the value of the parameter a equaled a_0, then the point M would coincide with M_0. If in the absence of noise the value of the parameter a in the transmitted signal were equal to a_1, then the point M would coincide with M_1. In reality, because of the presence of noise the point M does not coincide with either M_0 or M_1. But intuition suggests that if the point M is close to M_0, then the decision $a = a_0$ should be made; if the point M is close to M_1, then it should be reckoned that $a = a_1$. It is found that indeed the rule (5.65) has such a character. In fact, by taking (5.64) into account we will write it in the form

$$\begin{aligned} \sum_{i=1}^{k} [x_i - a_0 f(t_i)]^2 &> \sum_{i=1}^{k} [x_i - a_1 f(t_i)]^2 \rightarrow a = a_1 , \\ \sum_{i=1}^{k} [x_i - a_0 f(t_i)]^2 &< \sum_{i=1}^{k} [x_i - a_1 f(t_i)]^2 \rightarrow a = a_0 . \end{aligned} \tag{5.66}$$

The sum on the left side of these inequalities represents the square of the length of the vector M_0M (see Fig. 5.11), and the sum on the right side—the square of the length of the vector M_1M. Therefore conditions (5.66) mean that if the point M will be found closer to M_1 than M_0, then the decision $a = a_1$ should be made. Otherwise the decision $a = a_0$ is made.

For the application of the method of maximum *a posteriori* probability the rule (5.61) is used. Substituting (5.64) here, we arrive at the rule

$$F(X)\begin{cases} > 2\sigma^2 \ln(p_0/p_1) = h' \rightarrow a = a_1\,, \\ < 2\sigma^2 \ln(p_0/p_1) = h' \rightarrow a = a_0\,. \end{cases} \tag{5.67}$$

Thus the decision is determined by whether the quantity $F(X)$ will be greater or less than the threshold h'.

The methods derived can be put into the form of an algorithm in the optimal receiver, or more precisely the optimal detector. This algorithm can also be extended to the case of continuous reception of the signal. For the transition to the continuous case we will first stipulate that the discrete values $x_1, x_2, ..., x_k$ of the received signal are separated from one another by equal time intervals having a length Δt (Fig. 5.12). We will set

$$\Delta t = T/k, \tag{5.68}$$

where T = const—the observation time. If k tends to infinity, then $\Delta t \rightarrow 0$, the discrete values x_i will come infinitely close to one another and in the limit we will obtain continuous reception of the signal $x(t)$ $(0 \leqslant t \leqslant T)$.

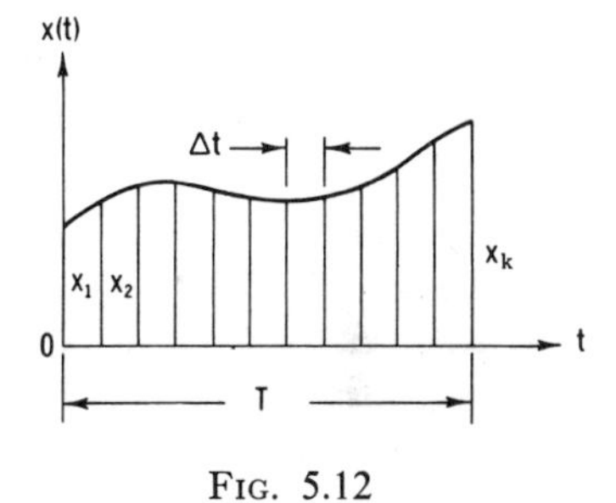

FIG. 5.12

Guided by the maximum likelihood method, we will consider conditions (5.66) as the initial ones. Multiplying both sides of the inequalities by Δt, we obtain

$$\begin{aligned} &\sum_{i=1}^{k} [x_i - a_0 f(t_i)]^2 \, \Delta t > \sum_{i=1}^{k} [x_i - a_1 f(t_i)]^2 \, \Delta t \rightarrow a = a_1, \\ &\sum_{i=1}^{k} [x_i - a_0 f(t_i)]^2 \, \Delta t < \sum_{i=1}^{k} [x_i - a_1 f(t_i)]^2 \, \Delta t \rightarrow a = a_0\,. \end{aligned} \tag{5.69}$$

Now if k tends to infinity and Δt to zero, then integrals appear in the inequalities (5.69) instead of sums and these conditions will take the form

$$\begin{aligned} &\int_0^T [x - a_0 f(t)]^2\, dt > \int_0^T [x - a_1 f(t)]^2\, dt \to a = a_1 , \\ &\int_0^T [x - a_0 f(t)]^2\, dt < \int_0^T [x - a_1 f(t)]^2\, dt \to a = a_0 . \end{aligned} \tag{5.70}$$

We transform these expressions, after restricting ourselves to the detection problem when $a_0 = 0$. For example, in this case the first of conditions (5.70) can be rewritten in the following way:

$$\int_0^T x^2(t)\, dt > \int_0^T x^2(t)\, dt - 2 \int_0^T x(t)[a_1 f(t)]\, dt + \int_0^T [a_1 f(t)]^2\, dt \to a = a_1 . \tag{5.71}$$

We will set

$$\int_0^T [a_1 f(t)]^2\, dt = E_0 . \tag{5.72}$$

This quantity can be called the energy of the signal $a_1 f(t)$. The quantity E_0 is regarded as assigned. Then from (5.71) after cancelling the integral $\int_0^T x^2(t)\, dt$ on both sides, we find the condition

$$\int_0^T x(t)[a_1 f(t)]\, dt > E_0/2 \to a = a_1 . \tag{5.73}$$

If inequality (5.73) is replaced by the opposite, then the decision $a = a_0 = 0$ is made, designating that the transmitted signal is absent.

From formula (5.73) a block diagram of the optimal detector can be constructed. This scheme is shown in Fig. 5.13. The integral on the

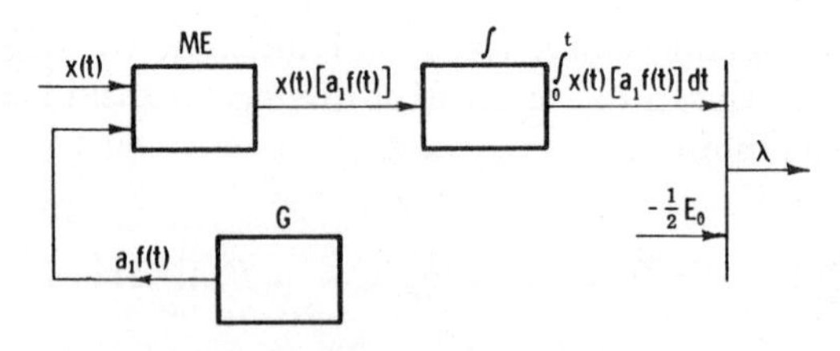

FIG. 5.13

left side of inequality (5.73) is obtained by multiplying in the multiplier element *ME* the input signal $x(t)$ and the signal $s(t) = a_1 f(t)$, produced by the local oscillator *G*, with subsequent integration in the integrating element. The latter is denoted by the sign $\int$. Let its output variable be equal to zero for $t = 0$. Then for $t > 0$ there is obtained at the output of the integrating element the integral

$$\int_0^t x(t)[a_1 f(t)]\,dt.$$

At the time instant $t = T$ this integral becomes equal to the left side of inequality (5.73). This value is compared with the constant magnitude $E_0/2$ by subtracting the latter from the integral. If the difference λ is positive, then the detector makes the decision $a = a_1$ about the presence of the signal. But if $\lambda < 0$, then the decision about the absence of the signal ($a = 0$) is made.

In the diagram of Fig. 5.13 the multiplier element *ME* accomplishes so-called synchronous detection of the received signal, and the integrating element accomplishes the integration of the product, or as this operation is sometimes called, coherent integration. As is shown below, the same unit is also encountered in optimal detectors under various types of optimality criterion.

Essentially the method of maximum *a posteriori* probability or the maximum likelihood method are prescriptions for obtaining a decision, but have not yet been proved at all to be optimal from some point of view. To obtain the optimal detection method some optimality criterion must first be established. The criteria related to error probabilities have received the greatest dissemination in the field under consideration.

Any decision method cannot be entirely without error. Errors in the decision are unavoidable, since the estimate of the value of the parameter a is made on the basis of observing the random variables x_1, x_2, ..., x_k. In order to find the error probability, we will examine the observation space (Fig. 5.14). To each observed realization x_1, x_2, ..., x_k there corresponds a vector X of observation space, and hence a point M in this space. As a result of the observation leading to a definite position of the point M, the observer or automatic machine replacing him must make a decision about what value, a_1 or a_0, the quantity a assumes. We will assume that some definite decision method has been chosen. Then to each observed point M there corresponds some definite decision. The decision $a = a_1$ corresponds to some points of observation space, the decision $a = a_0$ corresponds to other points of the space. Therefore all points of observation space are divided into two classes:

(a) The set of points corresponding to the decision $a = a_1$. This set forms the region Γ_1 of observation space (Fig. 5.14).

(b) The set of points corresponding to the decision $a = a_0$. This set forms the region Γ_0 of observation space.

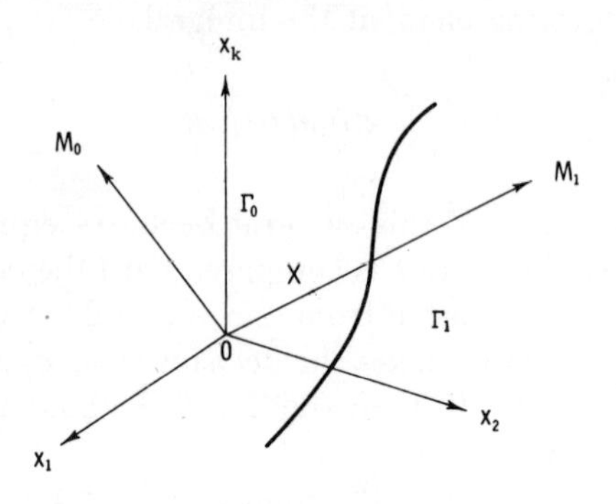

FIG. 5.14

For example, the point M_1 in Fig. 5.14 belongs to the region Γ_1, while the point M_0 belongs to the region Γ_0.

Any decision method represents a decomposition of all observation space into two regions: Γ_1 and Γ_0. Different decompositions of observation space into these regions correspond to different decision methods.

Two kinds of errors are peculiar to any decision method.

(a) An error of the first kind, when the decision $a = a_1$ is made when actually $a = a_0$. In particular, such an error for the solution of the detection problem means that the decision is made about the presence of the signal $s(t) \neq 0$ when in fact the signal is absent. This error is sometimes called false alarm error.

(b) An error of the second kind, when the decision $a = a_0$ is made when actually $a = a_1$. In the particular case of the detection problem this means that the decision is made about the absence of the signal at the time when in fact the signal exists. This error is sometimes called false ring error or signal gap error.

We will denote by the letter α the probability of an error of the first kind. This is the probability of the decision $a = a_1$ on condition that in fact $a = a_0$. Further, we will denote by the letter β the probability of an error of the second kind. This is the probability of making the decision $a = a_0$ on condition that in fact $a = a_1$.

From a geometric point of view α represents the probability of the end of the vector X occurring in the region Γ_1 (Fig. 5.14) on condition that in fact $a = a_0$. In other words, this is the probability of the event of the point M falling in the region Γ_1 on condition that $a = a_0$. In exactly the same way β is the probability of the point M falling in the region Γ_0 on condition that $a = a_1$.

Let $P(X \mid a_0)$ be the probability density of the vector X under the condition that $a = a_0$. Then the probability α of the end of the vector X falling in the region Γ_1 is the sum of occurrences over all infinitesimally small volumes of this region, i.e., the integral of $P(X \mid a_0)\, d\Omega(X)$ taken over the region Γ_1:

$$\alpha = \int_{\Gamma_1} P(X \mid a_0)\, d\Omega(X). \tag{5.74}$$

Let $P(X \mid a_1)$ be the probability density of the vector X under the condition that $a = a_1$. Then the probability β of its end falling in the region Γ_0 is equal to the integral of $P(X \mid a_1)\, d\Omega(X)$ taken over the region Γ_0:

$$\beta = \int_{\Gamma_0} P(X \mid a_1)\, d\Omega(X). \tag{5.75}$$

Evidently the probability ϵ of the correct decision in the presence of signal, i.e., for $a = a_1$, is expressed by the formula

$$\epsilon = \int_{\Gamma_1} P(X \mid a_1)\, d\Omega(X). \tag{5.76}$$

The validity of the latter expression follows from the fact that the integral of the density $P(X \mid a_1)$ over the entire observation space is equal to unity. Moreover, this expression follows directly from the definition of the quantity ϵ.

Let us now find the total error probability. Let p_0 and p_1 be the *a priori* probabilities of the values a_0 and a_1 of the parameter a respectively. Then the absolute probability of error of the first kind is the probability of the compound event consisting of the parameter a having the value a_0 (first event) and under the condition of the decision $a = a_1$ being made (second event). The probability of the first of these events is equal to p_0, the probability of the second under the condition that the first had occurred is equal to α.

Hence the probability of the compound event stated above is equal to $p_0\alpha$. By reasoning in the same way, the absolute probability of error of the second kind can be found, equal to $p_1\beta$. The total absolute probability of error q is equal to the sum of the absolute error probabilities of the first and second kinds:

$$q = p_0\alpha + p_1\beta. \tag{5.77}$$

With any decision method the probabilities of the various errors can be calculated from formulas (5.74), (5.75), and (5.77).

Errors of the first and second kinds are not always equally risky. In some cases the false alarm error implies such unpleasant consequences or such a large expenditure, that its probability is advisedly limited to some small assigned value beforehand. But an error of the second kind is also undesirable. An optimal decision method can be considered with such an approach, providing the "smallest signal gap probability for a prescribed false alarm probability." This criterion, proposed in the United States by Neyman and independently in England by Pearson, bears the name of the Neyman-Pearson criterion. For a given absolute false alarm probability

$$p_0\alpha = \text{const} = c_0 \tag{5.78}$$

the Neyman-Pearson criterion requires a minimum for the absolute signal gap probability:

$$p_1\beta = \min. \tag{5.79}$$

The substitution of expressions (5.74) and (5.75) into (5.78) and (5.79) leads to the conditions

$$p_0 \int_{\Gamma_1} P(X \mid a_0)\, d\Omega(X) = c_0 \tag{5.80}$$

and

$$p_1 \int_{\Gamma_0} P(X \mid a_1)\, d\Omega(X) = \min. \tag{5.81}$$

We will determine the regions Γ_1 and Γ_0 of observation space which are optimal according to the Neyman-Pearson criterion. For finding the conditional extremum of the integral (5.81) with the observance of condition (5.80), we will apply the method of indeterminate Lagrange multipliers. Let us find the minimum of the auxiliary expression

$$G = \lambda p_0 \int_{\Gamma_1} P(X \mid a_0)\, d\Omega(X) + p_1 \int_{\Gamma_0} P(X \mid a_1)\, d\Omega(X), \tag{5.82}$$

where λ is an indeterminate factor. Since the integral over all observation space Γ is decomposed into the sum of integrals over Γ_1 and Γ_0,

$$\int_{\Gamma} P(X \mid a_1)\, d\Omega(X) = \int_{\Gamma_1} P(X \mid a_1)\, d\Omega(X) + \int_{\Gamma_0} P(X \mid a_1)\, d\Omega(X) = 1. \tag{5.83}$$

From (5.82) and (5.83) it follows that

$$G = \lambda p_0 \int_{\Gamma_1} P(X \mid a_0)\, d\Omega(X) + p_1 \left[1 - \int_{\Gamma_1} P(X \mid a_1)\, d\Omega(X)\right]$$

$$= p_1 - \int_{\Gamma_1} [p_1 P(X \mid a_1) - \lambda p_0 P(X \mid a_0)]\, d\Omega(X). \tag{5.84}$$

The expression for G is minimal when the integral in formula (5.84) assumes the maximal value. This condition can be satisfied if the region Γ_1 is chosen in such a way that it includes all points of observation space at which the integrand is positive, and does not contain those points at which the integrand is negative. Hence those and only those points enter into the optimal region Γ_1 for which

$$p_1 P(X \mid a_1) - \lambda p_0 P(X \mid a_0) > 0 \tag{5.85}$$

or

$$\frac{P(X \mid a_1)}{P(X \mid a_0)} > \frac{\lambda p_0}{p_1} = h. \tag{5.86}$$

Here h is some threshold. This inequality describes the decision method which is optimal according to the Neyman-Pearson criterion. If it is satisfied, then the point under consideration belongs to the region Γ_1, i.e., the decision $a = a_1$ is made. If the opposite inequality is satisfied, then the decision $a = a_0$ is made.

On the left side of (5.86) is the likelihood ratio $\Lambda(X)$. Therefore the optimal decision rule can be written in the form

$$\Lambda(X) \begin{cases} > \lambda p_0/p_1 = h \rightarrow a = a_1, \\ < \lambda p_0/p_1 = h \rightarrow a = a_0. \end{cases} \tag{5.87}$$

Comparing (5.87) with the decision methods (5.61) and (5.62) found earlier, we are convinced that the decision algorithm is the same everywhere. The difference in the methods is only in the magnitude of the threshold h. In order to find this threshold for the Neyman-Pearson criterion, the λ appearing in conditions (5.87) must be computed. The way of determining the threshold h for a typical problem is shown below.

If the errors of the first and second kinds are equally risky, then for the optimality criterion the total probability of error q can be taken and its minimum be required in the optimal system. This criterion, assumed by V. A. Kotel'nikov as the basis for choosing the optimal receiver, bears the name of the Kotel'nikov criterion or the criterion of the ideal detector. Substituting expressions (5.74) and (5.75) into (5.77), we obtain the required condition in the form

$$q = p_0 \int_{\Gamma_1} P(X \mid a_0)\, d\Omega(X) + p_1 \int_{\Gamma_0} P(X \mid a_1)\, d\Omega(X) = \min. \tag{5.88}$$

Using Eq. (5.83), this condition can be written in the following way:

$$\begin{aligned} q &= p_0 \int_{\Gamma_1} P(X \mid a_0)\, d\Omega(X) + p_1 \left[1 - \int_{\Gamma_1} P(X \mid a_1)\, d\Omega(X)\right] \\ &= p_1 - \int_{\Gamma_1} [p_1 P(X \mid a_1) - p_0 P(X \mid a_0)]\, d\Omega(X). \end{aligned} \tag{5.89}$$

In order that q be minimal, the integral in this formula must be maximal. This can be achieved by choosing the region Γ_1 such that it includes all those and only those points of observation space at which the integrand is positive. Thus the points of the optimal region Γ_1 must satisfy the condition

$$p_1 P(X \mid a_1) - p_0 P(X \mid a_0) > 0. \tag{5.90}$$

In this case the decision $a = a_1$ is made. Otherwise the decision $a = a_0$ is made. Thus the optimal decision by the Kotel'nikov criterion has the following form:

$$\Lambda(X) \begin{cases} > p_0/p_1 = h \rightarrow a = a_1\,, \\ < p_0/p_1 = h \rightarrow a = a_0\,. \end{cases} \tag{5.91}$$

This algorithm belongs to the same type as was found earlier. It coincides completely with the algorithm (5.61) for the method of maximum *a posteriori* probability. But now this method is not arbitrary and is justified by weighty argument, since it has been proved that this method indeed ensures the minimum probability of error.

By substituting formula (5.63) for $\Lambda(X)$, for $a_0 = 0$ in the typical example being considered we obtain the rule

$$\exp\left\{\frac{-\sum_{i=1}^{k} [x_i - a_1 f(t_i)]^2 + \sum_{i=1}^{k} x_i^2}{2\sigma^2}\right\} > h \rightarrow a = a_1\,. \tag{5.92}$$

Taking the logarithm of both sides of the inequality, removing the brackets on its left side and making the cancellation, we arrive at the condition

$$a_1 \sum_{i=1}^{k} x_i f(t_i) - \tfrac{1}{2} \sum_{i=1}^{k} [a_1 f(t_i)]^2 > \sigma^2 \ln h \rightarrow a = a_1\,. \tag{5.93}$$

For the transition to the continuous case, in the same way as in the analogous example given above (see also Fig. 5.12), let us multiply both sides of inequality (5.93) by Δt:

$$\sum_{i=1}^{k} x_i[a_1 f(t_i)]\,\Delta t - \tfrac{1}{2}\sum_{i=1}^{k}[a_1 f(t_i)]^2\,\Delta t > \sigma^2 \ln h\,\Delta t \rightarrow a = a_1 . \qquad (5.94)$$

Now we let Δt tend to zero; then the sum on the left side of the inequality turns into an integral. If $\sigma^2 = \text{const}$ is set, then here the right side of the inequality will tend to zero. But the case of $\sigma^2 = \text{const}$ is not of interest. We will require that the sequence of independent random variables $n_i = n(t_i)$ in the limit become white noise with spectral density S_0. This can be achieved if $\sigma^2 = S_0/\Delta t$ is set.

Actually, we will consider the stationary random process for which the graph of the correlation function $K_n(\tau)$ has the form of the "triangular" curve depicted in Fig. 5.15. Values of the ordinates of this process which are separated from one another by an interval Δt are not correlated, since for $\tau \geqslant \Delta t$ the function $K_n(\tau)$ is equal to zero. Therefore it can be supposed that the independent random variables $n_i = n(t_i)$ indicated above are discrete values of the random process with correlation function $K_n(\tau)$. Now if we let Δt tend to zero, then the area of the curve in Fig. 5.15 remains fixed, since the height of the triangle increases in proportion to the decrease in its base. The area of this curve is equal to

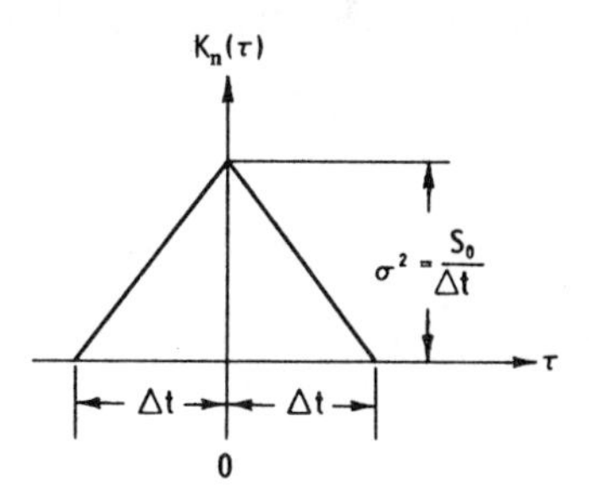

FIG. 5.15

$$\frac{1}{2}\frac{S_0}{\Delta t} 2\,\Delta t = S_0 .$$

From the curve of Fig. 5.15 in the limit we obtain an impulse function with area equal to S_0. According to formula (2.91) the spectral density for the random process is

$$\begin{aligned} S_n(\omega) &= \int_{-\infty}^{\infty} K_n(\tau) e^{-j\omega\tau}\, d\tau = 2\int_0^{\Delta t} K_n(\tau)\cos\omega\tau\, d\tau \\ &= 2\cos\omega\tau_m \int_0^{\Delta t} K_n(\tau)\, d\tau = S_0 \cos\omega\tau_m , \qquad 0 < \tau_m < \Delta t. \end{aligned} \qquad (5.95)$$

As $\Delta t \rightarrow 0$ the quantity $\tau_m \rightarrow 0$, and in the limit we obtain the equation

$S_n(\omega) = S_0 =$ const. Thus in the limit the process proves to be white noise with spectral density S_0.

If $\sigma^2 = S_0/\Delta t$, then the right side of inequality (5.94) will be equal to $S_0 \ln h$, and in the limit as $\Delta t \to 0$ condition (5.94) takes the form

$$I = \frac{1}{S_0}\int_0^T x(t)[a_1 f(t)]\, dt > \ln h + \frac{E_0}{2S_0} = h_1 \to a = a_1, \tag{5.96}$$

where as before E_0 is given by Eq. (5.72), and I is the notation for the integral on the left side of the inequality. Condition (5.96) gives the optimal detection method both from the Neyman-Pearson criterion and from the Kotel'nikov criterion. The difference is only in the magnitude of the threshold h. As is seen from (5.96), the block diagram of the optimal detector is the same as Fig. 5.13.

Let us find the false alarm probability α for the detection method given by the algorithm (5.96). In the absence of signal ($a = a_0 = 0$) the quantity $x(t) = n(t)$, and condition (5.96) acquires the form

$$(I)_{x(t)=n(t)} = I_0 = \frac{1}{S_0}\int_0^T n(t)[a_1 f(t)]\, dt > h_1 \to a = a_1. \tag{5.97}$$

The probability of realizing inequality (5.97) must be determined.

The integral I_0 is a random variable with a normal distribution law. In fact the quantity I_0 represents an integral, i.e., the sum of an infinite set of infinitesimally small terms. Each of these is normally distributed, since n_i is normally distributed noise and $a_1 f(t_i)$ is a factor which is constant for a given t_i. Hence the integral is also a normal random variable. Therefore for determining the distribution density $P(I_0)$ of the random variable I_0 it suffices to find the mean value $M\{I_0\}$ and dispersion σ_I^2 of this quantity. First we determine the value

$$M\{I_0\} = \frac{1}{S_0} M\left\{\int_0^T n(t)[a_1 f(t)]\, dt\right\} = \frac{1}{S_0}\int_0^T M\{n(t)\}[a_1 f(t)]\, dt = 0, \tag{5.98}$$

since $M\{n(t)\} = 0$. Further, in view of the fact that that the mean value was found to be equal to zero, we find the dispersion from the expression

$$\begin{aligned}\sigma_I^2 = M\{I_0^2\} &= M\left\{\frac{1}{S_0^2}\int_0^T n(t_1)[a_1 f(t_1)]\, dt_1 \int_0^T n(t_2)[a_1 f(t_2)]\, dt_2\right\} \\ &= M\left\{\frac{a_1^2}{S_0^2}\int_0^T\int_0^T [n(t_1)n(t_2)][f(t_1)f(t_2)]\, dt_1\, dt_2\right\} \\ &= \frac{a_1^2}{S_0^2}\int_0^T\int_0^T M\{n(t_1)n(t_2)\}[f(t_1)f(t_2)]\, dt_1\, dt_2.\end{aligned} \tag{5.99}$$

The mean value of the product $n(t_1)n(t_2)$ is the correlation function $K_n(\tau)$ of the noise, where $\tau = t_1 - t_2$ in accordance with (2.95).

Since for white noise with spectral density S_0

$$M\{n(t_1)n(t_2)\} = K_n(\tau) = S_0\delta(\tau) = S_0\delta(t_1 - t_2), \tag{5.100}$$

then

$$\begin{aligned}\sigma_I^2 &= \frac{a_1^2}{S_0^2}\int_0^T\int_0^T S_0\delta(t_2 - t_1)[f(t_1)f(t_2)]\,dt_1\,dt_2 \\ &= \frac{a_1^2}{S_0}\int_{t_1=0}^{t_1=T} f(t_1)\left[\int_{t_2=0}^{t_2=T}\delta(t_2 - t_1)f(t_2)\,dt_2\right]dt_1 \\ &= \frac{a_1^2}{S_0}\int_{t_1=0}^{t_1=T}[f(t_1)]^2\,dt_1 = \frac{E_0}{S_0}.\end{aligned} \tag{5.101}$$

Here the transformation of the integral in the middle brackets is carried out according to the known property of the δ-function, following from its definition

$$\int_{-\infty}^{\infty}\delta(t - \tau)\varphi(\tau)\,d\tau = \varphi(t). \tag{5.102}$$

This property is also suitable for the integral in the brackets of (5.101) if $0 < t_1 < T$. Therefore this integral is equal to $f(t_1)$.

Thus the distribution density $P(I_0)$ has the form

$$P(I_0) = \frac{1}{\sigma_I\sqrt{(2\pi)}}\exp\left\{-\frac{I_0^2}{2\sigma_I^2}\right\}, \tag{5.103}$$

where σ_I is defined by formula (5.101).

The quantity α represents the probability of realizing the inequality $I_0 > h_1$. This probability is easily found by knowing the distribution density $P(I_0)$, from the formula

$$\begin{aligned}\alpha &= \int_{h_1}^{\infty}P(I_0)\,dI_0 = \int_{h_1}^{\infty}\frac{1}{\sigma_I\sqrt{(2\pi)}}\exp\left\{-\frac{I_0^2}{2\sigma_I^2}\right\}dI_0 \\ &= \frac{1}{\sqrt{(2\pi)}}\int_{h_1/\sigma_I}^{\infty}\exp\left\{-\frac{z^2}{2}\right\}dz = \Phi(\infty) - \Phi\left(\frac{h_1}{\sigma_I}\right),\end{aligned} \tag{5.104}$$

where $z = I_0/\sigma_I$ and we set

$$\Phi(z) = \frac{1}{\sqrt{(2\pi)}}\int_0^z\exp\left\{-\frac{z^2}{2}\right\}dz. \tag{5.105}$$

The function $\Phi(z)$ has been tabulated; its tables are in handbooks on probability theory. Since $\Phi(\infty) = \frac{1}{2}$, then by taking (5.96) and (5.101) into account formula (5.104) for α can be rewritten in the form

$$\alpha = \frac{1}{2} - \Phi\left(\frac{\ln h + (E_0/2S_0)}{\sqrt{(E_0/S_0)}}\right). \tag{5.106}$$

The function Φ depends monotonically on its argument. Therefore the dependence of α or $p_0\alpha$ on h is likewise monotonic. It is shown in Fig. 5.16. As is seen from (5.106), for $h = 0$ the quantity $\alpha = 1$. This is also evident from physical considerations, since for a threshold h equal to zero the integral I exceeds this threshold for any arbitrarily small noise. The intersection of the curve and the horizontal straight line in Fig. 5.16 permits the threshold h^* to be determined. In an analogous

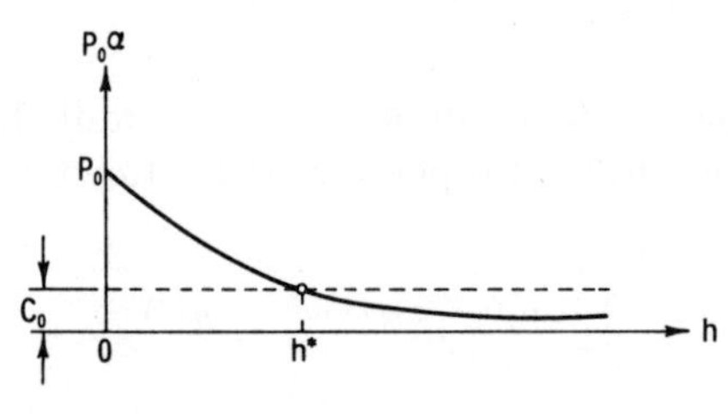

FIG. 5.16

fashion the quantity β, the probability of fulfilling the condition

$$(I)_{x(t)=n(t)+s(t)} = I_1 = \frac{1}{S_0}\int_0^T [a_1 f(t) + n(t)][a_1 f(t)]\, dt < h_1, \tag{5.107}$$

can be found.

This condition can be rewritten in the form

$$I_1 = \frac{1}{S_0}\int_0^T [a_1 f(t)]^2\, dt + \frac{1}{S_0}\int_0^T n(t)[a_1 f(t)]\, dt = \frac{E_0}{S_0} + I_0 < h_1. \tag{5.108}$$

Hence it follows that β is the probability that I_0 will be found to be less than $h_1 - (E_0/S_0)$:

$$\begin{aligned}\beta &= \int_{-\infty}^{h_1-(E_0/S_0)} P(I_0)\, dI_0 = \int_{-\infty}^{h_1-(E_0/S_0)} \frac{1}{\sigma_I\sqrt{(2\pi)}} \exp\left\{-\frac{I_0^2}{2\sigma_I^2}\right\} dI_0 \\ &= \frac{1}{\sqrt{(2\pi)}}\int_{-\infty}^{[h_1-(E_0/S_0)]/\sigma_I} \exp\left\{-\frac{z^2}{2}\right\} dz = \Phi\left(\frac{h_1-(E_0/S_0)}{\sigma_I}\right) - \Phi(-\infty) \\ &= \frac{1}{2} + \Phi\left(\frac{\ln h - (E_0/2S_0)}{\sqrt{(E_0/S_0)}}\right).\end{aligned} \tag{5.109}$$

For the optimal detector according to Kotel'nikov $h = p_0/p_1$. Therefore

$$\alpha = \frac{1}{2} - \Phi\left(\frac{\ln(p_0/p_1) + (E_0/2S_0)}{\sqrt{(E_0/S_0)}}\right) \tag{5.110}$$

and

$$\beta = \frac{1}{2} + \Phi\left(\frac{\ln(p_0/p_1) - (E_0/2S_0)}{\sqrt{(E_0/S_0)}}\right), \tag{5.111}$$

and by taking into account that $p_1 + p_0 = 1$, the total absolute probability of error q can be expressed by the formula

$$\begin{aligned} q &= p_0\alpha + p_1\beta \\ &= \frac{1}{2} + p_1\Phi\left(\frac{\ln(p_0/p_1) - (E_0/2S_0)}{\sqrt{(E_0/S_0)}}\right) - p_0\Phi\left(\frac{\ln(p_0/p_1) + (E_0/2S_0)}{\sqrt{(E_0/S_0)}}\right). \end{aligned} \tag{5.112}$$

Let us set

$$\rho = E_0/S_0\,. \tag{5.113}$$

The quantity ρ is the ratio of the energy of the signal E_0 to the spectral density of the noise S_0, i.e., the "signal-to-noise" ratio. For the example we will construct the curve $q = q(\rho)$ for the important special case $p_0 = p_1 = \frac{1}{2}$. Then $\ln(p_0/p_1) = 0$ and formula (5.112) takes the form

$$q = \frac{1}{2} + \frac{1}{2}\Phi\left(\frac{-\sqrt{\rho}}{2}\right) - \frac{1}{2}\Phi\left(\frac{\sqrt{\rho}}{2}\right) = \frac{1}{2} - \Phi\left(\frac{\sqrt{\rho}}{2}\right), \tag{5.114}$$

since Φ is an odd function. Since Φ is a monotonically increasing function, then with an increase in $\sqrt{\rho}$ the quantity q monotonically decreases. For $\sqrt{\rho} = 0$ the function $\Phi = 0$ and $q = \frac{1}{2}$. The dependence of q on ρ bears the name of "decision curve" (or "sampling curve"). After assigning the permitted value of the probability of error q, we can find the required signal-to-noise ratio from this curve. Usually the decision curve is represented in the form of the dependence of q on $\sqrt{\rho}$ with a logarithmic scale along the ordinate. The curve corresponding to Eq. (5.114) has been drawn in this way in Fig. 5.17.

We give an example of the application of this function to solve a specific problem. Let the signal be a high-frequency oscillation

$$s(t) = a \sin \omega_0 t \tag{5.115}$$

with period

$$T_0 = 2\pi/\omega_0\,. \tag{5.116}$$

As a simplification we will assume that the observation time T is equal to an integral number of periods (if $T \gg T_0$, then this can be achieved):

$$T = nT_0. \tag{5.117}$$

Then

$$E_0 = a_1^2 n \int_0^{T_0} f^2(t)\, dt = a_1^2 n \int_0^{T_0} \sin^2 \omega_0 t\, dt$$

$$= a_1^2 n \left[\int_0^{T_0} \left(\frac{1}{2} - \frac{1}{2}\cos 2\omega t\right) dt\right] = a_1^2 n \tfrac{1}{2} T_0 = \tfrac{1}{2} a_1^2 T. \tag{5.118}$$

Further,

$$\sqrt{\rho} = \sqrt{(E_0/S_0)} = \sqrt{(a_1^2 T/2S_0)} = a_1 \sqrt{(T/2S_0)}. \tag{5.119}$$

For example, let $a_1 = 1$ mv, $S_0 = 0.2$ mv² sec. What must the observation time T be in order that for $p_0 = p_1$ the probability of error q will be equal to 0.01 %? After setting $q = 10^{-4}$, we find from the curve of Fig. 5.17 the value $\sqrt{\rho} = 7.3$. Therefore from (5.119) we determine

$$\sqrt{\rho} = 7.3 = \frac{1}{\sqrt{(2 \cdot 0.2)}} \sqrt{T}. \tag{5.120}$$

Hence we obtain $T \approx 22$ sec. Thus in spite of a small signal-to-noise ratio, the decision can be made with a negligible probability of error in the problem under consideration. But a long observation time will serve as the price for the high accuracy of the decision.

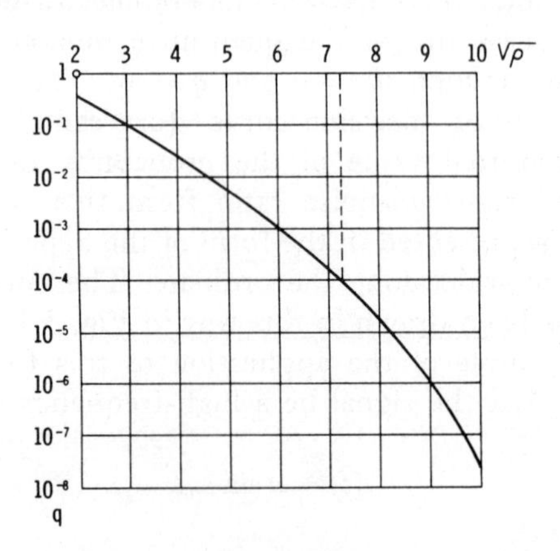

FIG. 5.17

3. Elements of General Statistical Decision Theory

The methods of the theory of two-alternative decisions can also be extended to more general classes of problems. For example, let the parameter a of the transmitted signal $s = af(t)$ be capable of having $m + 1$ possible values $a_0, a_1, a_2, \ldots, a_m$ with *a priori* probabilities $p_0, p_1, p_2, \ldots, p_m$, respectively. We will assume that the signal arrives at the receiver mixed with the noise $n(t)$. The problem is posed of deciding from the curve of $x(t)$ observed on the interval $0 \leqslant t \leqslant T$, what the value of the parameter a is in the transmitted signal. No matter what the decision method, it will consist of the observation space Γ being decomposed into $m + 1$ regions $\Gamma_0, \Gamma_1, \ldots, \Gamma_m$, corresponding to the decisions $a = a_0$, $a = a_1$, and so on. In this case as in the theory of two-alternative decisions, the methods of maximum likelihood, maximum *a posteriori* probability, the Kotel'nikov method (minimum absolute probability of error), or modifications of the Neyman-Pearson method can be applied. In the last case the probabilities of certain errors are prescribed, and the sum of the probabilities of the remaining errors are minimized.

The problems of parameter estimation represent a further generalization. Let the parameter a of the signal $s(t) = af(t)$ have a continuous *a priori* distribution with density $P(a)$. From an assumed sampling $X = (x_1, x_2, \ldots, x_k)$ of values of the received signal $x(t)$ or from a realization of the function $x(t)$ on the interval $0 \leqslant t \leqslant T$, it is required to decide what the value of the parameter a is in the transmitted signal. The methods considered above are extended to this case as well. For example, let $x_i = s_i + n_i$ $(i = 1, \ldots, k)$, and the same assumptions regarding the noise $n(t)$ are valid as in the typical example described in Section 2. Then the likelihood function for some value of the parameter a has the form [see (5.51) or (5.52)]

$$L(a) = P(X \mid a) = \frac{1}{(2\pi)^{k/2}\sigma^k} \exp\left\{-\frac{\sum_{i=1}^{k} [x_i - af(t_i)]^2}{2\sigma^2}\right\}. \qquad (5.121)$$

Guided by the maximum likelihood method, we select a value of a for which the likelihood function $L(a)$ is maximal, which is equivalent to the condition

$$F(a) = \sum_{i=1}^{k} [x_i - af(t_i)]^2$$

$$= \sum_{i=1}^{k} x_i^2 - 2a \sum_{i=1}^{k} x_i f(t_i) + a^2 \sum_{i=1}^{k} [f(t_i)]^2 = \min. \qquad (5.122)$$

We equate the derivative $dF(a)/da$ to zero:

$$-2\sum_{i=1}^{k} x_i f(t_i) + 2a \sum_{i=1}^{k} [f(t_i)]^2 = 0. \tag{5.123}$$

With condition (5.123) satisfied the function $F(a)$ is minimal. Hence we find the estimate $a = a^*$ from the maximum likelihood method:

$$a^* = \sum_{i=1}^{k} x_i f(t_i) \Big/ \sum_{i=1}^{k} [f(t_i)]^2. \tag{5.124}$$

In order to arrive at the case of a continuous measurement, the numerator and denominator of (5.124) can be multiplied by Δt. As $\Delta t \to 0$ and $k \to \infty$ the sums turn into integrals and formula (5.124) acquires the following form:

$$a^* = \frac{\int_0^T x(t) f(t)\, dt}{\int_0^T [f(t)]^2\, dt} = \frac{1}{e_0} \int_0^T x(t) f(t)\, dt, \tag{5.125}$$

where

$$e_0 = \int_0^T [f(t)]^2\, dt. \tag{5.126}$$

The quantity e_0 can be called the specific energy of the signal, since it is equal to the value of E_0 in formula (5.72) for $a_1 = 1$. From expression (5.125) it follows that the determination of a^* reduces to synchronous detection, i.e., the formation of the product $x(t)f(t)$ and then coherent integration—the obtaining of the integral of this product.

In the general case in the estimation problem for processes, the form of the transmitted signal $s(t)$ is subject to determination. It can be approximately known if, for example, the sequence of samples s_1, s_2, ..., s_k is given, where k is a sufficiently large number. The problem is posed of estimating these values.

Let us introduce the vector

$$S = (s_1, s_2, \ldots, s_k) \tag{5.127}$$

with Cartesian coordinates s_1, s_2, ..., s_k in k-dimensional space. We will call this space the "signal space." In the same way as for the spaces considered earlier, the *a priori* probability density $P(S)$ of this vector can be introduced. Then $P(S)\, d\Omega(S)$ is the probability of the end of the vector S falling in an infinitesimally small volume $d\Omega(S)$ of this space.

In the particular case of parameter estimation the signal $s(t)$ has the form $s(t, A)$, where $A = (a_1, ..., a_r)$ is a parameter vector, and in addition the form of the dependence of s on t and A is known. In this case the parameter space described in Section 1 can be used instead of the signal space. In the still more particular case of the theory of two-alternative decisions, there are only two possible values A_0 and A_1 of the vector A, which correspond to two points in parameter space. If the values a_0 and a_1 of a single parameter are considered in place of vectors, then the parameter space becomes one-dimensional and turns into a straight line, on which there are two points corresponding to the two possible values of the parameter.

From the preceding it is known that the receiver problem can be regarded as a decision problem, and this system itself is essentially a decision system. Its problem is to deliver to the output a "decision" about what signal is transmitted. If the problem of estimating a discrete process is posed, i.e., the vector $S = (s_1, s_2, ..., s_k)$, then estimates $d_1, ..., d_k$ must be given for each of the values $s_1, ..., s_k$. Hence in the general case the decision is a vector

$$D = (d_1, d_2, ..., d_k). \tag{5.128}$$

We will introduce a "decision space"—the k-dimensional space of the vector D constructed in the same way as the signal space of S. But its coordinates are now not the values s_i of the signal, but their estimates d_i made by the receiver. If the problem is posed of estimating the parameter vector A, then a decision about its coordinates $a_1, ..., a_r$ is made. In other words, r decisions $d_1, ..., d_r$ must be delivered to the output of the receiver, corresponding to the r-dimensional vector D. In this case the decision space is a replica of the parameter space of A.

Some decision D, i.e., a vector D in decision space, must be associated with each realization of the received signal, i.e., a point X of observation space. Any decision rule is a transformation rule for points of X-space (observation space) into points of D-space (decision space). Such a rule, being an algorithm of the receiver, is usually called the "strategy" of the receiver in statistical decision theory.

Two types of strategy are possible. In the first, one definite point of D-space corresponds to each fixed X. Such a strategy is called "determinate" or "regular." In the second type of strategy a certain probability distribution density $\Delta(D \mid X)$ of points of D-space corresponds to each fixed X. This means that the choice of decision is made in random fashion; but the statistical law regulating such a choice depends on the observed value of X. Strategies of the second type are called "random."

The second type of strategy is more general than the first type. In fact, a regular strategy can be regarded as a limit law for a random strategy, when the dispersion of points of D-space corresponding to the observed vector X_0 tends to zero. Here points of D-space are in practice concentrated in an infinitesimally small neighborhood of some definite point $D_0(X_0)$, corresponding to X_0 by a determinate, regular law. In the limit the probability density $\Delta(D \mid X_0)$ can be described by the expression

$$\Delta(D \mid X_0) = \delta[D - D_0(X_0)]. \tag{5.129}$$

Here δ is the notation for the unit impulse function, which is equal to zero everywhere at points $D \neq D_0(X_0)$, and at the point $D_0(X_0)$ becomes infinite. Here the integral

$$\int_D \delta[D - D_0(X_0)]\, d\Omega(D) \tag{5.130}$$

over all D-space equals unity, since in formula (5.129) the left and right sides are probability densities.

Thus, in general assigning the function $\Delta(D \mid X)$ is equivalent to prescribing a random strategy. In the case when Δ becomes a δ-function, the strategy is regular.

The function $\Delta(D \mid X)$ bears the name of "decision function" (see [5.15]). Indeed it is required to find this function, defining the strategy of the receiver. In statistical decision theory the problem is posed of finding the optimal decision function, where the optimality criterion is related to the errors in decision.

If the signal S is transmitted and the decision D made, then D may not be the correct estimate of S. We will assume that an incorrect decision stipulates a certain "loss" or "disadvantage," whose magnitude can be estimated by the "loss function" W, depending generally on S and D:

$$W = W(S, D). \tag{5.131}$$

Since for a correct decision the loss must be smallest, then

$$W(S, D = S) < W(S, D \neq S). \tag{5.132}$$

A set of various functions satisfies such a condition. For example, for estimating the coordinates $s_1, \ldots, s_k$ of the vector S we can select

$$W(S, D) = \text{const}[(s_1 - d_1)^2 + \cdots + (s_k - d_k)^2] \tag{5.133}$$

or

$$W(S, D) = \text{const}[| s_1 - d_1 | + \cdots + | s_k - d_k |], \tag{5.134}$$

and so on. Sometimes a simple and elementary function of the following form is designated as the loss function:

$$W(S, D) = \begin{cases} -\infty, & S = D, \\ 1, & S \neq D. \end{cases} \tag{5.135}$$

This function can be expressed by the formula

$$W(S, D) = 1 - \delta(S - D), \tag{5.136}$$

where the δ-function is denoted by the letter δ.

Let the transmitted signal S be fixed. Under this condition the decision D is generally random. Actually, in the first place the vector X, on the basis of whose value the decision is made, is random because of the presence of the noise N. In the second place, the very law for determining D for a given X can be random, if the receiver carries out a random strategy. But if D is random, then the loss function $W(S, D)$ is also a random variable. Therefore it cannot serve as a measure of the operation of the decision system. As a measure it is reasonable to choose the mathematical expectation of the loss function, which bears the name "risk."

We designate as the conditional risk $r(S, \Delta)$ the mathematical expectation of the loss function $W(S, D)$ for a fixed signal S and some fixed decision function Δ. According to the definition of the mathematical expectation, $W(S, D)$ must be averaged over different values of D. We will denote the conditional probability density of D for a given S by $P(D \mid S)$. Then

$$r(S, \Delta) = M\{W \mid S\} = \int_{\Omega(D)} W(S, D) P(D \mid S) \, d\Omega. \tag{5.137}$$

Here $\Omega(D)$ is the region of possible values of D, and $d\Omega$ is its infinitesimally small volume. From now on we will denote by $d\Omega$ an infinitesimally small volume of any region over which the integration is carried out.

By knowing the experimental conditions, i.e. the probabilistic characteristics of the noise and the method of combination of signal and noise in the communication channel, the probability density $P(X \mid S)$ can be found. For example, such a probability density has been determined in the typical problem considered in Section 2. If the operating law of the decision system is known i.e., the function $\Delta(D \mid X)$, then the function $P(D \mid S)$ appearing in formula (5.137) is defined by the expression

$$P(D \mid S) = \int_{\Omega(X)} \Delta(D \mid X) P(X \mid S) \, d\Omega. \tag{5.138}$$

In fact, the probability of making the decision D for a fixed S is equal to the sum of the probabilities of making this decision for arbitrary values of X which are possible for a given S, when the end of the vector X falls in the volume $d\Omega$ with probability $P(X \mid S) \, d\Omega$.

In the general case it is not known what signal is transmitted. Only the *a priori* probability density $P(S)$ of the transmitted signals is known. Therefore the conditional risk r must still be averaged over the entire region $\Omega(S)$ of possible signals. The result of such an averaging is called the "total" or "mean" risk and is denoted by the letter R. The quantity R is the mathematical expectation of the conditional risk r for various observations, when the signal source sends different signals S with probability density $P(S)$. The expression for R has the form

$$R = M\{r\} = \int_{\Omega(S)} r(S, \Delta) P(S) \, d\Omega. \tag{5.139}$$

In this case $d\Omega$ denotes an infinitesimally small volume of signal space $\Omega(S)$.

Now we can formulate the problem of finding the optimal strategy of the decision system as the problem of determining a decision function $\Delta(D \mid X)$ for which the mean risk is minimal.

All the problems of statistical decision theory considered above are special cases of the general problem of the minimization of risk. We illustrate this statement by an example from the theory of two-alternative decisions with a single parameter a. In this case a will appear in the formulas instead of S and the scalar quantity d will be found in place of D. Let the parameter a be capable of having two possible values a_1 and a_0 with *a priori* probabilities p_1 and p_0, respectively. Then d likewise can assume only two possible values:

$$d = \begin{cases} d_1 & (\text{decision} \quad a = a_1), \\ d_0 & (\text{decision} \quad a = a_0). \end{cases} \tag{5.140}$$

In the example under consideration let the loss function have the form

$$W = \begin{cases} 0, & d = a, \\ 1, & d \neq a. \end{cases} \tag{5.141}$$

We will write out the expression for the conditional risk $r(a, \Delta)$. Let the conditional probabilities of decisions d_1 and d_0 for a fixed a be $p(d_1 \mid a)$ and $p(d_0 \mid a)$. In this example the integral (5.137) is replaced by the sum

$$r(a, \Delta) = W(a, d_0)p(d_0 \mid a) + W(a, d_1)p(d_1 \mid a). \tag{5.142}$$

In particular

$$\begin{aligned} r(a_1, \Delta) &= W(a_1, d_0)p(d_0 \mid a_1) + W(a_1, d_1)p(d_1 \mid a_1) \\ &= W(a_1, d_0)p(d_0 \mid a_1) = p(d_0 \mid a_1) \end{aligned} \tag{5.143}$$

and

$$\begin{aligned} r(a_0, \Delta) &= W(a_0, d_0)p(d_0 \mid a_0) + W(a_0, d_1)p(d_1 \mid a_0) \\ &= W(a_0, d_1)p(d_1 \mid a_0) = p(d_1 \mid a_0). \end{aligned} \tag{5.144}$$

Further, the integral in formula (5.139) for the mean risk R is replaced in this example by the sum

$$R = M\{r\} = r(a_1, \Delta)p_1 + r(a_0, \Delta)p_0, \tag{5.145}$$

since p_1 and p_0 are the *a priori* probabilities of the values a_1 and a_0 respectively. Substituting (5.143) and (5.144) into (5.145), we obtain:

$$R = p_0 p(d_1 \mid a_0) + p_1 p(d_0 \mid a_1). \tag{5.146}$$

We will explain the meaning of the expression $p(d_1 \mid a_0)$ representing the probability of the decision that $a = a_1$ under the condition that in reality $a = a_0$. But in fact this is the probability of a "false alarm," which had been denoted earlier by the letter α. Thus,

$$p(d_1 \mid a_0) = \alpha. \tag{5.147}$$

In a quite analogous fashion we find that $p(d_0 \mid a_1)$ is the probability of the decision $a = a_0$ under the condition $a = a_1$, i.e., the probability of "false ring," denoted in Section 2 by the symbol β:

$$p(d_0 \mid a_1) = \beta. \tag{5.148}$$

After the substitution of (5.147) and (5.148) into formula (5.146), the latter acquires the form

$$R = p_0\alpha + p_1\beta = q, \tag{5.149}$$

i.e., in this case the mean risk coincides with the absolute probability of error q. Therefore in this case the condition $R = \min$ for the optimal strategy means the minimization of the absolute probability of error q. We have arrived at the Kotel'nikov criterion.

It can be shown (see, for example, [1.10], 2nd ed., pp. 604–611), that for another assignment of the loss function W, we can obtain conditions which coincide with other optimality criteria, in particular with the Neyman-Pearson criterion.

The problem formulated above of minimizing the total risk R can only be posed when the *a priori* probability density $P(S)$ of the signal S is known. Actually, the expression $P(S)$ appears in formula (5.139) for the risk R. The stated problem is designated as of Bayes' type, and the function $\Delta(D \mid X)$ obtained as a result of its solution is called the "Bayes' decision."

In many problems of practical importance, however, the *a priori* probability density $P(S)$ is not known. The "*a priori* difficulty" connected with this circumstance has already been discussed in this chapter. In such a case it remains only to apply formula (5.137) for the conditional risk r, whose calculation does not require a knowledge of the function $P(S)$. In statistical decision theory the following "pessimistic" method of argument is offered: we will find the "worst" signal S^* for which the conditional risk r is maximal (for a given Δ). This condition is written analytically as:

$$r(S^*, \Delta) = \max_{(S)} r(S, \Delta). \tag{5.150}$$

We will now compare the various decision functions Δ and select one of them (we will call it Δ^*) such that the "worst" conditional risk $r(S^*, \Delta)$ for it is minimal. Thus,

$$r(S^*, \Delta^*) = \min_{(\Delta)} r(S^*, \Delta) = \min_{(\Delta)} \max_{(S)} r(S, \Delta). \tag{5.151}$$

The decision Δ^* is called "minimax," and the strategy corresponding to it is called "minimax optimal," or "optimal in the minimax sense."

We will clarify in what sense this strategy can be regarded as best. This strategy in the worst case for it gives a better result (i.e., a smaller value of the risk r) than any other strategy in the worst case for this other strategy. The orientation to the worst possible case gives a guarantee that under any conditions the system will not operate worse at any rate.

However, such an orientation cannot be justified from a practical point of view if actually the worst conditions of operation are very rarely encountered. Therefore the minimax approach is by no means the only possible point of view. Even from Laplace's time it has sometimes been recommended, whenever the *a priori* distribution $P(S)$ is unknown, to regard it as uniform, i.e., to assume *a priori* that all possible values of S are equiprobable in the region $\Omega(S)$. It is interesting that such a point of view often leads to the choice of the same strategy as is obtained by using the minimax approach. The latter was introduced in game theory where the player, conducting a struggle with a reasonable opponent, expects the maximal injury to himself and must be prepared for the best actions under very unfavorable circumstances. But in the majority of communication and control problems nature, generating noise, is indifferent to the aims of man; it lacks a "malicious intent"—the deliberate tendency to cause injury. Therefore for the solution of communication and control problems which can be regarded as a "game" of man with nature, the minimax approach can prove to be unnecessarily pessimistic.

It should also be noted that in general the minimax optimal strategy Δ is not uniformly optimal at all for all signals S. For various types of signal it can give a worse result than other types of strategy.

The concepts of risk, optimal Bayes' and minimax strategies gain a considerable clarity if still another space is introduced—the "risk space." We will examine this space for the simplest case of the theory of two-alternative decisions, when it turns into a plane. First we investigate only regular strategies D_i. We will assume that some strategy D_1 is taken. In the case when $a = a_0$, let the conditional risk for the use of this strategy be equal to $r(a_0, D_1)$. We will recall that this expression is the mathematical expectation of the loss function under the condition that $a = a_0$ and the strategy D_1 is taken. But if $a = a_1$, then we will consider that with the use of the strategy D_1 the conditional risk is found to be equal to $r(a_1, D_1)$.

We will lay off values $r(a_0, D)$ and $r(a_1, D)$ of the conditional risk in the form of Cartesian coordinates in a plane, which we will call the risk plane (Fig. 5.18). The strategy D_1 corresponds to the point D_1 with coordinates $r(a_0, D_1)$ and $r(a_1, D_1)$ in the risk plane. For another strategy D_2 the conditional risks $r(a_0, D_2)$ and $r(a_1, D_2)$ will have other values in general. This strategy corresponds to another point D_2 in the risk plane. The smaller the risk, the better we consider the strategy. But a direct comparison of the strategies D_1 and D_2 in the risk plane does not always permit one of them to be expressed to advantage. For example, the value $r(a_1, D_1)$ of the risk for the strategy D_1 is smaller than the corresponding value $r(a_1, D_2)$ for D_2 (see Fig. 5.18). But on the other hand for the strategy D_2 the risk $r(a_0, D_2)$ is smaller. It is evident that a strategy D_3 must be rejected, since for this strategy both the conditional risks $r(a_0, D_3)$ and $r(a_1, D_3)$ are greater than for the strategy D_1.

Up to now we have designated as points in the plane only regular, or as they are sometimes called, "pure" strategies. But random strategies can also be represented in the risk plane. In fact, we will suppose that we choose either a strategy D_1 with probability q_1 or a strategy D_4 with probability q_2. We will denote such a random strategy by D_5'. Then the conditional risk r for $a = a_0$ is equal to the mean value of

$$r(a_0, D_5') = q_1 r(a_0, D_1) + q_2 r(a_0, D_4). \tag{5.152}$$

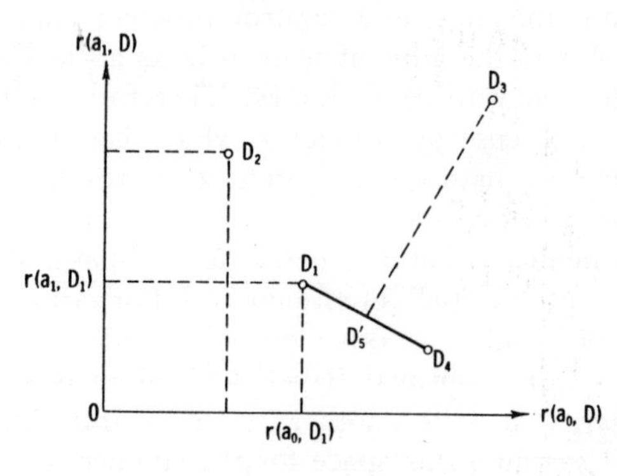

Fig. 5.18

In an analogous fashion

$$r(a_1, D_5') = q_1 r(a_1, D_1) + q_2 r(a_1, D_4). \tag{5.153}$$

It is not difficult to see that the point D_5' with coordinates $r(a_0, D_5')$ and $r(a_1, D_5')$ is on the straight line joining the points D_1 and D_4.

For $q_1 + q_2 = 1$ and a variation of q_1 from 0 to 1, the point D_5' can occupy any position on the segment D_1D_4, moving from D_4 to D_1. Such strategies, for which different regular strategies change in a random manner, are sometimes called "mixed." From the preceding it follows that any point of the segment between the two points of pure strategies corresponds to some mixed strategy.

Let us now join the points D_5' and D_3 with a straight line; it is shown in Fig. 5.18 by a dashed line. Then each point of the dashed segment corresponds to some mixed strategy. It is formed by the use of the strategies D_3 and D_5' with definite probabilities (which in the final result reduces to the use of the three regular strategies D_1, D_4, and D_3 with certain probabilities). Drawing straight lines in the same manner between the other points for the strategies obtained earlier, we can be convinced that any of the interior points of the convex polygon formed by the initial regular strategies—for example, D_1, D_2, D_3, D_4, D_5

in Fig. 5.19—corresponds to some mixed strategy. This polygon is shaded in Fig. 5.19. If a point corresponding to some initial regular strategy D_6 is inside the polygon, then it does not participate in its construction.

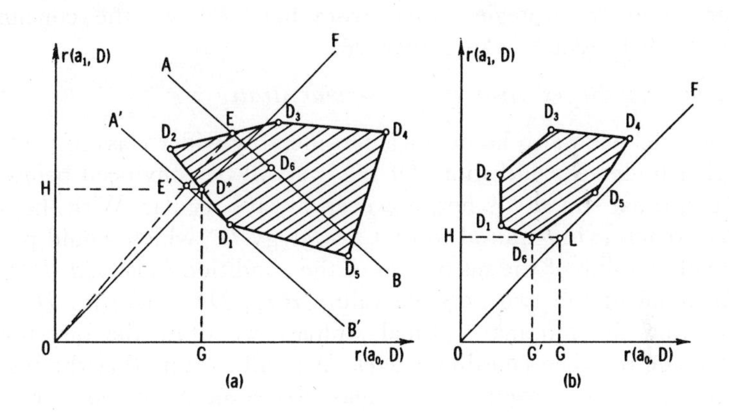

FIG. 5.19

How is the Bayes' strategy to be interpreted in the risk plane? In this case the *a priori* probabilities p_0 and p_1 of the values a_0 and a_1, respectively, are assigned. Then the mean risk R can be found from the formula

$$R = p_1 r(a_1, D) + p_0 r(a_0, D). \tag{5.154}$$

The curve $R =$ const on the risk plane is a straight line with slope p_0/p_1. If p_0 and p_1 are given, then the slope of the straight line $R =$ const is also given. Let this be the straight line AB in Fig. 5.19a. As is not difficult to see, the magnitude of the risk R for points on this straight line, i.e., the left side of Eq. (5.154), is proportional to the length of the perpendicular OE extended from the origin to the straight line AB. For example, if we take the strategy D_6 which corresponds to a point lying on the straight line AB, then the magnitude of the total risk R is determined by the length of the perpendicular OE. But from Fig. 5.19a it is seen that we can find a more successful strategy for which the risk R will turn out to be less. In order to find such a strategy, the straight line AB must be transferred parallel to itself along the direction to the origin. The last position $A'B'$ of this straight line corresponds to the smallest possible length OE' of the perpendicular, and hence to the smallest possible value of the total risk R. A further approach of the straight line AB to the origin is impossible, since all possible strategies D are situated within the limits of the shaded polygon. The end position $A'B'$ of the straight line corresponds to a unique strategy D_1, which is optimal as well. From the construction it is evident that the optimal strategy corresponds to a vertex of the rectangle. And since all vertices correspond to pure, regular strategies, then from here follows the conclusion, having great fundamental significance:

The optimal Bayes' strategy is a regular strategy.

This conclusion also holds for a considerably wider class of problems than that under consideration. Of this we will be convinced below.

The optimal strategy is not, however, always regular. With the minimax approach it is required to find a strategy D^* which would provide the smallest value of the maximum of the conditional risk $r(a, D^*)$, i.e., the maximal of the two possible values $r(a_0, D^*)$ and $r(a_1, D^*)$. For determining the minimax optimal strategy we draw the bisector OF of the angle between coordinate axes. We will assume that the bisector OF intersects the polygon of strategies. Let some point move from the origin along the straight line OF. The place of first contact of this point with the boundary of the polygon will correspond to the strategy D^*. In fact, for this point both values $OG = r(a_0, D^*)$ and $OH = r(a_1, D^*)$ of the risk are equal to one another and are less than the maximal risk r for any other point of the polygon. For example, for the point D_1 the maximal of the risks, $r(a_0, D_1)$, is greater than OG.

In the general case the bisector OF does not meet the polygon at its vertex. Therefore in the general case the strategy D^* is not pure, but mixed. Hence the important conclusion follows:

In general the minimax optimal strategy is a random strategy.

In a particular case the minimax optimal strategy can turn out to be regular. For example, let the bisector OF not meet the polygon $D_1D_2D_3D_4D_5D_6$ (Fig. 5.19b). In this case the vertical line LG and the horizontal line LH must be connected to the point L moving from the origin along the bisector. The optimal strategy corresponds to the point of the polygon first meeting one of these lines—in Fig. 5.19b this is the point D_6.

All these conclusions are extended to a wide class of problems. For example, the generalization to the theory of multi-alternative decisions consists of replacing the risk plane by an $(m + 1)$-dimensional space with coordinates $r(a_0, D)$, $r(a_1, D)$, ..., $r(a_m, D)$. The methods and results of the arguments conducted above on the plane are extended in their entirety to the case of a multidimensional space.

4. Statistical Decision Theory for Application to Automatic Control Systems

The application of statistical decision theory to open-loop control systems, or to systems reducing to open-loop, does not differ in principle from the application to problems in communication theory. But in this field there are few papers at present. In [5.23] the theory of an optimal system for character recognition was analyzed based on statistical decision theory. In [5.24] an open-loop system was analyzed as a preliminary before a presentation of the theory of closed-loop systems. In [5.25] certain basic concepts of statistical decision theory are presented and the simplest example given of the application of this theory to the design of one control system, where the latter reduces in practice to an open-loop system.

Let us consider the open-loop automatic control system depicted in Fig. 5.20 [5.24]. For simplicity we will assume that the variables in this circuit are scalars and are functions of discrete time $t = 0, 1, \ldots, s, \ldots, n$, where n is fixed. Let the driving action x^* have the form

$$x_s^* = x_s^*(s, \bar{\lambda}), \tag{5.155}$$

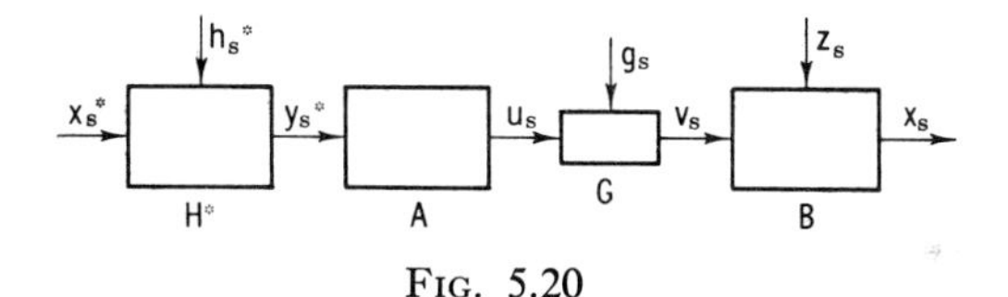

FIG. 5.20

where $\bar{\lambda}$ is a vector with random coordinates:

$$\bar{\lambda} = (\lambda_1, \ldots, \lambda_q). \tag{5.156}$$

Thus x_s^* is a discrete random process. We will mention specific examples of such a process:

$$x_s^* = \lambda_1, \tag{5.157}$$

$$x_s^* = \lambda_1 + \lambda_2 s + \lambda_3 s^2 + \lambda_4 s^3, \tag{5.158}$$

$$x_s^* = \lambda_1/(1 + \lambda_2 s), \tag{5.159}$$

$$x_s^* = \lambda_1 \exp\{-\lambda_2 s\}, \tag{5.160}$$

$$x_s^* = \sum_{i=1}^{q} \lambda_i f_i(s), \tag{5.161}$$

where the $f_i(s)$ are known functions.

In the particular case of (5.157) the random process degenerates into the random variable λ_1.

Let the *a priori* probability density $P(\bar{\lambda})$ of the vector $\bar{\lambda}$ be given, i.e., the joint *a priori* probability density of the variables $\lambda_1, \dots, \lambda_q$:

$$P(\bar{\lambda}) = P(\lambda_1, \dots, \lambda_q). \tag{5.162}$$

Let the probability density $P(h_s^*)$ of the noise h^* in the inertialess channel H^*, and also the method of combination of the signal x^* and the noise h^* in this channel be known. Then the conditional probability $P(y_s^* \mid x_s^*)$ can be found. This function is identical for all s, since the density $P(h_s^*)$ for the noise will be considered fixed. In addition, we will take x_s^*, h_s^* and any other external actions on the system as independent. Finally, let h_i^* ($i = 1, \dots, n$) be a sequence of variables independent of one another.

In the simplest case the controlled object does not have memory, and its operator is given by the expression

$$x_s = F(z_s, v_s), \tag{5.163}$$

where F is a known function, and the noise z_s depends on discrete time s and the random vector $\bar{\mu}$:

$$z_s = z_s(s, \bar{\mu}). \tag{5.164}$$

This expression is analogous to formula (5.155). The vector $\bar{\mu}$ has m random coordinates:

$$\bar{\mu} = (\mu_1, \dots, \mu_m). \tag{5.165}$$

Let the *a priori* distribution density $P(\bar{\mu})$ be given. The input variable v_s of the object B is generated at the output of the inertialess communication channel G. Passing through this channel, the control action u_s is mixed with the noise g_s whose distribution density $P(g_s)$ is fixed.

The values of g_s for various s are a sequence of independent random variables. Therefore the conditional probability density $P(v_s \mid u_s)$ can be found.

We will denote the specific loss function corresponding to the discrete instant s by $W_s(s, x_s^*, x_s)$. Let the total loss function W be equal to the sum of the specific loss functions

$$W = \sum_{s=0}^{n} W_s(s, x_s^*, x_s). \tag{5.166}$$

The problem consists of determining the optimal strategy of the controller A. In the general case this strategy can be regarded as random. We will denote by $\Gamma_s(u_s)$ the probability density for the control action u_s at the instant $t = s$. In the general case this quantity must depend on the entire input information accumulated by the controller A during the times $t = 0, 1, ..., s - 1$, i.e., on the values $y_0^*, y_1^*, ..., y_{s-1}^*$.

Let us introduce the temporal vectors

$$\begin{aligned} \mathbf{y}_s^* &= (y_0^*, y_1^*, ..., y_s^*), \\ \mathbf{x}_s^* &= (x_0^*, x_1^*, ..., x_s^*) \qquad (s = 0, 1, ..., n). \end{aligned} \tag{5.167}$$

The coordinates of the vector $\mathbf{y}_s^*$ are the sequence of values of the scalar variable y^* at the time instants $t = 0, 1, ..., s$. Analogous sequences of values of the variable x^* are combined into the form of the temporal vector $\mathbf{x}_s^*$. We will denote temporal vectors by boldface corresponding letters.

Since the optimal strategy, i.e., the optimal probability density $\Gamma_s(u_s)$ depends on the vector $\mathbf{y}_{s-1}^*$, it can be written in expanded form as the conditional probability density $\Gamma_s(u_s \mid \mathbf{y}_{s-1}^*)$. The problem consists of finding the optimal distributions Γ_s for all the time instants $t = 0, 1, ..., s, ..., n$. The functions Γ_s $(s = 0, 1, ..., n)$ must be selected such that a minimum of the mean risk R is ensured, i.e., the mathematical expectation of the loss function W. From (5.166) it follows that

$$R = M\{W\} = M\left\{\sum_{s=0}^{n} W_s\right\} = \sum_{s=0}^{n} M\{W_s\} = \sum_{s=0}^{n} R_s, \tag{5.168}$$

where the quantity

$$R_s = M\{W_s(s, x_s^*, x_s)\} \tag{5.169}$$

can be called the "specific risk." The specific risk R_s corresponds to the discrete time instant $t = s$.

We will denote by $P(x_s \mid u_s)$ the conditional probability density for x_s with a fixed u_s. For given $P(\bar{\mu})$ and $P(v_s \mid u_s)$ this function can be computed by using formulas (5.163) and (5.164). Further, let $P(\mathbf{y}_{s-1}^* \mid \mathbf{x}_{s-1}^*)$ be the conditional probability density for the vector $\mathbf{y}_{s-1}^*$ for a prescribed vector $\mathbf{x}_{s-1}^*$. With the properties of the channel H^* stated above, the individual values y_i^* are independent of one another $(i = 0, 1, ..., s - 1)$ since the individual values h_i^* are independent. Therefore

$$\begin{aligned} P(\mathbf{y}_{s-1}^* \mid \mathbf{x}_{s-1}^*) &= P(y_0^*, y_1^*, ..., y_{s-1}^* \mid \mathbf{x}_{s-1}^*) \\ &= \prod_{i=0}^{s-1} P(y_i^* \mid \mathbf{x}_{s-1}^*) = \prod_{i=0}^{s-1} P(y_i^* \mid x_i^*). \end{aligned} \tag{5.170}$$

The latter transformation is valid because the channel H^* is inertialess, and therefore $P(y_i^*)$ depends only on x_i^*, and not on the previous values x_j^* $(j < i)$.

Since the vector $\mathbf{x}_{s-1}^*$ depends on s and $\bar{\lambda}$, expression (5.170) also depends on s and $\bar{\lambda}$. Figure 5.21 depicts the calculation of the effect of random factors in each block of the system in the form of assigning conditional probability densities.

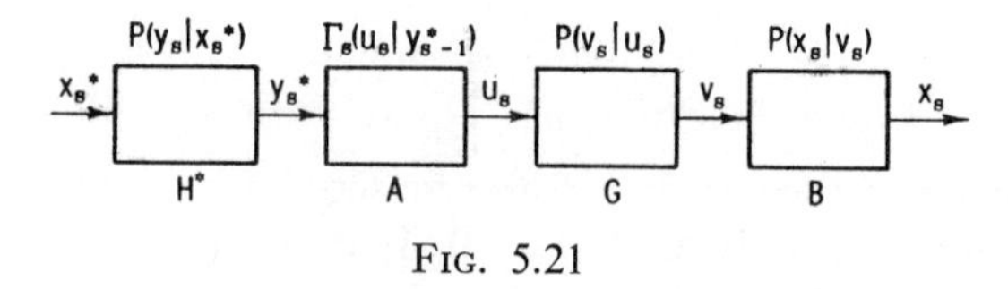

FIG. 5.21

We will denote by $\Omega(x_s, v_s, u_s, \mathbf{y}_{s-1}^*)$ the domain of variation of the parameters $x_s, v_s, u_s, y_0^*, \ldots, y_{s-1}^*$. It can be conceived of in the form of an $(s+3)$-dimensional space whose Cartesian coordinates are all the parameters stated above. Let $d\Omega(x_s, v_s, u_s, \mathbf{y}_{s-1}^*) = dx_s\, dv_s\, du_s\, dy_0^* \cdots dy_{s-1}^*$. An infinitesimally small volume of an arbitrary region is denoted below by $d\Omega$, if it is evident what region is in view.

We will first write the expression for the "conditional specific risk" r_s, meaning the value of the specific risk for a fixed vector $\mathbf{x}_s^*$, or, what is the same thing, for a fixed vector $\bar{\lambda}$. Then

$$r_s = M\{W_s \mid \mathbf{x}_s^*\} = \int_{\Omega(x_s)} W_s(s, x_s^*, x_s) P(x_s \mid \mathbf{x}_s^*)\, d\Omega. \tag{5.171}$$

Here the averaging of W_s is carried out with respect to the random variable x_s. Since $\mathbf{x}_s^*$ is considered fixed, then in averaging the conditional probability density of x_s for a fixed $\mathbf{x}_s^*$ must be used, i.e., the function $P(x_s \mid \mathbf{x}_s^*)$.

Since x_s depends on the input variable v_s of the object B, then by knowing the *a priori* probability characteristic of the noise z_s, $P(x_s \mid v_s)$ can be found (Fig. 5.21). But the quantity v_s is itself random, and in the end its distribution law depends on $\mathbf{x}_s^*$, i.e., it is the function $P(v_s \mid \mathbf{x}_s^*)$. The relation between these densities has the form

$$P(x_s \mid \mathbf{x}_s^*) = \int_{\Omega(v_s)} P(x_s \mid v_s) P(v_s \mid \mathbf{x}_s^*)\, d\Omega. \tag{5.172}$$

Here $d\Omega$ is the infinitesimally small volume of the region $\Omega(v_s)$ of all possible values of v_s. Formula (5.172) expresses the probability

$P(x_s \mid \mathbf{x}_s{}^*)\, dx_s$ in the form of the sum of probabilities for the output variable of the object falling in the range x_s to $x_s + dx_s$ for various values of v_s, but with a fixed $\mathbf{x}_s{}^*$.

Substituting (5.172) into (5.171) we find:

$$r_s = \int_{\Omega(x_s, v_s)} W_s(s, x_s{}^*, x_s) P(x_s \mid v_s) P(v_s \mid \mathbf{x}_s{}^*)\, d\Omega. \qquad (5.173)$$

Now the integration is carried out over the two-dimensional region $\Omega(x_s, v_s)$ of all possible values of x_s and v_s.

Let us continue the development of expression (5.173). By knowing $P(v_s \mid u_s)$ the probability density

$$P(v_s \mid \mathbf{x}_s{}^*) = \int_{\Omega(u_s)} P(v_s \mid u_s) P(u_s \mid \mathbf{x}_s{}^*)\, d\Omega \qquad (5.174)$$

can be found.

In an entirely analogous way we obtain the expression

$$P(u_s \mid \mathbf{x}_s{}^*) = \int_{\Omega(\mathbf{y}_{s-1}^*)} \Gamma_s(u_s \mid \mathbf{y}_{s-1}^*) P(\mathbf{y}_{s-1}^* \mid \mathbf{x}_{s-1}^*)\, d\Omega, \qquad (5.175)$$

where the integration is carried out over the region $\Omega(y_0{}^*, \ldots, y_{s-1}^*) = \Omega(\mathbf{y}_{s-1}^*)$.

Substituting (5.175) into (5.174), and (5.174) into (5.173), we arrive at the expression for the conditional specific risk

$$r_s = M\{W_s \mid \mathbf{x}_s{}^*\} = \int_{\Omega(x_s, v_s, u_s, \mathbf{y}_{s-1}^*)} W_s(s, x_s{}^*, x_s) P(x_s \mid v_s)$$
$$\times P(v_s \mid u_s) \Gamma_s(u_s \mid \mathbf{y}_{s-1}^*) P(\mathbf{y}_{s-1}^* \mid \mathbf{x}_{s-1}^*)\, d\Omega. \qquad (5.176)$$

As was shown above, the factor $P(\mathbf{y}_{s-1}^* \mid \mathbf{x}_{s-1}^*)$ in the integrand depends on the vector $\bar{\lambda}$. We will denote by $\Omega(\bar{\lambda})$ the domain of variation of the vector $\bar{\lambda}$. For different trials performed on the system, this vector can take on various values. Therefore the conditional specific risk r_s in different trials will also have various values, and thus is a random variable. The mean specific risk R_s can be found by averaging the conditional risk r_s with respect to $\bar{\lambda}$:

$$R_s = \int_{\Omega(\bar{\lambda})} r_s P(\bar{\lambda})\, d\Omega. \qquad (5.177)$$

Substituting (5.176) into (5.177) we find:

$$R_s = \int_{\Omega(x_s, u_s, v_s, \mathbf{y}^*_{s-1}, \bar{\lambda})} W_s(s, x_s{}^*, x_s) P(x_s \mid v_s) P(v_s \mid u_s)$$

$$\times \Gamma_s(u_s \mid \mathbf{y}^*_{s-1}) P(\mathbf{y}^*_{s-1} \mid \mathbf{x}^*_{s-1}) P(\bar{\lambda})\, d\Omega. \tag{5.178}$$

The problem consists of choosing the functions Γ_s such that the values of R_s, and hence the total risk $R = \Sigma_{s=0}^{n} R_s$ are minimal.

By considering the problem formulated above, it is not difficult to be convinced that it does not differ essentially from those communication theory problems examined in Sections 2 and 3 of this chapter. In fact, the risk formulation in the form of requiring the minimization of expression (5.168) does not differ from the general formulation given in Section 3. The element H^* in Fig. 5.20 can be regarded as a communication channel with noise h^*, and the elements A, G, B can be combined and called the receiver. However, in the latter problem there are certain differences from the problems solved in Sections 2 and 3, which though not fundamental do foster a known specific character:

(a) The prescribed elements G and B enter into the structure of the "receiving" part of the system. The problem consists of finding the algorithm of only the single part A of this system. We note that in communication theory as well analogous problems are encountered.

(b) There are the random noises g_s and z_s inside the same "receiving" part of the system. An analogous problem can also be formulated for a communication system in which there are internal noises.

(c) From the same source of the process, it is required to deliver to the output x_s of the object B a quantity that minimizes the mathematical expectation of the loss function. For example, if

$$W_s = (x_s{}^* - x_s)^2, \tag{5.179}$$

then evidently it is required to minimize a certain measure of the deviation of the "decision" x_s from the "transmitted signal" $x_s{}^*$. In the ideal case $x_s = x_s{}^*$ and W_s attains a minimal value equal to zero. But in view of the presence of random noise superimposed on the "received" signal $y_s{}^*$, and also because of the internal noises g_s and z_s of the "receiving" part of the system, errors in the decision are unavoidable, and in general x_s does not coincide with $x_s{}^*$. The difference in this problem statement from the conditions given in Sections 2 and 3 is that the earlier decision was given only at the end of the observation of the received signal. Meanwhile in this case it is required to effect a decision continuously as the signal $x_s{}^*$ enters (the current decision). This is a real problem in communication theory. For example, in television transmissions it is necessary that the receiver effect a current decision. In this case, it is true that a certain delay is allowable, while in control systems a delay in the decision is usually no more harmful than an error. But communication and control problems are very similar. In communication theory there are also real problems in which a decision is made right after the end of the observation of the entire received signal (for example, in the reception of stationary images). But in control theory cases can also be found when such a problem formulation is of interest.

Let us rewrite expression (5.178) in the following form:

$$R_s = \int_{\Omega(x_s, v_s, u_s, \mathbf{y}_{s-1}^*)} P(x_s \mid v_s) P(v_s \mid u_s) \Gamma_s(u_s \mid \mathbf{y}_{s-1}^*) \times \left\{ \int_{\Omega(\bar{\lambda})} W_s[s, x_s{}^*(s, \bar{\lambda}), x_s] P(\mathbf{y}_{s-1}^* \mid \mathbf{x}_{s-1}^*) P(\bar{\lambda}) d\Omega(\bar{\lambda}) \right\} \times d\Omega(x_s, u_s, v_s, \mathbf{y}_{s-1}^*). \tag{5.180}$$

We will introduce the function

$$\rho_s = \rho_s(x_s, \mathbf{y}_{s-1}^*) = \int_{\Omega(\bar{\lambda})} W_s[s, x_s{}^*(s, \bar{\lambda}), x_s] \times P(\mathbf{y}_{s-1}^* \mid \mathbf{x}_{s-1}^*) P(\bar{\lambda})\, d\Omega(\bar{\lambda}). \tag{5.181}$$

Then expression (5.180) can be rewritten in the more compact form:

$$R_s = \int_{\Omega(x_s, v_s, u_s, \mathbf{y}_{s-1}^*)} P(x_s \mid v_s) P(v_s \mid u_s) \Gamma_s(u_s \mid \mathbf{y}_{s-1}^*) \rho(x_s, \mathbf{y}_{s-1}^*)\, d\Omega. \tag{5.182}$$

From formula (5.182) it is seen that the choice of the function Γ_s for a fixed s affects only the one term R_s of the total risk R. Hence it suffices to select a function Γ_s in formula (5.182) such that the specific risk be minimal. This function will be optimal in the sense of minimizing the total risk R. Since Γ_s is a probability density, then $\Gamma_s > 0$ and moreover,

$$\int_{\Omega(u_s)} \Gamma_s(u_s \mid \mathbf{y}_{s-1}^*)\, d\Omega = 1. \tag{5.183}$$

Hence the chosen function Γ_s must satisfy condition (5.183), which together with the condition $\Gamma_s > 0$ is the constraint imposed on the choice of possible functions Γ_s.

We will rewrite expression (5.182) as

$$R_s = \int_{\Omega(u_s, \mathbf{y}_{s-1}^*)} \Gamma_s(u_s \mid \mathbf{y}_{s-1}^*) \left[\int_{\Omega(x_s, v_s)} P(x_s \mid v_s) P(v_s \mid u_s) \times \rho_s(x_s, \mathbf{y}_{s-1}^*)\, d\Omega(x_s, v_s)\right] d\Omega(u_s, \mathbf{y}_{s-1}^*). \tag{5.184}$$

The integral inside the brackets is a function of u_s and $\mathbf{y}_{s-1}^*$, which we will denote by $\xi_s(u_s, \mathbf{y}_{s-1}^*)$:

$$\xi_s(u_s, \mathbf{y}_{s-1}^*) = \int_{\Omega(x_s, v_s)} P(x_s \mid v_s) P(v_s \mid u_s) \rho_s(x_s, \mathbf{y}_{s-1}^*)\, d\Omega. \tag{5.185}$$

This function can be found from the conditions of the problem. The intermediate function ρ_s cannot be computed, but ξ_s can be found directly from the formula which is easily obtained by the substitution of ρ_s from (5.181) into (5.185).

From (5.185) and (5.184) we find

$$R_s = \int_{\Omega(u_s, \mathbf{y}_{s-1}^*)} \xi_s(u_s, \mathbf{y}_{s-1}^*) \Gamma_s(u_s \mid \mathbf{y}_{s-1}^*)\, d\Omega = \int_{\Omega(\mathbf{y}_{s-1}^*)} I(\mathbf{y}_{s-1}^*)\, d\Omega(\mathbf{y}_{s-1}^*), \tag{5.186}$$

where

$$I(\mathbf{y}_{s-1}^*) = \int_{\Omega(u_s)} \Gamma_s(u_s \mid \mathbf{y}_{s-1}^*) \xi_s(u_s, \mathbf{y}_{s-1}^*)\, d\Omega. \tag{5.187}$$

From formula (5.186) it follows that the risk R_s will be minimal, if for each value of $\mathbf{y}_{s-1}^*$ the function I will assume the minimal possible value. Hence Γ_s must be chosen in expression (5.187) such that the integral $I(\mathbf{y}_{s-1}^*)$ will be minimal for any value of the parameter $\mathbf{y}_{s-1}^*$.

On the basis of the mean value theorem and taking into account that the integrand in (5.187) is positive, we can write

$$I(\mathbf{y}_{s-1}^*) = (\xi_s)_m \int_{\Omega(u_s)} \Gamma_s(u_s \mid \mathbf{y}_{s-1}^*)\, d\Omega = (\xi_s)_m \geqslant (\xi_s)_{\min}, \tag{5.188}$$

where $(\xi_s)_m$ is a certain mean value, and $(\xi_s)_{\min}$ is the minimal value of ξ_s. The integral in (5.188) is equal to unity in accordance with (5.183). From expression (5.188) it follows that the minimal possible value of $I(\mathbf{y}_{s-1}^*)$ is equal to $(\xi_s)_{\min}$. It is not difficult to show that this value of I is attained if the function Γ_s is selected in the following way.

Let[†] u_s^* be the value of u_s corresponding to the minimum of the function $\xi_s(u_s)$ in the region $\Omega(u_s)$ of possible values of u_s. Here by assigning supplementary conditions the region $\Omega(u_s)$ can be restricted in some manner. It may turn out that $\xi_s(u_s^*)$ is the smallest of several local minima of the function $\xi_s(u_s)$. Like this or otherwise, we will consider that a value of u_s has been found for which

$$\xi_s(u_s^*) = \min_{(u_s)} \xi_s(u_s). \tag{5.189}$$

Let us now consider the function

$$\Gamma_s^*(u_s) = \delta(u_s - u_s^*), \tag{5.190}$$

where δ is the unit impulse function (Dirac function). Evidently the function Γ_s^* satisfies condition (5.183), since the integral of the Dirac function over the entire region $\Omega(u_s)$ equals unity. It turns out that expression (5.190) gives the optimal function Γ_s^*. In order to show this, we will note the known property of the δ-function: if $\varphi(x)$ is a continuous function, then

$$\int_{-\infty}^{\infty} \delta(x - x^*)\varphi(x)\,dx = \varphi(x^*). \tag{5.191}$$

This property is easily generalized to the case of multiple integration, i.e., integration over a multidimensional domain.

Let us substitute (5.190) into the left side of (5.187). Then

$$[I(\mathbf{y}_{s-1}^*)]_{\Gamma_s=\delta(u_s-u_s^*)} = \int_{\Omega(u_s)} \delta(u_s - u_s^*)\xi_s(u_s, \mathbf{y}_{s-1}^*)\,d\Omega \tag{5.192}$$

$$= \xi_s(u_s^*) = (\xi_s)_{\min} = (I)_{\min}.$$

Hence I attains its minimal value with the use of the strategy (5.190).

Thus, as is seen from formula (5.190), the optimal strategy Γ_s^* is found to be regular. However, it could be expected in this case since the problem is of Bayes' type and the *a priori* probabilities are given. In the previous section it was shown that in the theory of two-alternative decisions the optimal Bayes' strategy turns out to be regular. This rule has now been verified for a more complicated case as well.

[†] Here $\mathbf{y}_{s-1}^*$ is considered a parameter and is not calculated in explicit form.

The optimal algorithm of the controller consists of choosing the value u_s^* minimizing the function ξ_s:

$$\xi_s(u_s^*, \mathbf{y}_{s-1}^*) = \min_{(u_s)} \xi_s(u_s, \mathbf{y}_{s-1}^*). \tag{5.193}$$

Hence it follows that u_s^* is a function of $\mathbf{y}_{s-1}^*$:

$$u_s^* = u_s^*(\mathbf{y}_{s-1}^*). \tag{5.194}$$

Thus as should be expected, in general the optimal choice of the value u_s at the sth time instant $t = s$ depends on the entire "previous history" of the signals y_s^* observed by the device A, i.e. on the sequence of values $y_0^*, y_1^*, ..., y_{s-1}^*$.

If the function $\xi_s(u_s, \mathbf{y}_{s-1}^*)$ has been calculated in advance, then by substituting into it the vector $\mathbf{y}_{s-1}^*$ observed by the controller A and minimizing ξ_s with respect to u_s, we can find the desired value of the current optimal control action u_s^*. The block diagram of a possible alternate version of the controller A is shown in Fig. 5.22. The current

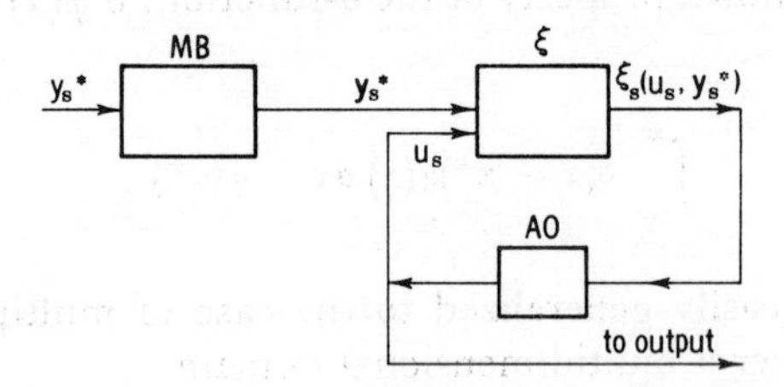

FIG. 5.22

values y_s^* act on the memory block MB, where the sequence $y_0^*, ..., y_{s-1}^*$, i.e., the vector $\mathbf{y}_{s-1}^*$, is stored. The value of this vector is transmitted from the memory block to the block ξ, where the function $\xi_s(u_s, \mathbf{y}_{s-1}^*)$ is formed. The value of u_s from the output of automatic optimizer AO acts on the other input to the block ξ.† The automatic optimizer selects the value $u_s = u_s^*$ minimizing the function ξ_s. The value u_s^* appears at the output of the block A. Of course, in specific schemes there is frequently no need to store each of the variables y_0^*, $y_1^*, ..., y_{s-1}^*$. Only a small number of functions $\psi_{s1}, ..., \psi_{sp}$ of these variables need be stored, where p is not too large, the so-called sufficient statistics (sufficient coordinates). An example of this kind is considered below.

† Descriptions of automatic optimizers of various types exist in the literature; see, for example, [3.25], chapter XIV.

Often the formula for $\xi_s(u_s, \mathbf{y}_{s-1}^*)$ in explicit form is obtained with difficulty. In this case the unit ξ can be constructed in the form of a computer in which the integration is automatically accomplished in accordance with expression (5.185). The numerical value of ξ_s appears at the output of this unit corresponding to the values of $\mathbf{y}_{s-1}^*$ and u_s at its inputs. As an example let us analyze the block diagram of the simple system depicted in Fig. 5.23.

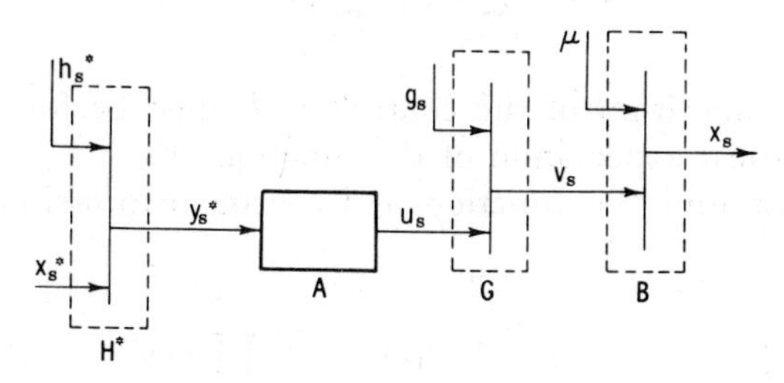

FIG. 5.23

The equations of the system have the form

$$y_s^* = x_s^* + h_s^*, \qquad v_s = u_s + g_s, \qquad x_s = v_s + \mu. \tag{5.195}$$

The first of these equations describes the properties of the channel H^*; the second, those of the channel G. The last equation is the equation of the object B.

Let h_s^* and g_s be sequences of independent random variables with zero mean values and the normal distributions

$$P(h_s^*) = \frac{1}{\sigma_h \sqrt{(2\pi)}} \exp\left\{-\frac{(h_s^*)^2}{2\sigma_h^2}\right\},$$
$$P(g_s) = \frac{1}{\sigma_g \sqrt{(2\pi)}} \exp\left\{-\frac{(g_s)^2}{2\sigma_g^2}\right\}. \tag{5.196}$$

In this example instead of the vector $\bar{\mu}$ the random variable μ appears with the probability density

$$P(\mu) = \frac{1}{\sigma_\mu \sqrt{(2\pi)}} \exp\left\{-\frac{\mu^2}{2\sigma_\mu^2}\right\}. \tag{5.197}$$

Let

$$x_s^* = \lambda, \tag{5.198}$$

where λ is a normal random variable with *a priori* probability density

$$P(\lambda) = \frac{1}{\sigma_\lambda \sqrt{(2\pi)}} \exp \left\{ -\frac{(\lambda - \lambda_0)^2}{2\sigma_\lambda^2} \right\}. \tag{5.199}$$

We will assign the loss function in the form of expression (5.179). Then

$$W = \sum_{s=0}^{n} W_s = \sum_{s=0}^{n} (x_s^* - x_s)^2. \tag{5.200}$$

The optimal algorithm of the controller A must be found minimizing the mathematical expectation of the function W.

Let us first find the function ρ_s by using expressions (5.181) and (5.170):

$$\rho_s = \rho_s(x_s, \mathbf{y}_{s-1}^*) = \int_{\Omega(\lambda)} [x_s^*(\lambda) - x_s]^2 \prod_{i=0}^{s-1} P(y_i^* \mid x_i^*) P(\lambda) \, d\Omega(\lambda)$$

$$= \int_{-\infty}^{\infty} (\lambda - x_s)^2 \prod_{i=0}^{s-1} P(y_i^* \mid \lambda) P(\lambda) \, d\lambda. \tag{5.201}$$

Since $h_i^* = y_i^* - x_i^* = y_i^* - \lambda$, the probability that the input variable to the device A will be found in the interval from y_i^* to $y_i^* + dy_i^*$ is equal to the probability that the noise in the channel H^* will be found in the interval from $y_i^* - \lambda$ to $y_i^* - \lambda + dy_i^*$. Hence according to the first of formulas (5.196) it follows that

$$P(y_i^* \mid \lambda) = \frac{1}{\sigma_h \sqrt{(2\pi)}} \exp \left\{ -\frac{(y_i^* - \lambda)^2}{2\sigma_h^2} \right\}. \tag{5.202}$$

Substituting (5.202) and (5.199) into expression (5.201), we obtain

$$\rho_s = \int_{-\infty}^{\infty} (\lambda - x_s)^2 \left[\prod_{i=0}^{s-1} \frac{1}{\sigma_h \sqrt{(2\pi)}} \exp \left\{ -\frac{(y_i^* - \lambda)^2}{2\sigma_h^2} \right\} \right]$$

$$\times \frac{1}{\sigma_\lambda \sqrt{(2\pi)}} \exp \left\{ -\frac{(\lambda - \lambda_0)^2}{2\sigma_\lambda^2} \right\} d\lambda = \frac{1}{(\sigma_h)^s \sigma_\lambda (2\pi)^{(s+1)/2}}$$

$$\times \int_{-\infty}^{\infty} (\lambda - x_s)^2 \exp \left\{ -\frac{\sum_{i=0}^{s-1} (y_i^* - \lambda)^2}{2\sigma_h^2} - \frac{(\lambda - \lambda_0)^2}{2\sigma_\lambda^2} \right\} d\lambda. \tag{5.203}$$

Let us introduce the notations

$$a = \frac{1}{(\sigma_h)^s \sigma_\lambda (2\pi)^{(s+1)/2}}, \qquad A_s = \frac{\sigma_h^2 + s\sigma_\lambda^2}{2\sigma_h^2\sigma_\lambda^2},$$

$$B_{s-1} = \frac{\sigma_\lambda^2 \sum_{i=0}^{s-1} y_i^* + \lambda_0\sigma_h^2}{\sigma_h^2\sigma_\lambda^2}, \qquad C_{s-1} = \frac{\sigma_\lambda^2 \sum_{i=0}^{s-1} (y_i^*)^2 + \lambda_0\sigma_h^2}{2\sigma_h^2\sigma_\lambda^2}. \tag{5.204}$$

Then formula (5.203) takes the form

$$\begin{aligned} \rho_s &= a \int_{-\infty}^{\infty} (\lambda - x_s)^2 \exp\{-A_s\lambda^2 + B_{s-1}\lambda - C_{s-1}\}\, d\lambda \\ &= a \exp\{-C_{s-1}\} \int_{-\infty}^{\infty} (\lambda - x_s)^2 \exp\{-A_s\lambda^2 + B_{s-1}\lambda\}\, d\lambda. \end{aligned} \tag{5.205}$$

We will make use of the known formula (see, for example, [5.26])

$$\int_{-\infty}^{\infty} x^2 \exp\{-px^2 + 2qx\}\, dx = \frac{1}{2p} \sqrt{\left(\frac{\pi}{p}\right)} \left(1 + \frac{2q^2}{p}\right) \exp\left\{\frac{q^2}{p}\right\}. \tag{5.206}$$

In applying this formula to (5.205), we will set

$$z = \lambda - x_s. \tag{5.207}$$

With such a substitution the integral of (5.205) is transformed into the form

$$\begin{aligned} \rho_s &= a \exp\{-C_{s-1} - A_s x_s^2 + B_{s-1}x_s\} \int_{-\infty}^{\infty} z^2 \exp\{-A_s z^2 \\ &\quad -(2A_s x_s - B_{s-1})z\}\, dz = a \exp\{-C_{s-1} - A_s x_s^2 + B_{s-1}x_s\} \\ &\quad \times \frac{1}{2A_s} \sqrt{\left(\frac{\pi}{A_s}\right)} \left[1 + \frac{2}{A_s}\left(A_s x_s - \frac{B_{s-1}}{2}\right)^2\right] \exp\left\{\frac{1}{A_s}\left(A_s x_s - \frac{B_{s-1}}{2}\right)^2\right\} \\ &= \frac{a\sqrt{\pi}}{2A_s^{3/2}} \left[1 + \frac{2}{A_s}\left(A_s x_s - \frac{B_{s-1}}{2}\right)^2\right] \exp\left\{-C_{s-1} + \frac{B_{s-1}}{4A_s}\right\}. \end{aligned} \tag{5.208}$$

Now let us find $\xi_s(u_s, \mathbf{y}_{s-1}^*)$. Since $\mu = x_s - v_s$, then

$$P(x_s \mid v_s) = P(\mu = x_s - v_s) = \frac{1}{\sigma_\mu \sqrt{(2\pi)}} \exp\left\{-\frac{(x_s - v_s)^2}{2\sigma_\mu^2}\right\}. \tag{5.209}$$

Further, since $g_s = v_s - u_s$, then

$$P(v_s \mid u_s) = P(g_s = v_s - u_s) = \frac{1}{\sigma_g \sqrt{(2\pi)}} \exp\left\{-\frac{(v_s - u_s)^2}{2\sigma_g^2}\right\}. \tag{5.210}$$

From (5.185) and (5.208) and taking (5.209) and (5.210) into account, we obtain

$$\xi_s = \int_{-\infty}^{\infty}\int_{-\infty}^{\infty} \frac{1}{\sigma_\mu \sqrt{(2\pi)}} \exp\left\{-\frac{(x_s - v_s)^2}{2\sigma_\mu{}^2}\right\} \frac{1}{\sigma_g \sqrt{(2\pi)}}$$

$$\times \exp\left\{-\frac{(v_s - u_s)^2}{2\sigma_g{}^2}\right\} \frac{a\sqrt{\pi}}{2A_s^{3/2}}\left[1 + \frac{2}{A_s}\left(A_s x_s - \frac{B_{s-1}}{2}\right)^2\right]$$

$$\times \exp\left\{-C_{s-1} + \frac{B_{s-1}^2}{4A_s}\right\} dx_s\, dv_s = b \exp\left\{-C_{s-1} + \frac{B_{s-1}^2}{4A_s}\right\}$$

$$\times \int_{x_s=-\infty}^{\infty} \left[1 + \frac{2}{A_s}\left(A_s x_s - \frac{B_{s-1}}{2}\right)^2\right]$$

$$\times \left\{\int_{-\infty}^{\infty} \exp\left\{-\frac{(x_s - v_s)^2}{2\sigma_\mu{}^2} - \frac{(v_s - u_s)^2}{2\sigma_g{}^2}\right\} dv_s\right\} dx_s\,, \tag{5.211}$$

where

$$b = \frac{a\sqrt{\pi}}{2A_s^{3/2}} \frac{1}{\sigma_\mu \sigma_g \cdot 2\pi}\,. \tag{5.212}$$

We will set

$$\alpha_s = \frac{\sigma_g{}^2 x_s{}^2 + \sigma_\mu{}^2 u_s{}^2}{2\sigma_\mu{}^2 \sigma_g{}^2}, \qquad \beta_s = \frac{\sigma_g{}^2 x_s + \sigma_\mu{}^2 u_s}{\sigma_\mu{}^2 \sigma_g{}^2}, \qquad \gamma = \frac{\sigma_g{}^2 + \sigma_\mu{}^2}{2\sigma_\mu{}^2 + \sigma_g{}^2}. \tag{5.213}$$

Then the inner integral in (5.211) inside the braces is equal to

$$I_0 = \int_{-\infty}^{\infty} \exp\{-\alpha_s + \beta_s v_s - \gamma v_s{}^2\}\, dv_s$$

$$= \exp\{-\alpha_s\} \int_{-\infty}^{\infty} \exp\{-\gamma v_s{}^2 + \beta_s v_s\}\, dv_s\,. \tag{5.214}$$

Using the formula (see [5.26])

$$\int_{-\infty}^{\infty} \exp\{-px^2 \pm qx\}\, dx = \sqrt{\left(\frac{\pi}{p}\right)} \exp\left\{\frac{q^2}{4p}\right\}, \tag{5.215}$$

we obtain as a result of calculation

$$I_0 = \exp\left\{-\alpha_s + \frac{\beta_s{}^2}{4\gamma}\right\} \sqrt{\left(\frac{\pi}{\gamma}\right)}. \tag{5.216}$$

Substituting (5.215) into (5.211), we arrive at the expression

$$\xi_s = \left[b\sqrt{\left(\frac{\pi}{\gamma}\right)} \exp\left\{-C_{s-1} + \frac{B_{s-1}^2}{4A_s}\right\}\right]\left(I_1 + \frac{2}{A_s} I_2\right), \tag{5.217}$$

where I_1 and I_2 are the integrals written out below. We will set

$$L = \frac{1}{2\sigma_\mu^2} - \frac{1}{4\gamma\sigma_\mu^4}, \qquad M = \frac{1}{2\gamma\sigma_\mu^2\sigma_g^2}, \qquad N = -\frac{1}{2\sigma_g^2} + \frac{1}{4\gamma\sigma_g^4}. \tag{5.218}$$

Then

$$\begin{aligned} I_1 &= \int_{-\infty}^{\infty} \exp\{-Lx_s^2 + Mx_s u_s + Nu_s^2\}\, dx_s \\ &= \sqrt{\left(\frac{\pi}{L}\right)} \exp\left\{\left(N + \frac{M^2}{4L}\right) u_s^2\right\}. \end{aligned} \tag{5.219}$$

This expression has been obtained by the application of formula (5.215). Further, by using (5.206) we obtain

$$\begin{aligned} I_2 &= \int_{-\infty}^{\infty} \left(A_s x_s - \frac{B_{s-1}}{2}\right)^2 \exp\{-Lx_s^2 + Mx_s u_s + Nu_s^2\}\, dx_s \\ &= \exp\{Nu_s^2\} \exp\left\{-\frac{LB_{s-1}^2}{4A_s^2} + \frac{Mu_s B_{s-1}}{2A_s}\right\} \\ &\quad \times \int_{-\infty}^{\infty} z^2 \exp\left\{-\frac{L}{A_s^2} z^2 + \left(-\frac{LB_{s-1}}{A_s^2} + \frac{Mu_s}{A_s}\right) z\right\} dz \\ &= \frac{A_s^2}{2L}\sqrt{\left(\frac{\pi A_s^2}{L}\right)}\left[1 + \frac{A_s^2}{2L}\left(-\frac{LB_{s-1}}{A_s^2} + \frac{Mu_s}{A_s}\right)^2\right] \\ &\quad \times \exp\left\{\left(N + \frac{M^2}{4L}\right) u_s^2\right\}. \end{aligned} \tag{5.220}$$

Substituting (5.219) and (5.220) into (5.217), we obtain after obvious transformations:

$$\xi_s = D_s[1 + \epsilon^2(\vartheta u_s - 1)^2], \tag{5.221}$$

where it is found that

$$N + (M^2/4L) = 0.$$

Here

$$D_s = \left(1 + \frac{A_s^2}{L}\right)\sqrt{\left(\frac{\pi}{L}\right)} b \sqrt{\left(\frac{\pi}{\gamma}\right)} \exp\left\{-C_{s-1} + \frac{B_{s-1}^2}{4A_s}\right\}, \qquad (5.222)$$

$$\epsilon^2 = \frac{LB_{s-1}^2}{2(L + A_s^2)}, \qquad \vartheta = \frac{2A_s}{B_{s-1}}.$$

The expressions for D_s and ϵ^2 do not depend on u_s. Therefore the minimum of ξ_s with respect to u_s is obtained if the parentheses in expression (5.221) are equated to zero:

$$\vartheta u_s - 1 = 0. \qquad (5.223)$$

Hence we obtain the optimal control action u_s^* in the form

$$u_s^* = 1/\vartheta = B_{s-1}/2A_s. \qquad (5.224)$$

Substituting the values of B_{s-1} and A_s from (5.204), we find

$$u_s^* = \frac{\lambda_0 + (\sigma_\lambda/\sigma_h)^2 \sum_{i=0}^{s-1} y_i^*}{1 + s(\sigma_\lambda/\sigma_h)^2}. \qquad (5.225)$$

The meaning of this formula is that it represents an estimate of the value of λ. In fact, from Fig. 5.23 it is seen that since the mean values of g_s and μ are equal to zero, the distributions $P(g_s)$ and $P(\mu)$ do not play any part in this case; therefore their parameters do not enter into expression (5.225). Of course, if the mean values of g_s and μ are not equal to zero, or for example the characteristics of the blocks G and B are nonlinear, then the parameters of the distributions $P(g_s)$ and $P(\mu)$ can enter into the formula for u_s^*. But in this example the value of u_s need only reproduce the value λ in the best form; then the value of x_s will correspond in the best manner to the value $x_s^* = \lambda$.

As is seen from expression (5.225), for small values of s and $(\sigma_\lambda/\sigma_h)^2$ (for example, for a large dispersion σ_h^2 of the noise) the basic estimate is the *a priori* mean value λ_0. But for sufficiently large s the value λ_0 now does not play a significant part, since after time s the sum $\sum_{i=0}^{s-1} y_i^*$ is stored and its magnitude in practice exceeds λ_0 considerably. In this case from formula (5.225) it follows that

$$u_s^* \cong \frac{\sum_{i=0}^{s-1} y_i^*}{s}, \qquad (5.226)$$

i.e., the estimate of the quantity λ reduces to obtaining the arithmetic mean of the values y_i^* $(i = 0, 1, \ldots, s - 1)$, measured at the input to the controller. This estimate fully conforms to the intuitive idea.

In an analogous fashion for the case examined a theory of optimal compensation can be constructed. For example, we investigate the system whose block diagram is depicted in Fig. 5.24.

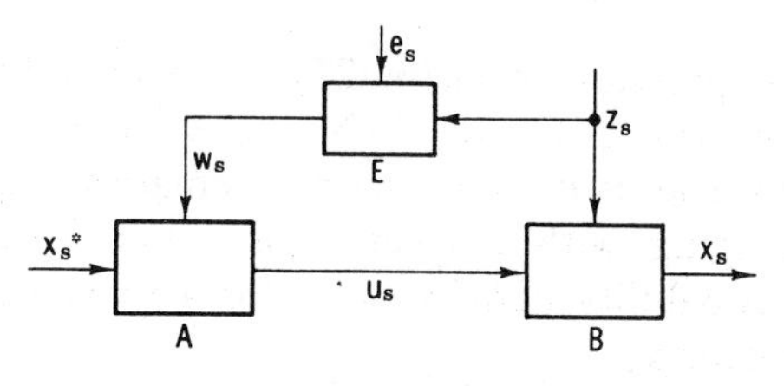

FIG. 5.24

For simplicity we will suppose that there is random noise only in the measurement channel E of the noise z_s. In addition, let x_s^* be prescribed in advance. Of course, it is not difficult to generalize the theory presented below to the case of noise present in the other channels as well, as was done above. Let the object not have memory and be described by the equation

$$x_s = F(z_s, u_s), \tag{5.227}$$

where as before

$$z_s = z_s(s, \bar{\mu}), \tag{5.228}$$

and the *a priori* probability density $P(\bar{\mu})$ of the vector $\bar{\mu}$ is given. The probability density $P(e_s)$ of the noise e_s in the measurement channel E of the noise z_s is also prescribed. Let the noise e_s be a sequence of independent variables with the same distribution law $P(e_s)$. By knowing this law and the method of combination of the signal z_s with the noise e_s in the channel E, the conditional probability density $P(w_s \mid z_s)$ can be found for the output variable w_s of this channel, supplied to the input of the controller A.

Let us introduce the temporal vectors

$$\mathbf{z}_k = (z_0, z_1, \ldots, z_k), \qquad \mathbf{w}_k = (w_0, w_1, \ldots, w_k),$$
$$\mathbf{x}_k^* = (x_0^*, x_1^*, \ldots, x_k^*). \tag{5.229}$$

The problem consists of determining the optimal strategies

$\Gamma_s(u_s \mid \mathbf{x}_s^*, \mathbf{w}_{s-1})$ of the controller A ($s = 0, 1, ..., n$). The expressions for the loss function W and risk R remain the same as in the theory presented above in this section.

We will first derive the formula for the conditional specific risk r_s for a fixed vector $\mathbf{z}_s$:

$$r_s = M\{W_s \mid \mathbf{z}_s\} = \int_{\Omega(u_s)} W_s[s, x_s^*, F(z_s, u_s)] P_s(u_s \mid \mathbf{x}_s^*, \mathbf{z}_{s-1}) \, d\Omega. \tag{5.230}$$

Instead of x_s its expression from (5.227) has been substituted into this formula inside the formula for W_s. Therefore the only random variable in the expression for w_s is u_s, and the mathematical expectation $M\{W_s\}$ is found by averaging with respect to u_s. Let $P_s(u_s \mid \mathbf{x}_s^*, \mathbf{z}_{s-1})$ be the conditional probability density of u_s for a fixed variable $\mathbf{z}_{s-1}$, and also for a fixed vector $\mathbf{x}_s^*$.

The function $P_s(u_s \mid \mathbf{x}_s^*, \mathbf{z}_{s-1})$ can be found from the formula which follows in an obvious manner from the block diagram shown in Fig. 5.24:

$$P_s(u_s \mid \mathbf{x}_s^*, \mathbf{z}_{s-1}) = \int_{\Omega(\mathbf{w}_{s-1})} \Gamma_s(u_s \mid \mathbf{x}_s^*, \mathbf{w}_{s-1}) P(\mathbf{w}_{s-1} \mid \mathbf{z}_{s-1}) \, d\Omega. \tag{5.231}$$

Here Γ_s is the strategy of the controller, and $P(\mathbf{w}_{s-1} \mid \mathbf{z}_{s-1})$ is the conditional probability density of the output vector of the channel E for a fixed vector $\mathbf{z}_{s-1}$ of the noise. Since the w_i are independent variables, then

$$P(\mathbf{w}_{s-1} \mid \mathbf{z}_{s-1}) = \prod_{i=0}^{s-1} P(w_i \mid z_i). \tag{5.232}$$

Substituting (5.232) into (5.231), and (5.231) into (5.230), we find

$$r_s = \int_{\Omega(u_s, \mathbf{w}_{s-1})} W_s\{s, x_s^*, F[z_s(s, \bar{\mu}), u_s]\} \Gamma_s(u_s \mid \mathbf{x}_s^*, \mathbf{w}_{s-1}) \times \left\{ \prod_{i=0}^{s-1} P[w_i \mid z_i(i, \bar{\mu})] \right\} d\Omega. \tag{5.233}$$

According to (5.228) the quantity z_i in this expression can be written in the form of a function of i and $\bar{\mu}$.

For different trials with the system different vectors $\mathbf{z}_s$ appear. If the probability density $P(\bar{\mu})$ is known, then the total specific risk R_s can be found from the formula

$$R_s = \int_{\Omega(\bar{\mu})} r_s(\bar{\mu}) P(\bar{\mu}) \, d\Omega. \tag{5.234}$$

After substituting expression (5.233) for r_s here, we obtain

$$R_s = \int_{\Omega(u_s, \mathbf{w}_{s-1}, \bar{\mu})} W_s\{s, x_s^*, F[z_s(s, \bar{\mu}), u_s]\}\Gamma_s(u_s \mid \mathbf{x}_s^*, \mathbf{w}_{s-1})$$

$$\times \left\{\prod_{i=0}^{s-1} P[w_i \mid z_i(i, \bar{\mu})]\right\} P(\bar{\mu})\, d\Omega. \qquad (5.235)$$

Since the risk R_s depends only on the single strategy Γ_s corresponding to the given instant $t = s$, then the optimal strategy Γ_s^* is the strategy minimizing the specific risk R_s. We will set

$$\eta_s(u_s, x_s^*, \mathbf{w}_{s-1}) = \int_{\Omega(\bar{\mu})} W_s\{s, x_s^*, F[z_s(s, \bar{\mu}), u_s]\}$$

$$\times \left\{\prod_{i=0}^{s-1} P[w_i \mid z_i(i, \bar{\mu})]\right\} P(\bar{\mu})\, d\Omega. \qquad (5.236)$$

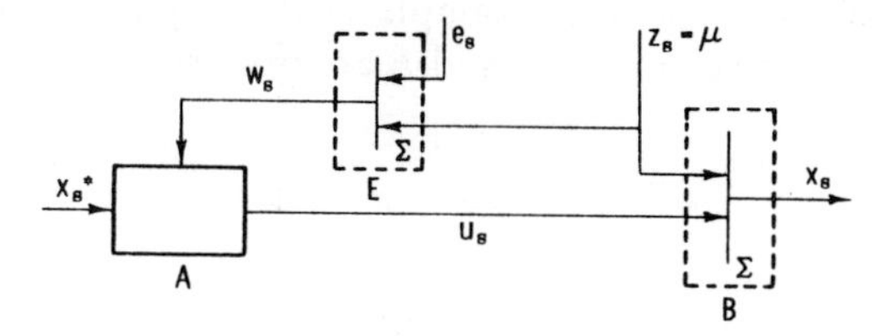

FIG. 5.25

Then

$$R_s = \int_{\Omega(\mathbf{w}_{s-1}, u_s)} \Gamma_s(u_s \mid \mathbf{x}_s^*, \mathbf{w}_{s-1})\eta_s(u_s, x_s^*, \mathbf{w}_{s-1})\, d\Omega$$

$$= \int_{\Omega(\mathbf{w}_{s-1})} I(\mathbf{w}_{s-1})\, d\Omega, \qquad (5.237)$$

where

$$I(\mathbf{w}_{s-1}) = \int_{\Omega(u_s)} \Gamma_s(u_s \mid \mathbf{x}_s^*, \mathbf{w}_{s-1})\eta_s(u_s, x_s^*, \mathbf{w}_{s-1})\, d\Omega. \qquad (5.238)$$

If Γ_s is chosen such that for any $\mathbf{w}_{s-1}$ the quantity $I(\mathbf{w}_{s-1})$ will be minimal, then R_s will also have a minimal value. This choice is made in the same way as in the theory presented above.

Let

$$\eta_s(u_s^*, x_s^*, \mathbf{w}_{s-1}) = \min_{(u_s)} \eta_s(u_s, x_s^*, \mathbf{w}_{s-1}). \qquad (5.239)$$

We will find the value u_s^* minimizing η_s. Evidently, in general u_s^* depends on x_s^* and $\mathbf{w}_{s-1}$:

$$u_s^* = u_s^*(x_s^*, \mathbf{w}_{s-1}). \tag{5.240}$$

This is the optimal control that depends on the current value x_s and on the entire "previous history" of the input variables w_i ($i = 0, 1, ..., s-1$), observed by the controller A at the time instants $t = 0, 1, ..., s-1$.

The optimal strategy Γ_s^* is regular:

$$\Gamma_s^* = \delta(u_s - u_s^*), \tag{5.241}$$

where u_s^* is determined from condition (5.239). The proof of the validity of this expression is the same as in the theory presented above.

Let us consider the simple example of the circuit given in Fig. 5.25. Let $z_s = \mu$, and the elements E and B are prescribed by the equations

$$w_s = e_s + \mu, \qquad x_s = \mu + u_s . \tag{5.242}$$

Further, let the normal random variables e_s and μ be characterized by the probability densities

$$\begin{aligned} P(\mu) &= \frac{1}{\sigma_\mu \sqrt{(2\pi)}} \exp\left\{-\frac{(\mu-\mu_0)^2}{2\sigma_\mu^2}\right\}, \\ P(e_s) &= \frac{1}{\sigma_e \sqrt{(2\pi)}} \exp\left\{-\frac{e_s^2}{2\sigma_e^2}\right\}. \end{aligned} \tag{5.243}$$

We assume a specific loss function in the form

$$W_s = (x_s^* - x_s)^2 = (x_s^* - \mu - u_s)^2. \tag{5.244}$$

Since $e_i = w_i - \mu$, then from (5.242) and (5.243) it follows that:

$$\begin{aligned} P(w_i \mid z_s) &= P(w_i \mid \mu) = P(e_i = w_i - \mu) \\ &= \frac{1}{\sigma_e \sqrt{(2\pi)}} \exp\left\{-\frac{(w_i-\mu)^2}{2\sigma_e^2}\right\}. \end{aligned} \tag{5.245}$$

In accordance with formula (5.236) we find the expression for η_s:

$$\eta_s = \int_{-\infty}^{\infty} (x_s^* - \mu - u_s)^2 \left[\prod_{i=0}^{s-1} \frac{1}{\sigma_e \sqrt{(2\pi)}} \exp\left\{-\frac{(w_i - \mu)^2}{2\sigma_e^2}\right\}\right]$$

$$\times \frac{1}{\sigma_\mu \sqrt{(2\pi)}} \exp\left\{-\frac{(\mu - \mu_0)^2}{2\sigma_\mu^2}\right\} d\mu$$

$$= a \int_{-\infty}^{\infty} (x_s^* - \mu - u_s)^2 \exp\left\{-\frac{\sum_{i=0}^{s-1}(w_i - \mu)^2}{2\sigma_e^2} - \frac{(\mu - \mu_0)^2}{2\sigma_\mu^2}\right\} d\mu, \tag{5.246}$$

where a is a constant. We transform the expression in the braces under the integral by substituting

$$\alpha = \mu + u_s - x_s^* = \mu - y_s, \tag{5.247}$$

where

$$y_s = x_s^* - u_s. \tag{5.248}$$

After the transformations expression (5.246) takes the form

$$\eta_s = a \int_{-\infty}^{\infty} \alpha^2 \exp\{-A_s - B_s\alpha - C_s\alpha^2\}\, d\alpha$$

$$= a \exp\{-A_s\} \int_{-\infty}^{\infty} \alpha^2 \exp\{-C_s\alpha^2 - B_s\alpha\}\, d\alpha, \tag{5.249}$$

where

$$A_s = \frac{(\sigma_\mu^2 \sum_{i=0}^{s-1} w_i^2 + \sigma_e^2\mu_0^2) - y_s(2\sigma_\mu^2 \sum_{i=0}^{s-1} w_i + 2\sigma_e^2\mu_0) + (s\sigma_\mu^2 + \sigma_e^2)y_s^2}{2\sigma_e^2\sigma_\mu^2}, \tag{5.250}$$

$$B_s = \frac{-\sigma_\mu^2 \sum_{i=0}^{s-1} w_i - \sigma_e^2\mu_0 + y_s(s\sigma_\mu^2 + \sigma_e^2)}{\sigma_e^2\sigma_\mu^2}, \qquad C_s = \frac{s\sigma_\mu^2 + \sigma_e^2}{2\sigma_e^2\sigma_\mu^2}.$$

The integral (5.249) is determined from formula (5.206). As a result we finally find:

$$\eta_s = b \exp\{-A_s + (B_s^2/4C_s)\} \cdot [1 + (B_s^2/2C_s)], \tag{5.251}$$

where b is constant, also including certain factors depending on s. Setting

$$\beta_s = \frac{\sigma_\mu^2 \sum_{i=0}^{s-1} w_i + \sigma_e^2\mu_0}{\sigma_e^2\sigma_\mu^2}, \tag{5.252}$$

and comparing the latter two formulas in (5.250), the relation

$$B_s = 2y_s C_s - \beta_s \tag{5.253}$$

can be obtained. By examining the expression under the exponential sign in (5.251), we can be convinced that it does not contain y_s, and hence does not depend on u_s. In fact, from (5.253) we find

$$\frac{B_s^2}{4C_s} = \frac{(2y_s C_s - \beta_s)^2}{4C_s} = y_s^2 C_s - y_s \beta_s + \frac{\beta_s^2}{4C_s}. \tag{5.254}$$

Let us compare expression (5.254) with the formula for A_s in (5.250). The terms containing y_s^2 in these expressions are identical. In exactly the same way the terms containing y_s are also identical. Hence these terms vanish under the exponential sign, and the difference

$$\{-A_s + (B_s^2/4C_s)\}$$

does not depend on y_s, and so in accordance with (5.248) not on u_s either. Therefore the minimization of η_s with respect to u_s in formula (5.251) reduces to the minimization of the factor

$$[1 + (B_s^2/2C_s)],$$

in which B_s depends on u_s. This factor is minimal for $B_s = 0$. From expression (5.253) it follows that this condition is satisfied by the value

$$y_s = \beta_s/2C_s. \tag{5.255}$$

After substituting the value of y_s from (5.248) and β_s from (5.252), we find the optimal control law:

$$u_s^* = x_s^* - \frac{\sigma_\mu^2 \sum_{i=0}^{s-1} w_i + \sigma_e^2 \mu_0}{s\sigma_\mu^2 + \sigma_e^2} = x_s^* - \frac{\mu_0 + (\sigma_\mu/\sigma_e)^2 \sum_{i=0}^{s-1} w_i}{1 + (\sigma_\mu/\sigma_e)^2 s}. \tag{5.256}$$

The second term in this formula represents an estimate of the quantity μ. In fact, as is seen from Fig. 5.25, the value

$$u_s = x_s^* - \mu \tag{5.257}$$

would be ideal, since for just this condition x_s equals x_s^* and the specific loss function W_s vanishes. Comparing (5.257) and (5.256), we are easily convinced that the second term in (5.256) gives precisely the estimate of μ which is most advantageous in the light of the optimality criterion applied.

From formula (5.256) it follows that for s not large and small ratios σ_μ/σ_e, for example with a large dispersion of the noise e_s, the *a priori* mean value μ_0 plays a fundamental role in the estimate of μ. But for sufficient large s, when the significant sum $\sum_{i=0}^{s-1} w_i$ is accumulated, formula (5.256) can be rewritten approximately in the form

$$u_s^* \cong x_s^* - \frac{\sum_{i=0}^{s-1} w_i}{s}, \tag{5.258}$$

and the estimate of μ reduces to the arithmetic mean of the values w_i.

In Section 1 of this chapter the fundamental possibility was shown of replacing the direct measurement of the noise z_s by an indirect means of measuring the input and output of the object B. For example, let us consider the feedback scheme depicted in Fig. 5.26. In this diagram

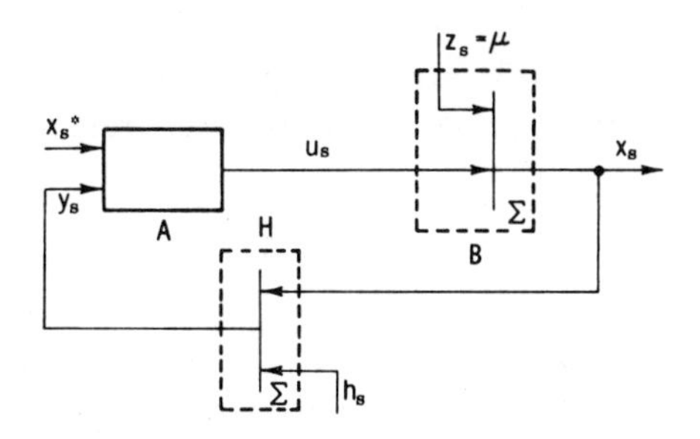

FIG. 5.26

the direct measurement of the noise $z_s = \mu$ is missing, and in the feedback circuit there is a channel H in which the noise, a sequence of independent normal random variables, is mixed additively with the signal x_s. Let the equations of the elements B and H have the form

$$x_s = u_s + z_s = u_s + \mu, \qquad y_s = x_s + h_s. \tag{5.259}$$

The controller A receives the output variable y_s of the channel H. From (5.259) it follows that

$$y_s - u_s = \mu + h_s; \tag{5.260}$$

therefore the difference $y_s - u_s$, which can be obtained in the controller, represents whatever the result of the measurement of the quantity $z_s = \mu$ would be with an error h_s. Thus the schemes depicted in Fig. 5.24 and 5.25 are completely equivalent, if only the condition

$$e_s = h_s \tag{5.261}$$

is observed. Thus the scheme in Fig. 5.25 is reduced to the circuit of an open-loop system, and hence the theory presented above, constructed for the system of Fig. 5.24, is entirely suitable for the analysis of the diagram of Fig. 5.26. We will call closed-loop systems reducible to open-loop ones, "reducible" systems for short.

In a similar fashion, the system with feedback which appears as an example in [5.25] can be reduced to an open-loop system.

This system is depicted in Fig. 5.27a.

Let the object B be linear with a transfer function

$$K_B(p) = 1/p(p + a). \tag{5.262}$$

The quantity

$$\epsilon = x^* - x \tag{5.263}$$

acts on the block A_3 of the controller, representing an amplifier element with amplification coefficient k. Therefore the transfer function of the open-loop circuit is equal to

$$K(p) = k/p(p + a). \tag{5.264}$$

We will assume that the parameter a of the object B can vary slowly in a manner not expected beforehand. This has been reflected in Fig. 5.27a by arrow a, acting on the object B. By measuring the input u and output x variables of the object B, the value of the parameter a can be determined in an indirect way. The block A_1 accomplishes this function. We will assume that the result of measurement a^* is not exactly equal to a. This value a^* acts on another block A_2, into which is put in advance the algorithm of the proper setting of the varying value of k of the block A_3 as a function of the quantity a^*. Thus with the aid of the scheme of Fig. 5.27a, the amplification coefficient k can be changed with the slow variation of the parameter a in a corresponding manner. By changing k, we can obtain a character of the processes in the system which is optimal in some sense, or in any case sufficiently favorable, even under conditions of significant variations in the parameter a.

The system depicted in Fig. 5.27a appears very complicated. But with the aid of elementary transformations it can be reduced to a simple enough equivalent open-loop system. For this reason the investigation of the circuit and the construction of its theory do not cause difficulties. Actually, the indirect measurement of the parameter a by means of the variables u and x is equivalent to its direct measurement with a certain error with the aid of the block A_1' (Fig. 5.27b). This unit must be such that the very same quantity a^* be obtained at its output, which is observed at the output of the block A_1 in the scheme of Fig. 5.27a. Thus the scheme of joining the blocks A_1 and A_2 without any change in the essence of the matter reduces to an equivalent open-loop circuit. Further, the closed-loop circuit of the basic part of the system in Fig. 5.27b, including the blocks A_3, B and the inverter with amplification coefficient -1, can be regarded as some equivalent object O. Since this part of the system is linear, it can be replaced by a system without feedback which is equivalent to it (Fig. 5.27c). The transfer function $K_0(p)$ of this system is easily determined from the formula

$$K_0(p) = \frac{K(p)}{1 + K(p)}, \tag{5.265}$$

where $K(p)$ is the transfer function of the open-loop system. Substituting expression (5.264) into (5.265), we find

$$K_0(p) = k/(p^2 + ap + k). \tag{5.266}$$

Thus the circuit of the system under consideration reduces to an equivalent scheme not containing any feedback connections (Fig. 5.27c). The quantity a is the external perturbing action (noise), operating on the coefficient to the first power of p in the denominator of formula (5.266). This action enters the equivalent object O at the point N. The other action—the variation in the parameter k— acts on the object O from the block A_2 at the point M.

In [5.25] it is assumed that the parameter a can have three possible values: $a_1 = 1$, $a_2 = 2$, $a_3 = 4$. Therefore a^* can likewise assume only the same possible values. But noise in the measurement channel causes the possibility of error. Since the values of the quantities a and a^* at the input and output of the channel are discrete, then in place of a conditional probability density the measurement channel A_1' is characterized by a matrix of conditional probabilities $p(a_j^* \mid a_i)$ ($j = 1, 2, 3$, $i = 1, 2, 3$). These probabilities are written out in the accompanying table. An examination of this table shows that the probabilities of a

	a^*		
a	1	2	4
1	0.6	0.2	0.2
2	0.1	0.6	0.3
4	0.0	0.4	0.6

correct measurement are the largest; all of them are equal to 0.6. The probabilities of various errors are smaller. For example, the probability $p\ (a^* = 1 \mid a = 2)$ of detecting $a^* = 1$ when $a = 2$ is equal to 0.1. The matrix of the $p(a_j^* \mid a_i)$ completely characterizes the properties of the measurement channel in the same way that the probability density $P(a^* \mid a)$ completely characterizes the properties of the analogous channel with variables which are continuous with respect to level.

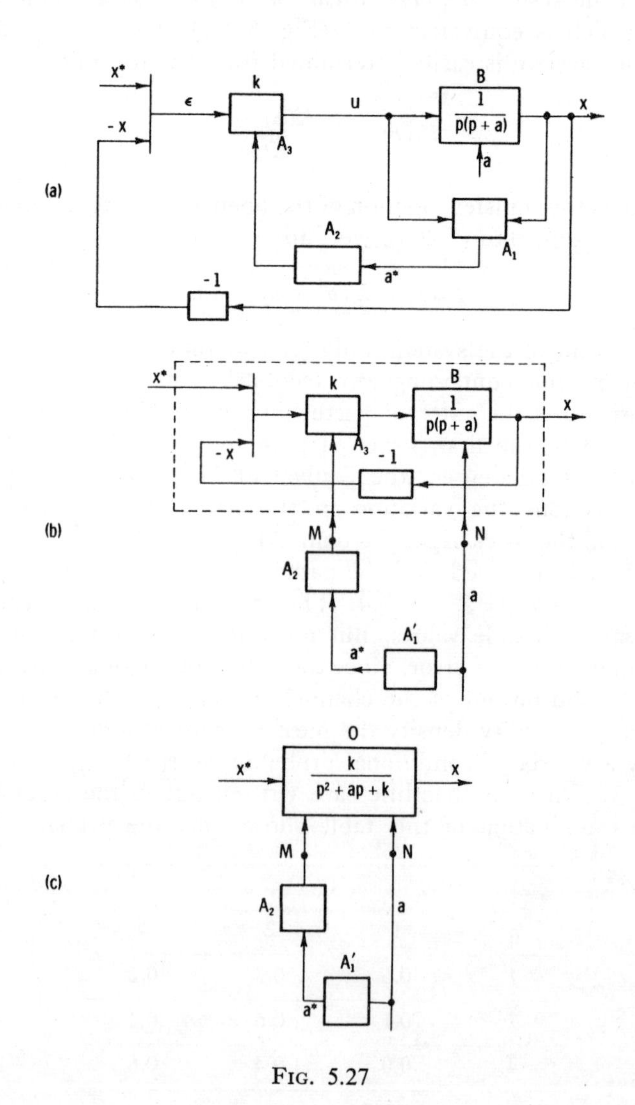

FIG. 5.27

We will assume that the *a priori* probabilities of possible values of the parameter a are given:

$$p(a = a_1 = 1) = 0.3, \qquad p(a = a_2 = 2) = 0.5, \qquad p(a = a_3 = 4) = 0.2.$$

Thus the problem turns out to be of Bayes' type. If the value of a were measured exactly, then for each a the value of k which is best in some sense could be determined in advance, and afterwards the corresponding algorithm put into the unit A_2.

In [5.25] values of k were chosen for each possible value of a, for which the transfer function of the equivalent object O has the most favorable character in a deterministic sense (in [5.25] by applying an integral criterion of quality). The calculations showed that under the condition of exact measurement of a, the block A_2 must realize the dependence of k on a as shown in the following table.

a	$a = a_1 = 1$	$a = a_2 = 2$	$a = a_3 = 4$
k	$k = k_1 = 0.4$	$k = k_2 = 1.5$	$k = k_3 = 6$

But because of error in the measurements it may turn out that for a equal to a_i, the setting of the value $k = k_j$ takes place, where $j \neq i$. In [5.25] for all possible combinations of i and j, the matrix of the values of the loss function $W(a_i, k_j) = W(i, j)$ is given in the form as shown in the accompanying table.

	k		
a	$k_1 = 0.4$	$k_2 = 1.5$	$k_3 = 6$
$a_1 = 1$	2.4	6	7.2
$a_2 = 2$	4.6	1.2	2.7
$a_3 = 4$	10.8	2.3	0.6

From this table it is seen that in general, the correct setting of the value $k = k_i$ when $a = a_i$ gives a smaller value of the loss function $W(i, j)$, while incorrect settings increase the value of $W(i, j)$. For example, if $a = a_1 = 1$ and the value $k = k_1 = 0.4$ is set, then the corresponding value $W(1, 1) = 2.4$. But if $a = a_3 = 4$ and $k = k_1 = 0.4$ is set, then $W(3,1) = 10.8$.

Let it be required to determine the algorithm of the device A_2, i.e., the method of setting the values k_j from the measured data a_j^*. Such an algorithm must be found for which the risk is minimal.

Let a_j^* be the measured value, a_i the actual value of the parameter a, and k_ν the setting of the value of k ($j = 1, 2, 3$; $i = 1, 2, 3$; $\nu = 1, 2, 3$). Then the loss is equal to $W(i, \nu)$. With the appearance of a_j^* let the value k_ν be chosen. We find, under the condition that a_j^* appeared, the value of the conditional risk

$$r(a_j^*, k_\nu) = M\{W(i, \nu) \mid a_j^*\} = \sum_{i=1}^{3} W(i, \nu) p(a_i \mid a_j^*).$$

From Bayes' formula

$$p(a_i \mid a_j^*) = \frac{p(a_i) p(a_j^* \mid a_i)}{p(a_j^*)} = \frac{p(a_i) p(a_j^* \mid a_i)}{\sum_{\nu=1}^{3} p(a_j^* \mid a_\nu) p(a_\nu)};$$

hence it follows that

$$r(a_j^*, k_\nu) = \left[\sum_{i=1}^{3} W(i, \nu) p(a_i) p(a_j^* \mid a_i)\right] \cdot \frac{1}{p(a_j^*)}. \tag{5.267}$$

We will look at what value k_ν is most advantageous to set, if the measured value a^* is equal to $a_3^* = 4$. We will first use $k_\nu = k_1 = 0.4$. Then the expression in the brackets is equal to

$$\sum_{i=1}^{3} W(i, 1) p(a_3^* \mid a_i) p(a_i)$$
$$= (2.4)(0.2)(0.3) + (4.6)(0.3)(0.5) + (10.8)(0.6)(0.2) = 2.13. \tag{5.268}$$

If in an analogous fashion this expression is calculated for the settings $k = k_2$ and $k = k_3$, then the values shown in the accompanying table can be obtained. From this table it is seen that the smallest value of the risk is obtained for $k = k_2 = 1.5$. Thus for $a = a_3^* = 4$ the best

k	$k_1 = 0.4$	$k_2 = 1.5$	$k_3 = 6$
$r \cdot p(a_j^*)$	2.13	0.816	0.909

setting is $k = k_2 = 1.5$. In this case, as is easily calculated, $p(a_j^*)$ equals 0.33. However, it is not required to compute this value for determining the optimal k_ν. The values of the risk r are obtained by dividing the lower row of the last table by 0.33. In an analogous fashion the most advantageous values of the setting k can also be found for other measured values of a^*.

Similar examples of the computation have been used in [5.29].

CHAPTER VI

Optimal Systems with Active Information Storage

1. Statement of the Simplest Problem for an Optimal Dual Control System

Fig. 6.1 shows the block diagram of a closed-loop automatic control system. The driving action $\bar{x}^*$ enters the input of the controller A through the channel H^*, where it is mixed with the noise $\bar{h}^*$. Therefore in general the action $\bar{y}^*$ fed directly to the input of A is not equal to the actual value of the driving action $\bar{x}^*$. In an analogous fashion the mixing of the signal $\bar{x}$ concerning the state of the controlled object B with the noise $\bar{h}$ takes place in the channel H. The latter is in the feedback circuit; its output variable $\bar{y}$ acts on the input to the controller A.

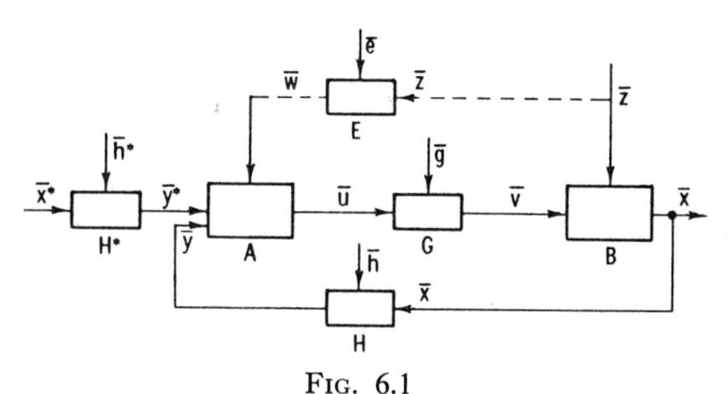

FIG. 6.1

Further, the control action $\bar{u}$ gets into the input to the object B, also traversing the channel G where it is mixed with the noise $\bar{g}$. Therefore the action $\bar{v}$ directly entering the object B is in general not equal to $\bar{u}$. The compensation circuit which can be added to the system is shown by a dotted line. The value of the noise $\bar{z}$ acting on the object B is measured and transmitted to the controller A through the channel E, where it is mixed with the noise $\bar{e}$. As a result the action $\bar{w}$ gets into the input of the device A, generally not equal to the quantity $\bar{z}$.

In the closed-loop scheme depicted in Fig. 6.1, processes are possible which do not have an analog in the open-loop systems studied in the previous chapter. In the open-loop diagram, shown for example in Fig. 5.20 or Fig. 5.23, the study of the actions $\bar{x}^*$ or $\bar{z}$ is only possible by passive observation. Meanwhile in the closed-loop diagram of Fig. 6.1 the process of studying the noise $\bar{z}$ can bear an active character. With a view to studying the object B better and obtaining more complete information about the noise $\bar{z}$, i.e., essentially information about the characteristics of the object which change in an unexpected manner, it can "feel" the test actions, having a perceptive character.

But control actions are necessary not only for studying the object, but also for bringing it to the required behavior. Therefore in the scheme of Fig. 6.1 the control actions must bear a "dual" character. They must be investigators to a known degree, but also directors to a known degree.

The control for which the control actions bear the dual character mentioned above has been called dual control [6.1]. This chapter is devoted to the theory of dual control.

Dual control is appropriate or even necessary in those cases when the operator F of the object B and the perturbing action $\bar{z}$ have a complicated character, and thus the object is distinguished by the complexity and changeability of its characteristics. Examples of dual control systems are automatic search systems, and in particular automatic optimization systems [3.25, 6.2–6.6]. In the usual systems of this type it is often easy to separate the investigating or "testing" part of the action from the directing or "working" part of the action, either on account of the difference in their frequency ranges or because they alternate in time. But in the general case such a division is not compulsory; the same action can have a dual character and be partially perceptive and partially directive.

In dual control systems a contradiction arises between the two aspects of the control action mentioned above. In fact, successful control is only possible with simultaneous action on the object. A delayed action makes the quality of the control process worse. But we can control successfully only when the properties of the object are known accurately enough. Meanwhile the study of the object requires time. Too "hurried" a controller will perform unjustified control actions, which will not be substantiated properly by the information obtained as a result of the study of the object. Too "cautious" a system will wait for an unnecessarily long time, accumulating information, and will not be able to direct the object in time to the required condition. Both in this and the other case the control process may turn out to be not the best, or even not successful.

Partial information about the object will be contained in the probability distributions of its characteristics. In general these distributions, as a measure of the study of the object, will characterize its properties more and more precisely. Indeed, a gradual change in the *a posteriori* probability distributions, their concentration close to the actual characteristics of the object, is an estimate of the rate of its study. A distinctive property of dual control systems is the dependence of this rate on the strategy of the controller.

We will now formulate the problem of constructing the optimal controller. In the formulation it is convenient to use certain concepts of statistical decision theory. We will first formulate a problem of a particular type; it will be generalized in subsequent sections.

We will examine the circuit depicted in Fig. 6.2 under the following conditions:

(1) A discrete-continuous system is investigated. All the quantities appearing in the diagram are considered only at the discrete time instants $t = 0, 1, ..., n$, where n is fixed. The value of any of the variables at the time instant $t = s$ is provided with a subscript s (for example, x_s^*, x_s, y_s).

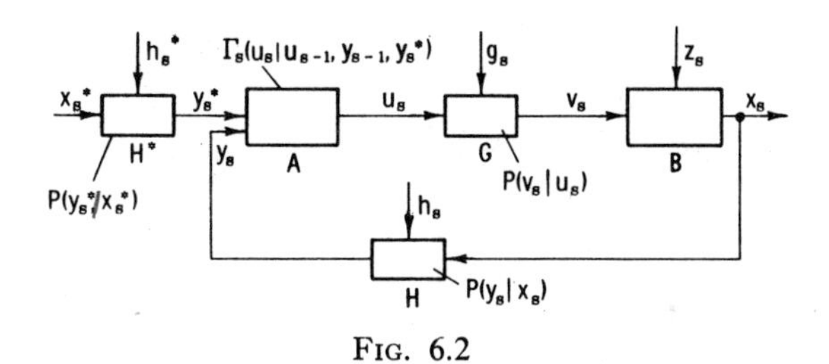

FIG. 6.2

(2) The Bayes' problem is considered, in which the *a priori* probability distributions are given. Let h_s^*, h_s, g_s be sequences of independent random variables with fixed probability densities $P(h_s^*)$, $P(h_s)$, $P(g_s)$, respectively. Further, let

$$z_s = z(s, \bar{\mu}), \tag{6.1}$$

where $\bar{\mu}$ is a random vector,

$$\bar{\mu} = (\mu_1, ..., \mu_m). \tag{6.2}$$

We regard the *a priori* probability density $P(\bar{\mu})$ of the vector $\bar{\mu}$ as assigned. In an analogous fashion we set the driving action x_s^* in the form

$$x_s^* = x_s^*(s, \bar{\lambda}), \tag{6.3}$$

where $\bar{\lambda}$ is a random vector,

$$\bar{\lambda} = (\lambda_1, ..., \lambda_q) \tag{6.4}$$

with a prescribed *a priori* probability density $P(\bar{\lambda})$. We consider that all the external actions entering the circuit—z_s, x_s^*, h_s^*, h_s, and g_s—are statistically independent.

(3) We consider that the object B does not have memory.† The operator characterizing this object turns into the relation

$$x_s = F(z_s, v_s), \tag{6.5}$$

relating the values of x_s, z_s, v_s at the same time instant. We regard the function F as finite, single-valued, and differentiable, but not necessarily one-to-one at all. Thus for a given value of z_s several values of v_s can correspond to certain values of x_s. Some forms of functions $F(v_s)$ are given in Fig. 6.3.

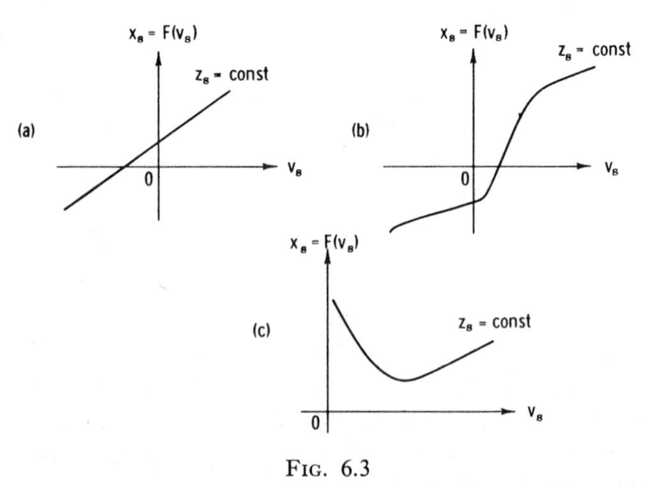

FIG. 6.3

(4) We consider the methods of combining the signal and noise in the blocks H^*, H, and G known and fixed, and these units themselves do not have memory. Hence,

$$v_s = v_s(u_s, g_s), \qquad y_s^* = y_s^*(h_s^*, x_s^*), \qquad y_s = y_s(h_s, x_s). \tag{6.6}$$

† Such objects are very important from a practical point of view, since a dynamical object will behave like an object without memory, if a steady-state value x_s is determined at each trial with the input parameter v_s.

Instead of assigning these functions and probabilistic characteristics of the noises h_s^*, h_s, g_s, the conditional probability densities $P(v_s \mid u_s)$, $P(y_s^* \mid x_s^*)$ and $P(y_s \mid x_s)$ can be prescribed at once (Fig. 6.2).

(5) We will introduce the specific loss function

$$W_s = W_s(s, x_s, x_s^*). \tag{6.7}$$

Let the total loss function W have the form

$$W = \sum_{s=0}^{n} W_s(s, x_s, x_s^*). \tag{6.8}$$

We will call optimal the system for which the total risk R, the mathematical expectation of the loss function:

$$R = M\{W\} = \sum_{s=0}^{n} M\{W_s\} = \sum_{s=0}^{n} R_s, \tag{6.9}$$

is minimal. Here R_s is the specific risk.

(6) We will consider that in the general case the controller A has memory. In addition, for generality we will assume that the device A is characterized by a random strategy. Let us introduce the temporal vectors

$$\begin{aligned} \mathbf{u}_s &= (u_0, u_1, ..., u_s), & \mathbf{x}_s^* &= (x_0^*, x_1^*, ..., x_s^*), \\ \mathbf{v}_s &= (v_0, v_1, ..., v_s), & \mathbf{y}_s^* &= (y_0^* y_1^*, ..., y_s^*), \\ \mathbf{x}_s &= (x_0, x_1, ..., x_s), & \mathbf{y}_s &= (y_0, y_1, ..., y_s) \end{aligned} \tag{6.10}$$
$$(0 \leqslant s \leqslant n).$$

The value u_s is a random function of the variables y_i, acting on the input to the device A at previous time instants ($i < s$); further, u_s is a function of the values y_j^* ($j \leqslant s$). Finally, u_s can also depend on the values u_ν of the output of the device A in the past ($\nu < s$). The previous values u_ν can be stored in this device and enter the input to the computer unit determining the current value u_s, in exactly the same way that the other input variables y_i and y_j^* act on the input to A.

The problem consists of determining the optimal random strategy of the device A, i.e., the optimal probability densities

$$P_s(u_s) = \Gamma_s(u_s \mid \mathbf{u}_{s-1}, \mathbf{y}_{s-1}, \mathbf{y}_s^*) \qquad (0 \leqslant s \leqslant n), \tag{6.11}$$

for which the total risk R is minimal. Since Γ_s is a probability density, then $\Gamma_s \geqslant 0$ and the functions Γ_s must satisfy the constraint

$$\int_{\Omega(u_s)} \Gamma_s(u_s)\, d\Omega = 1. \tag{6.12}$$

Here $\Omega(u_s)$ is the region of possible values of u_s, and $d\Omega$ its infinitesimally small element. We will call the Γ_i $(i = 0, 1, ..., n)$ specific strategies, and their totality the complete strategy.

The first step in the solution of the problem consists of deriving a formula for the risk R. We will first find the expression for the conditional specific risk r_s. We will define the latter as the term of the risk corresponding to the sth time (i.e., the time instant $t = s$), for "fixed values of the inputs to the controller A." Thus in the derivation of the formula for r_s, the temporal vectors $\mathbf{y}_s^*$, $\mathbf{u}_{s-1}$, $\mathbf{y}_{s-1}$ are considered given. Hence,

$$\begin{aligned} r_s &= M\{W_s \mid \mathbf{y}_s^*, \mathbf{u}_{s-1}, \mathbf{y}_{s-1}\} \\ &= \int_{\Omega(\bar{\lambda}, x_s)} W_s[s, x_s^*(s, \bar{\lambda}), x_s] \cdot P(\bar{\lambda}, x_s \mid \mathbf{y}_s^*, \mathbf{u}_{s-1}, \mathbf{y}_{s-1})\, d\Omega. \end{aligned} \tag{6.13}$$

We will explain this formula. Here $\Omega(\bar{\lambda}, x_s)$ is the region containing the set of possible values of the vector $\bar{\lambda}$ and the variable x_s. This region can be represented as a $(q + 1)$-dimensional space with Cartesian coordinates $\lambda_1, \lambda_2, ..., \lambda_q, x_s$. In expression (6.13) the notation $d\Omega$ is used for an infinitesimally small element of the region $\Omega(\bar{\lambda}, x_s)$. Since in the expression for W_s the variables $\bar{\lambda}$ and x_s are random, in formula (6.13) for the mathematical expectation, the function W_s must be multiplied by the joint conditional probability density of $\bar{\lambda}$ and x_s, i.e., by the function $P(\bar{\lambda}, x_s \mid \mathbf{y}_s^*, \mathbf{u}_{s-1}, \mathbf{y}_{s-1})$, under the condition that $\mathbf{y}_s^*, \mathbf{u}_{s-1}, \mathbf{y}_{s-1}$ are assigned. The result of the multiplication is integrated over the region $\Omega(\bar{\lambda}, x_s)$; as a result the expression on the right side of (6.13) is obtained.

Let us find the expression for the joint conditional probability density of $\bar{\lambda}$ and x_s. According to the theorem on multiplication of probabilities [see (2.4)] we find

$$\begin{aligned} &P(\bar{\lambda}, x_s \mid \mathbf{y}_s^*, \mathbf{u}_{s-1}, \mathbf{y}_{s-1}) \\ &\quad = P(\bar{\lambda} \mid \mathbf{y}_s^*, \mathbf{u}_{s-1}, \mathbf{y}_{s-1}) \cdot P(x_s \mid \bar{\lambda}, \mathbf{y}_s^*, \mathbf{u}_{s-1}, \mathbf{y}_{s-1}). \end{aligned} \tag{6.14}$$

But the probability density for $\bar{\lambda}$ is related only to $\mathbf{y}_s^*$ (see Fig. 6.2) and will not vary, if in addition to the constant vector $\mathbf{y}_s^*$ the vectors $\mathbf{u}_{s-1}$ and $\mathbf{y}_{s-1}$ are still fixed as well. In fact, the circuit containing the block H^* in Fig. 6.2 does not enter into the closed-loop scheme and is separate. The properties of this circuit permit the function $P(\bar{\lambda} \mid \mathbf{y}_s^*)$

to be determined, which does not depend on the values $\mathbf{u}_{s-1}$ and $\mathbf{y}_{s-1}$. Therefore

$$P(\bar{\lambda} \mid \mathbf{y}_s^*, \mathbf{u}_{s-1}, \mathbf{y}_{s-1}) = P(\bar{\lambda} \mid \mathbf{y}_s^*) = P_s(\bar{\lambda}). \tag{6.15}$$

Here the notation $P_s(\bar{\lambda})$ is used for the function $P(\bar{\lambda} \mid \mathbf{y}_s^*)$; this expression is the *a posteriori* probability density for $\bar{\lambda}$ computed at the time instant $t = s$ on the basis of the observed values $y_0^*, y_1^*, ..., y_s^*$.

The second factor on the right side of Eq. (6.14) can also be simplified. In fact, the probability density for x_s with the fixed inputs $\mathbf{y}_s^*$, $\mathbf{y}_{s-1}$, $\mathbf{u}_{s-1}$ to the device A will not vary, if $\bar{\lambda}$ is still fixed. Hence,

$$P(x_s \mid \bar{\lambda}, \mathbf{y}_s^*, \mathbf{u}_{s-1}, \mathbf{y}_{s-1}) = P(x_s \mid \mathbf{y}_s^*, \mathbf{u}_{s-1}, \mathbf{y}_{s-1}). \tag{6.16}$$

Substituting (6.14) into (6.13) with (6.15) and (6.16) taken into account, we obtain

$$r_s = \int_{\Omega(\bar{\lambda}, x_s)} W_s[s, x_s^*(s, \bar{\lambda}), x_s] \cdot P_s(\bar{\lambda}) P(x_s \mid \mathbf{y}_s^*, \mathbf{u}_{s-1}, \mathbf{y}_{s-1}) \, d\Omega. \tag{6.17}$$

From Fig. 6.2 it is seen that for a fixed value of $\bar{\mu}$ and hence of the variable z_s as well, and also for a fixed u_s, the output x_s of the system will be a random variable in view of the presence of the random noise g_s. We will denote by $P(x_s \mid \bar{\mu}, s, u_s)$ the conditional probability density of x_s for fixed $\bar{\mu}$ and u_s, taken at the time instant $t = s$. Also let $P_s(\bar{\mu})$ be the *a posteriori* probability density of the vector $\bar{\mu}$ at the time instant $t = s$. This function differs from the *a priori* density $P(\bar{\mu}) = P_0(\bar{\mu})$ of the vector $\bar{\mu}$, since the information arriving at the controller A at the times $t = 0, 1, ..., s - 1$, permits the value of the vector $\bar{\mu}$ to be refined. The probability density $P(x_s \mid \mathbf{y}_s^*, \mathbf{u}_{s-1}, \mathbf{y}_{s-1})$ can be found from the formula

$$P(x_s \mid \mathbf{y}_s^*, \mathbf{u}_{s-1}, \mathbf{y}_{s-1}) = \int_{\Omega(\bar{\mu}, u_s)} P(x_s \mid \bar{\mu}, s, u_s) \cdot P_s(\bar{\mu}) \times \Gamma_s(u_s \mid \mathbf{y}_s^*, \mathbf{u}_{s-1}, \mathbf{y}_{s-1}) \, d\Omega. \tag{6.18}$$

It is not difficult to be convinced of the validity of this formula, since for obtaining $P(x_s \mid \mathbf{y}_s^*, \mathbf{u}_{s-1}, \mathbf{y}_{s-1})$ the averaging must be carried out in the expression for $P(x_s \mid \bar{\mu}, s, u_s)$ over the two independent variables $\bar{\mu}$ and u_s. They are independent whenever the inputs to the controller are fixed. In a certain sense this condition is equivalent to breaking the closed-loop circuit of the scheme[†] depicted in Fig. 6.2.

[†] The value x_s is not related to $\mathbf{y}_{s-1}$ directly through the object, since the latter does not have memory. The binding of x_s and $\mathbf{y}_{s-1}$ is realized only through the strategy of the controller A, in which $\mathbf{y}_{s-1}$ is stored.

Substituting (6.18) into (6.17) we find that

$$r_s = \int_{\Omega(\bar{\lambda}, \bar{\mu}, x_s, u_s)} W_s[s, x_s^*(s, \bar{\lambda}), x_s] \cdot P_s(\bar{\lambda}) \cdot P(x_s \mid \bar{\mu}, s, u_s)$$

$$\times P_s(\bar{\mu}) \cdot \Gamma_s(u_s \mid \mathbf{y}_s^*, \mathbf{u}_{s-1}, \mathbf{y}_{s-1}) \, d\Omega. \tag{6.19}$$

We will now find the *a posteriori* probability density $P_s(\bar{\lambda})$. Since

$$P(\bar{\lambda}, \mathbf{y}_s^*) = P(\bar{\lambda}) \cdot P(\mathbf{y}_s^* \mid \bar{\lambda}) = P(\bar{\lambda} \mid \mathbf{y}_s^*) \cdot P(\mathbf{y}_s^*), \tag{6.20}$$

then

$$P_s(\bar{\lambda}) = P(\bar{\lambda} \mid \mathbf{y}_s^*) = P(\bar{\lambda}) \frac{P(\mathbf{y}_s^* \mid \bar{\lambda})}{P(\mathbf{y}_s^*)}. \tag{6.21}$$

Here $P(\bar{\lambda})$ is the *a priori* probability density for $\bar{\lambda}$, $p(\mathbf{y}_s^*)$ the *a priori* (absolute) probability density of $\mathbf{y}_s^*$, and $P(\mathbf{y}_s^* \mid \bar{\lambda})$ the conditional probability density for $\mathbf{y}_s^*$ for a fixed value of $\bar{\lambda}$. The latter function (it has the form of a likelihood function) can be found by knowing the properties of the channel H^* (Fig. 6.2).

Since the values h_i^* are independent for different i, and the channel H^* is inertialess, then[†]

$$P(\mathbf{y}_s^* \mid \bar{\lambda}) = \prod_{i=0}^{s} P(y_i^* \mid \bar{\lambda}) = \prod_{i=0}^{s} P(y_i^* \mid i, \bar{\lambda}). \tag{6.22}$$

It is considerably more difficult to find the *a posteriori* probability density $P_s(\bar{\mu})$. The calculation of this quantity can be performed in the controller A, in which values of the variables u_i and y_i are stored ($i = 0, ..., s - 1$). Thus the input and output of the object B are studied. The measurement of the v_i and x_i—the input and output variables of the object—takes place with errors in view of the presence of the noises g_i and h_i. But in general the study of the object gives the possibility of refining the probability density of the parameter vector $\bar{\mu}$, which determines z_s—the noise acting on the object B. The noise z_s is actually the designation for the changing characteristic of the object. The more exactly the vector $\bar{\mu}$ is known and hence the characteristic of the object B as well, the more definitively the best method of controlling the object can be explained. Therefore the calculation of the *a posteriori* probability density $P_s(\bar{\mu})$ is very important, because in it all the information about the object accumulated in the device A is concentrated.

[†] It is emphasized here that P can depend on the number i of the time.

To determine $P_s(\bar{\mu})$ we will also use a formula of Bayes' type. We will examine the joint probability density $P(\bar{\mu}, \mathbf{u}_{s-1}, \mathbf{y}_{s-1} \mid \mathbf{y}_s^*)$ of the vectors $\bar{\mu}$, $\mathbf{u}_{s-1}$, and $\mathbf{y}_{s-1}$ for fixed $\mathbf{y}_s^*$. Let $P(\mathbf{u}_{s-1}, \mathbf{y}_{s-1} \mid \bar{\mu}, \mathbf{y}_s^*)$ be the joint conditional probability density for the vectors $\mathbf{u}_{s-1}$ and $\mathbf{y}_{s-1}$ for fixed $\bar{\mu}$, and of course for fixed $\mathbf{y}_s^*$, and $P(\bar{\mu} \mid \mathbf{u}_{s-1}, \mathbf{y}_{s-1}, \mathbf{y}_s^*)$ be the conditional probability density for $\bar{\mu}$ for the fixed vectors $\mathbf{u}_{s-1}$, $\mathbf{y}_{s-1}$ and $\mathbf{y}_s^*$. The latter function is also the desired *a posteriori* probability density of the vector $\bar{\mu}$.

Then according to the theorem on multiplication of probabilities,

$$\begin{aligned} P(\bar{\mu}, \mathbf{u}_{s-1}, \mathbf{y}_{s-1} \mid \mathbf{y}_s^*) &= P(\mathbf{u}_{s-1}, \mathbf{y}_{s-1} \mid \bar{\mu}, \mathbf{y}_s^*) \cdot P(\bar{\mu}) \\ &= P(\bar{\mu} \mid \mathbf{u}_{s-1}, \mathbf{y}_{s-1}, \mathbf{y}_s^*) P(\mathbf{u}_{s-1}, \mathbf{y}_{s-1} \mid \mathbf{y}_s^*). \end{aligned} \tag{6.23}$$

In this formula the function $P(\bar{\mu}) = P_0(\bar{\mu})$ is the *a priori* (absolute) probability density of $\bar{\mu}$, and $P(\mathbf{u}_{s-1}, \mathbf{y}_{s-1})$ is the joint absolute probability density of the vectors $\mathbf{u}_{s-1}$ and $\mathbf{y}_{s-1}$. Since $\bar{\mu}$ does not depend on $\bar{\lambda}$ or $\mathbf{y}_s^*$, the dependence on $\mathbf{y}_s^*$ is missing in the expression for $P(\bar{\mu})$.

From (6.23) we find

$$\begin{aligned} P_s(\bar{\mu}) = P(\bar{\mu} \mid \mathbf{u}_{s-1}, \mathbf{y}_{s-1}, \mathbf{y}_s^*) &= \frac{P(\bar{\mu}) \cdot P(\mathbf{u}_{s-1}, \mathbf{y}_{s-1} \mid \bar{\mu}, \mathbf{y}_s^*)}{P(\mathbf{u}_{s-1}, \mathbf{y}_{s-1} \mid \mathbf{y}_s^*)} \\ &= \frac{P(\bar{\mu}) \cdot P(\mathbf{u}_{s-1}, \mathbf{y}_{s-1} \mid \bar{\mu}, \mathbf{y}_s^*)}{\int_{\Omega(\bar{\mu})} P(\mathbf{u}_{s-1}, \mathbf{y}_{s-1} \mid \bar{\mu}, \mathbf{y}_s^*) P(\bar{\mu})\, d\Omega}. \end{aligned} \tag{6.24}$$

Thus in order to determine $P_s(\bar{\mu})$, the expression for the likelihood function $P(\mathbf{u}_{s-1}, \mathbf{y}_{s-1} \mid \bar{\mu}, \mathbf{y}_s^*)$ in the closed-loop system depicted in Fig. 6.2 must be found. It is not difficult to see that the probability density of the compound event, consisting of the occurrence first of the pair of values u_0, y_0, then u_1, y_1, afterwards u_2, y_2 and so on, for a fixed value of $\bar{\mu}$ is the product of several factors: the probability density $P(u_0, y_0 \mid \bar{\mu}, \mathbf{y}_0^*)$ of the first of these events, the probability density $P(u_1, y_1 \mid \bar{\mu}, u_0, y_0, \mathbf{y}_1^*)$ of the second of the events under the condition that the first occurred, then the probability density $P(u_2, y_2 \mid \bar{\mu}, u_1, y_1, \mathbf{y}_2^*)$ of the third of the events under the condition that the first and second occurred, and so on:

$$\begin{aligned} P(\mathbf{u}_{s-1}, \mathbf{y}_{s-1} \mid \bar{\mu}, \mathbf{y}_s^*) = {} & P(u_0, y_0 \mid \bar{\mu}, y_0^*) P(u_1, y_1 \mid \bar{\mu}, u_0, y_0, \mathbf{y}_1^*) \\ & \times P(u_2, y_2 \mid \bar{\mu}, \mathbf{u}_1, \mathbf{y}_1, \mathbf{y}_2^*) \\ & \cdots P(u_{s-1}, y_{s-1} \mid \bar{\mu}, \mathbf{u}_{s-2}, \mathbf{y}_{s-2}, \mathbf{y}_{s-1}^*). \end{aligned} \tag{6.25}$$

Let us consider the ith factor of this expression $(0 < i \leqslant s - 1)$ and transform it:

$$P(u_i, y_i \mid \bar{\mu}, \mathbf{u}_{i-1}, \mathbf{y}_{i-1}, \mathbf{y}_i^*)$$
$$= P(y_i \mid \bar{\mu}, u_i, \mathbf{u}_{i-1}, \mathbf{y}_{i-1}, \mathbf{y}_i^*)\, P(u_i \mid \bar{\mu}, \mathbf{u}_{i-1}, \mathbf{y}_{i-1}, \mathbf{y}_i^*). \quad (6.26)$$

The first factor on the right side of (6.26) is the conditional probability density of y_i at the time instant $t = i$, under the condition that $\bar{\mu}$ and u_i are fixed. From Fig. 6.2 it is seen that it will not change if the vectors $\mathbf{y}_i^*$, $\mathbf{u}_{i-1}$, and $\mathbf{y}_{i-1}$ are also fixed; hence this density does not depend on the vectors $\mathbf{u}_{i-1}$, $\mathbf{y}_i^*$, and $\mathbf{y}_{i-1}$. It can be denoted by $P(y_i \mid \bar{\mu}, i, u_i)$, where the i to the right of the vertical line means that in general this function also depends on the number i of the time. Further, the conditional probability density $P(u_i \mid \bar{\mu}, \mathbf{u}_{i-1}, \mathbf{y}_{i-1}, \mathbf{y}_i^*)$ does not depend on $\bar{\mu}$ whenever $\mathbf{u}_{i-1}$ and $\mathbf{y}_{i-1}$ are fixed. This function is the random strategy Γ_i of the controller A. Therefore it can be written[†] as $\Gamma_i(u_i \mid \mathbf{y}_i^*, \mathbf{u}_{i-1}, \mathbf{y}_{i-1})$. Thus formula (6.26) can be rewritten:

$$P(u_i, y_i \mid \bar{\mu}, \mathbf{u}_{i-1}, \mathbf{y}_{i-1}, \mathbf{y}_i^*) = P(y_i \mid \bar{\mu}, i, u_i)\Gamma_i, \quad (6.27)$$

where Γ_i is the abbreviated notation for

$$\Gamma_i(u_i \mid \mathbf{y}_i^*, \mathbf{u}_{i-1}, \mathbf{y}_{i-1}).$$

Substituting (6.27) into (6.26), and (6.26) into (6.25), we obtain the expression for the likelihood function in the form

$$P(\mathbf{u}_{s-1}, \mathbf{y}_{s-1} \mid \bar{\mu}, \mathbf{y}_s^*) = \left[\prod_{i=0}^{s-1} P(y_i \mid \bar{\mu}, i, u_i)\right] \cdot \prod_{i=0}^{s-1} \Gamma_i, \quad (6.28)$$

where the notation

$$\Gamma_0 = P_0(u_0) \quad (6.29)$$

is introduced. This probability density does not depend on the observations which did not yet exist at the initial instant $t = 0$, and is determined only by *a priori* data.

The substitution of (6.28) into expression (6.24) leads to the formula

$$P_s(\bar{\mu}) = P(\bar{\mu} \mid \mathbf{u}_{s-1}, \mathbf{y}_{s-1}, \mathbf{y}_s^*)$$
$$= \frac{P(\bar{\mu}) \cdot [\prod_{i=0}^{s-1} P(y_i \mid \bar{\mu}, i, u_i)] \cdot \prod_{i=0}^{s-1} \Gamma_i}{P(\mathbf{u}_{s-1}, \mathbf{y}_{s-1} \mid \mathbf{y}_s^*)}. \quad (6.30)$$

[†] It does not depend on y_i if storage of the values y_j $(j < i)$ passing earlier is made in A. Therefore $\mathbf{y}_{i-1}$ can be written instead of $\mathbf{y}_i$.

We will turn our attention to the fundamental difference between the latter formula and expression (6.21). The storage of information about the characteristics of the object is expressed in the fact that the *a priori* probability density $P(\bar{\mu})$ is replaced at each new time by the *a posteriori* densities $P_s(\bar{\mu})$, characterizing the vector $\bar{\mu}$ all the more precisely.

From formula (6.30) it is seen that the form of the function $P_s(\bar{\mu})$ and hence the rate of information storage about the object depend in general on all the previous functions Γ_i of the strategy ($i < s$). In other words, the rate of object study depends on the extent to which experiments on the study of this object, the input of actions u_i to it and the analysis of the reactions y_i of the object to these actions were rationally formulated.

Meanwhile in formula (6.21) for $P_s(\bar{\lambda})$, in the same way as in analogous formulas of the previous chapter, the dependence of the rate of information storage on the strategy Γ_i ($i = 0, 1, ..., s - 1$) is missing. Therefore the process of information storage about the action x_s^* in the open-loop circuit is passive or independent.

The comparison of formulas (6.30) and (6.21) permits the basic difference between dual and nondual control to be explained. With dual control the rate of object study depends on the character of the control actions, i.e., on the strategy of the controller, while with nondual control this dependence is missing.

After substituting expressions (6.21) and (6.30) into (6.19), we arrive at the definitive formula for the conditional specific risk:

$$r_s = \int_{\Omega(\bar{\lambda},\bar{\mu},x_s,u_s)} W_s[s, x_s^*(s, \bar{\lambda}), x_s] \times \frac{P(\bar{\lambda}) \cdot P(\mathbf{y}_s^* \mid \bar{\lambda})}{P(\mathbf{y}_s^*)} P(x_s \mid \bar{\mu}, s, u_s) \times \frac{P(\bar{\mu}) \cdot \prod_{i=0}^{s-1} P(y_i \mid \bar{\mu}, i, u_i)}{P(\mathbf{y}_{s-1}, \mathbf{u}_{s-1} \mid \mathbf{y}_s^*)} \cdot \prod_{i=0}^{s} \Gamma_i \, d\Omega. \tag{6.31}$$

If the values of r_s are examined for different trials performed with the system, then in general the vectors $\mathbf{y}_s^*$, $\mathbf{u}_{s-1}$, and $\mathbf{y}_{s-1}$, not known in advance, can assume different values. Let $P(\mathbf{y}_s^*, \mathbf{u}_{s-1}, \mathbf{y}_{s-1})$ be the joint distribution density of these vectors. Then the mean specific risk R_s, the mean value of the conditional specific risk r_s for a mass production of trials, is defined by the formula

$$R_s = \int_{\Omega(\mathbf{y}_s^*,\mathbf{u}_{s-1},\mathbf{y}_{s-1})} r_s \cdot P(\mathbf{y}_s^*, \mathbf{u}_{s-1}, \mathbf{y}_{s-1}) \, d\Omega. \tag{6.32}$$

We will take into account that

$$P(\mathbf{y}_s^*, \mathbf{u}_{s-1}, \mathbf{y}_{s-1}) = P(\mathbf{u}_{s-1}, \mathbf{y}_{s-1} \mid \mathbf{y}_s^*)P(\mathbf{y}_s^*). \tag{6.33}$$

After substituting (6.31) and (6.33) into (6.32), with (6.22) taken into account we obtain the fundamental formula—the expression for R_s in the form

$$R_s = \int_{\Omega(\bar{\lambda}, \bar{\mu}, x_s, \mathbf{y}_s^*, \mathbf{u}_s, \mathbf{y}_{s-1})} W_s[s, x_s^*(s, \bar{\lambda}), x_s]$$

$$\times P(\bar{\lambda}) \cdot \left[\prod_{i=0}^{s} P(y_i^* \mid i, \bar{\lambda})\right] \cdot P(x_s \mid \bar{\mu}, s, u_s)$$

$$\times P(\bar{\mu}) \left[\prod_{i=0}^{s-1} P(y_i \mid \bar{\mu}, i, u_i)\right]\left[\prod_{i=0}^{s} \Gamma_i\right] d\Omega. \tag{6.34}$$

Although in this problem the object B does not have memory, still the risk R_s corresponding to the time instant $t = s$ depends on the complete strategy, i.e., on the totality of specific strategies—the functions Γ_i at the time instants $t = 0, 1, \ldots, s$. The physical reason for this occurrence, missing in the open-loop system, is just that of dual control. Hence it follows that the control for $t = k$ $(k < n)$ must be calculated not only in order to reduce the value of R_k—the specific risk corresponding to this time instant—but also by a better study of the object to assist in reducing the values R_i $(i > k)$ of the risks at subsequent time instants. We will examine S_k, the part of the total risk R depending on the strategy Γ_k:

$$S_k = \sum_{i=k}^{n} R_i = R_k + \sum_{i=k+1}^{n} R_i. \tag{6.35}$$

With respect to the specific strategy Γ_k, the first term of the right side of (6.35) can be called the "operating risk," and the second the "probing risk." The primitive strategy of selecting the action u_k (or its probability density Γ_k) such that only the operating risk will be minimized is not optimal. On the other hand, disregarding the operating risk and minimizing only the probing risk, i.e., choosing u_k (or Γ_k) only with the aim of the best study of the object to use this information in subsequent actions, also will not be the optimal behavior. The optimal strategy for dual control must minimize the sum S_k of the operating and probing risks.

2. Solution of the Problem and Very Simple Examples

For determining the optimal strategy of dual control we will use the method of dynamic programming [6.1, 6.7].

Let us introduce the auxiliary functions ($k = 0, ..., n$)

$$\alpha_k = \alpha_k(\mathbf{u}_k, \mathbf{y}_{k-1}, \mathbf{y}_k^*)$$
$$= \int_{\Omega(\bar{\lambda},\bar{\mu},x_k)} W_k[k, x_k^*(k, \bar{\lambda}), x_k] \cdot P(\bar{\lambda}) \cdot \left[\prod_{i=0}^{k} P(y_i^* \mid i, \bar{\lambda})\right]$$
$$\times P(x_k \mid \bar{\mu}, k, u_k) \cdot P(\bar{\mu}) \cdot \left[\prod_{i=0}^{k-1} P(y_i \mid \bar{\mu}, i, u_i)\right] d\Omega \tag{6.36}$$

and

$$\beta_k = \prod_{i=0}^{k} \Gamma_i \,. \tag{6.37}$$

We will first examine the risk R_n for the last time instant $t = n$, considering that the strategies $\Gamma_0 \cdots \Gamma_{n-1}$ have been stipulated in some manner. From (6.34) with (6.37) and (6.36) taken into account we obtain:

$$R_n = \int_{\Omega(\mathbf{u}_n, \mathbf{y}_{n-1}, \mathbf{y}_n^*)} \alpha_n(\mathbf{u}_n, \mathbf{y}_{n-1}, \mathbf{y}_n^*) \cdot \beta_{n-1} \cdot \Gamma_n \, d\Omega$$
$$= \int_{\Omega(\mathbf{u}_{n-1}, \mathbf{y}_{n-1}, \mathbf{y}_n^*)} \beta_{n-1} \cdot \kappa_n(\mathbf{u}_{n-1}, \mathbf{y}_{n-1}, \mathbf{y}_n^*) \, d\Omega, \tag{6.38}$$

where

$$\kappa_n(\mathbf{u}_{n-1}, \mathbf{y}_{n-1}, \mathbf{y}_n^*) = \int_{\Omega(u_n)} \alpha_n(u_n, \mathbf{u}_{n-1}, \mathbf{y}_{n-1}, \mathbf{y}_n^*)$$
$$\times \Gamma_n(u_n \mid \mathbf{u}_{n-1}, \mathbf{y}_{n-1}, \mathbf{y}_n^*) \, d\Omega. \tag{6.39}$$

On the basis of the mean value theorem with (6.12) taken into account we can write:

$$\kappa_n = (\alpha_n)_m \int_{\Omega(u_n)} \Gamma_n \, d\Omega = (\alpha_n)_m \geqslant (\alpha_n)_{\min} \,. \tag{6.40}$$

It is necessary to choose Γ_n such that R_n will be minimized. The latter is realized if for any $\mathbf{u}_{n-1}$, $\mathbf{y}_{n-1}$, $\mathbf{y}_n^*$, Γ_n is chosen such that the function κ_n will be minimal. With this aim, as in the previous chapter we will find the value $u_n = u_n^*$ minimizing the function α_n. Let $\alpha_n = \gamma_n$ and

$$\gamma_n^* = \alpha_n(u_n^*, \mathbf{u}_{n-1}, \mathbf{y}_{n-1}, \mathbf{y}_n^*) = \min_{u_n \in \Omega(u_n)} \alpha_n(u_n, \mathbf{u}_{n-1}, \mathbf{y}_{n-1}, \mathbf{y}_n^*). \tag{6.41}$$

The quantity u_n^* is evidently a function of $\mathbf{u}_{n-1}, \mathbf{y}_{n-1}, \mathbf{y}_n^*$:

$$u_n^* = u_n^*(\mathbf{u}_{n-1}, \mathbf{y}_{n-1}, \mathbf{y}_n^*). \tag{6.42}$$

Then the optimal strategy Γ_n^* is given by the expression

$$\Gamma_n^* = \delta(u_n - u_n^*), \tag{6.43}$$

where δ is the unit impulse function. This means that Γ_n^* is a regular strategy, where in accordance with (6.42) the optimal value of u_n is equal to u_n^*. From expression (6.42) it is seen that u_n^* is a function of the values u_s, y_s ($s = 0, 1, ..., n-1$) which acted earlier on the controller A, and also of the values y_i^* ($i = 0, ..., n$).

It is not difficult to prove that the strategy expressed by formulas (6.41)–(6.43) is optimal. In the same way as in Chapter V, let us substitute expression (6.43) into relation (6.39). Then we will obtain

$$\kappa_n = (\alpha_n)_{u_n=u_n^*} = (\alpha_n)_{\min}. \tag{6.44}$$

But according to (6.40) this value is the minimal possible one for κ_n. Hence Γ_n^* is the optimal strategy.

In order to find the optimal strategies Γ_i^* for $i < n$, it is necessary to shift successively from the final instant $t = n$ to the beginning. For example, for the purpose of determining Γ_{n-1}^* we will consider the last two terms entering into the composition of the total risk R, i.e., the quantity

$$\begin{aligned} S_{n-1} &= R_{n-1} + R_n \\ &= \int_{\Omega(\mathbf{u}_{n-1}, \mathbf{y}_{n-2}, \mathbf{y}_{n-1}^*)} \alpha_{n-1}\beta_{n-1}\, d\Omega + \int_{\Omega(\mathbf{u}_n, \mathbf{y}_{n-1}, \mathbf{y}_n^*)} \alpha_n\beta_n\, d\Omega \\ &= \int_{\Omega(\mathbf{u}_{n-2}, \mathbf{y}_{n-2}, \mathbf{y}_{n-1}^*)} \beta_{n-2}\kappa_{n-1}(\mathbf{u}_{n-2}, \mathbf{y}_{n-2}, \mathbf{y}_{n-1}^*)\, d\Omega, \end{aligned} \tag{6.45}$$

where

$$\begin{aligned} &\kappa_{n-1}(\mathbf{u}_{n-2}, \mathbf{y}_{n-2}, \mathbf{y}_{n-1}^*) \\ &= \int_{\Omega(u_{n-1})} \left\{ \Gamma_{n-1}\alpha_{n-1} + \Gamma_{n-1} \int_{\Omega(y_{n-1}, y_n^*)} \alpha_n^*\, d\Omega(y_{n-1}, y_n^*) \right\} d\Omega(u_{n-1}) \\ &= \int_{\Omega(u_{n-1})} \Gamma_{n-1} \left\{ \alpha_{n-1} + \int_{\Omega(y_{n-1}, y_n^*)} \alpha_n^*\, d\Omega(y_{n-1}, y_n^*) \right\} d\Omega(u_{n-1}). \end{aligned} \tag{6.46}$$

Here the function Γ_n is not written out in explicit form, since from the preceding it is known to be contained in the replacement of u_n by expression (6.42). We will regard that such a substitution has been made; then from now on Γ_n will not appear in our formulas, and u_n^* turns out to be a known function of $\mathbf{u}_{n-1}, \mathbf{y}_{n-1}, \mathbf{y}_n^*$. Let us examine the function

$$\gamma_{n-1} = \gamma_{n-1}(\mathbf{u}_{n-1}, \mathbf{y}_{n-2}, \mathbf{y}_{n-1}^*) = \alpha_{n-1} + \int_{\Omega(y_{n-1}, y_n^*)} \gamma_n^*(u_n^*, \mathbf{u}_{n-1}, \mathbf{y}_{n-1}, \mathbf{y}_n^*)\, d\Omega. \tag{6.47}$$

We will find the value u_{n-1}^* giving this function a minimum. It is evident that this value in general depends on all the remaining variables appearing in the expression for γ_{n-1}:

$$u_{n-1}^* = u_{n-1}^*(\mathbf{u}_{n-2}, \mathbf{y}_{n-2}, \mathbf{y}_{n-1}^*). \tag{6.48}$$

Then the optimal strategy is

$$\Gamma_{n-1}^* = \delta(u_{n-1} - u_{n-1}^*), \tag{6.49}$$

i.e., Γ_{n-1}^* is also a regular strategy. The proof of the validity of this formula is analogous to that given above for Γ_n^*. Continuing this process further by passing to the determination of $\Gamma_{n-2}^*, \ldots, \Gamma_0^*$, the following prescription for obtaining the optimal strategy can be found. Let $\gamma_n = \alpha_n$ and

$$\gamma_{n-k} = \alpha_{n-k} + \int_{\Omega(y_{n-k}, y_{n-k+1}^*)} \gamma_{n-k+1}^*(u_{n-k+1}^*, \mathbf{u}_{n-k}, \mathbf{y}_{n-k}, \mathbf{y}_{n-k+1}^*)\, d\Omega, \tag{6.50}$$

and u_{n-k}^* is the value of u_{n-k} minimizing the function γ_{n-k}, i.e.,

$$\gamma_{n-k}^* = \min_{u_{n-k} \in \Omega(u_{n-k})} \gamma_{n-k} = \gamma_{n-k}(u_{n-k}^*). \tag{6.51}$$

It is evident that

$$u_{n-k}^* = u_{n-k}^*(\mathbf{y}_{n-k}^*, \mathbf{u}_{n-k-1}, \mathbf{y}_{n-k-1}). \tag{6.52}$$

Then the optimal strategy Γ_{n-k}^* is given by the expression

$$\Gamma_{n-k}^* = \delta(u_{n-k} - u_{n-k}^*), \tag{6.53}$$

i.e., the optimal strategy is regular and consists of choosing $u_{n-k} = u_{n-k}^*$ according to formula (6.52). As is seen from this formula, in general the optimal control u_{n-k}^* at the instant $t = n - k$ depends on the entire "previous history" of the input variables acting on A, i.e., on all the values u_i, y_i ($i = 0, 1, ..., n - k - 1$), and also on the values y_j^* ($j \leqslant n - k$).

It is not difficult to show that in the expression for α_0 we should put

$$\prod_{i=0}^{s-1} P(y_i \mid \bar{\mu}, i, u_i) = 1. \tag{6.54}$$

Then it turns out that

$$u_0^* = u_0^*(y_0^*), \tag{6.55}$$

i.e., the initial control action depends on the initial value received at the input to the device A, and in addition, of course, on the *a priori* data.

It should be noted that, in the optimal algorithm (6.52) which must be put into the controller A, the value u_{n-k}^* depends on the variables u_i, y_i, y_i^* fed to the input of A in the past and also on the current value y_{n-k}^*, but not on the future values of these quantities. Therefore the optimal controller is physically realizable.

If the values x_i^* ($i = 0, 1, ..., n$) are known beforehand, then the formula for the optimal control is simplified. Such a case can be encountered in automatic stabilization systems or in programmed control systems. In this case the element H^* in Fig. 6.2 turns out to be unnecessary, and the values $x_i^* = y_i^*$ can be put in advance into the control law in the form of known parameters. Formulas for problems of this type can be obtained as a special case of formulas derived above, if we set

$$P(\bar{\lambda}) = \delta(\bar{\lambda} - \bar{\lambda}^*), \qquad P(y_i^* \mid i, \bar{\lambda}) = \delta[y_i^* - x_i^*(\bar{\lambda}^*)], \tag{6.56}$$

where $\bar{\lambda}$ is a fixed vector known in advance. Since x_k^* can also be regarded as known in advance, then formula (6.34) takes the form

$$R_s = \int_{\Omega(\bar{\mu}, x_s, \mathbf{u}_s, \mathbf{y}_{s-1})} W_s(s, x_s^*, x_s) \cdot P(x_s \mid \bar{\mu}, s, u_s) \cdot P(\bar{\mu}) \times \left[\prod_{i=0}^{s-1} P(y_i \mid \bar{\mu}, i, u_i)\right] \cdot \left[\prod_{i=0}^{s} \Gamma_i\right] d\Omega. \tag{6.57}$$

Considering x_k^* as a parameter and not writing it out explicitly, if we now put

$$\alpha_k = \alpha_k(\mathbf{u}_k, \mathbf{y}_{k-1}) = \int_{\Omega(\bar{\mu}, x_k)} W_k(k, x_k^*, x_k) \cdot P(x_k \mid \bar{\mu}, k, u_k) \times P(\bar{\mu}) \cdot \left[\prod_{i=0}^{k-1} P(y_i \mid \bar{\mu}, i, u_i)\right] d\Omega, \tag{6.58}$$

then we arrive at a formula of the same type as expression (6.38):

$$R_k = \int_{\Omega(\mathbf{u}_k, \mathbf{y}_{k-1})} \alpha_k(\mathbf{u}_k, \mathbf{y}_{k-1})\beta_k \, d\Omega. \tag{6.59}$$

Carrying out the same arguments for this case as above, we arrive at the following prescription for finding the optimal strategy: let $\gamma_n = \alpha_n$ and

$$\gamma_{n-k} = \alpha_{n-k} + \int_{\Omega(y_{n-k})} \gamma_{n-k+1}(u^*_{n-k+1}, u_{n-k}, \mathbf{y}_{n-k}) \, d\Omega. \tag{6.60}$$

Further, let u^*_{n-k} be the value of u_{n-k} minimizing the function γ_{n-k} according to expression (6.51). It is evident that

$$u^*_{n-k} = u^*_{n-k}(\mathbf{u}_{n-k-1}, \mathbf{y}_{n-k-1}). \tag{6.61}$$

Here the dependence on the parameter $\mathbf{x}_n{}^*$ is not written out in explicit form. Then the optimal strategy Γ^*_{n-k} is given by expression (6.53) where, however, u^*_{n-k} is computed according to relations (6.60) and (6.61).

Let us first consider some comparatively simple examples [6.1, 6.7]. A block diagram of the simplest system is shown in Fig. 6.4. The output variable x_s of the controlled object B is brought to the input of the controller A. The driving action $x_s{}^*$, assumed known, is fed to the other input of this device. The output variable u_s of the device A is the control action which is fed to the input of the object B through the communication channel G, where it is mixed with the noise g_s. In this example the noise z_s reduces to the random variable μ. The object B has an operator of the following form:

$$x_s = v_s + \mu, \tag{6.62}$$

where

$$v_s = u_s + g_s. \tag{6.63}$$

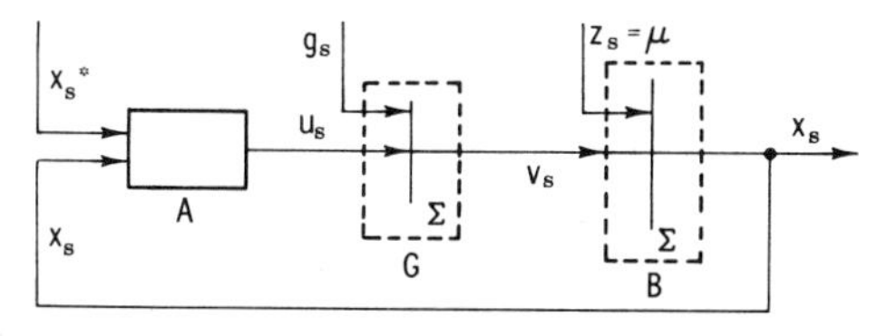

FIG. 6.4

The latter expression is the law of additive composition of the noise and signal in the communication channel G. The specific loss function is defined by the expression

$$W_s = (x_s - x_s^*)^2. \tag{6.64}$$

Let $P_0(\mu)$ and $P(g_s) = q(g_s)$ be normal distributions with zero mean values and dispersions σ_μ^2 and σ_g^2, respectively:

$$\begin{aligned} P_0(\mu) &= \frac{1}{\sigma_\mu \sqrt{(2\pi)}} \exp\left\{-\frac{\mu^2}{2\sigma_\mu^2}\right\}, \\ P(g_s) = q(g_s) &= \frac{1}{\sigma_g \sqrt{(2\pi)}} \exp\left\{-\frac{g_s^2}{2\sigma_g^2}\right\}. \end{aligned} \tag{6.65}$$

No constraints are imposed on the values of the variables.

It is required to realize the synthesis of an optimal controller A, for which the risk R—the mathematical expectation of the loss function $W = \sum_{s=0}^{n} W_s$—is minimal.

According to the general theory, the function α_k defined by (6.58) must first be found. For this it is required to find out the expressions $P(x_k \mid \bar{\mu}, k, u_k)$ and $P(y_i \mid \bar{\mu}, i, u_i)$ as a preliminary. Since in this problem $y = x$, μ is a scalar variable and the desired probability density does not depend on i or k, we will denote it by $P(x_i \mid \mu, u_i)$.

From (6.62) and (6.63) it follows that:

$$x_i = (u_i + \mu) + g_i\,. \tag{6.66}$$

Therefore the probability that the output variable of the object lies in the interval between the fixed values x_i and $x_i + dx_i$ is equal to the probability that g_i will be found between the two fixed values $x_i - (u_i + \mu)$ and $x_i - (u_i + \mu) + dx_i$.

Hence,

$$P(x_i \mid \mu, u_i) = q(x_i - u_i - \mu), \tag{6.67}$$

where q is the density of the noise g_s, which is given by the latter of expressions (6.65). Substituting this expression into (6.58), we find:

$$\begin{aligned} \alpha_k &= \int_{\Omega(x_k,\mu)} W_k P_0(\mu) \cdot \prod_{i=0}^{k} P(x_i \mid \mu, u_i)\, d\Omega \\ &= \int_{-\infty}^{\infty} \int_{-\infty}^{\infty} (x_k^* - x_k)^2 \cdot P_0(\mu) \prod_{i=0}^{k} q(x_i - u_i - \mu)\, dx_k\, d\mu. \end{aligned} \tag{6.68}$$

After substituting expressions (6.65) here, we obtain

$$\alpha_k = \frac{1}{(\sigma_g)^{k+1}\sigma_\mu(2\pi)^{k/2+1}} \int_{\mu=-\infty}^{\infty} \exp\left\{-\frac{\mu^2}{2\sigma_\mu^2} - \sum_{i=0}^{k-1} \frac{(x_i - u_i - \mu)^2}{2\sigma_g^2}\right\}$$

$$\times \left[\int_{x_k=-\infty}^{\infty} (x_k^* - x_k)^2 \exp\left\{-\frac{(x_k - u_k - \mu)^2}{2\sigma_g^2}\right\} dx_k\right] d\mu. \tag{6.69}$$

We will make use of the formula ($p > 0$)

$$\int_{-\infty}^{\infty} x^2 \exp\{-px^2 + 2qx\}\, dx = \frac{1}{2p}\sqrt{\left(\frac{\pi}{p}\right)}\left(1 + \frac{2q^2}{p}\right) \exp\left\{\frac{q^2}{p}\right\}. \tag{6.70}$$

We will find the integral I_k in the brackets of formula (6.69):

$$I_k = \int_{-\infty}^{\infty} (x_k^* - x_k)^2 \exp\left\{-\frac{(x_k - u_k - \mu)^2}{2\sigma_g^2}\right\} dx_k$$

$$= \exp\left\{-\frac{l_k^2}{2\sigma_g^2}\right\} \int_{-\infty}^{\infty} z^2 \exp\left\{-\frac{z^2}{2\sigma_g^2} + 2\left(-\frac{l_k}{2\sigma_g^2}\right) z\right\} dz, \tag{6.71}$$

where

$$x_k - x_k^* = z, \qquad x_k^* - u_k - \mu = l_k\,. \tag{6.72}$$

After using formula (6.70), we find from (6.71) and (6.72)

$$I_k = \sigma_g \sqrt{(2\pi)}[\sigma_g^2 + (x_k^* - u_k - \mu)^2]. \tag{6.73}$$

Then, according to (6.69),

$$\alpha_k = \frac{1}{\sigma_g^k \sigma_\mu (2\pi)^{(k+1)/2}} \left[\sigma_g^2 \int_{-\infty}^{\infty} \exp\left\{-\frac{\mu^2}{2\sigma_\mu^2}\right.\right.$$

$$\left. - \sum_{i=0}^{k-1} \frac{(x_i - u_i)^2 - 2\mu(x_i - u_i) + \mu^2}{2\sigma_g^2}\right\} d\mu + \int_{-\infty}^{\infty} \exp\left\{-\frac{\mu^2}{2\sigma_\mu^2}\right.$$

$$\left.\left. - \sum_{i=0}^{k-1} \frac{(x_i - u_i)^2 - 2\mu(x_i - u_i) + \mu^2}{2\sigma_g^2}\right\} (x_k^* - u_k - \mu)^2\, d\mu\right]. \tag{6.74}$$

The first of the integrals inside the brackets in expression (6.74) can be found from using the formula ($p > 0$):

$$\int_{-\infty}^{\infty} \exp\{-px^2 \pm qx\}\, dx = \sqrt{\left(\frac{\pi}{p}\right)} \exp\left\{\frac{q^2}{4p}\right\}. \tag{6.75}$$

The second integral in the brackets can be determined after setting $\mu + u_k - x_k^* = z$ and using formula (6.70). As a result the formula for α_k is obtained in final form.

Let us introduce the notations

$$u_k - x_k^* = w_k, \qquad \sum_{i=0}^{k-1} (x_i - u_i) = \Sigma_{k-1},$$

$$\epsilon_k = \frac{1}{2\sigma_\mu^2} + \frac{k}{2\sigma_g^2}, \qquad \sum_{i=0}^{k-1} (x_i - u_i)^2 = \theta_{k-1},$$

$$a_k = \frac{\sigma_g^2}{\sqrt{2}\, \sigma_g^k \sigma_\mu (2\pi)^{k/2} \sqrt{(\epsilon_k)}}, \qquad b_k = \frac{1}{2\sigma_g^k \sigma_\mu (2\pi)^{k/2} (2\epsilon_k)^{3/2}}. \tag{6.76}$$

Then

$$\alpha_k = \left\{ a_k + b_k + \frac{2b_k}{\epsilon_k} \left(w_k \epsilon_k + \frac{\Sigma_{k-1}}{2\sigma_g^2} \right)^2 \right\} \exp\left\{ -\frac{\theta_{k-1}}{2\sigma_g^2} + \frac{\Sigma_{k-1}^2}{4\sigma_g^4 \epsilon_k} \right\}. \tag{6.77}$$

In this formula the quantity u_k contained in the expression for w_k is only present in the factor in front of the exponential, but u_k or a function of this variable is missing under the exponential sign. This condition greatly simplifies the minimization of α_k with respect to u_k.

We will now perform the successive calculations for determining Γ_i^*. In formula (6.77) we will set $k = n$ and find the minimum of the function $\gamma_n = \alpha_n$ with respect to u_n. The minimum condition coincides with the vanishing of the parentheses in expression (6.77):

$$w_n \epsilon_n + (\Sigma_{n-1}/2\sigma_g^2) = 0. \tag{6.78}$$

Hence by substituting the values of w_n and Σ_{n-1} from (6.76), we find the optimal control:

$$u_n^* = x_n^* - \frac{\sum_{i=0}^{n-1} (x_i - u_i)}{2\sigma_g^2 \epsilon_n}. \tag{6.79}$$

Further, we determine from (6.77)

$$\gamma_n^* = \alpha_n^* = (a_n + b_n) \exp\left\{ -\frac{\theta_{n-1}}{2\sigma_g^2} + \frac{\Sigma_{n-1}^2}{4\sigma_g^4 \epsilon_n} \right\}. \tag{6.80}$$

Then we find the function γ_{n-1}:

$$\gamma_{n-1} = \alpha_{n-1} + \int_{-\infty}^{\infty} \alpha_n^* \, dx_{n-1}$$

$$= \left\{ a_{n-1} + b_{n-1} + \frac{2b_{n-1}}{\epsilon_{n-1}} \left[w_{n-1}\epsilon_{n-1} + \frac{\Sigma_{n-2}}{2\sigma_g^2} \right]^2 \right\}$$

$$\times \exp\left\{ - \frac{\theta_{n-2}}{2\sigma_g^2} + \frac{\Sigma_{n-2}^2}{4\sigma_g^4 \epsilon_{n-1}} \right\} + \int_{-\infty}^{\infty} (a_n + b_n)$$

$$\times \exp\left\{ - \frac{(x_{n-1} - u_{n-1})^2 + \theta_{n-2}}{2\sigma_g^2} + \frac{[\Sigma_{n-2} + (x_{n-1} - u_{n-1})]^2}{4\sigma_g^4 \epsilon_n} \right\} dx_{n-1}. \tag{6.81}$$

The integral in this formula does not depend on u_{n-1}. In fact, the quantity u_{n-1} enters into α_n^* only in the form of the difference $x_{n-1} - u_{n-1}$. Therefore by replacing x_{n-1} by the new variable $x_{n-1} - u_{n-1} = z$, $dx_{n-1} = dz$, we will obtain an integral in which neither the integrand nor the limits of integration $(-\infty, \infty)$ depend on u_{n-1}.

Hence, only α_{n-1} in expression (6.81) depends on u_{n-1}. This means that, as should be expected, the operating risk for $t = n - 1$ depends on the control action u_{n-1} while the probing risk is independent of u_{n-1}, although different from zero. Such systems in which the process of studying the object takes place identically with any control have been called neutral [6.1]. The system just under consideration is neutral. The physical meaning of this phenomenon for this example is explained below.

Minimizing α_{n-1} with respect to u_{n-1}, we find the minimum condition in the form of equating the parentheses in (6.77) to zero:

$$w_{n-1}\epsilon_{n-1} + (\Sigma_{n-2}/2\sigma_g^2) = 0; \tag{6.82}$$

hence

$$u_{n-1}^* = x_{n-1}^* - \frac{\Sigma_{i=0}^{n-2} (x_i - u_i)}{2\sigma_g^2 \epsilon_{n-1}}. \tag{6.83}$$

Reasoning in an entirely analogous fashion, we find that generally γ_s^* also does not depend on w_s and depends only on the difference $x_{s-1} - u_{s-1}$. Therefore the substitution $z = x_{s-1} - u_{s-1}$ eliminates the dependence on u_{s-1} in the integral $\int \gamma_s^* \, dx_{s-1}$. Hence for any $s - 1$ only the first term of α_{s-1} depends on u_{s-1} in the formula for γ_{s-1}.

The system is neutral at each time. Hence it follows that the formula for determining u_s has the same form as both (6.79) and (6.83) for any s:

$$u_s^* = x_s^* - \frac{\sum_{i=0}^{s-1}(x_i - u_i)}{2\sigma_g^2\epsilon_s} \qquad (s = 0, 1, ..., n). \tag{6.84}$$

Substituting the value of ϵ_s from (6.76), we arrive at the following optimal control law:

$$u_s^* = x_s^* - \frac{\sum_{i=0}^{s-1}(x_i - u_i)}{s + (\sigma_g/\sigma_\mu)^2} \qquad (s = 1, 2, ..., n). \tag{6.85}$$

For $s = 0$ we obtain $u_0^* = x_0^*$.

Formula (6.85) has a simple meaning. If the value of μ were known, then it would be required to set the variable $u_s = x_s^* - \mu$ in order to satisfy the ideal condition $x_s = x_s^*$ (in the absence of the noise g_s). Hence the second term in (6.85) gives an estimate of the variable μ. If the noise were absent ($g_s = 0$), then with $u_s = v_s$ the value of μ could be determined easily: $\mu = x_i - u_i$. But from Fig. 6.4 it is seen that

$$x_i - u_i = (\mu + v_i) - u_i = \mu + g_i . \tag{6.86}$$

Hence the difference $x_i - u_i$ gives the value of μ measured with an error g_i. Thus the estimate of μ is constructed from the several measured values of the differences $x_i - u_i$ for $i = 0, 1, ..., s - 1$. For sufficiently large values of s the second term of expression (6.85) acquires the form of the arithmetic mean of the values $x_i - u_i$ for $i = 0, 1, ..., s - 1$, which corresponds to the intuitive idea.

Let us turn our attention to the similarity between formula (6.85) and expression (5.225). If $\mu_0 = 0$ is set, and this is just what was assumed in the solution of the latter problem, then these expressions become identical. The reason is that the circuit of Fig. 6.4 is reducible. It reduces to an open-loop system of the type depicted in Fig. 5.23. In fact, from formula (6.86) it follows that the measurement of x_s and u_s is equivalent to the measurement of the noise $z_s = \mu$ with an error g_s. Therefore under the condition $e_s = g_s$, i.e., $\sigma_e = \sigma_g$, the circuits depicted in Figs. 5.23 and 6.4 are equivalent from a definite point of view. The information brought to the controller is the same in them.

The example considered above was given only for illustrating general formulas. Of course, it would be much simpler to first transform the circuit to the open-loop equivalent, and then use the far simpler formulas of the previous chapter. However, in the first place the example examined above shows how the general formulas of this chapter are used. Second, it shows that the same results are obtained from these formulas as with the aid of other methods, which confirms the correctness of the formulas. In addition, it should be noted that a reducible circuit need not always be transformed first into an open-loop one, and afterwards investigated. Sometimes the reduction process itself turns out to be cumbersome. And finally the essential one is that there exists a class of irreducible systems, whose investigation is possible only with the use of the formulas obtained in this chapter.

The example given above shows that in a closed-loop system the rate of information storage about the object can turn out to be independent of the values of the control actions. Systems of such a kind have been called "neutral." For example, the system depicted in Fig. 6.4 was found to be neutral. Of course, all reducible systems are neutral, since in the open-loop systems equivalent to them the rate of information storage about the object does not depend on the strategy of the controller.

The extrapolation of the results of the example given above to the case of a continuous system is of interest. Let the time length be

$$\Delta t = t/s;$$

maintaining $t =$ const, we let s tend to infinity. Then Δt tends to zero. We will assume that the problem is posed of synthesizing a control system for the case when $g(t)$ is stationary white noise with spectral density S_0. In Chapter V it was shown that the sequence of independent normal variables g_s will turn out to be such noise in the limit, if we set

$$\sigma_g^2 = S_0/\Delta t. \tag{6.87}$$

Let us now rewrite formula (6.85), with (6.87) taken into account, in the following manner:

$$u_s^* = x_s^* - \frac{\sum_{i=0}^{s-1} (x_i - u_i)\,\Delta t}{s\,\Delta t + (\sigma_g^2\,\Delta t/\sigma_\mu^2)}. \tag{6.88}$$

Since according to (6.87) the quantity $\sigma_g^2\,\Delta t = S_0$, and also $s\,\Delta t = t$, then from (6.88) in the limit as $\Delta t \to 0$ the expression

$$u^*(t) = x^*(t) - \frac{\int_0^t (x - u)\,dt}{t + (S_0/\sigma_\mu^2)} = x^*(t) - \frac{\int_0^t (x - u)\,dt}{t + a} \tag{6.89}$$

can be obtained, where $a = S_0/\sigma_\mu^2$. The second term of this expression can be obtained at the output of a filter with a transient response ($u_{\text{in}} = 1$ for $t > 0$):

$$u_{\text{out}} = \frac{1}{t+a}\int_0^t u_{\text{in}}\, dt = \frac{t}{t+a}, \tag{6.90}$$

if the difference $x - u$ is fed to the input of the filter. The transient response (6.90) is not exponential; but it can be approximately replaced by an exponential with a time constant $\tau \approx 1.1a$. Then the system, close to the optimal, takes the form shown in Fig. 6.5a. By transforming this circuit, we obtain a series of schemes equivalent to it, depicted in Fig. 6.5b and c.

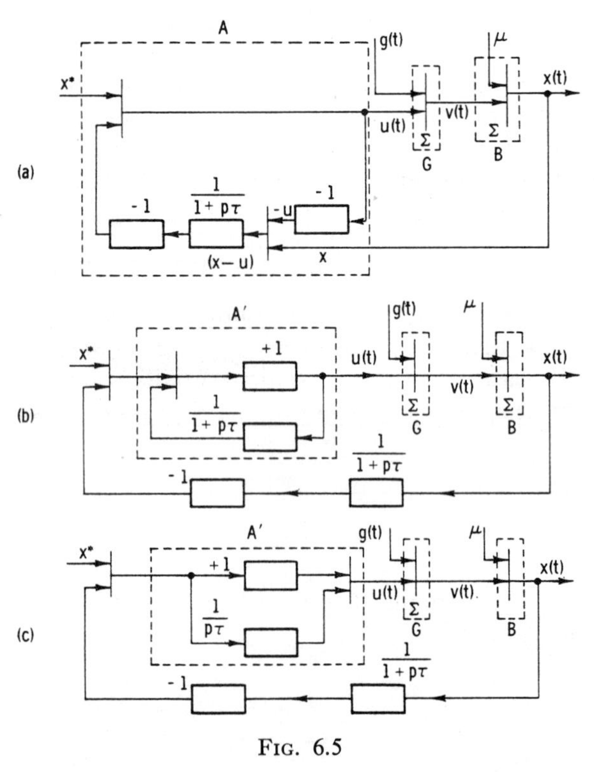

FIG. 6.5

Let us consider a somewhat more complicated example, when information storage about the driving action x_s^* and the noise z_s, acting on the object B, takes place simultaneously. Figure 6.6 shows a circuit in which the blocks G and B are the same as in Fig. 6.4, but the driving action x_s^* enters the controller A through the channel H^* where it is mixed with the noise h_s^*. The equations of the channels H^* and G respectively have the form

$$y_s^* = x_s^* + h_s^* \tag{6.91}$$

and

$$v_s = u_s + g_s\,. \tag{6.92}$$

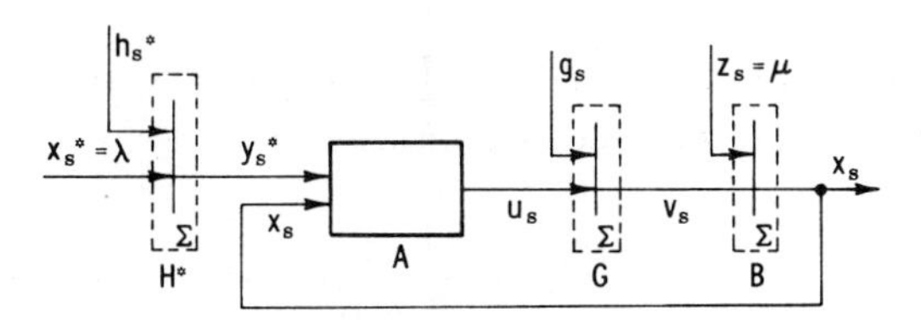

FIG. 6.6

The equation of the object B has the form

$$x_s = v_s + \mu = u_s + g_s + \mu. \tag{6.93}$$

Let

$$x_s^* = \lambda = \text{const}, \qquad z_s = \mu = \text{const}, \tag{6.94}$$

where the random variables λ, μ, g_s, and h_s^* have normal distribution laws:

$$\begin{gathered}
P(\mu) = \frac{1}{\sigma_\mu \sqrt{(2\pi)}} \exp\left\{-\frac{\mu^2}{2\sigma_\mu^2}\right\}, \\
P(\lambda) = \frac{1}{\sigma_\lambda \sqrt{(2\pi)}} \exp\left\{-\frac{(\lambda-\lambda_0)^2}{2\sigma_\lambda^2}\right\}, \\
P(g_s) = q(g_s) = \frac{1}{\sigma_g \sqrt{(2\pi)}} \exp\left\{-\frac{g_s^2}{2\sigma_g^2}\right\}, \\
P(h_s^*) = \rho(h_s^*) = \frac{1}{\sigma_h \sqrt{(2\pi)}} \exp\left\{-\frac{h_s^{*2}}{2\sigma_h^2}\right\}.
\end{gathered} \tag{6.95}$$

The specific loss function is given by the expression

$$W_s = (x_s - x_s^*)^2 = (x_s - \lambda)^2. \tag{6.96}$$

We will find the optimal controller A, for which in accordance with (6.36) we will compute the function

$$\alpha_k = \int_{\Omega(\lambda,\mu,x_k)} (x_k - \lambda)^2 P(\lambda) \left[\prod_{i=0}^{k} P(y_i^* \mid \lambda)\right] \cdot P(\mu) \times \left[\prod_{i=0}^{k} P(x_i \mid \mu, u_i)\right] d\Omega \tag{6.97}$$

as a preliminary. In this problem $y_i = x_i$, and in the same way as in formula (6.67) in the previous example,

$$P(x_i \mid \mu, u_i) = q(x_i - u_i - \mu) = \frac{1}{\sigma_g \sqrt{(2\pi)}} \exp\left\{-\frac{(x_i - u_i - \mu)^2}{2\sigma_g^2}\right\}. \tag{6.98}$$

From (6.91) it follows that

$$h_s^* = y_s^* - x_s^* = y_s^* - \lambda. \tag{6.99}$$

Therefore the probability that the output variable of the channel H^* will be found in the interval between the fixed values y_s^* and $y_s^* + dy_s^*$, is equal to the probability that the noise in the channel H^* assumes a value lying in the interval between the values $y_s^* - \lambda$ and $y_s^* - \lambda + dy_s^*$.

Hence it follows that

$$P(y_i^* \mid \lambda) = \rho(y_i^* - \lambda) = \frac{1}{\sigma_h \sqrt{(2\pi)}} \exp\left\{-\frac{(y_i^* - \lambda)^2}{2\sigma_h^2}\right\}. \tag{6.100}$$

Substituting (6.98) and (6.100) into (6.97), we find

$$\alpha_k = \int_{\Omega(\lambda,\mu)} P(\lambda) \cdot P(\mu) \left[\prod_{i=0}^{k} \rho(y_i^* - \lambda)\right] \cdot \left[\prod_{i=0}^{k-1} q(x_i - u_i - \mu)\right] \times \left[\int_{\Omega(x_k)} (x_k - \lambda)^2 q(x_k - u_k - \mu)\, d\Omega(x_k)\right] d\Omega(\lambda, \mu). \tag{6.101}$$

The integral in the brackets of expression (6.101) can be calculated. It is equal to

$$I_k = \int_{-\infty}^{\infty} (x_k - \lambda)^2 q(x_k - u_k - \mu)\, dx_k = \sigma_g \sqrt{(2\pi)}[\sigma_g^2 + (\lambda - u_k - \mu)^2]. \tag{6.102}$$

This integral is computed in the same way as the integral (6.73). Therefore according to (6.101) and (6.102),

$$\alpha_k = \int_{-\infty}^{\infty}\int_{-\infty}^{\infty} \sigma_g \sqrt{(2\pi)}[\sigma_g^2 + (\lambda - u_k - \mu)^2]P(\lambda)\cdot P(\mu)$$

$$\times \left[\prod_{i=0}^{k} \rho(y_i^* - \lambda)\right]\left[\prod_{i=0}^{k-1} q(x_i - u_i - \mu)\right] d\mu\, d\lambda$$

$$= \int_{-\infty}^{\infty} P(\lambda)\left[\prod_{i=0}^{k} \rho(y_i^* - \lambda)\right]\left\{a_k + b_k + \frac{2b_k}{\epsilon_k}\left(w_k\epsilon_k\right.\right.$$

$$\left.\left. + \frac{\Sigma_{k-1}}{2\sigma_g^2}\right)^2\right\} \exp\left\{-\frac{\theta_{k-1}}{2\sigma_g^2} + \frac{\Sigma_{k-1}^2}{4\sigma_g^4\epsilon_k}\right\} d\lambda. \qquad (6.103)$$

Here the same notations are used as in formulas (6.76) and (6.77).

In the formula for α_n we perform the minimization with respect to the variable u_n, which is contained only in the term w_n. In this case both u_n^* and the quantity $\gamma_n^* = \alpha_n^* = \min \alpha_n$ are found to be dependent on the variables $u_0, u_1, \ldots, u_{n-1}$, where u_{n-1} is contained in the terms Σ_{n-1} and θ_{n-1}. But the dependence on u_{n-1} in these terms represents a dependence on the difference $x_{n-1} - u_{n-1} = \nu_{n-1}$. Hence u_n^* and γ_n^* depend on just this difference. For the subsequent integration with respect to x_{n-1} this variable can be replaced by another variable ν_{n-1}. Then the dependence on u_{n-1} vanishes in the integral

$$\int_{-\infty}^{\infty} \gamma_n^*(x_{n-1} - u_{n-1})\, dx_{n-1} = \int_{-\infty}^{\infty} \gamma_n^*(\nu_{n-1})\, d\nu_{n-1}. \qquad (6.104)$$

Hence in the expression (6.60) for γ_{n-1} the second term does not depend on u_{n-1}; in this case only α_{n-1} depends on u_{n-1}. Therefore instead of minimizing γ_{n-1} with respect to u_{n-1}, it suffices to carry out the minimization of α_{n-1} with respect to u_{n-1}; as a result u_{n-1}^* is found. Continuing this reasoning further, we can conclude that any u_s^* can be found as a minimization of α_s with respect to u_s. From formula (6.103) for α_s it suffices to examine only that part of it which depends on $w_s = u_s - \lambda$, and that is,

$$J_s = \int_{-\infty}^{\infty} [u_s\epsilon_s + (\Sigma_{s-1}/2\sigma_g^2) - \epsilon_s\lambda]^2 P(\lambda)\left[\prod_{i=0}^{s} \rho(y_i^* - \lambda)\right] d\lambda. \quad (6.105)$$

We will explain how this integral was obtained. Certainly the exponential in (6.103) can be taken out of the integral sign, since it does not depend on λ. In general it can be left out of consideration, since this does not affect the result of minimization with respect to u_s. Further, the term of formula (6.103) containing $a_s + b_s$ also can be left out of consideration, since here there is no dependence on u_s. Hence only the integral (6.105) remains to be considered.

Let us set

$$J_{m,s} = \int_{-\infty}^{\infty} \lambda^m P(\lambda) \left[\prod_{i=0}^{s} p(y_i^* - \lambda)\right] d\lambda. \tag{6.106}$$

Then

$$J_s = (u_s\epsilon_s + \Sigma_{s-1}/2\sigma_g^2) J_{0,s} - 2\epsilon_s(u_s\epsilon_s + \Sigma_{s-1}/2\sigma_g^2) J_{1,s} + \epsilon_s^2 J_{2,s}. \tag{6.107}$$

From the condition

$$\partial J_s / \partial u_s = 0 \tag{6.108}$$

we find the value u_s^* minimizing J_s:

$$u_s^* = \frac{J_{1,s}}{J_{0,s}} - \frac{\Sigma_{s-1}}{2\epsilon_s\sigma_g^2} = \frac{J_{1,s}}{J_{0,s}} - \frac{\Sigma_{i=0}^{s-1}(x_i - u_i)}{s + (\sigma_g/\sigma_\mu)^2}. \tag{6.109}$$

Dividing the numerator and denominator of the first term of formula (6.109) by $P(\mathbf{y}_s^*)$ [compare with (6.21)], we are easily able to discover that this term is the mathematical expectation of the quantity λ at the time instant $t = s$:

$$\frac{J_{1,s}}{J_{0,s}} = \frac{\int_{-\infty}^{\infty} \lambda P_s(\lambda)\, d\lambda}{\int_{-\infty}^{\infty} P_s(\lambda)\, d\lambda} = \int_{-\infty}^{\infty} \lambda P_s(\lambda)\, d\lambda = M\{\lambda \mid \mathbf{y}_s^*\}. \tag{6.110}$$

Carrying out the integration with the aid of formulas (6.70) and (6.75), we find

$$M\{\lambda \mid \mathbf{y}_s^*\} = \frac{J_{1,s}}{J_{0,s}} = \frac{\lambda_0}{1 + (\sigma_\lambda/\sigma_h)^2(s+1)} + \frac{\Sigma_{r=0}^{s} y_r^*}{(\sigma_h/\sigma_\lambda)^2 + s + 1}. \tag{6.111}$$

Substituting (6.111) into (6.109), we arrive at the final formula determining the optimal strategy:

$$u_s^* = \frac{\lambda_0}{1 + (\sigma_\lambda/\sigma_h)^2(s+1)} + \frac{\Sigma_{r=0}^{s} y_r^*}{(\sigma_h/\sigma_\lambda)^2 + (s+1)} - \frac{\Sigma_{i=0}^{s-1}(x_i - u_i)}{s + (\sigma_g/\sigma_\mu)^2}. \tag{6.112}$$

If the noises g_s and h_s^* were absent, then for ensuring the ideal condition $x_s = x_s^*$ it would be necessary to set the value $u_s = x_s^* - \mu = \lambda - \mu$.

In formula (6.112) the first two terms give the estimate of λ, and the last one gives the estimate of μ. For small values of s the *a priori* mean value λ_0 can have an important influence on the estimate of λ. But for large s the first term now does not play a part, while the second approaches the arithmetic mean of the values y_i^*. In the same way as in (6.85), for large values of s the last term of formula (6.112) gives the estimate of μ in the form of the arithmetic mean of the values $x_i - u_i$.

When the dispersion σ_h^2 of the noise h_s^* in the channel H^* tends to zero, then the first term in expression (6.112) vanishes, and the second one assumes the form of the arithmetic mean of the constant $y_i^* = x_s^* = \lambda$, i.e., it becomes the constant x_s^*. In this case formula (6.112) turns into (6.85).

3. Examples of Irreducible Systems

Up to this point we have considered only examples of reducible systems, which naturally turned out to be neutral. But an important class of systems is not neutral. Systems that are not neutral are certainly irreducible. In fact, if the system were reducible, then according to what was presented above it could be replaced by an open-loop system and an open-loop system is neutral. Hence it suffices to show that the system is not neutral; by the same token it will be shown that it is irreducible.

By the simplest examples it is easy to be convinced that systems which are not neutral and hence irreducible exist. For example, let us consider the inertialess object B, for which the output variable x_s is related to the input v_s by the dependence

$$x_s = \mu_1 v_s + \mu_2 , \tag{6.113}$$

where μ_1 and μ_2 are random variables with a prescribed joint *a priori* probability density. A rough graph of the dependence (6.113) is shown in Fig. 6.7a. Let the specific loss function have the form

$$W_s = (x_s - x_s^*)^2 \tag{6.114}$$

and $x_s^* \neq 0$. Then for setting the best value of v_s it is necessary to have information about the variables μ_1 and μ_2. From an examination of the graph of Fig. 6.7a it turns out that there exist both the best and worst sequences of values of v_s, providing the best or worst study of the object. For example, if for any s the same value $v_s = v^1$ is set, then we will only obtain information about a single point M of the object characteristic, which does not give the possibility of determining the slope of the characteristic. But if various values of v_s are set, then in principle the values of μ_1 and μ_2 can be ascertained completely.

Let the admissible values of v_s be in the interval $|v_s| \leqslant 1$, and x_s be measured with a random error h_s. Then setting the values v^1 and v^2 will be unsuccessful, since the ordinates of the points M_1 and M_2 measured with error can cause a large error in the determination of the parameters μ_1 and μ_2 (the dotted line passing through the points N_1 and N_2). In this case intuition suggests that the points v^1 and v^2 possibly must be spread farther apart, placing them at the edges of the admissible range. Then the same errors in the measurements of the values x_s will give smaller errors in the determination of the parameters (the dashed line $N_1'N_2'$), and the process of studying the object for the same number of times will prove to be more successful.

An analogous situation exists with the characteristic of the form

$$x_s = \mu_1 v_s^2 + \mu_2 v_s + \mu_3 . \tag{6.115}$$

This characteristic is shown in Fig. 6.7b. In this case even an exact determination of the two points M_1 and M_2 of the characteristic still

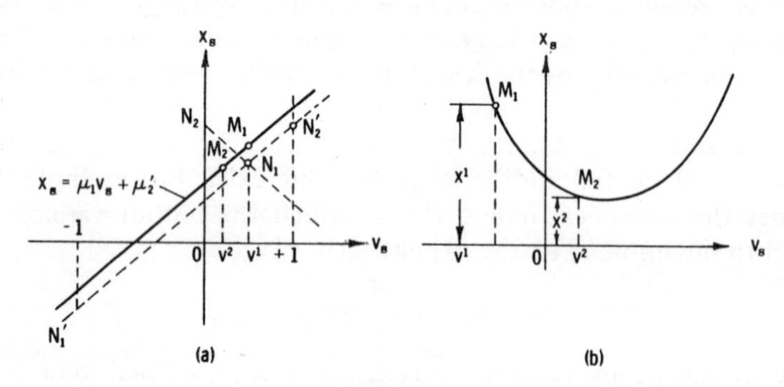

FIG. 6.7

does not permit it to be ascertained completely. Let it be required to establish the minimum x_s; then the specific loss function W_s has the form

$$W_s = x_s . \tag{6.116}$$

The coordinate v_s of the minimum x_s for $\mu_1 > 0$ has the value

$$v_s^* = -\mu_2 / 2\mu_1 . \tag{6.117}$$

But for determining μ_1 and μ_2 it is necessary to solve a system of three linear equations (two equations are insufficient since after eliminating μ_3 one equation in all is obtained), obtained as a result of three measurements, with abscissas v^1, v^2, v^3 and ordinates x^1, x^2, x^3:

$$\begin{aligned} x^1 &= \mu_1(v^1)^2 + \mu_2 v^1 + \mu_3 , \\ x^2 &= \mu_1(v^2)^2 + \mu_2 v^2 + \mu_3 , \\ x^3 &= \mu_1(v^3)^2 + \mu_2 v^3 + \mu_3 . \end{aligned} \tag{6.118}$$

Hence the strategy consisting of a sequence of identical values of v_s or even of setting one of the two possible values of v_s several times will be bad. The object will be studied well only when the number of possible values of v_s is larger than two, and these values are arranged in the most successful way. From this simple argument it now follows that in this case it suffices to give values of v_s on the boundaries of the admissible range.

For example, let us examine the circuit depicted in Fig. 6.8. Let the equation of the object B have the form

$$x_s = \mu u_s , \tag{6.119}$$

where μ is a normal random variable. The feedback loop passes through the channel H with noise h_s. The equation of the channel is

$$y_s = x_s + h_s . \tag{6.120}$$

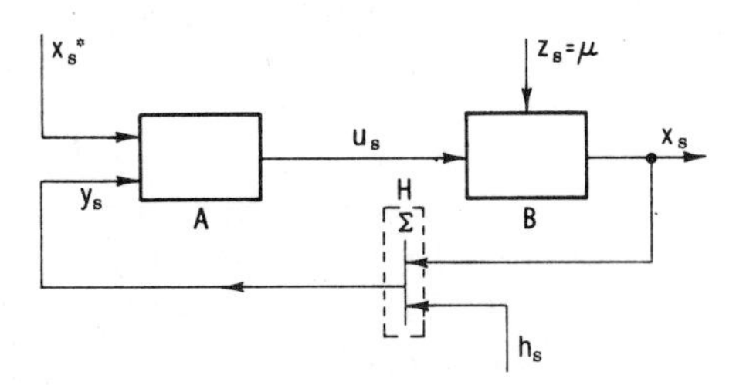

FIG. 6.8

Let h_s be a sequence of independent normal random variables with the same probability density

$$P(h_s) = q(h_s) = \frac{1}{\sigma_h \sqrt{(2\pi)}} \exp\left\{-\frac{h_s^2}{2\sigma_h^2}\right\}. \tag{6.121}$$

The *a priori* probability density of μ is expressed by the formula

$$P(\mu) = \frac{1}{\sigma_\mu \sqrt{(2\pi)}} \exp\left\{-\frac{(\mu-\mu_0)^2}{2\sigma_\mu^{\ 2}}\right\}. \tag{6.122}$$

Let the specific loss function be given by the expression

$$W_s = (x_s - x_s^*)^2, \tag{6.123}$$

where x_s^* is a prescribed sequence of variables ($s = 0, 1, \ldots, n$). It is required to find the optimal strategy of the controller A minimizing the risk R, where

$$R = M\left\{\sum_{s=0}^{n} W_s\right\} = \sum_{s=0}^{n} R_s\,. \tag{6.124}$$

In Fig. 6.9 the characteristic $x_s = \mu u_s$ of the object B is shown. The characteristic is the straight line OM_0 with unknown slope μ. If the value of μ were known, then the loss function W_s could be reduced to zero by setting $x_s = \mu u_s = x_s^*$, from which it follows that the required value u_0' would be equal to x_s^*/μ. But the value of μ is unknown; it can be determined approximately by feeding the values u_i to the input of the object B and measuring the corresponding quantities y_i at the output of the channel H. Let y_i be the ordinate of the point N_i not equal to the actual value x_s—the ordinate of the point M_i. For the same mean square deviation of N_i from M_i, the value of μ, and with it the required value u_0' as well will be found, the more precisely the larger u_i is chosen in absolute value. But if u_i increases significantly, then the difference $x_s - x_s^*$ grows, and with it the magnitude of the loss function W_s as well, in accordance with (6.123). Hence the point M_i must not be too far from M_0. For small i, apparently larger values of u_i are rational since the probing risk has a big influence, and a significant part of the total risk depends on how well the object will be studied at the first times. As i increases the part of the probing risk in the total value R drops; therefore it is rational to reduce the quantity u_i, nevertheless leaving its value larger than the expected value u_0'.

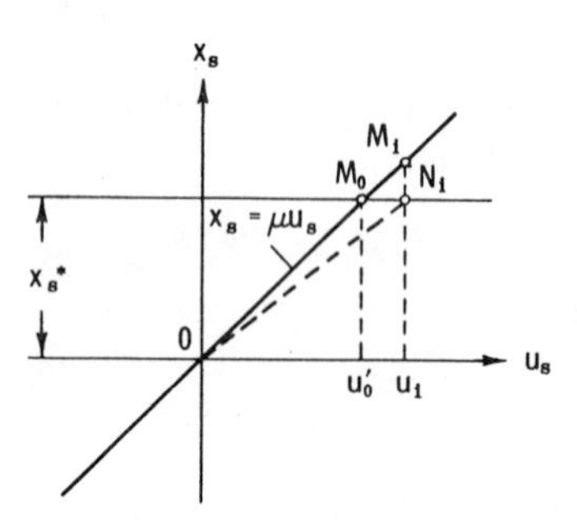

FIG. 6.9

To solve the problem the function α_k must first be determined from formula (6.58). Since $x_k = \mu u_k$, then by making this substitution we obtain α_k in the form of an integral with respect to μ only. Since $P(x_k \mid \mu, u_k) = \delta(x_k - \mu u_k)$, we find

$$\begin{aligned}\alpha_k &= \alpha_k(\mathbf{u}_k, \mathbf{y}_{k-1}) \\ &= \int_{\Omega(\mu, x_k)} W_k(x_k^*, x_k) P(x_k \mid \mu, u_k) P(\mu) \left[\prod_{i=0}^{k-1} P(y_i \mid \mu, u_i)\right] d\Omega \\ &= \int_{\mu=-\infty}^{\infty} (\mu u_k - x_k^*)^2 P(\mu) \left[\prod_{i=0}^{k-1} \frac{1}{\sigma_h \sqrt{(2\pi)}} \exp\left\{-\frac{(y_i - \mu u_i)^2}{2\sigma_h^2}\right\}\right] d\mu. \end{aligned} \tag{6.125}$$

Discarding the unessential constant factor and denoting proportionality by the symbol $\equiv$, and then substituting $P(\mu)$ from (6.122), we arrive at the expression

$$\begin{aligned}\alpha_k &\equiv \int_{-\infty}^{\infty} (\mu u_k - x_k^*)^2 \exp\left\{-\frac{(\mu - \mu_0)^2}{2\sigma_\mu^2} - \frac{1}{2\sigma_h^2} \sum_{i=0}^{k-1} (y_i - \mu u_i)^2\right\} d\mu \\ &= \int_{-\infty}^{\infty} (\mu u_k - x_k^*)^2 \exp\{-A_{k-1}\mu^2 + B_{k-1}\mu - C_{k-1}\} \, d\mu \\ &= \exp\{-C_{k-1}\} \int_{-\infty}^{\infty} (\mu u_k - x_k^*)^2 \exp\{-A_{k-1}\mu^2 + B_{k-1}\mu\} \, d\mu, \end{aligned} \tag{6.126}$$

where

$$\begin{aligned} A_{k-1} &= \frac{1}{2\sigma_\mu^2} + \frac{\sum_{i=0}^{k-1} u_i^2}{2\sigma_h^2} > 0, \\ B_{k-1} &= \frac{\mu_0}{\sigma_\mu^2} + \frac{1}{\sigma_h^2} \sum_{i=0}^{k-1} u_i y_i, \\ C_{k-1} &= \frac{\mu_0^2}{2\sigma_\mu^2} + \frac{1}{2\sigma_h^2} \sum_{i=0}^{k-1} y_i^2 > 0. \end{aligned} \tag{6.127}$$

Let us set

$$\mu u_k - x_k^* = z. \tag{6.128}$$

Then the integral in expression (6.126) takes the form

$$\begin{aligned} I_k &= \frac{1}{u_k} \int_{-\infty}^{\infty} z^2 \exp\left\{-A_{k-1} \frac{(z + x_k^*)^2}{u_k^2} + B_{k-1} \frac{z + x_k^*}{u_k}\right\} dz \\ &= \frac{1}{u_k} \exp\left\{-A_{k-1} \left(\frac{x_k^*}{u_k}\right)^2 + B_{k-1} \left(\frac{x_k^*}{u_k}\right)\right\} \\ &\quad \times \int_{-\infty}^{\infty} z^2 \exp\left\{-\frac{A_{k-1}}{u_k^2} z^2 + \left(\frac{B_{k-1}}{u_k} - \frac{2A_{k-1}x_k^*}{u_k^2}\right) z\right\} dz. \end{aligned} \tag{6.129}$$

With the aid of formula (6.70) this integral can be found in explicit form. Substituting the integral I_k into (6.126), after simple transformations we obtain

$$\alpha_k \equiv \exp\left\{\frac{\sigma_\mu^2 D_{k-1} - \sigma_h^2 E_{k-1}}{4\sigma_\mu^2 \sigma_h^4 A_{k-1}}\right\} \frac{1}{2A_{k-1}} \sqrt{\left(\frac{\pi}{A_{k-1}}\right)} \times \left[u_k^2 + \frac{(2A_{k-1}x_k^* - 2B_{k-1}u_k)^2}{2A_{k-1}}\right]. \tag{6.130}$$

Here we set

$$D_{k-1} = \left(\sum_{i=0}^{k-1} y_i u_i\right)^2 - \left(\sum_{i=0}^{k-1} u_i^2\right)\left(\sum_{i=0}^{k-1} y_i^2\right),$$
$$E_{k-1} = \mu_0^2 \sum_{i=0}^{k-1} u_i^2 - 2\mu_0 \sum_{i=0}^{k-1} y_i u_i + \sum_{i=0}^{k-1} y_i^2. \tag{6.131}$$

The first step in determining the optimal strategy—finding u_n^*—is carried out without difficulty. Since the dependence on u_n is hidden only in the last factor of (6.130)—the brackets—then for obtaining $\gamma_n^* = \alpha_n^* = (\alpha_n)_{\min}$ it is necessary to minimize the expression

$$u_n^2 + \frac{(2A_{n-1}x_n^* - 2B_{n-1}u_n)^2}{2A_{n-1}}. \tag{6.132}$$

Equating the derivative of this expression with respect to u_n to zero, we find the optimal value

$$u_n^* = \frac{2A_{n-1}B_{n-1}x_n^*}{A_{n-1} + 2B_{n-1}^2}. \tag{6.133}$$

Substituting this value into (6.130) for $k = n$, we obtain the optimal value of α_n, denoted by α_n^* or γ_n^*:

$$\gamma_n^* = \alpha_n^* = \min_{u_n} \alpha_n = \frac{(x_n^*)^2 \sqrt{(\pi A_{n-1})}}{(A_{n-1} + 2B_{n-1}^2)} \exp\left\{\frac{\sigma_\mu^2 D_{n-1} - \sigma_h^2 E_{n-1}}{4\sigma_\mu^2 \sigma_h^4 A_{n-1}}\right\}. \tag{6.134}$$

This expression must be integrated with respect to y_{n-1} in the interval $-\infty < y_{n-1} < \infty$ to obtain the second term in the formula for γ_{n-1} (the first term is α_{n-1}). As follows from (6.127), A_{n-1} does not depend on y_{n-1}. But B_{n-1} depends on y_{n-1}, similarly D_{n-1} and E_{n-1}—the latter is seen from (6.131). By considering expression (6.134) together with (6.131) and (6.127) the following features of the problem can be observed.

(a) D_{n-1} and E_{n-1} are quadratic functions of y_{n-1}, and B_{n-1} depends linearly on y_{n-1}. Hence the expression for $\gamma_n{}^*$ is a complicated function of y_{n-1}; there is no similar function in a table of integrals (an exponential of a quadratic function divided by a quadratic function is found under the integral sign). Apparently, in general this integral cannot be represented in the form of a finite combination of parameters. In principle such a difficulty is surmountable if some of the formulas of approximate integration are used. But because of this the approximate solution to the problem becomes much more cumbersome.

(b) Another fact is more important from a fundamental point of view. The dependence on u_{n-1} enters both into A_{n-1} and into B_{n-1}, D_{n-1}, and E_{n-1}, and moreover u_{n-1} does not enter in the form of the difference $y_{n-1} - u_{n-1}$ (then the dependence on u_{n-1} could be suppressed by the substitution $z = y_{n-1} - u_{n-1}$). Thus the result of the integration will depend on u_{n-1}. Hence the system is not neutral, which was predicted above with the aid of elementary considerations. Thus this system is irreducible.

The impossibility of accomplishing the operations of integration and minimization in general form hampers the solution to the problem exceedingly. At first glance it seems that in general this problem is unsolvable even by modern universal digital computers. In fact, $\gamma_n{}^*$ is a function of $\mathbf{u}_{n-1}$ and $\mathbf{y}_{n-1}$, and the integral $\int \gamma_n{}^* \, dy_{n-1}$ is a function of $\mathbf{u}_{n-1}$ and $\mathbf{y}_{n-2}$ (not counting $x_n{}^*$), i.e., a function of $2n - 1$ variables. For large values of n the storage and minimization of a function of such a large number of variables may prove to be unrealizable in practice even in the largest of modern digital computers. But in many specific examples the computation can be organized in such a way that it is necessary to store only functions of a small number of variables. These functions are sometimes called sufficient coordinates, sufficient statistics, or sufficient estimators. Such a designation is used in mathematical statistics [6.8–6.10]. The number of variables on which sufficient estimators depend is not a function of n here.

We will find sufficient estimators for the problem under consideration. Let us set

$$L_k = \sum_{i=0}^{k} u_i^2, \qquad M_k = \sum_{i=0}^{k} u_i y_i \qquad (k = 0, 1, ..., n-1). \tag{6.135}$$

As is shown below, in this example the quantities L_{k-1} and M_{k-1} are sufficient estimators. They have the following property:

$$L_k = L_{k-1} + u_k^2, \qquad M_k = M_{k-1} + u_k y_k . \tag{6.136}$$

The quantities A_{k-1} and B_{k-1} are easily expressed in terms of L_{k-1} and M_{k-1}:

$$A_{k-1} = \frac{1}{2\sigma_\mu^2} + \frac{L_{k-1}}{2\sigma_h^2}, \qquad B_{k-1} = \frac{\mu_0}{\sigma_\mu^2} + \frac{M_{k-1}}{\sigma_h^2}. \tag{6.137}$$

Therefore in accordance with (6.126) α_k can also be represented in the form of a function of the variables L_{k-1}, M_{k-1}, C_{k-1}, u_k (the remaining quantities, as known parameters, are not written out as arguments). Carrying out the integration with respect to μ in (6.126), we can find α_k in the following form:

$$\alpha_k = \exp\{-C_{k-1}\}[u_k^2\varphi_{2k}(L_{k-1}, M_{k-1}) - u_k x_k^* \varphi_{1k}(L_{k-1}, M_{k-1}) + (x_k^*)^2\varphi_{0k}(L_{k-1}, M_{k-1})], \tag{6.138}$$

where

$$\varphi_{2k}(L_{k-1}, M_{k-1}) = \int_{-\infty}^{\infty} \mu^2 \exp\{-A_{k-1}\mu^2 + B_{k-1}\mu\}\, d\mu,$$

$$\varphi_{1k}(L_{k-1}, M_{k-1}) = \int_{-\infty}^{\infty} \mu \exp\{-A_{k-1}\mu^2 + B_{k-1}\mu\}\, d\mu, \tag{6.139}$$

$$\varphi_{0k}(L_{k-1}, M_{k-1}) = \int_{-\infty}^{\infty} \exp\{-A_{k-1}\mu^2 + B_{k-1}\mu\}\, d\mu.$$

From (6.138), with expression (6.127) for C_{k-1} taken into account, it follows that

$$\alpha_k \equiv \exp\left\{\frac{1}{2\sigma_h^2}\sum_{i=0}^{k-1} y_i^2\right\}[u_k^2\varphi_{2k}(L_{k-1}, M_{k-1}) - u_k x_k^*\varphi_{1k}(L_{k-1}, M_{k-1}) + (x_k^*)^2\varphi_{0k}(L_{k-1}, M_{k-1})], \tag{6.140}$$

where the sign $\equiv$ denotes proportionality.

Now the procedure can be carried out for determining the optimal strategy by storing only a function of two variables at each step. Actually, we will first write α_n in the form

$$\begin{aligned}\alpha_n &\equiv \exp\left\{\frac{1}{2\sigma_h^2}\sum_{i=0}^{n-1} y_i^2\right\}[u_n^2\varphi_{2n}(L_{n-1}, M_{n-1}) \\ &\qquad - u_n x_n^*\varphi_{1n}(L_{n-1}, M_{n-1}) + (x_n^*)^2\varphi_{0n}(L_{n-1}, M_{n-1})] \\ &= \exp\left\{\frac{1}{2\sigma_h^2}\sum_{i=0}^{n-1} y_i^2\right\}\psi(u_n, L_{n-1}, M_{n-1}).\end{aligned} \tag{6.141}$$

Here the expression in the second brackets is denoted by $\psi(u_n, L_{n-1}, M_{n-1})$. The parameter x_n^* is not written out in explicit form. Minimizing α_n with respect to u_n, we find u_n^* and α_n^*:

$$u_n^* = u_n^*(L_{n-1}, M_{n-1}) \tag{6.142}$$

and

$$\alpha_n^* \equiv \exp\left\{\frac{1}{2\sigma_h^2}\sum_{i=0}^{n-1} y_i^2\right\}\psi^*(L_{n-1}, M_{n-1}), \tag{6.143}$$

where

$$\psi^*(L_{n-1}, M_{n-1}) = \min_{u_n \in \Omega(u_n)} \psi(u_n, L_{n-1}, M_{n-1}). \tag{6.144}$$

Then we find the function

$$\begin{aligned}
\gamma_{n-1} &= \alpha_{n-1} + \int_{-\infty}^{\infty} \gamma_n^*\, dy_{n-1} \\
&\equiv \exp\left\{\frac{1}{2\sigma_h^2}\sum_{i=0}^{n-2} y_i^2\right\}\psi(u_{n-1}, L_{n-2}, M_{n-2}) \\
&\quad + \int_{-\infty}^{\infty} \exp\left\{\frac{1}{2\sigma_h^2}\sum_{i=0}^{n-1} y_i^2\right\}\psi^*(L_{n-1}, M_{n-1})\, dy_{n-1} \\
&= \exp\left\{\frac{1}{2\sigma_h^2}\sum_{i=0}^{n-2} y_i^2\right\}\Bigg\{\psi(u_{n-1}, L_{n-2}, M_{n-2}) \\
&\quad + \int_{-\infty}^{\infty} \exp\left\{\frac{1}{2\sigma_h^2} y_{n-1}^2\right\}\psi^*(L_{n-2} + u_{n-1}^2, M_{n-2} \\
&\quad + u_{n-1}y_{n-1})\, dy_{n-1}\Bigg\}.
\end{aligned} \tag{6.145}$$

The latter transformation was done with (6.136) taken into account. We will set

$$\begin{aligned}
\varphi(u_k, L_{k-1}, M_{k-1}) &= \psi(u_k, L_{k-1}, M_{k-1}) \\
&\quad + \int_{-\infty}^{\infty} \exp\left\{\frac{1}{2\sigma_h^2} y_k^2\right\}\psi^*(L_{k-1} + u_k^2, M_{k-1} + u_k y_k)\, dy_k.
\end{aligned} \tag{6.146}$$

By knowing the function ψ^* of two variables and carrying out the integration by any approximate method, the integral on the right side of (6.146) can also be found in the form of a function of two variables. Since ψ is a simple combination of functions of two variables, then the function φ is determined by assigning functions of two variables. Thus,

$$\gamma_{n-1} = \exp\left\{\frac{1}{2\sigma_h^2}\sum_{i=0}^{n-2} y_i^2\right\}\varphi(u_{n-1}, L_{n-2}, M_{n-2}). \tag{6.147}$$

We find the minimum of γ_{n-1} with respect to u_{n-1}. Let us set

$$\varphi^*(L_{k-1}, M_{k-1}) = \min_{u_k\in\Omega(u_k)} \varphi(u_k, L_{k-1}, M_{k-1}). \tag{6.148}$$

Then evidently u_{n-1}^* will be found in the form

$$u_{n-1}^* = u_{n-1}^*(L_{n-2}, M_{n-2}) \tag{6.149}$$

by the minimization of φ. An analogous procedure permits an arbitrary

$$u_{n-k}^* = u_{n-k}^*(L_{n-k-1}, M_{n-k-1}) \tag{6.150}$$

to be determined as well by the minimization of $\varphi(u_{n-k}, L_{n-k-1}, M_{n-k-1})$. We will note that storage of functions of not more than two variables is required in the course of the entire procedure.

Let us consider still another example, in which an inertialess object has a quadratic characteristic (Fig. 6.10). The equations of the system have the form

$$x_s = \eta_s^2 = (u_s + \mu)^2, \qquad y_s = x_s + h_s. \tag{6.151}$$

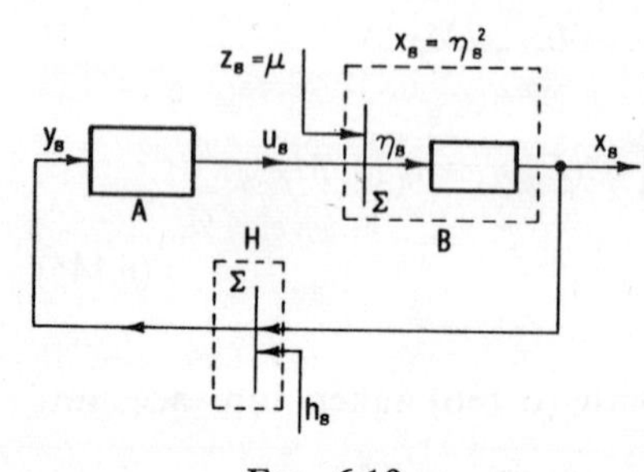

FIG. 6.10

Here μ is a random variable with *a priori* probability density $P_0(\mu)$. The noise h_s has a normal distribution with zero mean value and dispersion σ_h^2. The variable x_s^* is missing in explicit form (for example, we can set $x_s^* = 0$). All the values of W_s are equal to zero except the last one:

$$W_n = x_n. \tag{6.152}$$

Let the function $P_0(\mu)$ differ from zero only for $|\mu| \leqslant 1$. Analogously we assume that the admissible values are also limited by the condition $|u_s| \leqslant 1$.

The problem consists of determining the optimal algorithm of the controller satisfying the condition:

$$R = M\{W_n\} = M\{x_n\} = \min. \tag{6.153}$$

We will clarify the meaning of this problem. We will assume that there is an object with the parabolic characteristic

$$x = (u + \mu)^2, \tag{6.154}$$

where the value of μ is unknown, and the output variable x of the object is measured with a random error h_s. First for $i = 0, 1, ..., n - 1$ let trial values of the input variable u_i of the object be set and the corresponding quantities $y_i = x_i + h_i$ be measured. Then for $i = n$ a value u_n must be set such that the mathematical expectation of the output x_n of the object is minimal. The solution to this problem gives the optimal method of searching for the minimum of a parabolic function of the form (6.154) after $n + 1$ times $t = 0, 1, ..., n$. This method can underlie the construction of an optimal automatic search system.

For this problem the probability that the value of the output variable of the controller A lies in the range from y_i to $y_i + dy_i$ is equal to the probability that the quantity h_s is in the range from $y_i - x_i$ to $(y_i - x_i) + dy_i$. The probability density of the noise is expressed by the formula

$$P(h_i) = \frac{1}{\sigma_h \sqrt{(2\pi)}} \exp\left\{-\frac{h_i^2}{2\sigma_h^2}\right\}, \tag{6.155}$$

then

$$\begin{aligned} P(y_i \mid \mu, u_i) &= \frac{1}{\sigma_h \sqrt{(2\pi)}} \exp\left\{-\frac{(y_i - x_i)^2}{2\sigma_h^2}\right\} \\ &= \frac{1}{\sigma_h \sqrt{(2\pi)}} \exp\left\{-\frac{1}{2\sigma_h^2}[y_i - (u_i + \mu)^2]^2\right\} \\ &= \frac{1}{\sigma_h \sqrt{(2\pi)}} \exp\{a_i + b_i\mu + c_i\mu^2 + d_i\mu^3 + \mu^4\}, \end{aligned} \tag{6.156}$$

where

$$\begin{aligned} a_i &= -\frac{(y_i - u_i^2)^2}{2\sigma_h^2}, & b_i &= -\frac{2u_i(u_i^2 - y_i)}{\sigma_h^2}, \\ c_i &= -\frac{1}{\sigma_h^2}(3u_i^2 - y_i), & d_i &= -\frac{2u_i}{\sigma_h^2}. \end{aligned} \tag{6.157}$$

Since all the W_k are equal to zero for $k < n$, then also all $\alpha_k = 0$ for $k < n$. It is necessary to find only the function α_n from formula (6.58). Taking into account that in this problem

$$P(x_k \mid \mu, u_k) = \delta[x_k - (u_k + \mu)^2], \tag{6.158}$$

where δ is the unit impulse function, the integration with respect to

x_n can be replaced by the substitution $x_n = (u_n + \mu)^2$. Then to within a constant factor

$$\alpha_n \equiv \int_{-1}^{1} (u_n + \mu)^2 P_0(\mu) \exp\left\{\sum_{i=0}^{n-1} (a_i + b_i\mu + c_i\mu^2 + d_i\mu^3 + \mu^4\right\} d\mu. \tag{6.159}$$

In the same way as in the previous example, sufficient estimators can be introduced, which here will be

$$\sum_{i=0}^{n-1} b_i, \qquad \sum_{i=0}^{n-1} c_i, \qquad \sum_{i=0}^{n-1} d_i$$

(the exponential of $\sum_{i=0}^{n-1} a_i$ is taken out of the integration sign), and a computation scheme can be constructed in which only functions of a small number of variables appear. Therefore it is convenient, if this is possible, to resort to some still simpler approximate method. As the calculations carried out by I.V. Tim showed, we can expand the function $b_i\mu + c_i\mu^2 + d_i\mu^3 + \mu^4$ in the so-called series of Pike and Silverberg [6.13, 6.14], i.e., represent it in the following form:

$$b_i\mu + c_i\mu^2 + d_i\mu^3 + \mu^4 \cong \varphi_1(y_i, u_i) f_1(\mu) + \varphi_2(y_i, u_i) f_2(\mu). \tag{6.160}$$

Here f_1, f_2, φ_1, φ_2 are certain functions which can be found with the aid of the procedure presented in [6.13] and [6.14]. If Eq. (6.160) is approximately correct, then from (6.159) we find

$$\begin{aligned} \alpha_n \cong \int_{-1}^{1} (u_n + \mu)^2 P_0(\mu) \exp\left\{A_{n-1} + \sum_{i=0}^{n-1} [\varphi_1(y_i, u_i) f_1(\mu)\right. \\ \left. + \varphi_2(y_i, u_i) f_2(\mu)]\right\} d\mu = \exp\{A_{n-1}\} \int_{-1}^{1} (u_n + \mu)^2 P_0(\mu) \\ \times \exp\{E_{n-1} f_1(\mu) + F_{n-1} f_2(\mu)\}\, d\mu, \end{aligned} \tag{6.161}$$

where

$$\begin{aligned} A_s &= \sum_{i=0}^{s} a_i = A_{s-1} + a_s, \\ E_s &= \sum_{i=0}^{s} \varphi_1(y_i, u_i) = E_{s-1} + \varphi_1(y_s, u_s), \\ F_s &= \sum_{i=0}^{s} \varphi_2(y_i, u_i) = F_{s-1} + \varphi_2(y_s, u_s). \end{aligned} \tag{6.162}$$

Carrying out the integration with respect to μ in (6.161) by the approximate method, we obtain the function

$$\gamma_n{}^* = \alpha_n{}^* = \min_{u_n \in \Omega(u_n)} \alpha_n = \exp\{A_{n-1}\}\theta_n{}^*(E_{n-1}, F_{n-1}), \tag{6.163}$$

where $\theta_n{}^*$ is a function of the two variables E_{n-1} and F_{n-1}. Minimizing $\alpha_n{}^*$ with respect to u_n, we find the optimal control action

$$u_n{}^* = u_n{}^*(E_{n-1}, F_{n-1}). \tag{6.164}$$

Further, we determine the function

$$\begin{aligned}\gamma_{n-1} = \int_{-\infty}^{\infty} \gamma_n{}^* \, dy_{n-1} &= \exp\{A_{n-2}\} \int_{-\infty}^{\infty} \exp\{a_{n-1}\}\theta_n{}^*[E_{n-2} \\ &+ \varphi_1(y_{n-1}, u_{n-1}), F_{n-2} + \varphi_2(y_{n-1}, u_{n-1})] \, dy_{n-1} \\ &= \exp\{A_{n-2}\}\theta_{n-1}(E_{n-2}, F_{n-2}, u_{n-1}),\end{aligned} \tag{6.165}$$

where the result of the integration is denoted by θ_{n-1}—a function depending on the three variables E_{n-2}, F_{n-2}, u_{n-1}. Let us set

$$\theta_{n-1}^* = \theta_{n-1}^*(E_{n-2}, F_{n-2}) = \min_{u_{n-1} \in \Omega(u_{n-1})} \theta_{n-1}. \tag{6.166}$$

Minimizing θ_{n-1} with respect to u_{n-1}, we will find the optimal control

$$u_{n-1}^* = u_{n-1}^*(E_{n-2}, F_{n-2}) \tag{6.167}$$

and the function θ_{n-1}^*. Here

$$\gamma_{n-1}^* = \exp\{A_{n-2}\}\theta_{n-1}^*(E_{n-2}, F_{n-2}). \tag{6.168}$$

In an analogous fashion we obtain

$$\gamma_{n-k}^* = \exp\{A_{n-k}\}\theta_{n-k}^*(E_{n-k-1}, F_{n-k-1}) \tag{6.169}$$

($k = n, n - 1, ..., 0$) and the corresponding optimal controls

$$u_{n-k}^* = u_{n-k}^*(E_{n-k-1}, F_{n-k-1}). \tag{6.170}$$

Since the minimization with respect to u_{n-k} can be replaced by a comparison of the series of integrals $\int_{-\infty}^{\infty} \gamma_{n-k+1}^* \, dy_{n-k}$ for the various values u_{n-k}, then as a result it proves necessary to store only the functions θ_{n-k}^* at each step, and also the functions u_{n-k}^* obtained, i.e., only functions of two variables. The obtaining and storing of functions of two variables is entirely achievable in practice with the use of modern digital machines. The comparison of the integrals can be facilitated if the extreme values $u = \pm 1$ are checked.

It should be noted that the separation of systems into reducible and irreducible is not related at all to the linearity or nonlinearity of the characteristic of the object B, and to the additive or nonadditive inclusion of the coordinates μ_i in the formula for the object operator. On one hand, a reducible system with a linear object (6.62) and an irreducible system with a linear object (6.119) were considered above. On the other hand, in the last example an irreducible system with a nonlinear object was analyzed; it is also easy to give an example of a reducible system with a nonlinear object. For example, let the equation of the object B have the form

$$x_i = F(\mu + v_i), \tag{6.171}$$

where F is a nonlinear monotonic characteristic. Then the inverse function $\varphi = F^{-1}$ can be constructed. Let x_i and $v_i = u_i + g_i$ be measured. Since

$$\mu + g_i = \varphi(x_i) - u_i , \tag{6.172}$$

then the measurement of x_i and u_i is equivalent to the measurement of μ with a random error g_i. Thus the closed-loop system can be reduced to an open-loop system equivalent to it, where the result of the measurement of the noise μ passes through a channel E with an error $e_s = g_s$.

We will now give a precise formulation of the concept of reducible systems for the class of inertialess objects. Let us compare the two circuits depicted in Fig. 6.11a and b—closed-loop and open-loop. We

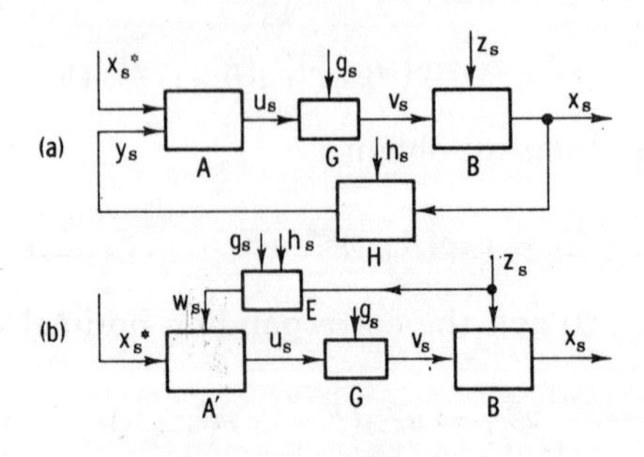

FIG. 6.11

can find the optimal strategy of the controller A for the scheme of Fig. 6.11a from the formulas of dual control theory. At the same time the optimal strategy of the controller A' for the scheme of Fig. 6.11b can be found by the considerably simpler methods mentioned in Chapter V. It is asked, is it impossible to replace the circuit of Fig. 6.11a by the circuit of Fig. 6.11b? It turns out that for a definite class of systems such a substitution can be made; such systems are called reducible. A precise definition of a reducible system is formulated in the following manner. Let all the external actions and the operators G and B for the circuits of Fig. 6.11a and b be identical. Further, let each of the systems be optimal. If a function

$$w_s = w_s(g_s, h_s, z_s) \tag{6.173}$$

can be chosen for the scheme of Fig. 6.11b such that the output actions u_s of the optimal controllers A and A' in both systems are optimal, then the circuit of Fig. 6.11a is called reducible to the circuit of Fig. 6.11b, or more briefly is called reducible. Here the algorithms of A and A' cannot coincide, if only because different variables act at the inputs to these devices.

It was stated above that all reducible systems are neutral, and hence all systems which are not neutral are irreducible. Therefore there exists a class of irreducible systems which cannot be replaced by equivalent circuits of the type depicted in Fig. 6.11b. Such systems can only be analyzed by the methods of dual control theory; for them the methods presented in the previous chapter are inadequate.

At first glance it appears that an equivalent scheme of the open-loop system type can be constructed for any case. For example, for the circuit depicted in Fig. 6.8 and Eqs. (6.119) and (6.120) of the object B and the channel H, respectively, we can argue in the following way: since from these equations it follows that

$$y_s = \mu u_s + h_s, \tag{6.174}$$

then an equivalent of the variable y_s can be constructed as is shown in Fig. 6.12a. This quantity can be obtained by multiplying out $z_s = \mu$ and u_s in the multiplier element ME, and then summing the product μu_s and h_s in the summing element Σ. But it is not difficult to be convinced that the circuit depicted in Fig. 6.12a is not at all identical to the circuit of Fig. 6.11b. In the scheme of Fig. 6.12a there is local feedback from u_s to y_s, which does not exist in Fig. 6.11b. If the elements ME and Σ are thought of as "pulled" inside the controller A (where a new device A' is now formed), then in this case the situation is as though μ is simply fed to the input of A'. But such a device A' will not be optimal. In fact, if the quantity μ were simply fed to the input of A', then a better device could be made which would at once establish the ideal value u_s. The controller A' obtained by "pulling" ME and Σ inside A cannot be optimal if only because here the random variable h_s is found to be inside of it. Meanwhile it was proved above that for the class of systems under consideration the optimal strategy is regular.

The circuit depicted in Fig. 6.12a differs fundamentally from the circuit of Fig. 6.11b. In fact, the influence of the values of u_s on the course of the measurement of the noise $z_s = \mu$ is clearly seen in the scheme of Fig. 6.12a. Depending on one or another value of u_s the noise μ is measured with a high or low relative accuracy. The larger u_s is, the larger μu_s turns out to be compared to the noise h_s; hence the more accurately the measurement of the value μu_s is accomplished, and so the value of μ as well.

An analogous argument can also be made for the circuit depicted in Fig. 6.10, with Eqs. (6.151). Here

$$y_s = (u_s + \mu)^2 + h_s . \tag{6.175}$$

Therefore the scheme of Fig. 6.10 can be replaced by the circuit depicted in Fig. 6.12b, where the sum $u_s + \mu$ is squared in the square-law generator SL and added to the noise h_s, forming the quantity y_s. This scheme also does not reduce to the circuit of Fig. 6.11b by virtue of the same considerations as stated regarding the scheme of Fig. 6.12a. In the circuit of Fig. 6.12b the influence of the variable u_s on the process of measuring the noise μ is also clearly seen, a characteristic of systems which are not neutral.

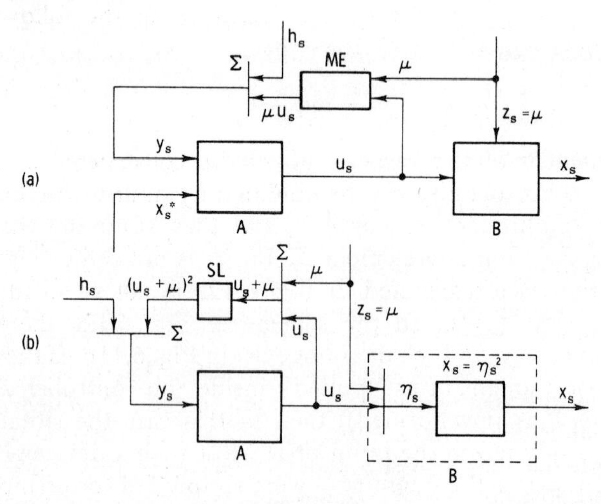

FIG. 6.12

From what was presented above it does not follow at all that all neutral systems are reducible. For example, for the object described by Eqs. (6.119) and (6.120) the system is irreducible, but for the admissible values $u = \pm 1$ it is neutral. In this example the calculation of u_s^* can be reduced to a definitive formula.

4. Generalization to the Problem with Inertial Objects

The formulation presented above of the problem in the theory of dual control and the methods of solving it can be generalized in various directions. The most important of them can be considered:

(a) The generalization to systems with several outputs of the object B, and also with several control actions.

(b) The generalization to the case of control systems with objects with memory, i.e., to the case of control by dynamical systems.

(c) The generalization to a wider class of noises z_i and driving actions x_i^*. For example, the random parameter vectors $\bar{\mu}$ and $\bar{\lambda}$, entering into the expressions for z_i and x_i respectively, can be replaced by Markov random processes $\bar{\mu}_i$ and $\bar{\lambda}_i$.

(d) The generalization to systems with discrete time with n not fixed in advance. For example, the problem can be considered of finding the optimal strategy of the controller A, minimizing the mathematical expectation $M\{n\}$ of the number of times, under the condition that the specific risk R_n is found to be a sufficiently small quantity $R_n \leqslant \epsilon$, where ϵ is an assigned value (however, it must not be too small, otherwise the solution to the problem proves to be impossible). This is the analog of the problem of maximal speed of response.

(e) The generalization to the case of continuous systems.

Some of these generalizations are considered below.

Let us consider now the generalization to the case of objects with memory and with several inputs and outputs [6.1].

The block diagram of the system under consideration is given in Fig. 6.13. We will regard all variables as functions of discrete time $t = s$ ($s = 0, 1, ..., n$), where n is fixed. The control actions $u_{1s}, u_{2s}, ..., u_{rs}$ from the output of the controller A act on the controlled object B through inertialess channels $G_1, G_2, ..., G_r$, in which the useful signals are mixed with the noises $g_{1s}, g_{2s}, ..., g_{rs}$ respectively. Therefore not the variables u_{js} but the output variables $v_{1s}, v_{2s}, ..., v_{rs}$ of the channels G_j ($j = 1, ..., r$) render the actions on the object B. In addition the noise

$\bar{z}_s$ operates on the object B, being a vector in general, i.e., a collection of scalar noises z_{is} $(i = 1, ..., w)$ applied to perhaps different parts of the object B. We will prescribe for the vector $\bar{z}_s$ the following form:

$$\bar{z}_s = \bar{z}_s(s, \bar{\mu}), \tag{6.176}$$

where $\bar{\mu}$ is the random parameter vector:

$$\bar{\mu} = (\mu_1, \mu_2, ..., \mu_m). \tag{6.177}$$

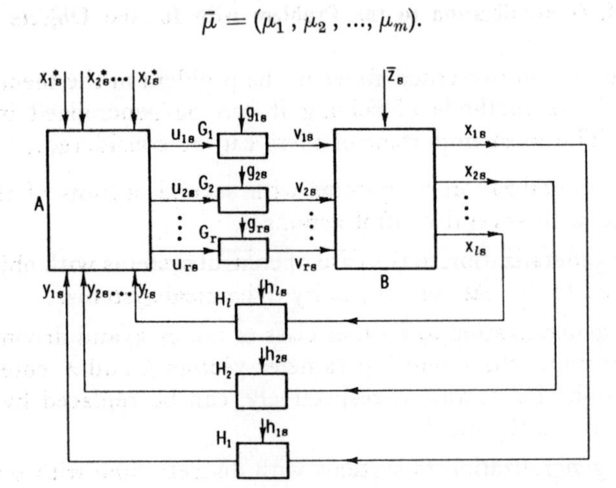

FIG. 6.13

The output variables $x_{1s}, x_{2s}, ..., x_{ls}$ of the object B are fed through the channels $H_1, H_2, ..., H_l$, respectively, to the inputs of the controller A, being mixed with the noises $h_{1s}, h_{2s}, ..., h_{ls}$. Therefore in general the input variables $y_{1s}, y_{2s}, ..., y_{ls}$ of the controller A differ from the outputs of the object B. We will consider that the driving actions $x_{1s}^*, x_{2s}^*, ..., x_{ls}^*$ act directly on the inputs to the controller A and are prescribed. The generalization to the case when these quantities are not known exactly and act through channels with noises, can be made in the same way as was done in Section 1 of this chapter. In certain cases, for example for finding the minimum of some quantity x, the driving actions are generally not necessary. Such a case was analyzed above (see the circuit depicted in Fig. 6.10).

In particular cases there is only one scalar variable x_s at the output of the object even with several inputs; such a scheme is encountered in automatic optimization systems. In other types of systems the control action u_s is transmitted to the object only along a single channel $(r = 1)$, while several variables x_{is} $(i = 1, ..., l)$ can be measured at the output.

Let us introduce the vectors composed of values of the corresponding variables at the same time instant $t = s$ (vectors of aggregates or spatial vectors). We will denote these vectors by letters with a bar:

$$\begin{aligned} \bar{x}_s^* &= (x_{1s}^*, x_{2s}^*, ..., x_{ls}^*), & \bar{x}_s &= (x_{1s}, x_{2s}, ..., x_{ls}), \\ \bar{y}_s &= (y_{1s}, y_{2s}, ..., y_{ls}), & \bar{u}_s &= (u_{1s}, u_{2s}, ..., u_{rs}), \\ \bar{v}_s &= (v_{1s}, v_{2s}, ..., v_{rs}), & \bar{g}_s &= (g_{1s}, g_{2s}, ..., g_{rs}), \\ & & \bar{h}_s &= (h_{1s}, h_{2s}, ..., h_{ls}). \end{aligned} \tag{6.178}$$

These vectors must differ from the temporal vectors, denoted by boldface letter:

$$\begin{aligned} \mathbf{x}_{is}^* &= (x_{i0}^*, x_{i1}^*, ..., x_{is}^*), & \bar{\mathbf{x}}_{is} &= (x_{i0}, x_{i1}, ..., x_{is}), \\ \mathbf{y}_{is} &= (y_{i0}, y_{i1}, ..., y_{is}), & \bar{\mathbf{u}}_{js} &= (u_{j0}, u_{j1}, ..., u_{js}), \\ & & \bar{\mathbf{v}}_{js} &= (v_{j0}, v_{j1}, ..., v_{js}) \\ & & & (i = 1, 2, ..., l, \quad j = 1, 2, ..., r). \end{aligned} \tag{6.179}$$

If the vector notations stated above are applied, then the circuit of Fig. 6.13 can be represented in a more compact form (Fig. 6.14). Here the set of channels $G_1, ..., G_r$ is replaced by a single vector channel G, and the set of channels $H_1, H_2, ..., H_l$ is replaced by the vector channel H.

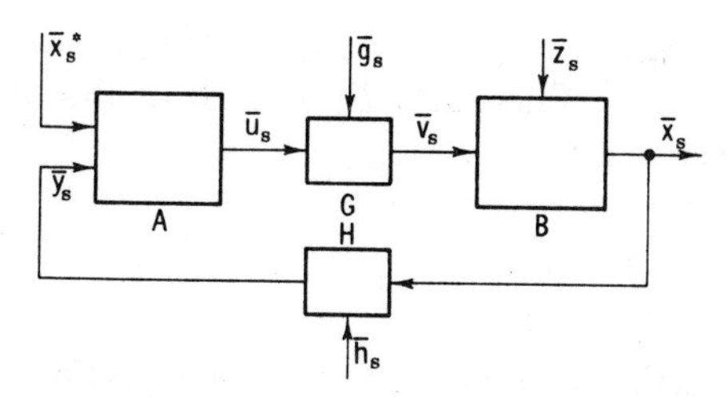

FIG. 6.14

We will restrict the form of the equations of the object B by the following condition: we will assume that these equations can be solved, and thus the vector $\bar{x}_s$ can be determined as a function of the vector $\bar{\mu}$ and the vectors $\bar{v}_k$ $(k = 0, 1, ..., s)$:

$$\bar{x}_s = \bar{x}_s(\bar{\mu}, \bar{v}_0, \bar{v}_1, ..., \bar{v}_s). \tag{6.180}$$

The initial conditions, if they are known, are included in the set of parameters of the solution and are not written out in explicit form.

But if only the probability distribution of the initial conditions is given, then they can be regarded as coordinates of the vector $\bar{\mu}$.

Let us examine various special cases which may be encountered under such a problem formulation. For example, let the motion of the object B be described in general by the nonlinear finite difference equations:

$$x_{i,s+1} = F_i(s, \bar{z}_s, x_{1s}, \ldots, x_{ls}, v_{1s}, \ldots, v_{rs}) = F_i(s, \bar{\mu}, \bar{x}_s, \bar{v}_s) \quad (6.181)$$

or

$$\bar{x}_{s+1} = \bar{F}(s, \bar{\mu}, \bar{x}_s, \bar{v}_s), \quad (6.182)$$

where $\bar{F}$ is a vector with coordinates $F_1, \ldots, F_l$.

After prescribing the initial values $\bar{x}_0$ and $\bar{v}_0$ of the vectors $\bar{x}$ and $\bar{v}$, the vector

$$\bar{x}_1 = \bar{F}(0, \bar{\mu}, \bar{x}_0, \bar{v}_0) = \Phi_1(\bar{\mu}, \bar{v}_0) \quad (6.183)$$

can be found from this formula. Here $\bar{x}_0$, as a known parameter, can be omitted from the general notation. By knowing $\bar{x}_1$ the following vector can be found:

$$\bar{x}_2 = \bar{F}(1, \bar{\mu}, \bar{x}_1, \bar{v}_1) = \Phi_2(\bar{\mu}, \bar{v}_0, \bar{v}_1). \quad (6.184)$$

In this equation $\bar{x}_1$ has been replaced by expression (6.183). By continuing this procedure we find a formula of the type (6.180) for the vector $\bar{x}_s$.

As another example the block diagram of the object B depicted in Fig. 6.15 can be cited. The parts B_1 and B_3 of the object B are linear with transient responses

$$x_{1s} = \sum_{k=0}^{s} a_{1k}(s, \bar{\mu}) v_{s-k}, \qquad x_{3s} = \sum_{k=0}^{s} a_{3k}(s, \bar{\mu}) x_{2,s-k}. \quad (6.185)$$

The part B_2 is inertialess nonlinear with the equation

$$x_{2s} = F(s, \bar{\mu}, x_{1s}). \quad (6.186)$$

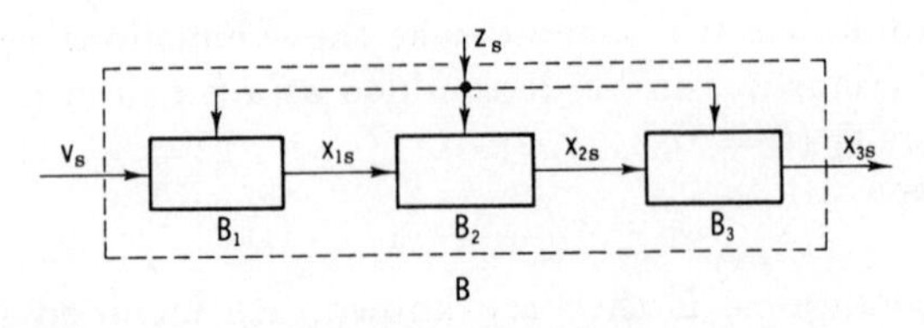

FIG. 6.15

The dependence of x_{2s} on x_{1s} must be single-valued, but not necessarily one-to-one. Thus a dependence of the type shown in Fig. 6.7b is allowed.

From Eqs. (6.185) and (6.186) it is seen that the variables x_{1s}, x_{2s}, x_{3s} can be expressed as functions of the vector $\bar{\mu}$ and the vector $\mathbf{v} = (v_0, v_1, \ldots, v_s)$, i.e., a dependence of the type (6.180) can be found.

Thus in very important and sufficiently general classes of cases the characteristic of the controlled object B can be reduced to the form (6.180). Later on we will consider that such a reduction has been made and the function (6.180) has been prescribed.

We will now introduce space-time matrices, which we will denote by a bar above the boldface letter. On the one hand, each such matrix is a set of temporal vectors, i.e., a column vector whose "coordinates" are temporal vectors. But it can be considered as a set of spatial vectors, i.e., a row vector whose "coordinates" are spatial vectors. For example,

$$\bar{\mathbf{u}}_s = \begin{vmatrix} u_{10}, & u_{11}, \ldots, u_{1s} \\ u_{20}, & u_{21}, \ldots, u_{2s} \\ \vdots & \vdots \qquad \vdots \\ u_{r0}, & u_{r1}, \ldots, u_{rs} \end{vmatrix} = \begin{vmatrix} \mathbf{u}_{1s} \\ \mathbf{u}_{2s} \\ \vdots \\ \mathbf{u}_{rs} \end{vmatrix} = | \, \bar{u}_0, \quad \bar{u}_1, \ldots, \bar{u}_s \, |. \tag{6.187}$$

In the same manner other space-time matrices can be introduced:

$$\bar{\mathbf{v}}_s = \begin{vmatrix} v_{10}, \ldots, v_{1s} \\ \vdots \qquad \vdots \\ v_{r0}, \ldots, v_{rs} \end{vmatrix} = \begin{vmatrix} \mathbf{v}_{1s} \\ \vdots \\ \mathbf{v}_{rs} \end{vmatrix} = | \, \bar{v}_0, \ldots, \bar{v}_s \, |, \tag{6.188}$$

$$\bar{\mathbf{x}}_s = \begin{vmatrix} x_{10}, \ldots, x_{1s} \\ \vdots \qquad \vdots \\ x_{l0}, \ldots, x_{ls} \end{vmatrix} = \begin{vmatrix} \mathbf{x}_{1s} \\ \vdots \\ \mathbf{x}_{ls} \end{vmatrix} = | \, \bar{x}_0, \ldots, \bar{x}_s \, |, \tag{6.189}$$

and so on. Let us select the loss function in the system in the form

$$W = \sum_{s=0}^{n} W_s, \tag{6.190}$$

where

$$W_s = W_s(s, \bar{\mu}, \bar{x}_s{}^*, \bar{x}_s). \tag{6.191}$$

Cases can be encountered where an explicit dependence of W_s on $\bar{\mu}$ occurs. For example, in the circuit of Fig. 6.13 let the dependence of a single output x_s on the inputs be defined by the formula

$$x_s = \mu_{r+1} + \sum_{i,j=1}^{r} a_{ij}(v_{is} - \mu_i)(v_{js} - \mu_j), \tag{6.192}$$

where $a_{ij} = \text{const}$, and the v_{js} are the input variables of the object. Further, let the second term on the right side of (6.192) be incapable of being negative. Then the minimal possible value of x_s is μ_{r+1}, and the values $v_{is} = \mu_i$ ensure the minimum x_s. The deviation of x_s from the minimum is expressed by the difference $x_s - \mu_{r+1}$. Therefore in this case if the automatic finding of the minimum x_s is required, we can set

$$W_s = x_s - \mu_{r+1} \tag{6.193}$$

and require the minimization of the mathematical expectation of the corresponding function W.

Let the *a priori* probability density of the vector $\bar{\mu}$ be given, which we will denote by $P(\bar{\mu}) = P_0(\bar{\mu})$, and also the probability densities $P(g_{js})$ $(j = 1, 2, ..., r)$ of the random variables g_{js}. For fixed j the quantities g_{js} represent a sequence of independent random variables with the same probability density $P(g_{js})$ not depending on time. For different channels (different j) the probability densities can be distinct.

The same conditions hold for the probability densities $P(h_{is})$ of the variables h_{is} $(i = 1, 2, ..., l)$.

We regard all the external actions on the system as mutually independent.

Let the channels G and H not have memory. Then

$$v_{js} = v_j(g_{js}, u_{js}), \qquad y_{is} = y_i(h_{is}, x_{is}). \tag{6.194}$$

By knowing the data shown above, the conditional probability densities $P(y_{is} \mid x_{is})$, $P(v_{js} \mid u_{js})$, and hence $P(\bar{y}_s \mid \bar{x}_s)$ and $P(\bar{v}_s \mid \bar{u}_s)$ as well can be found. In this case the conditional probability densities

$$P(\mathbf{y}_{is} \mid \mathbf{x}_{is}) = \prod_{\nu=0}^{s} P(y_{i\nu} \mid x_{i\nu}), \qquad P(\mathbf{v}_{js} \mid \mathbf{u}_{js}) = \prod_{\nu=0}^{s} P(v_{j\nu} \mid u_{j\nu}) \tag{6.195}$$

can be discovered. By knowing these probability densities, the conditional probability densities $P(\bar{\mathbf{y}}_s \mid \bar{\mathbf{x}}_s)$ and $P(\bar{\mathbf{v}}_s \mid \bar{\mathbf{u}}_s)$ for the matrices can be found.

The algorithm of the controller A is characterized by the probability density Γ_s of its output vector $\bar{u}_s$, which in general depends on all the previous values of the input variables to the controller and on the current values of the driving actions:

$$P_s(\bar{u}_s) = \Gamma_s(\bar{u}_s \mid \bar{\mathbf{x}}_s^*, \bar{\mathbf{y}}_{s-1}, \bar{\mathbf{u}}_{s-1}). \tag{6.196}$$

The problem is posed of finding a sequence of functions Γ_i $(i = 1, ..., n)$ and also an initial probability density $\Gamma_0 = P(u_0)$, such that the risk R, the mathematical expectation of the loss function W defined by (6.190) and (6.191), is minimal.

From (6.191) and (6.180) we find

$$W_s = W_s[s, \bar{\mu}, \bar{x}_s^*, \bar{x}_s(s, \bar{\mu}, \bar{\mathbf{v}}_s)] = W_s(s, \bar{\mu}, \bar{x}_s^*, \bar{\mathbf{v}}_s). \tag{6.197}$$

We will confine ourselves to the derivation of the expression† for the risk R with a fixed matrix $\bar{\mathbf{x}}_s^*$. Let the joint conditional probability density

$$P(\bar{\mu}, \bar{\mathbf{u}}_s, \bar{\mathbf{v}}_s, \bar{\mathbf{y}}_{s-1} \mid \bar{\mathbf{x}}_s^*)$$

of the correlated random vectors and matrices be known.

The matrix $\bar{\mathbf{y}}_{s-1}$ appears in this expression, since according to (6.196) the matrix $\bar{\mathbf{u}}_s$ is indeed related to $\bar{\mathbf{y}}_{s-1}$.

The specific risk R_s corresponding to the instant $t = s$ is equal to

$$\begin{aligned} R_s &= M\{W_s \mid \bar{\mathbf{x}}_s^*\} \\ &= \int_{\Omega(\bar{\mu}, \bar{\mathbf{u}}_s, \bar{\mathbf{v}}_s, \bar{\mathbf{y}}_{s-1})} W_s(s, \bar{\mu}, \bar{x}_s^*, \bar{\mathbf{v}}_s)\, P(\bar{\mu}, \bar{\mathbf{u}}_s, \bar{\mathbf{v}}_s, \bar{\mathbf{y}}_{s-1} \mid \bar{\mathbf{x}}_s^*)\, d\Omega. \end{aligned} \tag{6.198}$$

On the basis of the multiplication theorem for probabilities we can write

$$P(\bar{\mu}, \bar{\mathbf{u}}_s, \bar{\mathbf{v}}_s, \bar{\mathbf{y}}_{s-1} \mid \bar{\mathbf{x}}_s^*) = P(\bar{\mu}) P(\bar{\mathbf{v}}_s, \bar{\mathbf{u}}_s, \bar{\mathbf{y}}_{s-1} \mid \bar{\mu}, \bar{\mathbf{x}}_s^*). \tag{6.199}$$

This formula is valid if only $\bar{\mu}$ and $\bar{\mathbf{x}}_s^*$ are independent.

Let us consider the second factor on the right side of Eq. (6.199) and transform it:

$$\begin{aligned} P(\bar{\mathbf{v}}_s, \bar{\mathbf{u}}_s, \bar{\mathbf{y}}_{s-1} \mid \bar{\mu}, \bar{\mathbf{x}}_s^*) &= P(\bar{\mathbf{u}}_s, \bar{\mathbf{y}}_{s-1} \mid \bar{\mu}, \bar{\mathbf{x}}_s^*)\, P(\bar{\mathbf{v}}_s \mid \bar{\mathbf{u}}_s, \bar{\mathbf{y}}_{s-1}, \bar{\mu}, \bar{\mathbf{x}}_s^*) \\ &= P(\bar{\mathbf{u}}_s, \bar{\mathbf{y}}_{s-1} \mid \bar{\mu}, \bar{\mathbf{x}}_s^*)\, P(\bar{\mathbf{v}}_s \mid \bar{\mathbf{u}}_s, \bar{\mathbf{y}}_{s-1}, \bar{\mu}). \end{aligned} \tag{6.200}$$

The latter transformation holds since the probability density of the matrix $\bar{\mathbf{v}}_s$ is completely determined by setting $\bar{\mathbf{u}}_s$, $\bar{\mathbf{y}}_{s-1}$, $\bar{\mu}$, and additional information about $\bar{\mathbf{x}}_s^*$ cannot change it.

Further, it is not difficult to show by analogy with (6.25) that

$$\begin{aligned} P(\bar{\mathbf{u}}_s, \bar{\mathbf{y}}_{s-1} \mid \bar{\mu}, \bar{\mathbf{x}}_s^*) &= P(\bar{u}_0, \bar{y}_0 \mid \bar{\mu}, \bar{x}_0^*) \cdot P(\bar{u}_1, \bar{y}_1 \mid \bar{\mu}, \bar{u}_0, \bar{y}_0, \bar{\mathbf{x}}_1^*) \\ &\times P(\bar{u}_2, \bar{y}_2 \mid \bar{\mu}, \bar{\mathbf{u}}_1, \bar{\mathbf{y}}_1, \bar{\mathbf{x}}_2^*) \times P(\bar{u}_i, \bar{y}_i \mid \bar{\mu}, \bar{\mathbf{u}}_{i-1}, \bar{\mathbf{y}}_{i-1}, \bar{\mathbf{x}}_i^*) \times \cdots \\ &\times P(\bar{u}_{s-1}, \bar{y}_{s-1} \mid \bar{\mu}, \bar{\mathbf{u}}_{s-2}, \bar{\mathbf{y}}_{s-2}, \bar{\mathbf{x}}_{s-1}^*)\, P(\bar{u}_s \mid \bar{\mu}, \bar{\mathbf{u}}_{s-1}, \bar{\mathbf{y}}_{s-1}, \bar{\mathbf{x}}_s^*). \end{aligned} \tag{6.201}$$

† This derivation can be carried out in the same way as for objects without memory. Another derivation is given below.

We will examine the ith factor of this product $(0 < i < s)$:

$$\begin{aligned} P(\bar{u}_i, \bar{y}_i \mid \bar{\mu}, \bar{\mathbf{u}}_{i-1}, \bar{\mathbf{y}}_{i-1}, \bar{\mathbf{x}}_i^*) &= P(\bar{y}_i \mid \bar{\mu}, \bar{\mathbf{u}}_i, \bar{\mathbf{y}}_{i-1}, \bar{\mathbf{x}}_i^*) \cdot P(\bar{u}_i \mid \bar{\mu}, \bar{\mathbf{u}}_{i-1}, \bar{\mathbf{y}}_{i-1}, \bar{\mathbf{x}}_i^*) \\ &= P(\bar{y}_i \mid i, \bar{\mu}, \bar{\mathbf{y}}_{i-1}, \bar{\mathbf{u}}_i) \cdot \Gamma_i(\bar{u}_i \mid \bar{\mathbf{u}}_{i-1}, \bar{\mathbf{y}}_{i-1}, \bar{\mathbf{x}}_i^*). \end{aligned} \tag{6.202}$$

Actually, the probability density of $\bar{y}_i$ is defined for fixed $\bar{\mu}$, $\bar{\mathbf{y}}_{i-1}$, and $\bar{\mathbf{u}}_i$, and the additional setting of $\bar{\mathbf{x}}_i^*$ will not change it. In the formula for the probability density of $\bar{y}_i$ it is emphasized that it can depend explicitly on the time instant i.

The second factor in (6.202) is Γ_i, the probability density of $\bar{u}_i$, which in accordance with (6.196) depends only on $\bar{\mathbf{u}}_{i-1}$, $\bar{\mathbf{y}}_{i-1}$, $\bar{\mathbf{x}}_i^*$.

Substituting (6.202) into (6.201), we find

$$P(\bar{\mathbf{u}}_s, \bar{\mathbf{y}}_{s-1} \mid \bar{\mu}, \bar{\mathbf{x}}_s^*) = \prod_{i=0}^{s} \Gamma_i \cdot \prod_{i=0}^{s-1} P(\bar{y}_i \mid i, \bar{\mu}, \bar{\mathbf{y}}_{i-1}, \bar{\mathbf{u}}_i). \tag{6.203}$$

Here we set

$$P(\bar{y}_0 \mid i, \bar{\mu}, \bar{\mathbf{y}}_{-1}, \bar{\mathbf{u}}_0) = P(\bar{y}_0 \mid i, \bar{\mu}, \bar{u}_0); \qquad \Gamma_0 = P_0(\bar{u}_0, \bar{\mathbf{x}}_0^*). \tag{6.204}$$

Let us substitute expression (6.203) into (6.200), then (6.200) into (6.199) and (6.199) into (6.198). Then the expression for R_s takes the form

$$\begin{aligned} R_s = \int_{\Omega(\bar{\mu}, \bar{\mathbf{v}}_s, \bar{\mathbf{u}}_s, \bar{\mathbf{y}}_{s-1})} & W(s, \bar{\mu}, \bar{x}_s^*, \bar{\mathbf{v}}_s) \cdot P(\bar{\mu}) \cdot P(\bar{\mathbf{v}}_s \mid \bar{\mathbf{u}}_s, \bar{\mathbf{y}}_{s-1}, \bar{\mu}) \\ & \times \prod_{i=0}^{s-1} P(\bar{y}_i \mid i, \bar{\mu}, \bar{\mathbf{y}}_{i-1}, \bar{\mathbf{u}}_i) \cdot \prod_{i=0}^{s} \Gamma_i(\bar{u}_i \mid \bar{\mathbf{u}}_{i-1}, \bar{\mathbf{y}}_{i-1}, \bar{\mathbf{x}}_i^*)\, d\Omega. \end{aligned} \tag{6.205}$$

The total risk R is defined as before by the equation

$$R = \sum_{s=0}^{n} R_s. \tag{6.206}$$

In the special case when there is one input and one output for the object B, the space-time matrices in formula (6.205) are replaced by temporal vectors, and the spatial vectors by scalars. Here expression (6.205) takes the following form:

$$\begin{aligned} R_s = \int_{\Omega(\bar{\mu}, \mathbf{v}_s, \mathbf{u}_s, \mathbf{y}_{s-1})} & W_s(s, \bar{\mu}, x_s^*, \mathbf{v}_s) \cdot P(\bar{\mu}) \cdot P(\mathbf{v}_s \mid \mathbf{u}_s, \mathbf{y}_{s-1}, \bar{\mu}) \\ & \times \prod_{i=0}^{s-1} P(y_i \mid i, \bar{\mu}, \mathbf{y}_{i-1}, \mathbf{u}_i) \prod_{i=0}^{s} \Gamma_i(u_i \mid \mathbf{u}_{i-1}, \mathbf{y}_{i-1}, \mathbf{x}_i^*)\, d\Omega. \end{aligned} \tag{6.207}$$

This expression is more general than (6.57), since it takes into account the presence of memory in the object B. Therefore in formula (6.207), W_s depends not on the scalar v_s but on the temporal vector $\mathbf{v}_s$, which causes the necessity for calculating the probability density $P(\mathbf{v}_s \mid \mathbf{u}_s, \mathbf{y}_{s-1}, \bar{\mu})$. Moreover, the conditional probability density for y_i now depends not only on the value of u_i, but also on the entire vector $\mathbf{u}_i$, i.e., on the "previous history" of the output variable of the object B, and also on $\mathbf{y}_{i-1}$. In the absence of the noise g_s, for example with $u_s = v_s$, the dependence of $P(y_i)$ on $\mathbf{y}_i$ is missing.

For determining the optimal strategy we will examine the specific risk R_n, as in Section 2 of this chapter, regarding the densities Γ_i $(i = 0, 1, ..., n-1)$ as fixed. We will set

$$\alpha_k = \alpha_k(\bar{\mathbf{u}}_k, \bar{\mathbf{y}}_{k-1}, \bar{x}_k^*) = \int_{\Omega(\bar{\mu}, \bar{\mathbf{v}}_k)} W_k(k, \bar{\mu}, \bar{x}_k^*, \bar{\mathbf{v}}_k)$$

$$\times P(\bar{\mu}) \cdot P(\bar{\mathbf{v}}_k \mid \bar{\mathbf{u}}_k, \bar{\mathbf{y}}_{k-1}, \bar{\mu}) \cdot \prod_{i=0}^{k-1} P(\bar{y}_i \mid i, \bar{\mu}, \bar{\mathbf{y}}_{i-1}, \bar{\mathbf{u}}_i)\, d\Omega. \quad (6.208)$$

For $k = 0$ the quantity $\prod_{i=0}^{k-1} P(\bar{y}_i)$ must be set equal to unity. We will also write

$$\beta_k = \beta_k(\bar{\mathbf{u}}_k, \bar{\mathbf{x}}_k^*, \bar{\mathbf{y}}_{k-1}) = \prod_{i=0}^{k} \Gamma_i. \quad (6.209)$$

Then

$$R_n = \int_{\Omega(\bar{\mathbf{u}}_n, \bar{\mathbf{y}}_{n-1})} \alpha_n(\bar{u}_n, \bar{\mathbf{u}}_{n-1}, \bar{y}_{n-1}, \bar{x}_n^*) \cdot \beta_{n-1} \cdot \Gamma_n \, d\Omega$$

$$= \int_{\Omega(\bar{\mathbf{u}}_{n-1}, \bar{\mathbf{y}}_{n-1})} \beta_{n-1} \kappa_n(\bar{\mathbf{u}}_{n-1}, \bar{\mathbf{y}}_{n-1}, \bar{\mathbf{x}}_n^*)\, d\Omega, \quad (6.210)$$

where

$$\kappa_n(\bar{\mathbf{u}}_{n-1}, \bar{\mathbf{y}}_{n-1}, \bar{\mathbf{x}}_n^*) = \int_{\Omega(\bar{u}_n)} \alpha_n(\bar{u}_n, \bar{\mathbf{u}}_{n-1}, \bar{\mathbf{y}}_{n-1}, \bar{x}_n^*) \Gamma_n(\bar{u}_n \mid \bar{\mathbf{u}}_{n-1}, \bar{\mathbf{y}}_{n-1}, \bar{\mathbf{x}}_n^*)\, d\Omega. \quad (6.211)$$

The choice of the function Γ_n, which is a probability density, is restricted by the condition

$$\int_{\Omega(\bar{u}_n)} \Gamma_n(\bar{u}_n \mid \bar{\mathbf{u}}_{n-1}, \bar{\mathbf{y}}_{n-1}, \bar{\mathbf{x}}_n^*)\, d\Omega = 1. \quad (6.212)$$

The function Γ_n satisfying condition (6.212) must be selected such that the quantity R_n is minimal. But this variable will be minimal if for any $\bar{\mathbf{u}}_{n-1}$, $\bar{\mathbf{y}}_{n-1}$ the quantity κ_n is minimal, which in turn will be ensured if

$$\Gamma_n = \delta(\bar{u}_n - \bar{u}_n^*) \tag{6.213}$$

is chosen. Here δ is the unit impulse function, and the value $\bar{u}_n^*$ is defined by the condition

$$\gamma_n^* = \alpha_n(\bar{u}_n^*, \bar{\mathbf{u}}_{n-1}, \bar{\mathbf{y}}_{n-1}, \bar{x}_n^*) = \min_{\bar{u}_n \in \Omega(\bar{u}_n)} \alpha_n(\bar{u}_n, \bar{\mathbf{u}}_{n-1}, \bar{\mathbf{y}}_{n-1}, \bar{x}_n^*). \tag{6.214}$$

From (6.214) it follows that the optimal control u_n^* depends on $\bar{\mathbf{u}}_{n-1}$, $\bar{\mathbf{y}}_{n-1}$, $\bar{x}_n^*$:

$$\bar{u}_n^* = \bar{u}_n^*(\bar{\mathbf{u}}_{n-1}, \bar{\mathbf{y}}_{n-1}, \bar{x}_n^*). \tag{6.215}$$

Thus the optimal strategy Γ_n turns out not to be random, but regular. The choice of the optimal control action $\bar{u}_n^*$ is given by formulas (6.214) and (6.215).

In an analogous fashion, passing to the examination of the sum of terms $R_{n-1} + R_n$, then $R_{n-2} + R_{n-1} + R_n$, and so on, the specific optimal strategies Γ_i $(i = n-1, \ldots, n-2, \ldots)$ can be found. An analogous proof has been carried out in Section 1. The result consists of the following: let us introduce the function

$$\gamma_{n-k} = \alpha_{n-k} + \int_{\Omega(\bar{y}_{n-k})} \gamma_{n-k+1}^* \, d\Omega, \tag{6.216}$$

where $\gamma_n = \alpha_n$ and

$$\gamma_{n-k}^* = (\gamma_{n-k})_{\bar{u}_{n-k} = \bar{u}_{n-k}^*}, \tag{6.217}$$

and the value $\bar{u}_{n-k}^*$ is obtained from the condition

$$\begin{aligned} &\gamma_{n-k}(\bar{u}_{n-k}^*, \bar{\mathbf{u}}_{n-k-1}, \bar{\mathbf{y}}_{n-k-1}, \bar{x}_{n-k}^*) \\ &\quad = \min_{\bar{u}_{n-k} \in \Omega(\bar{u}_{n-k})} \gamma_{n-k}(\bar{u}_{n-k}, \bar{\mathbf{u}}_{n-k-1}, \bar{\mathbf{y}}_{n-k-1}, \bar{x}_{n-k}^*). \end{aligned} \tag{6.218}$$

Evidently,

$$\bar{u}_{n-k}^* = \bar{u}_{n-k}^*(\bar{\mathbf{u}}_{n-k-1}, \bar{\mathbf{y}}_{n-k-1}, \bar{x}_{n-k}^*). \tag{6.219}$$

Then the optimal strategy Γ_{n-k}^* is defined by the expression

$$\Gamma_{n-k}^* = \delta(\bar{u}_{n-k} - \bar{u}_{n-k}^*), \tag{6.220}$$

i.e., it is regular, where the optimal control $\bar{u}_{n-k}^*$ determined from

(6.219) depends on the "previous history" of the input variables to the controller A, and also on the current value $\bar{x}^*_{n-k}$.

It is not difficult to generalize the formula derived above to the case when the driving action x_s^* enters the input to the controller through a channel with noise. The derivation is analogous to the one carried out in Section 1 of this chapter.

As an example of the use of the formulas derived, let us consider the problem of synthesizing the algorithm of the optimal controller A for an automatic stabilization system. The block diagram of the system is shown in Fig. 6.16. The controlled object B consists of a part B_1 having memory, and a part B_2 without memory. The equations of these parts have the form

$$w_s = \sum_{k=0}^{s} a_k v_{s-k}, \qquad x_s = \mu + w_s. \tag{6.221}$$

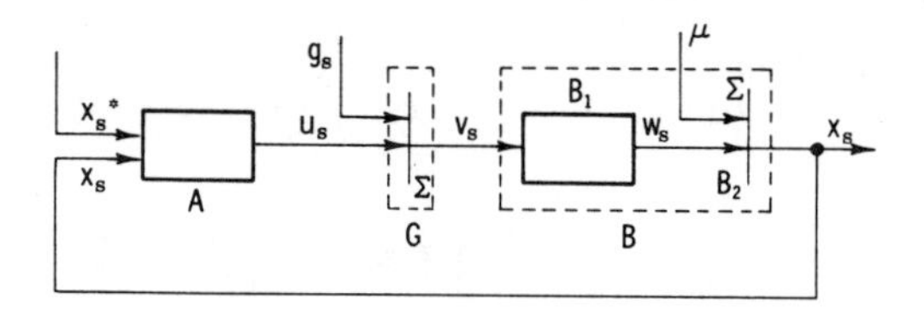

FIG. 6.16

Here the a_k are prescribed constants, μ is a random variable with a given *a priori* distribution density $P(\mu) = P_0(\mu)$. The overall equation of the object B can be written in the form

$$x_s = \mu + \sum_{k=0}^{s} a_k v_{s-k}. \tag{6.222}$$

The sequence g_s of independent random variables has the same probability density $q(g_s)$. The functions $P(\mu)$ and $q(g_s)$ are normal distribution densities

$$P(\mu) = \frac{1}{\sigma_\mu \sqrt{(2\pi)}} \exp\left\{-\frac{\mu^2}{2\sigma_\mu^2}\right\}, \qquad q(g_s) = \frac{1}{\sigma_g \sqrt{(2\pi)}} \exp\left\{-\frac{g_s^2}{2\sigma_g^2}\right\}. \tag{6.223}$$

The equation of the channel G has the form

$$v_s = g_s + u_s. \tag{6.224}$$

The specific loss function is defined by the expression

$$W_s = (x_s - x_s^*)^2. \tag{6.225}$$

Here the total loss function takes the form

$$W = \sum_{s=0}^{n} W_s = \sum_{s=0}^{n} (x_s - x_s^*)^2. \tag{6.226}$$

In the particular case of systems with one input and one output, we obtain from formula (6.208) the expression for α_k in the form

$$\begin{aligned} \alpha_k &= \alpha_k(\mathbf{u}_k, \mathbf{y}_{k-1}, x_k^*) \\ &= \int_{\Omega(\bar{\mu}, \mathbf{v}_k)} W(k, \bar{\mu}, x_k^*, \mathbf{v}_k)\, P(\bar{\mu}) \cdot P(\mathbf{v}_k \mid \mathbf{u}_k, \mathbf{y}_{k-1}, \bar{\mu}) \\ &\quad \times \prod_{i=0}^{k-1} P(y_i \mid i, \bar{\mu}, \mathbf{y}_{i-1}, \mathbf{u}_i)\, d\Omega. \end{aligned} \tag{6.227}$$

In the example being considered†

$$\begin{aligned} \alpha_k = \int_{\Omega(\mu, \mathbf{v}_k)} \left(\mu - x_k^* + \sum_{\rho=0}^{k} a_\rho v_{k-\rho}\right)^2 \frac{1}{\sigma_\mu \sqrt{(2\pi)}} \exp\left\{-\frac{\mu^2}{2\sigma_\mu^2}\right\} \\ \times P(\mathbf{v}_k \mid \mathbf{u}_k, \mathbf{x}_{k-1}, \bar{\mu}) \cdot \prod_{i=0}^{k-1} P(y_i \mid i, \mu, \mathbf{y}_{i-1}, \mathbf{u}_i). \end{aligned} \tag{6.228}$$

Let us find the expressions for $P(\mathbf{v}_k \mid \mathbf{u}_k, \mathbf{x}_{k-1}, \bar{\mu})$ and $P(y_i \mid i, \mu, \mathbf{y}_{i-1}, \mathbf{u}_i)$. From formula (6.224) it follows that

$$P(v_i \mid u_i) = q(v_i - u_i) = \frac{1}{\sigma_g \sqrt{(2\pi)}} \exp\left\{-\frac{(v_i - u_i)^2}{2\sigma_g^2}\right\}. \tag{6.229}$$

Since the random variables g_i ($i = 0, 1, ..., k$) are independent, then v_k depends only on u_k. Therefore

$$\begin{aligned} P(\mathbf{v}_k \mid \mathbf{u}_k, \mathbf{x}_{k-1}, \bar{\mu}) &= P(v_k \mid u_k) \cdot P(\mathbf{v}_{k-1} \mid \mathbf{u}_{k-1}, \mathbf{x}_{k-1}, \bar{\mu}) \\ &= \frac{1}{\sigma_g \sqrt{(2\pi)}} \exp\left\{-\frac{(v_k - u_k)^2}{2\sigma_g^2}\right\} \cdot P(\mathbf{v}_{k-1} \mid \mathbf{u}_{k-1}, \mathbf{x}_{k-1}, \bar{\mu}). \end{aligned} \tag{6.230}$$

First we will find $P(y_i \mid i, \mu, \mathbf{y}_{i-1}, \mathbf{u}_i) = P(y_i \mid \mu, \mathbf{y}_{i-1}, \mathbf{u}_i)$.

The index i to the right of the vertical line can be removed, since the equations of the object B and the channel G do not depend on i.

† In the paper [6.1] the dependence of $P(y_i)$ and $P(\mathbf{v}_i)$ on $\mathbf{y}_{i-1}$ was not taken into account in this example, which simplifies the formulas but can give a substantial deviation from optimality. Here the exact formulas have been given.

From formula (6.222) it follows that

$$\begin{aligned}
x_i &= \mu + \sum_{\rho=0}^{i} a_\rho v_{i-\rho} = \mu + \sum_{\rho=0}^{i} a_\rho (u_{i-\rho} + g_{i-\rho}),\\
x_{i-1} &= \mu + \sum_{\rho=0}^{i-1} a_\rho v_{i-1-\rho} = \mu + \sum_{\rho=0}^{i-1} a_\rho (u_{i-1-\rho} + g_{i-1-\rho}),\\
&\vdots\\
x_0 &= \mu + a_0 v_0 = \mu + a_0(u_0 + g_0).
\end{aligned} \tag{6.231}$$

This is a system of linear equations with respect to $g_0, g_1, \ldots, g_i$. Let us rewrite it in the following form:

$$\begin{aligned}
\sum_{\rho=0}^{i} a_\rho g_{i-\rho} &= x_i - \mu - \sum_{\rho=0}^{i} a_\rho u_{i-\rho},\\
\sum_{\rho=0}^{i-1} a_\rho g_{i-1-\rho} &= x_{i-1} - \mu - \sum_{\rho=0}^{i-1} a_\rho u_{i-1-\rho},\\
&\vdots\\
a_0 g_0 &= x_0 - \mu - a_0 u_0 .
\end{aligned} \tag{6.232}$$

We will denote the right sides of these equations by b_k:

$$\begin{aligned}
x_i - \mu - \sum_{\rho=0}^{i} a_\rho u_{i-\rho} &= b_i,\\
x_{i-1} - \mu - \sum_{\rho=0}^{i-1} a_\rho u_{i-1-\rho} &= b_{i-1},\\
&\vdots\\
x_0 - \mu - a_0 u_0 &= b_0 .
\end{aligned} \tag{6.233}$$

If the variables $g_0, g_1, \ldots, g_{i-1}$ are eliminated from the system (6.232), then we will obtain the relation between x_i and g_i for fixed $\mathbf{x}_{i-1}$, μ, and $\mathbf{u}_i$. Let us denote the determinant of the system (6.232) by the letter Δ:

$$\Delta = \begin{vmatrix} a_i & a_{i-1} & \cdots & a_1 & a_0 \\ a_{i-1} & a_{i-2} & \cdots & a_0 & 0 \\ \cdot & \cdot & \cdots & \cdot & \cdot \\ a_1 & a_0 & \cdots & 0 & 0 \\ a_0 & 0 & \cdots & 0 & 0 \end{vmatrix}. \tag{6.234}$$

Since a_0 is not equal to zero [otherwise from Eq. (6.222) it would follow that $x_0 = \mu$, i.e., the parameter μ is known exactly], the determinant Δ is likewise not equal to zero.

Solving the system of equations, we find g_i:

$$g_i = \sum_{j=0}^{i} \kappa_{ij} b_j \,, \tag{6.235}$$

where the κ_{ij} are constants expressed in terms of $a_0, \ldots, a_i$ from the known formulas:

$$\kappa_{ij} = \frac{1}{\Delta} \begin{vmatrix} a_i & a_{i-1} & \cdots & a_1 & 0 \\ a_{i-1} & a_{i-2} & \cdots & a_0 & 0 \\ \cdot & \cdot & \cdots & \cdot & 1 \\ a_1 & a_0 & \cdots & \cdot & 0 \\ a_0 & 0 & \cdots & \cdot & 0 \end{vmatrix} \quad (j\text{th row}). \tag{6.236}$$

From formula (6.233) it is seen that only b_i depends on the variable x_i; the latter enters linearly into expression (6.235) with the coefficient κ_{ij}.

Therefore the probability density $P(x_i \mid \mu, \mathbf{u}_i, \mathbf{x}_{i-1})$ can be written in the following form:

$$P(x_i \mid \mu, \mathbf{u}_i, \mathbf{x}_{i-1}) = |\kappa_{ii}| \frac{1}{\sqrt{(2\pi\sigma_g^2)}} \exp\left\{ - \frac{(\sum_{j=0}^{i} \kappa_{ij} b_j)^2}{2\sigma_g^2} \right\}. \tag{6.237}$$

Now the final expression for $P(\mathbf{v}_{k-1} \mid \mathbf{u}_{k-1}, \mathbf{x}_{k-1}, \mu)$ can be written into formula (6.230). The values $v_0, \ldots, v_{k-1}$ are determined directly from the given μ, $x_0, \ldots, x_{k-1}$, as is seen from (6.231). The solutions to these equations are written analogously to (6.235), but with g_i replaced by v_i; now by b_j is meant the variables $(x_j - \mu)$ $(j = 0, \ldots, k-1)$. Thus,

$$v_i = \sum_{j=0}^{i} \kappa_{ij}(x_j - \mu) \qquad (i = 0, \ldots, k-1). \tag{6.238}$$

Hence $P(\mathbf{v}_{k-1} \mid \mathbf{u}_{k-1}, \mathbf{x}_{k-1}, \mu)$ is converted into an impulse function, and in expression (6.228) for α_k the integration with respect to $\mathbf{v}_{k-1}$ is replaced by the substitution of expressions (6.238). We find to within a constant factor:

$$\begin{aligned} \alpha_k \equiv \int_{\mu=-\infty}^{\infty} \int_{v_k=-\infty}^{\infty} & \left[\mu - x_k{}^* + a_0 v_k + \sum_{\rho=1}^{k} a_\rho \sum_{j=0}^{k-\rho} \kappa_{k-\rho,j}(x_j - \mu) \right]^2 \\ & \times \exp\left\{ - \frac{\mu^2}{2\sigma_\mu^2} - \frac{(v_k - u_k)^2}{2\sigma_g^2} \right. \\ & \left. - \frac{\sum_{i=0}^{k-1} [\sum_{j=0}^{i} \kappa_{ij}(x_j - \mu - \sum_{\rho=0}^{j} a_\rho u_{j-\rho})]^2}{2\sigma_g^2} \right\} d\mu \, dv_k \,. \end{aligned} \tag{6.239}$$

Let us introduce the notations:

$$x_k^{(0)*} = x_k{}^* - a_0 u_k - \sum_{\rho=1}^{k} a_\rho \sum_{j=0}^{k-\rho} \kappa_{k-\rho,j} x_j;$$

$$x_i^{(0)} = \sum_{j=0}^{i} \kappa_{ij} \left(x_j - \sum_{\rho=0}^{j} a_\rho u_{j-\rho}\right); \qquad \sum_{j=0}^{i} \kappa_{ij} = d_i;$$

$$F_k = 1 - \sum_{\rho=1}^{k} a_\rho \sum_{j=0}^{k-\rho} \kappa_{k-\rho,j}; \qquad B_{k-1} = \sum_{i=0}^{k-1} \frac{d_i x_i^{(0)}}{\sigma_g{}^2}; \tag{6.240}$$

$$C_{k-1} = \frac{1}{2}\left(\sum_{i=0}^{k-1} \frac{d_i}{\sigma_g{}^2}\right) + \frac{1}{2\sigma_\mu{}^2} > 0; \qquad D_{k-1} = \sum_{i=0}^{k-1} \frac{(x_i^{(0)})^2}{2\sigma_g{}^2} > 0;$$

$$E_k = \frac{B_{k-1} F_k}{2C_{k-1}} - x_k^{(0)*}.$$

Then after transformations expression (6.239) can be rewritten in the following form to within constant factor:

$$\alpha_k = \alpha_k(\mathbf{u}_k, \mathbf{x}_{k-1}, x_k{}^*) \equiv \int_{\mu=-\infty}^{\infty} \exp\left\{-\frac{\mu^2}{2\sigma_\mu{}^2} - \frac{\sum_{i=0}^{k-1} [x_i^{(0)} - d_i\mu]^2}{2\sigma_g{}^2}\right\}$$

$$\times \left[\int_{z_k=-\infty}^{\infty} [F_k\mu - x_k^{(0)*} + a_0 z_k]^2 \exp\left\{-\frac{z_k{}^2}{2\sigma_g{}^2}\right\} dz_k\right] d\mu, \tag{6.241}$$

where $z_k = v_k - u_k$. The integral I_k in the square brackets is to within a constant factor equal to

$$I_k \equiv 1 + \frac{(x_k^{(0)*} - F_k\mu)^2}{a_0{}^2\sigma_g{}^2}. \tag{6.242}$$

Therefore

$$\alpha_k \equiv \int_{\mu=-\infty}^{\infty} \left[1 + \frac{(x_k^{(0)*} - F_k\mu)^2}{a_0{}^2\sigma_g{}^2}\right]$$

$$\times \exp\left\{-\frac{\mu^2}{2\sigma_\mu{}^2} - \frac{\sum_{i=0}^{k-1} [x_i^{(0)} - d_i\mu]^2}{2\sigma_g{}^2}\right\} d\mu \tag{6.243}$$

or after transformations,

$$\alpha_k \equiv \exp\{-D_{k-1}\} \left[\int_{-\infty}^{\infty} \exp\{-C_{k-1}\mu^2 + B_{k-1}\mu\} \, d\mu\right.$$

$$\left. + \int_{-\infty}^{\infty} \frac{(x_k^{(0)*} - F_k\mu)^2}{a_0{}^2\sigma_g{}^2} \exp\{-C_{k-1}\mu^2 + B_{k-1}\mu\} \, d\mu. \right. \tag{6.244}$$

By determining the integrals in the brackets (see [5.26]), we arrive at the expression

$$\alpha_k \equiv \exp\left\{-D_{k-1} + \frac{B_{k-1}^2}{4C_{k-1}}\right\}\left(1 + \frac{F_k{}^2}{2a_0{}^2\sigma_g{}^2C_{k-1}} + \frac{E_k{}^2}{a_0{}^2\sigma_g{}^2}\right). \tag{6.245}$$

In this formula only E_k depends on u_k, which is equal to

$$E_k = \frac{B_{k-1}F_k}{2C_{k-1}} - x_k^{(0)*}$$
$$= \sum_{i=0}^{k-1}\left(\frac{d_iF_k}{2\sigma_g{}^2C_{k-1}} + a_{k-i}\right)\sum_{j=0}^{i}\kappa_{ij}\left(x_j - \sum_{\rho=0}^{j}a_\rho u_{j-\rho}\right) - x_k{}^* + \sum_{j=0}^{k}a_{k-j}u_j\,. \tag{6.246}$$

Hence from (6.245) it follows that the minimization of α_n with respect to u_n reduces to the minimization of E_n with respect to u_n, and the latter will be obtained if $u_n{}^*$ is chosen such that $E_n = 0$. Therefore

$$u_n{}^* = \frac{1}{a_0}\left[x_n{}^* - \sum_{i=0}^{n-1}\left(\frac{d_iF_n}{2\sigma_g{}^2C_{n-1}} + a_{n-i}\right)\sum_{j=0}^{i}\kappa_{ij}\left(x_j - \sum_{\rho=0}^{j}a_\rho u_{j-\rho}\right)\right.$$
$$\left. - \sum_{j=0}^{n-1}a_{n-j}u_j\right] = \frac{K_n}{a_0}, \tag{6.247}$$

where the square brackets are denoted by the letter K_n.

This is the optimal strategy at the nth time.

Passing from n to $n-1$, $\gamma_n{}^* = \alpha_n{}^*$ must be integrated with respect to x_{n-1}, and then the function α_{n-1} added to the integral, after which the value u_{n-1}^* minimizing the expression obtained can be found. From formula (6.245) it is seen that for $E_n = 0$ the function $\alpha_n = \alpha_n{}^*$ depends on u_{n-1}, since this variable enters into the expressions for D_{n-1} and B_{n-1}. However, after using the substitution

$$x_{n-1} - \sum_{j=0}^{n-1}a_{n-1-j}u_j = \lambda_{n-1} \tag{6.248}$$

and integrating with respect to the new variable λ_{n-1} within the limits from $-\infty$ to $+\infty$, we can be convinced that $\int_{-\infty}^{\infty}\alpha_n{}^*\,dx_{n-1}$ does not depend on u_{n-1}. Therefore the minimization of γ_{n-1} with respect to u_{n-1} reduces to the minimization of α_{n-1} with respect to u_{n-1}. Analogous arguments show that for any k, $u_k{}^*$ is found by the minimization of α_k, and this reduces to choosing $u_k{}^*$ such that the condition $E_k = 0$ be ensured. Hence we find the optimal strategy at an arbitrary sth time:

$$u_s{}^* = \frac{1}{a_0}\left[x_s{}^* - \sum_{i=0}^{s-1}\left(\frac{d_iF_s}{2\sigma_g{}^2C_{s-1}} + a_{s-i}\right)\sum_{j=0}^{i}\kappa_{ij}\left(x_j - \sum_{\rho=0}^{j}a_\rho u_{j-\rho}\right)\right.$$

$$\left. - \sum_{j=0}^{s-1}a_{s-j}u_j\right] = \frac{K_s}{a_0}, \tag{6.249}$$

where the expression in the brackets is denoted by K_s.

The physical meaning of the result obtained becomes clearer if it is taken into account that

$$\mu = x_i - w_i = x_i - \sum_{j=0}^{i}a_{i-j}v_j = x_i - \sum_{j=0}^{i}[a_{i-j}(u_j + g_j)]$$

$$= \left(x_i - \sum_{j=0}^{i}a_{i-j}u_j\right) - \sum_{j=0}^{i}a_{i-j}g_j\,. \tag{6.250}$$

The last term in this expression is random, and its dispersion depends on i. The expression in the brackets is the mean value. Thus the second term in the formula for K_s gives an estimate of the random variable μ. Different results of measurements of μ enter into this estimate with different weights, since the dispersions are unequal for different measurements.

The last term in the formula for K_s is the mean value of the results of previous actions u_j ($j < s$), remaining at the instant $t = s$ due to the presence of memory. It is natural that for determining $u_s{}^*$ the calculation of this residual is necessary. Thus all the terms of formula (6.249) have a clear physical meaning.

It can be shown under certain auxiliary conditions that for a sufficiently large value of s, the mean term in the formula for K_s reduces to the value

$$\frac{1}{s}\sum_{i=0}^{s-1}\left(x_i - \sum_{j=0}^{i}a_{i-j}u_j\right), \tag{6.251}$$

i.e., to the arithmetic mean of the results of measurements of μ.

In principle the result obtained could be extended to the case of a continuous system. If in the continuous case the relation between $w(t)$ and $v(t)$ is described, for example by the equation of an inertial element with a transfer function

$$K(p) = 1/(1 + pT_0), \tag{6.252}$$

then

$$w(t) = \int_0^t b(\tau)v(t - \tau)\, d\tau, \tag{6.253}$$

where the weighting function is

$$b(t) = (1/T_0) \exp\{-t/T_0\}. \tag{6.254}$$

We make Eq. (6.253) discrete by setting

$$w(t) \approx \sum_{\rho=0}^{s} b(t_\rho)v(t - t_\rho)\, \Delta t = \sum_{\rho=0}^{s} a_\rho v(t - t_\rho) = \sum_{\rho=0}^{s} a_\rho v_{s-\rho}, \tag{6.255}$$

where

$$a_\rho = b(t_\rho)\, \Delta t = (\Delta t/T_0) \exp\{-\rho\, \Delta t/T_0\}, \qquad a_0 = \Delta t/T_0\,. \tag{6.256}$$

Substituting the values a_ρ into formula (6.249) for the optimal strategy, and by letting Δt tend to zero after setting $s\Delta t = t$, in the limit the optimal strategy for the continuous case could be obtained in the same way as in the example presented in Section 2 of this chapter. However, as is also clear from physical considerations, here $a_0 \to 0$ and for example, the value of u_n^* in formula (6.247) becomes infinitely large. The same is obtained for other u_s as well. Meanwhile for a real system only the solution corresponding to bounded values of u_s has physical meaning. Therefore an auxiliary condition can still be introduced into the discrete case, for example

$$| u_s | \leqslant M, \tag{6.257}$$

where M is some constant.

Here, however, the optimal strategy changes. For example, instead of (6.247) we must now write

$$u_n^* = \begin{cases} K_n/a_0 & \text{for} \quad | K_n/a_0 | \leqslant M, \\ M \operatorname{sign} (K_n/a_0) & \text{for} \quad | K_n/a_0 | > M. \end{cases} \tag{6.258}$$

The investigation of this problem shows that now $\int \gamma_n^*\, dx_{n-1}$ depends on u_{n-1}, and the system cannot be regarded as neutral. Therefore the calculation is made significantly more complex.

Is the system under consideration reducible ? We can answer affirmatively to this question. The definition of a reducible system given in the previous paragraph can also be carried over to systems in which the objects have memory. Instead of the function $f(s, z_s, g_s, h_s)$ characterizing the block E in Fig. 6.11b, it is only necessary to have in view the more complicated function $f(s, \mathbf{z}_s, \mathbf{g}_s, \mathbf{h}_s)$, where $\mathbf{z}_s$, $\mathbf{g}_s$, $\mathbf{h}_s$ are the corresponding temporal vectors. In the example under consideration, from (6.222) and (6.224) with (6.250) taken into account it follows that:

$$x_i - \sum_{k=0}^{i} a_k u_{i-k} = \mu + \sum_{k=0}^{i} a_k g_{i-k} . \tag{6.259}$$

Hence the combination of measured variables appearing on the left side of this equation is equivalent to measuring the quantity μ with a random error $\sum_{k=0}^{s} a_k g_{s-k}$. Thus the formation of the quantities $x_i{}^*$ used in the control algorithm is completely equivalent to measuring the value of μ in the equivalent open-loop diagram depicted in Fig. 6.17. This circuit corresponds to Eq. (6.259). The block C is characterized by the equation

$$g_s{}' = \sum_{k=0}^{s} a_k g_{s-k} , \tag{6.260}$$

where g_s is the input and $g_s{}'$ is the output variable of the block. This block is identical to the block B of the object.

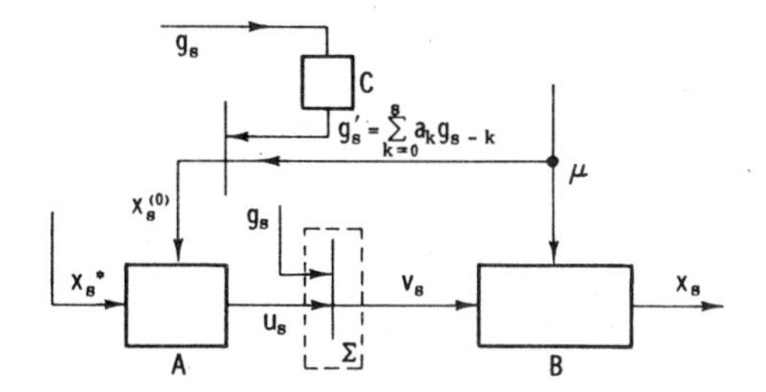

FIG. 6.17

Since the circuit under consideration is reducible, it must be neutral in the sense that the rate of study of the object B must not depend on the values of u_s. But why in such a case does the integral

$$\int_{-\infty}^{\infty} \gamma_{n-k}^{*} \, dx_{n-k-1}$$

in the presence of the constraint (6.257) nevertheless turn out to depend on u_{n-k}, which greatly complicates the calculations? The reason, it turns out, lies in the fact that in the case of objects with memory, this integral now characterizes not only the probing risk but also the known part of the operating risk, more precisely the "aftereffect" risk. In fact, now the control action u_s produced at $t = s$ will also have an influence on x_k at subsequent times ($k > s$), where the aftereffect will also occur in the equivalent open-loop circuit depicted in Fig. 6.17. As long as the constraint (6.257) did not play a part, these aftereffects could be neglected. Actually, in the optimal control law (6.249) a term will be contained (the last one in the brackets), which just compensates, neutralizes all "traces" of the previous actions remaining at the output of the system, on account of the fact that the object B has memory. But in those cases when u_s is bounded in absolute value, such a compensation may prove to be impossible. In this case the control action u_s affects the subsequent values of x_k even in the open-loop system (as also in the closed-loop system of Fig. 6.16 equivalent to it). This influence must not be confused with the probing risk; in reducible systems it enters in the form of a component of the operating risk, which can be called the aftereffect risk.

5. Generalization to the Problem with Markov Objects

Let us now consider the generalization to the wider class of random processes z_i and x_i^* [6.11, 6.12]. Let

$$z_s = z_s(s, \bar{\mu}_s) \tag{6.261}$$

and

$$x_i^* = x_i^*(s, \bar{\lambda}_s), \tag{6.262}$$

where in contrast to the preceding $\bar{\mu}_s$ and $\bar{\lambda}_s$ are not random variables, but discrete random vector Markov processes. We will call objects for which $\bar{\mu}_s$ is a Markov process Markov objects. The vectors

$$\bar{\mu}_s = (\mu_s^1, \ldots, \mu_s^m) \tag{6.263}$$

and

$$\bar{\lambda}_s = (\lambda_s^1, \ldots, \lambda_s^l) \tag{6.264}$$

represent the sets of scalar variables μ_s^i and λ_s^j ($i = 1, \ldots, m; j = 1, \ldots, l$). Here μ_s^i and μ_s^k are discrete scalar Markov processes related to each other in the general case. The same also holds with respect to λ_s^j and λ_s^r. But we regard the vectors $\bar{\mu}_s$, $\bar{\lambda}_s$, and also the noises h_s^*, h_s, g_s as independent of each other.

Let the probabilistic characteristics of the Markov processes $\bar{\mu}_s$ and $\bar{\lambda}_s$ be described. This means that both the initial probability densities $P_0(\bar{\mu}_0)$ and $P_0(\bar{\lambda}_0)$ at $t = 0$, and the transition probability densities, i.e., $P(\bar{\mu}_{i+1} \mid \bar{\mu}_i)$ and $P(\bar{\lambda}_{i+1} \mid \bar{\lambda}_i)$, are given. For the rest the problem remains the same; only for simplicity we will confine ourselves to the case when the object B has a single input and a single output. We will consider that the object does not have memory. The generalization to objects with memory and with several inputs and outputs can be made in the same way as above. In the generalization under consideration the matrices $\bar{\boldsymbol{\mu}}_s$ and $\bar{\boldsymbol{\lambda}}_s$ must be introduced, composed of the column vectors $\bar{\mu}_s$ and $\bar{\lambda}_s$:

$$\bar{\boldsymbol{\mu}}_s = (\bar{\mu}_0, \bar{\mu}_1, ..., \bar{\mu}_s), \qquad \bar{\boldsymbol{\lambda}}_s = (\bar{\lambda}_0, \bar{\lambda}_1, ..., \bar{\lambda}_s). \tag{6.265}$$

The specific risk r_s is given by a formula very close to the previous one, and differing from it only in that $\bar{\lambda}_s$ is in place of $\bar{\lambda}$. In addition, $\bar{\mu}_s$ appears instead of $\bar{\mu}$ in subsequent formulas. Thus,

$$\begin{aligned} r_s &= M\{W_s \mid \mathbf{y}_s{}^*, \mathbf{y}_{s-1}, \mathbf{u}_{s-1}\} \\ &= \int_{\Omega(\bar{\lambda}_s, x_s)} W_s[s, x_s, x_s{}^*(s, \bar{\lambda}_s)] P(\bar{\lambda}_s, x_s \mid \mathbf{y}_s{}^*, \mathbf{u}_{s-1}, \mathbf{y}_{s-1})\, d\Omega. \end{aligned} \tag{6.266}$$

Here $P(\bar{\lambda}_s, x_s \mid \mathbf{y}_s{}^*, \mathbf{u}_{s-1}, \mathbf{y}_{s-1})$ is the joint conditional probability density of $\bar{\lambda}_s$ and x_s for fixed vectors $\mathbf{y}_s{}^*, \mathbf{y}_{s-1}, \mathbf{u}_{s-1}$. According to the multiplication theorem

$$\begin{aligned} P(\bar{\lambda}_s, x_s \mid \mathbf{y}_s{}^*, \mathbf{u}_{s-1}, \mathbf{y}_{s-1}) &= P(x_s \mid \mathbf{y}_s{}^*, \mathbf{u}_{s-1}, \mathbf{y}_{s-1}, \bar{\lambda}_s) \cdot P(\bar{\lambda}_s \mid \mathbf{y}_s{}^*, \mathbf{u}_{s-1}, \mathbf{y}_{s-1}) \\ &= P(\bar{\lambda}_s \mid \mathbf{y}_s{}^*) \cdot P(x_s \mid \mathbf{y}_s{}^*, \mathbf{u}_{s-1}, \mathbf{y}_{s-1}). \end{aligned} \tag{6.267}$$

The latter transformation holds since the probability density of $\bar{\lambda}_s$ for fixed $\mathbf{y}_s{}^*$ does not change if $\mathbf{u}_{s-1}, \mathbf{y}_{s-1}$ are still fixed (see Fig. 6.2). Further, the probability density of x_s for fixed $\mathbf{y}_s{}^*$ does not change if $\bar{\lambda}_s$ is still fixed as well. We will rewrite the second factor of (6.267) in expanded form:

$$\begin{aligned} &P(x_s \mid \mathbf{y}_s{}^*, \mathbf{u}_{s-1}, \mathbf{y}_{s-1}) \\ &\quad = \int_{\Omega(\bar{\mu}_s, u_s)} P(x_s \mid \bar{\mu}_s, u_s) \cdot P_s(\bar{\mu}_s) \cdot \Gamma_s(u_s \mid \mathbf{y}_s{}^*, \mathbf{u}_{s-1}, \mathbf{y}_{s-1})\, d\Omega. \end{aligned} \tag{6.268}$$

Here $P_s(\bar{\mu}_s)$ is the *a posteriori* probability density for $\bar{\mu}_s$ at the sth time:

$$\begin{aligned} P_s(\bar{\mu}_s) &= P(\bar{\mu}_s \mid \mathbf{y}_{s-1}, \mathbf{u}_{s-1}, \mathbf{y}_s{}^*) \\ &= \int_{\Omega(\bar{\mu}_{s-1})} P(\bar{\mu}_s \mid \bar{\mu}_{s-1}) \cdot P(\bar{\mu}_{s-1} \mid \mathbf{y}_{s-1}, \mathbf{u}_{s-1}, \mathbf{y}_s{}^*)\, d\Omega. \end{aligned} \tag{6.269}$$

Since

$$P(\bar{\mu}_{s-1} \mid \mathbf{y}_{s-1}, \mathbf{u}_{s-1}, \mathbf{y}_s^*) = \int_{\Omega(\bar{\boldsymbol{\mu}}_{s-2})} P(\bar{\boldsymbol{\mu}}_{s-1} \mid \mathbf{y}_{s-1}, \mathbf{u}_{s-1}, \mathbf{y}_s^*)\, d\Omega, \tag{6.270}$$

then it is required to find the conditional probability $P(\bar{\boldsymbol{\mu}}_{s-1} \mid \mathbf{y}_{s-1}, \mathbf{u}_{s-1}, \mathbf{y}_s^*)$ for the matrix $\bar{\boldsymbol{\mu}}_{s-1}$. From the equation

$$\begin{aligned} P(\bar{\boldsymbol{\mu}}_{s-1}, \mathbf{u}_{s-1}, \mathbf{y}_{s-1} \mid \mathbf{y}_s^*) &= P(\mathbf{u}_{s-1}, \mathbf{y}_{s-1} \mid \bar{\boldsymbol{\mu}}_{s-1}, \mathbf{y}_s^*) \cdot P(\bar{\boldsymbol{\mu}}_{s-1}) \\ &= P(\bar{\boldsymbol{\mu}}_{s-1} \mid \mathbf{u}_{s-1}, \mathbf{y}_{s-1}, \mathbf{y}_s^*) \cdot P(\mathbf{u}_{s-1}, \mathbf{y}_{s-1} \mid \mathbf{y}_s^*) \end{aligned} \tag{6.271}$$

we find:

$$\begin{aligned} P_{s-1}(\bar{\boldsymbol{\mu}}_{s-1}) &= P(\bar{\boldsymbol{\mu}}_{s-1} \mid \mathbf{u}_{s-1}, \mathbf{y}_{s-1}, \mathbf{y}_s^*) \\ &= \frac{P(\mathbf{u}_{s-1}, \mathbf{y}_{s-1} \mid \bar{\boldsymbol{\mu}}_{s-1}, \mathbf{y}_s^*) \cdot P(\bar{\boldsymbol{\mu}}_{s-1})}{P(\mathbf{u}_{s-1}, \mathbf{y}_{s-1} \mid \mathbf{y}_s^*)}. \end{aligned} \tag{6.272}$$

Here $P(\mathbf{u}_{s-1}, \mathbf{y}_{s-1} \mid \mathbf{y}_s^*)$ is the joint conditional *a priori* density of the vectors $\mathbf{u}_{s-1}, \mathbf{y}_{s-1}$; $P(\bar{\boldsymbol{\mu}}_{s-1})$ is the *a priori* probability density of the matrix $\bar{\boldsymbol{\mu}}_{s-1}$, and $P(\mathbf{u}_{s-1}, \mathbf{y}_{s-1} \mid \bar{\boldsymbol{\mu}}_{s-1}, \mathbf{y}_s^*)$ is the joint conditional probability density of $\mathbf{u}_{s-1}$ and $\mathbf{y}_{s-1}$ for a fixed matrix $\bar{\boldsymbol{\mu}}_{s-1}$ and fixed $\mathbf{y}_s^*$ (the likelihood function). Calculations analogous to that carried out in Section 1 permit the expression

$$P_{s-1}(\bar{\boldsymbol{\mu}}_{s-1}) = \frac{P_0(\bar{\mu}_0) \cdot \prod_{i=1}^{s-1} P(\bar{\mu}_i \mid \bar{\mu}_{i-1}) [\prod_{i=0}^{s-1} P(y_i \mid \bar{\mu}_i, i, u_i)] \cdot [\prod_{i=0}^{s-1} \Gamma_i]}{P(\mathbf{u}_{s-1}, \mathbf{y}_{s-1} \mid \mathbf{y}_s^*)} \tag{6.273}$$

to be found. The substitution of (6.273) into (6.270), then (6.270) into (6.269), (6.269) into (6.268), (6.268) into (6.267) permits the second factor in (6.267) to be determined. Let us now examine the first factor of this expression—the *a posteriori* probability density of the vector $\bar{\lambda}_s$:

$$P_s(\bar{\lambda}_s) = P(\bar{\lambda}_s \mid \mathbf{y}_s^*) = \int_{\Omega(\bar{\boldsymbol{\lambda}}_{s-1})} P(\bar{\boldsymbol{\lambda}}_s \mid \mathbf{y}_s^*)\, d\Omega. \tag{6.274}$$

Since

$$P(\bar{\boldsymbol{\lambda}}_s, \mathbf{y}_s^*) = P(\bar{\boldsymbol{\lambda}}_s) \cdot P(y_s^* \mid \bar{\boldsymbol{\lambda}}_s) = P(\bar{\boldsymbol{\lambda}}_s \mid \mathbf{y}_s^*) \cdot P(\mathbf{y}_s^*), \tag{6.275}$$

then

$$P_s(\bar{\boldsymbol{\lambda}}_s) = P(\bar{\boldsymbol{\lambda}}_s) \cdot \frac{P(\mathbf{y}_s^* \mid \bar{\boldsymbol{\lambda}}_s)}{P(\mathbf{y}_s^*)}. \tag{6.276}$$

The *a priori* probability density $P(\bar{\boldsymbol{\lambda}}_s)$ of the matrix $\bar{\boldsymbol{\lambda}}_s$ is determined from a formula which holds for a Markov process:

$$P(\bar{\boldsymbol{\lambda}}_s) = P(\bar{\lambda}_0, \bar{\lambda}_1, ..., \bar{\lambda}_s) = P_0(\bar{\lambda}_0) \cdot P(\bar{\lambda}_1 \mid \bar{\lambda}_0) \cdot P(\bar{\lambda}_2 \mid \bar{\lambda}_1) \cdots$$
$$\cdots P(\bar{\lambda}_s \mid \bar{\lambda}_{s-1}) = P_0(\bar{\lambda}_0) \cdot \prod_{i=1}^{s} P(\bar{\lambda}_i \mid \bar{\lambda}_{i-1}). \tag{6.277}$$

Further, in view of the channel H^* being inertialess, the conditional probability density of $\mathbf{y}_s{}^*$ for fixed $\bar{\boldsymbol{\lambda}}_s$ is defined in the following way:

$$P(\mathbf{y}_s{}^* \mid \bar{\boldsymbol{\lambda}}_s) = P(y_0{}^* \mid \bar{\lambda}_0) \cdot P(y_1{}^* \mid \bar{\lambda}_1) \cdot \cdots \cdot P(y_s{}^* \mid \bar{\lambda}_s) = \prod_{i=0}^{s} P(y_i{}^* \mid \bar{\lambda}_i). \tag{6.278}$$

From (6.276)–(6.278) we obtain

$$P_s(\bar{\boldsymbol{\lambda}}_s) = \frac{P_0(\bar{\lambda}_0) \cdot \prod_{i=1}^{s} P(\bar{\lambda}_i \mid \bar{\lambda}_{i-1}) \cdot \prod_{i=0}^{s} P(y_i{}^* \mid \bar{\lambda}_i)}{P(\mathbf{y}_s{}^*)}. \tag{6.279}$$

Substituting (6.279) into (6.274), we arrive at a definitive formula for $P_s(\bar{\lambda}_s)$.

There exists a fundamental difference between formulas (6.279) and (6.273). While $P_s(\bar{\boldsymbol{\lambda}}_s)$ does not depend on the strategies Γ_i of the controller A, the expression for $P_s(\bar{\boldsymbol{\mu}}_s)$ does depend on them. Hence the storage of information about $\bar{\mu}_s$, i.e., about the unexpectedly varying characteristics of the object B, in general depends on the strategy of the controller. This dependence is the fact underlying the theory of dual control.

Making all the substitutions indicated above and then after substituting (6.267) into (6.266), a definitive formula for the conditional specific risk r_s can be obtained. If the values of r_s are examined for various trials made with the system, then in general the vectors $\mathbf{y}_s{}^*$, $\mathbf{u}_{s-1}$ and $\mathbf{y}_{s-1}$, not known in advance, can assume different values and be random. Their joint probability density is

$$P(\mathbf{y}_s{}^*, \mathbf{u}_{s-1}, \mathbf{y}_{s-1}) = P(\mathbf{u}_{s-1}, \mathbf{y}_{s-1} \mid \mathbf{y}_s{}^*)P(\mathbf{y}_s{}^*). \tag{6.280}$$

Then the specific risk R_s, the mean value of r_s after a mass production of trials, is defined by the formula

$$R_s = M\{r_s\} = \int_{\Omega(\mathbf{y}_s{}^*, \mathbf{u}_{s-1}, \mathbf{y}_{s-1})} r_s \cdot P(\mathbf{y}_s{}^*, \mathbf{u}_{s-1}, \mathbf{y}_{s-1})\, d\Omega$$
$$= \int_{\Omega(\mathbf{y}_s{}^*, \mathbf{u}_{s-1}, \mathbf{y}_{s-1})} r_s P(\mathbf{u}_{s-1}, \mathbf{y}_{s-1} \mid \mathbf{y}_s{}^*)P(\mathbf{y}_s{}^*)\, d\Omega. \tag{6.281}$$

Here after substituting the expression for r_s, we arrive at the formula

$$R_s = \int_{\Omega(\bar{\lambda}_s, x_s, \bar{\mu}_s, \mathbf{y}_s^*, u_s, \mathbf{y}_{s-1})} W_s[s, x_s^*(s, \bar{\lambda}_s), x_s] \cdot P_0(\bar{\lambda}_0)$$

$$\times \prod_{i=1}^{s} P(\bar{\lambda}_i \mid \bar{\lambda}_{i-1}) \cdot \prod_{i=0}^{s} P(y_i^* \mid \bar{\lambda}_i) \cdot P(x_s \mid s, \bar{\mu}_s, u_s) \cdot P_0(\bar{\mu}_0)$$

$$\times \prod_{i=1}^{s} P(\bar{\mu}_i \mid \bar{\mu}_{i-1}) \cdot \prod_{i=0}^{s-1} P(y_i \mid \bar{\mu}_i, i, u_i) \cdot \prod_{i=0}^{s} \Gamma_i \, d\Omega. \tag{6.282}$$

We will introduce the auxiliary functions α_k $(0 \leqslant k \leqslant n)$:

$$\alpha_k = \alpha_k(\mathbf{y}_k^*, u_k, \mathbf{u}_{k-1}, \mathbf{y}_{k-1}) = \int_{\Omega(\bar{\lambda}_k, \bar{\mu}_k, x_k)} W_k[k, x_k^*(k, \bar{\lambda}_k), x_k]$$

$$\times P(x_k \mid k, \bar{\mu}_k, u_k) \cdot P_0(\bar{\lambda}_0) \cdot \prod_{i=1}^{k} P(\bar{\lambda}_i \mid \bar{\lambda}_{i-1}) \cdot \prod_{i=0}^{k} P(y_i^* \mid \bar{\lambda}_i)$$

$$\times P_0(\bar{\mu}_0) \cdot \prod_{i=1}^{k} P(\bar{\mu}_i \mid \bar{\mu}_{i-1}) \cdot \prod_{i=0}^{k-1} P(y_i \mid \bar{\mu}_i, i, u_i) \, d\Omega$$

$$(k = 0, 1, \ldots, n). \tag{6.283}$$

Also let

$$\beta_k = \prod_{i=0}^{k} \Gamma_i . \tag{6.284}$$

Then, for example, the formula for the risk R_n corresponding to the time instant $t = n$ will take the same form as (6.38). Repeating the same arguments as in Section 1, we arrive at an analogous procedure for determining the optimal strategy: let $\gamma_n = \alpha_n$ and

$$\gamma_{n-k} = \gamma_{n-k}(\mathbf{y}_{n-k}^*, \mathbf{u}_{n-k}, \mathbf{y}_{n-k-1})$$

$$= \alpha_{n-k} + \int_{\Omega(y_{n-k}, y_{n-k+1}^*)} \gamma_{n-k+1}^*(u_{n-k+1}^*, \mathbf{y}_{n-k+1}^*, \mathbf{u}_{n-k}, \mathbf{y}_{n-k}) \, d\Omega$$

$$(k = 0, 1, \ldots, n). \tag{6.285}$$

Here

$$\gamma_{n-k}^* = \min_{u_{n-k} \in \Omega(u_{n-k})} \gamma_{n-k} = \gamma_{n-k}(u_{n-k}^*). \tag{6.286}$$

It is evident that

$$u_{n-k}^{*} = u_{n-k}^{*}(\mathbf{y}_{n-k}^{*}, \mathbf{u}_{n-k-1}, \mathbf{y}_{n-k-1}). \tag{6.287}$$

Then the optimal strategy is given by the expression

$$\Gamma_{n-k}^{*} = \delta(u_{n-k} - u_{n-k}^{*}), \tag{6.288}$$

i.e., it turns out to be regular and consists of choosing $u_{n-k} = u_{n-k}^{*}$.

In the particular case when the process $\bar{\lambda}_s$ turns into the random variable $\bar{\lambda}$, and $\bar{\mu}_s$ into the random variable $\bar{\mu}$, formula (6.283) for α_k is simplified and takes the form

$$\alpha_k = \int_{\Omega(\bar{\lambda},\bar{\mu},x_k)} W_k[k, x_k^*(k, \bar{\lambda}), x_k] \cdot P_0(\bar{\lambda}) \cdot \prod_{i=0}^{k} P(y_i^* \mid \bar{\lambda})$$

$$\times P(x_k \mid k, \bar{\mu} \;\;, u_k) \cdot P_0(\bar{\mu}_0) \cdot \prod_{i=0}^{k-1} P(y_i \mid \bar{\mu}, i, u_i) \, d\Omega. \tag{6.289}$$

This formula coincides with (6.36).

If the function $\bar{\mu}$ does not reduce to a constant and is a Markov random process, then the solution to the problem of dual control in the steady-state process acquires meaning. Otherwise if the quantity μ is constant during the process, with an infinite number of measurements to its end, the value of μ can be ascertained with an arbitrarily small error, and the information about the object becomes complete.

Let $n \to \infty$. In the steady-state process, the selection of a strategy can be required for which the minimum mean risk necessary at the time is ensured, i.e., the quantity

$$\rho = \lim_{n\to\infty} \frac{1}{n} \sum_{i=0}^{n} R_i. \tag{6.290}$$

If the limit

$$R_\infty = \lim_{\substack{i\to\infty \\ n\to\infty \\ n-i\to\infty}} R_i \tag{6.291}$$

exists, then $\rho = R_\infty$.

As the simplest example illustrating the theory presented above, let us examine the system whose block diagram is given in Fig. 6.18. Let the object B and the channel G be characterized by the equations

$$x_s = \mu_s + v_s \tag{6.292}$$

and

$$v_s = u_s + g_s \tag{6.293}$$

respectively. Here μ_s is a discrete Gaussian Markov random process with the initial probability density

$$P_0(\mu_0) = \frac{1}{\sigma_0 \sqrt{(2\pi)}} \exp\left\{-\frac{\mu_0^2}{2\sigma_0^2}\right\} \tag{6.294}$$

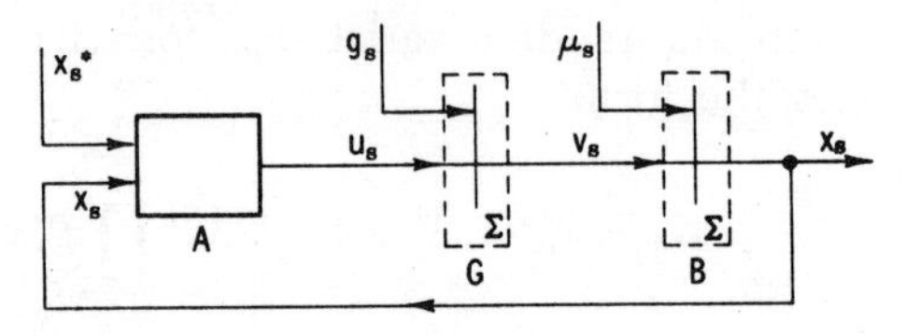

FIG. 6.18

and the transition probability density

$$P(\mu_k \mid \mu_{k-1}) = \frac{1}{\sigma_1 \sqrt{(2\pi)}} \exp\left\{-\frac{(\mu_k - \mu_{k-1})^2}{2\sigma_1^2}\right\}. \tag{6.295}$$

The variables g_s are independent, and the probability density is

$$P(g_s) = q(g_s) = \frac{1}{\sigma_g \sqrt{(2\pi)}} \exp\left\{-\frac{g_s^2}{2\sigma_g^2}\right\}. \tag{6.296}$$

Further, let $x_s^* = x^* =$ const and

$$W_k = (x^* - x_k)^2. \tag{6.297}$$

Since $h_s = 0$, then $y_s = x_s$. In this example

$$\begin{aligned} P(y_i \mid \mu_i, i, u_i) &= P(x_i \mid \mu_i, u_i) \\ &= q(x_i - u_i - \mu_i) = \frac{1}{\sigma_g \sqrt{(2\pi)}} \exp\left\{-\frac{(x_i - u_i - \mu_i)^2}{2\sigma_g^2}\right\}. \end{aligned} \tag{6.298}$$

From the general expression (6.283) we find α_k for this example. Here it must be taken into account that $h_s^* = 0$; therefore $P_0(\bar{\lambda}_0)$, $P(\bar{\lambda}_i \mid \bar{\lambda}_{i-1})$,

$P(y_i^* \mid \bar{\lambda}_i)$ degenerate into δ-functions and expression (6.283) can be rewritten in the form

$$\alpha_k = \alpha_k(x_k^*, \mathbf{u}_k, \mathbf{y}_{k-1}) = \int_{\Omega(\bar{\lambda}_k, \bar{\mu}_k, x_k)} W_k(k, x_k^*, x_k)$$

$$\times P(x_k \mid k, \bar{\mu}_k, u_k) \cdot P_0(\bar{\mu}_0) \cdot \prod_{i=1}^{k} P(\bar{\mu}_i \mid \bar{\mu}_{i-1})$$

$$\times \prod_{i=0}^{k-1} P(y_i \mid \bar{\mu}_i, i, u_i)\, d\Omega. \tag{6.299}$$

Here after substituting the appropriate expressions from (6.294), (6.295), (6.298), we find:

$$\alpha_k = C \int_{x_k=-\infty}^{\infty} \int_{\mu_0=-\infty}^{\infty} \cdots \int_{\mu_k=-\infty}^{\infty} (x^* - x_k)^2$$

$$\times \exp\left\{-\frac{\mu_0^2}{2\sigma_0^2} - \frac{(\mu_1 - \mu_0)^2 + (\mu_2 - \mu_1)^2 + \cdots + (\mu_k - \mu_{k-1})^2}{2\sigma_1^2}\right\}$$

$$\times \exp\left\{-\sum_{i=0}^{k} \frac{(x_i - u_i - \mu_i)^2}{2\sigma_g^2}\right\} d\mu_0 \cdot d\mu_1 \cdots d\mu_k \cdot dx_k, \tag{6.300}$$

where C is a constant.

After computation the integral (6.300) takes the following form [6.11]:

$$\frac{\alpha_k}{C'} = \exp\{-\bar{A}^* K \bar{A}\}\left\{\left[A_k^* - \sum_{i=0}^{k-1} e_{ik} A_i\right]^2 + d_k\right\}. \tag{6.301}$$

Here $C' =$ const, and the coefficients are obtained from the equations

$$\frac{x_i - u_i}{\sigma_g} = A_i, \qquad \frac{\mu_i}{\sigma_g} = \nu_i \qquad (i = 0, 1, \ldots, k),$$

$$\frac{x^* - u_k}{\sigma_g} = A_k^*, \qquad \frac{\sigma_0}{\sigma_g} = \theta_0, \qquad \frac{\sigma_1}{\sigma_g} = \theta_1 \tag{6.302}$$

and

$$a_1 = \frac{\theta_1^2}{\theta_0^2} + \theta_1^2 + 1,$$

$$a = 2 + \theta_1^2, \qquad b = 1 + \theta_1^2. \tag{6.303}$$

In addition, we introduce the notations for the $(k + 1)$-dimensional vectors A and ν:

$$A = (A_0, \ldots, A_k), \qquad \nu = (\nu_0, \ldots, \nu_k), \tag{6.304}$$

and we denote by (A, A) the scalar product of the vector with itself, by (A, ν) the scalar product of the vectors A and ν, and finally by A^* the vector transposed from A (the vector A is a column vector, and A^* is a row vector). Further, let

$$\bar{A} = (A_0, \ldots, A_{k-1}) \tag{6.305}$$

be a k-dimensional vector. The expression $\{\bar{A}^* K \bar{A}\}$ represents the scalar product of $\bar{A}^*$ and $K\bar{A}$, where K is some matrix with coefficients k_{ij} not containing the A_k^*. These coefficients are not needed later on and therefore are not written out. The coefficients e_{ik} and d_k are defined by the formulas

$$e_{ik} = \frac{C_{kk}}{1 - C_{kk}^2} C_{ik}, \qquad d_k = \frac{1}{1 - C_{kk}^2}, \tag{6.306}$$

where the C_{ik} are the coefficients of the triangular matrix C (from C_{00} to C_{kk}), which are defined by the expressions

$$C_{jk} = \frac{\Delta_j \theta_1}{\sqrt{[\Delta_k (b\,\Delta_k - \Delta_{k-1})]}} = \frac{\Delta_j}{\Delta_k} C_{kk} \qquad (0 \leqslant j \leqslant k). \tag{6.307}$$

Here we take $\Delta_0 = 1$ and

$$C_{kk} = \theta_1 \sqrt{\left(\frac{\Delta_k}{b\,\Delta_k - \Delta_{k-1}}\right)}, \tag{6.308}$$

and Δ_i $(i = 1, \ldots, k)$ is the determinant with i rows and columns:

$$\Delta_i = \begin{vmatrix} a_1 & -1 & 0 & \cdots & 0 \\ & \ddots & & & \\ -1 & a & -1 & \cdots & 0 \\ 0 & & \ddots & & \vdots \\ \vdots & & & \ddots & 0 \\ 0 & \cdots & 0\ -1 & a & -1 \\ & & & & \ddots \\ 0 & \cdots & 0 & -1 & a \end{vmatrix}. \tag{6.309}$$

After substituting (6.307) into the first of formulas (6.306), we find

$$e_{ik} = \left(\frac{(b\,\Delta_k - \Delta_{k-1})}{\theta_1^2\,\Delta_k} - 1\right)^{-1} \cdot \frac{\Delta_i}{\Delta_k}. \tag{6.310}$$

Let us set

$$Q_i = \frac{\Delta_i}{\Delta_{i-1}}, \qquad q = \frac{b\,\Delta_k - \Delta_{k-1}}{\Delta_k}. \tag{6.311}$$

It is evident that $Q_1 = \Delta_1/\Delta_0 = \Delta_1 = a_1 > 1$.

With (6.311) taken into account formula (6.310) can be rewritten in the form

$$e_{ik} = \left(\frac{q}{\theta_1^2} - 1\right)^{-1} \cdot \frac{\Delta_i}{\Delta_k}. \tag{6.312}$$

As is not difficult to show, from (6.309) it follows that

$$\Delta_i = a\Delta_{i-1} - \Delta_{i-2} \qquad (i = 2, \ldots, k). \tag{6.313}$$

Hence we find:

$$Q_i = a - (Q_{i-1})^{-1} \qquad (i = 2, \ldots, k). \tag{6.314}$$

Since $Q_1 > 1$ and $a = 2 + \theta_1^2 > 2$, then from relation (6.314) it follows that

$$Q_i > 1 \qquad (i = 1, \ldots, k). \tag{6.315}$$

The quantity

$$q = b - (Q_k)^{-1} > 0, \tag{6.316}$$

since $b = 1 + \theta_1^2 > 1$; but q can be both larger and smaller than unity.

If the expressions for A_i and A_k^* according to (6.302) are substituted into (6.301), then we will obtain, for example, for $k = n$:

$$\alpha_n = \left\{\left[\frac{x^* - u_n}{\sigma_g} - \sum_{i=0}^{n-1} e_{in}\left(\frac{x_i - u_i}{\sigma_g}\right)\right]^2 + d_n\right\}$$

$$\times \exp\left\{-\sum_{i,j=0}^{n-1} \frac{k_{ij}}{2} \cdot \frac{x_i - u_i}{\sigma_g} \cdot \frac{x_j - u_j}{\sigma_g}\right\}. \tag{6.317}$$

The variable u_n only enters into the composition of the brackets. The minimum of α_n with respect to u_n is obtained if these brackets are equated to zero. From this condition we find:

$$u_n^* = x^* - \sum_{i=0}^{n-1} e_{in}(x_i - u_i). \tag{6.318}$$

According to the general prescription for determining the optimal strategy, for finding u_{n-1}^* it is first necessary to ascertain γ_{n-1}, where

$$\gamma_{n-1} = \alpha_{n-1} + \int_{-\infty}^{\infty} \alpha_n^* \, dx_{n-1} = \alpha_{n-1} + \sigma_g \int_{-\infty}^{\infty} \alpha_n^* \, dA_{n-1}. \tag{6.319}$$

The integrand in the latter integral has the form

$$\alpha_n^* = \min_{u_n} \alpha_n = d_n \exp\left\{ - \sum_{i,j=0}^{n-1} \frac{k_{ij}}{2} A_i A_j \right\}. \tag{6.320}$$

If this expression is substituted into formula (6.319) and the integration carried out, then it will turn out that the second term in (6.319) does not depend on A_{n-1}, and hence not u_{n-1} either. Therefore the optimal value u_{n-1}^* can be obtained by minimizing only the component α_{n-1} of the expression for γ_{n-1} with respect to u_{n-1}. In the formula for α_{n-1} only the term A_{n-1} in the brackets depends on u_{n-1} [see (6.302)]. Equating the square brackets to zero, we find that

$$u_{n-1}^* = x^* - \sum_{i=0}^{n-2} e_{i,n-1}(x_i - u_i). \tag{6.321}$$

Analogous arguments for $k = n - 2, n - 3, \ldots$ lead to the general formula for the optimal strategy:

$$u_k^* = x^* - \sum_{i=0}^{k-1} e_{ik}(x_i - u_i). \tag{6.322}$$

We will explain the physical meaning of this formula. From Fig. 6.18 it is seen that the mathematical expectation of $(x^* - x_k)^2$ will be minimal if $u_k = x^* - \mu_k$ is set. The second term in formula (6.322) also represents an estimate of μ_k based on the measurement of the differences $x_i - u_i$ for $i < k$. The coefficients e_{ik} entering into this estimate depend on $k - i$. The values of these coefficients are defined by formula (6.310) or (6.312).

Let us examine the ratio

$$\frac{e_{ik}}{e_{jk}} = \frac{\Delta_i}{\Delta_j} < 1 \qquad (0 \leqslant i < j \leqslant k). \tag{6.323}$$

The physical meaning of the property of the optimal strategy, expressed by the ratio (6.323), is that lesser weight is attached to information of older origin—it "becomes obsolete." Thus not only the storage of new information, but also the process of degradation of obsolete information, takes place in the controller.

Let us set

$$i = k - \nu \tag{6.324}$$

and

$$e_{ik} = \left(\frac{q}{\theta_1^2} - 1\right)^{-1} \cdot \frac{\Delta_{k-\nu}}{\Delta_k} = f_{k\nu} \qquad (0 \leqslant i \leqslant k, \nu = 0, 1, .., k). \tag{6.325}$$

We will find the values of the coefficients for steady-state behavior:

$$\lim_{k\to\infty} f_{k\nu} = f_\nu \,. \tag{6.326}$$

Let us examine formula (6.314) as a preliminary. We take into account that $Q_1 = a_1 > a/2$. In general, let $Q_{i-1} > a/2$. Then

$$Q_i = a - \frac{1}{Q_{i-1}} > a - \frac{2}{a} = \frac{a}{2}\left(2 - \frac{4}{a^2}\right) > \frac{a}{2}\,. \tag{6.327}$$

Thus all the Q_i satisfy the condition $Q_i > a/2$. According to (6.314) and (6.327), the relation between Q_i and Q_{i-1} can be represented geometrically in the form of a hyperbola with equation $-1/Q_{i-1}$, shifted upwards by an amount a (curve 1 in Fig. 6.19). The bisector

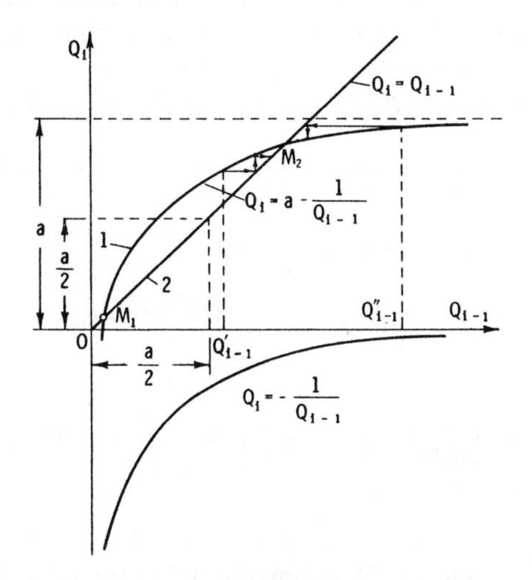

FIG. 6.19

of the first quadrant, i.e., the line $Q_i = Q_{i-1}$ (line 2), intersects curve 1 at the two points M_1 and M_2. The abscissas (or ordinates) x_1 and x_2 of these points represent the roots of the quadratic equation

$$x = a - x^{-1}, \tag{6.328}$$

i.e., the values

$$x_{1,2} = a/2 \pm \sqrt{[(a^2/4) - 1]}. \tag{6.329}$$

Hence it follows that the ordinate of the point M_1 is less than $a/2$, which is also shown in Fig. 6.19. We will note that the construction in this figure is valid for arbitrary $i = 1, 2, \ldots$.

Let us assign some definite value of $Q_{i-1} > a/2$.

In Fig. 6.19 this is Q'_{i-1} or Q''_{i-1}. Then the appropriate values Q_i, Q_{i+1}, Q_{i+2}, and so on can be found graphically with the aid of the "staircase" shown in Fig. 6.19. The value Q_{i+1} corresponding to Q_i' is obtained if a dashed ordinate from the point Q_i' to curve 1 is drawn. Then the ordinate of the intersection point obtained (we will call it the first point of the "staircase") is equal to Q_{i+1}. But this value must be laid off on the abscissa if we wish to find Q_{i+2}. However, the transfer of the value Q_{i+1} on the abscissa can be replaced by drawing a horizontal line (it is provided with an arrow) from the first point of the "staircase" to the bisector, i.e., to line 2. The intersection point obtained is the second point of the "staircase." It is not difficult to see that the value Q_{i+2} is obtained if a vertical line is drawn from the second point of the "staircase" to the intersection with curve 1. Analogous transfers permit Q_{i+3}, Q_{i+4}, and so on, to be found. As $i \to \infty$ the value Q_i converges to the quantity Q_∞ equal to the largest of the values of x, as is seen from the construction in Fig. 6.19. Thus from (6.329) we obtain:

$$Q_\infty = \lim_{i\to\infty} Q_i = a/2 + \sqrt{[(a^2/4) - 1]} > a/2 > 1. \tag{6.330}$$

Further, as $k \to \infty$ the quantity

$$q = b - (Q_k)^{-1} \to b - (Q_\infty)^{-1} = q_\infty . \tag{6.331}$$

Hence,

$$f_0 = \left(\frac{q_\infty}{\theta_1^2} - 1\right)^{-1} = \left[\frac{1}{\theta_1^2}\left(b - \frac{1}{Q_\infty}\right) - 1\right]^{-1}. \tag{6.332}$$

Further,

$$\frac{f_{\nu+1}}{f_\nu} = \lim_{k\to\infty} \frac{\Delta_{k-\nu-1}}{\Delta_{k-\nu}} = \lim_{k\to\infty} (Q_{k-\nu})^{-1} = (Q_\infty)^{-1} < 1. \tag{6.333}$$

If e_{ik} is replaced by $f_{k\nu}$ according to (6.325), and we pass to the limit as $k \to \infty$, then expression (6.322) is replaced by an infinite series in which u_0 is the current value, and $x_{-\nu}$ and $u_{-\nu}$ are the values measured ν times ago:

$$u_0 = x^* - \sum_{\nu=1}^{\infty} f_\nu (x_{-\nu} - u_{-\nu}). \tag{6.334}$$

The weights f_ν, with which the differences $x_{-\nu} - u_{-\nu}$ enter into the sum (6.334), decrease with an increase in ν by a geometric progression law; this follows from (6.333).

Formula (6.334) can be given a simpler form. From (6.334) and (6.333) it follows that:

$$u_{-1} = x^* - \sum_{\nu=2}^{\infty} f_{\nu-1}(x_{-\nu} - u_{-\nu}) = x^* - Q_\infty \sum_{\nu=2}^{\infty} f_\nu (x_{-\nu} - u_{-\nu}). \tag{6.335}$$

But according to (6.334) we can write:

$$\sum_{\nu=2}^{\infty} f_\nu (x_{-\nu} - u_{-\nu}) = x^* - u_0 - f_1(x_{-1} - u_{-1}). \tag{6.336}$$

Substituting (6.336) into (6.335), we arrive at the equation

$$Q_\infty u_0 = (Q_\infty - 1)x^* + u_{-1} - f_1(x_{-1} - u_{-1}). \tag{6.337}$$

From (6.333) it follows that

$$f_1 = \frac{f_0}{Q_\infty}.$$

Therefore from (6.337) we find:

$$u_0 = (Q_\infty)^{-1} \left[(Q_\infty - 1)x^* + \left(1 + \frac{f_0}{Q_\infty}\right) u_{-1} - \frac{f_0}{Q_\infty} x_{-1} \right]. \tag{6.338}$$

The block diagram corresponding to formula (6.338) is shown in Fig. 6.20. In this circuit τ is a single-time delay element. Therefore at the output of the element there is, for example, the value u_{-1} if u_0 acts at the input. The remaining elements of the scheme are amplifiers and one sum type. The circuit of the optimal controller A, serving only for steady-state behavior, contains internal positive feedback through the amplifier element $1 + f_0/Q_\infty$ and delay element.

For example, let $\theta_1 = 1$. Then $b = 1 + \theta_1^2 = 2$; $a = 2 + \theta_1^2 = 3$. The value Q_∞ computed from Eq. (6.330) is found to be equal to 2.615. Then

$$f_0 = \left[\frac{1}{1}\left(2 - \frac{1}{2.615}\right) - 1\right]^{-1} = 1.62,$$

$$f_1 = \frac{f_0}{Q_\infty} = \frac{1.62}{2.615} = 0.618,$$

$$f_2 = \frac{f_1}{Q_\infty} = \frac{0.618}{2.615} = 0.235 \quad \text{and so on.}$$

Thus the series (6.334) takes the form

$$u_0 = x^* - 0.618(x_{-1} - u_{-1}) - 0.235(x_{-2} - u_{-2}) - \cdots .$$

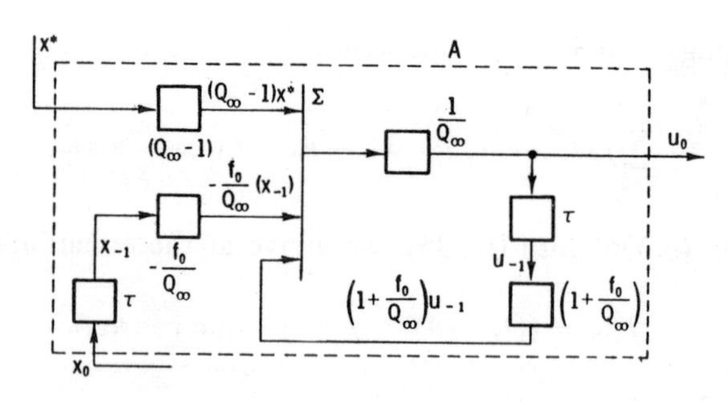

FIG. 6.20

Let us pass to continuous measurement.

Let the time $T = n\,\Delta t$ be fixed and divided into n equal intervals of length Δt. We first assume that measurements are made only at the ends of the intervals, and the value of σ_g^2 depends on Δt in the following way:

$$\sigma_g^2 = \frac{S_0}{\Delta t}. \tag{6.339}$$

Then as is known from Chapter V, in the limit as $\Delta t \to 0$ the sequence of random variables g_s turns into white noise with spectral density S_0. Further, let the dispersion σ_1^2 of the transition probability density depend on Δt according to the law:

$$\sigma_1^2 = \sigma_2^2\,\Delta t \tag{6.340}$$

where $\sigma_2^2 = \text{const}$. Then formula (6.295) assumes the form of the probability density for a normal continuous Markov process:

$$P(\mu_0 + \Delta\mu, t_0 + \Delta t \mid \mu_0, t_0) = \frac{1}{\sigma_2 \sqrt{(2\pi \Delta t)}} \exp\left\{-\frac{(\Delta\mu)^2}{2\sigma_2^2 \Delta t}\right\}, \quad (6.341)$$

where μ_0 is the value of μ at $t = t_0$, and $\Delta\mu$ is the increment of μ during the time Δt. In this case as $\Delta t \to 0$

$$\theta_1^2 = \frac{\sigma_2^2(\Delta t)^2}{S_0} \to 0, \qquad \theta_0^2 = \frac{\sigma_0^2 \Delta t}{S_0} \to 0. \quad (6.342)$$

Since for small values of Δt and θ_1 the value $Q_\infty = 1 + \theta_1$, then from formula (6.332) we find for this case:

$$f_0 \cong \left[\frac{1}{\theta_1^2}\left(1 - \frac{1}{1+\theta_1}\right) - 1\right]^{-1} \cong \theta_1 = (\sigma_2/\sqrt{S_0})\,\Delta t = L\,\Delta t, \quad (6.343)$$

where

$$L = \frac{\sigma_2}{\sqrt{S_0}}. \quad (6.344)$$

Expression (6.334) takes the form

$$u_0 = x^* - \sum_{\nu=1}^{\infty} \frac{L\,\Delta t}{(1 + L\,\Delta t)^\nu}(x_{-\nu} - u_{-\nu}). \quad (6.345)$$

After taking the notations $u(t)$ and $x(t)$ for current values, we obtain in the limit as $\Delta t \to 0$:

$$u(t) = x^* - L\int_{\tau=0}^{\infty} \exp\{-L\tau\}[x(t-\tau) - u(t-\tau)]\,d\tau. \quad (6.346)$$

Thus to obtain the optimal control law the difference $x - u$ must be fed to the input of an inertial element with a time constant L^{-1}. The output variable of this element is defined by the second term in expression (6.346).

6. On Block Diagrams of Optimal Controllers

A comparison of open-loop and closed-loop automatic systems with inertial controlled objects is of interest. With this aim let us examine a simple particular case of a closed-loop system, whose block diagram is shown in Fig. 6.21. Let the noise z_s depend both on s and the parameter vector $\bar{\mu}$:

$$z_s = z_s(s, \bar{\mu}), \tag{6.347}$$

and the equation of the discrete inertial object B can be written in the form

$$x_s = F_0(\mathbf{z}_s, \mathbf{u}_s) = F(s, \bar{\mu}, \mathbf{u}_s). \tag{6.348}$$

Here we regard the initial conditions as known and entering into expression (6.348) as parameters. If they are not known, then they can be added to the coordinates of the vector $\bar{\mu}$. We will set the specific loss function (taking x_s^* as a known parameter) in the form

$$W_s = W_s[s, x_s, x_s^*] = W_s[s, F(s, \bar{\mu}, \mathbf{u}_s), x_s^*] = W_s[s, \bar{\mu}, \mathbf{u}_s]. \tag{6.349}$$

The characteristic of the inertialess channel H in the feedback path is defined by the expression

$$y_s = y_s(h_s, x_s). \tag{6.350}$$

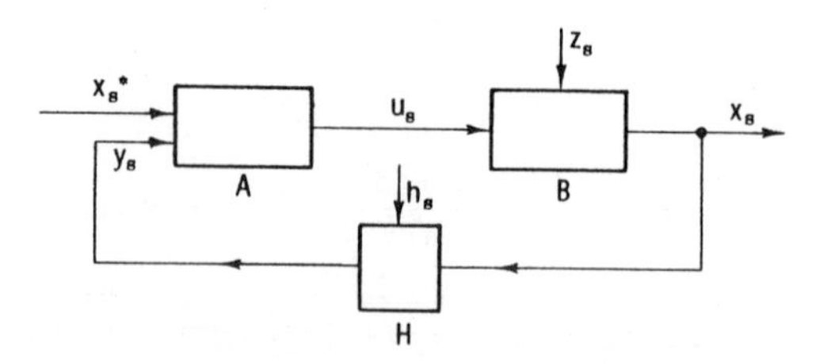

FIG. 6.21

Let the *a priori* probability densities $P(\bar{\mu})$ of the vector $\bar{\mu}$ and $P(h_s)$ for the independent random variables h_s ($s = 0, 1, 2, ..., n$) be prescribed. We take $\bar{\mu}$ and h_s independent. It is required to find the optimal strategy

$$P_s(u_s) = \Gamma_s(u_s \mid \mathbf{u}_{s-1}, \mathbf{y}_{s-1}) \tag{6.351}$$

of the device A. The specific risk R_s is expressed by this formula

$$R_s = M\{W_s\} = \int_{\Omega(\bar{\mu}, \mathbf{u}_s, \mathbf{y}_{s-1})} W_s[s, \bar{\mu}, \mathbf{u}_s] \cdot P(\bar{\mu}) \times \prod_{i=0}^{s-1} P(y_i \mid i, \bar{\mu}, \mathbf{u}_i) \cdot \prod_{i=0}^{s} \Gamma_i \, d\Omega. \tag{6.352}$$

Since for given u_i the information about y_i will not increase if $\mathbf{y}_{i-1}$ is prescribed, then the dependence on $\mathbf{y}_{i-1}$ is missing. The function α_k takes the form

$$\alpha_k(\mathbf{u}_k, \mathbf{y}_{k-1}) = \int_{\Omega(\bar{\mu})} W_k[k, \bar{\mu}, \mathbf{u}_k] \cdot P(\bar{\mu}) \prod_{i=0}^{k-1} P(y_i \mid i, \bar{\mu}, \mathbf{u}_i)\, d\Omega. \quad (6.353)$$

If $\gamma_n = \alpha_n$ is set and

$$\gamma_{n-k}(\mathbf{u}_{n-k}, \mathbf{y}_{n-k-1}) = \alpha_{n-k} + \int_{\Omega(y_{n-k})} \gamma_{n-k+1}^{*}\, d\Omega, \quad (6.354)$$

then the determination of the optimal value u_s^* reduces to the minimization of γ_s with respect to u_s, as was shown above repeatedly.

Can the closed-loop diagram of Fig. 6.21 be replaced by some equivalent open-loop diagram, not containing a feedback path but having a measurement circuit for the noise z_s?

In the equivalent diagram the values of u_s must be the same as in the actual scheme; therefore the behavior of the object B in both diagrams must be completely identical.

The circuit of Fig. 6.21 is easily transformed by using the formula following from (6.350) and (6.348):

$$y_s = y_s[h_s, x_s] = y_s[h_s, F_0(\mathbf{z}_s, \mathbf{u}_s)]. \quad (6.355)$$

In fact, if the operator F_0 of the object is given, then its artificial model B' can be constructed having the same operator, and z_s and u_s can be fed to the input of the model (Fig. 6.22). Let the initial conditions of the model B' and the object B coincide. Then at the output of the model we will obtain a quantity x_s' which is identical to the actual output variable of the object.

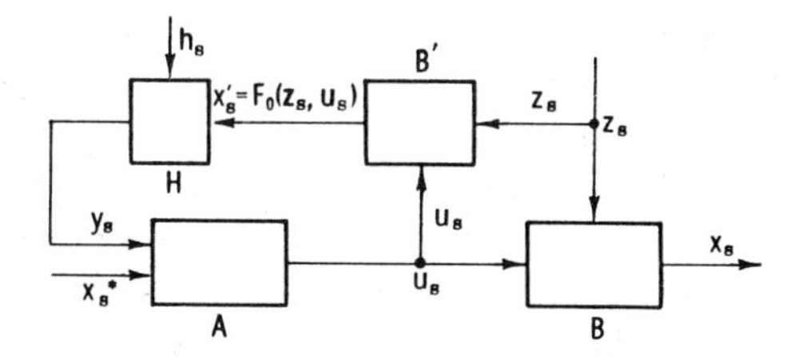

FIG. 6.22

However, the circuit depicted in Fig. 6.22 is not open loop in the general case, since it contains local feedback from the output of the device A to its input through the elements B' and H. The question is posed in the following manner: can such an open-loop system be found without any external feedbacks, as shown in Fig. 6.23, which would be equivalent to the optimal system depicted in Fig. 6.21 ? The block E in the scheme of Fig. 6.23 can also be a dynamical system,

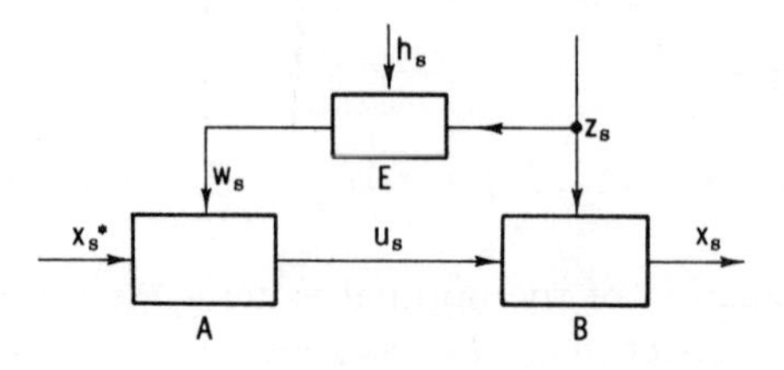

FIG. 6.23

i.e., its output w_s can depend not only on the current values h_s and z_s, but also on the "previous history" of the input variables. If the system E is inertialess with respect to h_s, then

$$w_s = w_s(h_s, \mathbf{z}_s) = w_s(h_s, s, \bar{\mu}). \tag{6.356}$$

The circuit shown in Fig. 6.23 is of independent interest. Let us find the optimal strategy of its controller A in the form

$$P_s(u_s) = \Gamma_s(u_s \mid \mathbf{u}_{s-1}, \mathbf{w}_{s-1}). \tag{6.357}$$

We will explain the meaning of the vectors to the right of the vertical line in expression (6.357). The optimal value or distribution of the variable u_s must depend on all the values $w_0, w_1, ..., w_{s-1}$, i.e., on the vector $\mathbf{w}_{s-1}$. Actually, these values give the possibility of computing z_s, and with it the unknown parameter vector $\bar{\mu}$ as well. The system E can be regarded as an inertial channel for measuring z_s, in which the noise h_s is also present. The more values w_i $(i = 0, 1, ..., s - 1)$ observed in the past, the more precisely the value of the unknown parameter vector $\bar{\mu}$ can in general be determined.

The vector $\mathbf{u}_{s-1}$ has an entirely different meaning in expression (6.357). Its calculation when u_s is computed is necessary with a view to the correct control of the object B. Since this object is inertial, then in general at the instant $t = s$ the consequence of all those actions $u_0, u_1, ..., u_{s-1}$ which have acted on it in the past manifests itself. Therefore the control action u_s must be determined with the state of the object taken into account, carrying upon itself the imprint of the entire "previous history" of its input actions along the channel of u. Thus u_s must depend on the vector $\mathbf{u}_{s-1}$.

We will find the strategy Γ_s that is optimal in the sense that the risk

$$R = M\{W\} = M\left\{\sum_{s=0}^{n} W_s\right\} = \sum_{s=0}^{n} R_s \tag{6.358}$$

is minimal. Here as before n is fixed.

Let us first find the conditional specific risk. Considering that x_s^* has been given and is a parameter, it cannot be written out in explicit form. Then

$$r_s = M\{W_s \mid \mathbf{w}_{s-1}\} = \int_{\Omega(\bar{\mu}, \mathbf{u}_s)} W_s[s, F_0(\mathbf{z}_s, \mathbf{u}_s)] \cdot P(\bar{\mu}, \mathbf{u}_s \mid \mathbf{w}_{s-1})\, d\Omega. \tag{6.359}$$

The joint probability density of $\bar{\mu}$ and $\mathbf{u}_s$ can be found from the expression

$$P(\bar{\mu}, \mathbf{u}_s \mid \mathbf{w}_{s-1}) = P(\bar{\mu} \mid \mathbf{w}_{s-1}) \cdot P(\mathbf{u}_s \mid \bar{\mu}, \mathbf{w}_{s-1})$$

$$= P(\bar{\mu} \mid \mathbf{w}_{s-1}) \cdot \prod_{i=0}^{s} \Gamma_i(u_i \mid \mathbf{u}_{i-1}, \mathbf{w}_{i-1}). \tag{6.360}$$

The latter transformation holds since $\mathbf{u}_s$ depends only on $\mathbf{w}_{s-1}$ (see Fig. 6.23) and the information about $\bar{\mu}$ does not change the probability density $P(\mathbf{u}_s \mid \mathbf{w}_{s-1})$. Moreover, $P(\mathbf{u}_s \mid \mathbf{w}_{s-1})$ is evidently the product of functions Γ_i.

Further, the joint probability density is

$$P(\mathbf{w}_{s-1}, \bar{\mu}) = P(\mathbf{w}_{s-1} \mid \bar{\mu}) \cdot P(\bar{\mu}) = P(\bar{\mu} \mid \mathbf{w}_{s-1}) \cdot P(\mathbf{w}_{s-1}), \tag{6.361}$$

where $P(\mathbf{w}_{s-1})$ is the absolute probability density of $\mathbf{w}_{s-1}$. Therefore the *a posteriori* probability density is

$$P_s(\bar{\mu}) = P(\bar{\mu} \mid \mathbf{w}_{s-1}) = \frac{P(\bar{\mu}) \cdot P(\mathbf{w}_{s-1} \mid \bar{\mu})}{P(\mathbf{w}_{s-1})}. \tag{6.362}$$

Substituting Bayes' formula (6.362) into (6.360) and taking into account that

$$P(\mathbf{w}_{s-1} \mid \bar{\mu}) = \prod_{i=0}^{s-1} P(w_i \mid i, \bar{\mu}), \tag{6.363}$$

we arrive at the expression

$$P(\bar{\mu}, \mathbf{u}_s \mid \mathbf{w}_{s-1}) = \frac{P(\bar{\mu}) \cdot \prod_{i=0}^{s-1} P(w_i \mid i, \bar{\mu})}{P(\mathbf{w}_{s-1})} \cdot \prod_{i=0}^{s} \Gamma_i. \tag{6.364}$$

This expression can be substituted into (6.359).

According to (6.359) and (6.364), the specific risk after a mass production of trials is expressed by the formula

$$R_s = \int_{\Omega(\mathbf{w}_{s-1})} r_s \cdot P(\mathbf{w}_{s-1})\, d\Omega = \int_{\Omega(\bar{\mu}, \mathbf{u}_s, \mathbf{w}_{s-1})} W_s(s, \bar{\mu}, \mathbf{u}_s) \cdot P(\bar{\mu})$$

$$\times \prod_{i=0}^{s-1} P(w_i \mid i, \bar{\mu}) \prod_{i=0}^{s} \Gamma_i(u_i \mid \mathbf{u}_{i-1}, \mathbf{w}_{i-1})\, d\Omega. \qquad (6.365)$$

Now the optimal strategy can be found by the same method as in the theory of dual control. We will introduce the auxiliary function

$$\alpha_k(\mathbf{u}_k, \mathbf{w}_{k-1}) = \int_{\Omega(\bar{\mu})} W_k(k, \bar{\mu}, \mathbf{u}_k) \cdot P(\bar{\mu}) \times \prod_{i=0}^{k-1} P(w_i \mid i, \bar{\mu})\, d\Omega. \qquad (6.366)$$

Then reasoning in exactly the same way as in dual control theory, an analogous result can likewise be obtained in this case. The optimal strategy Γ_{n-k}^* turns out to be regular. In order to obtain it the function

$$\gamma_{n-k} = \alpha_{n-k} + \int_{\Omega(w_{n-k})} \gamma_{n-k+1}^*\, d\Omega \qquad (6.367)$$

must be formed, where $\gamma_n = \alpha_n$ and

$$\gamma_i^* = \min_{u_i \in \Omega(u_i)} \gamma_i = \gamma_i(u_i^*). \qquad (6.368)$$

By minimizing γ_{n-k} with respect to u_{n-k}, we obtain the optimal control

$$u_{n-k}^* = u_{n-k}^*(\mathbf{u}_{n-k-1}, \mathbf{w}_{n-k-1}). \qquad (6.369)$$

The optimal control u_{n-k}^* depends on $\mathbf{w}_{n-k-1}$, since the observation of all the previous values of w_i ($i = 0, 1, \ldots, n-k-1$) gives the possibility of estimating the parameter vector $\bar{\mu}$. To disregard some of the values of w_i ($i < n - k$) observed earlier would lead to the loss of information about $\bar{\mu}$, and hence to an increase in uncertainty in estimating the parameters of the object, which can only make the control worse. In addition, u_{n-k}^* depends on $\mathbf{u}_{n-k-1}$, since all the past controls u_i ($i = 0, 1, \ldots, n-k-1$) leave their own "traces" in the form of the measurement of the state of the object B, and the calculation of this state is required in determining the optimal control at the instant $t = n - k$.

Thus, formally, the operations in determining the optimal strategy in the closed-loop circuit of Fig. 6.21 and in the open-loop circuit of Fig. 6.23 have proved to be almost identical.

If, however, the formulas are looked at closely, then it will turn out that there is an essential difference between them. In fact, let us compare formulas (6.353) and (6.366) for α_k in the closed-loop and open-loop systems respectively. We see that the dependence on $\mathbf{u}_k$ in expression (6.353) is fundamentally more complex than in (3.366). The coordinates of this vector enter into the integrand of (6.353) not only in the factor $W_k(k, \bar{\mu}, \mathbf{u}_k)$, but also in the factors $P(y_i \mid i, \bar{\mu}, \mathbf{u}_i)$. This property reflects the multiform character of the effect of the control u_k on the specific risk R_s for $s > k$. On one hand, in the dynamical system of the object a "trace" of the action u_k remains at the instant $t = s$. Therefore W_s depends on the entire vector $\mathbf{u}_s$ and not only on the scalar u_s. Moreover, the *a posteriori* probability $P_s(\bar{\mu})$ also depends on how successful the control u_k proved to be in the sense of refining the object data. Therefore the factors $P(y_i \mid i, \bar{\mu}, \mathbf{u}_i)$ enter into expression (6.353). This second channel of influence of u_k on R_s is missing in the open-loop system. In formula (6.366) for α_k in the open-loop system, the factors $P(w_i \mid i, \bar{\mu})$ do not depend on $\mathbf{u}_i$.

But cases exist when the system depicted in Fig. 6.21 turns out to be reducible. This is possible when formula (6.353) can be given a form analogous to (6.366). If a function

$$w_i = w_i(y_i, i, \mathbf{u}_i) \tag{6.370}$$

can be found such that the equation

$$P(y_i \mid i, \bar{\mu}, \mathbf{u}_i) = P(w_i \mid i, \bar{\mu}) \tag{6.371}$$

is realized, then (6.353) reduces to the form of (6.366) and the closed-loop system proves to be reducible. Here the condition

$$\int_{\Omega(w_i)} P(w_i \mid i, \bar{\mu})\, d\Omega = 1 \tag{6.372}$$

must be realized, since the integrand in (6.372) is the probability density of w_i.

An example of a reducible system with an inertial object has been considered [see Eq. (6.222) and so on]. From formula (6.259) it is seen that the measurement of $\mathbf{x}_i$ and $\mathbf{u}_i$ in this system is equivalent to the measurement of μ with a certain error. But the class of reducible systems is very restricted compared to the class of irreducible systems, for which conditions of the type (6.370) and (6.371), for example, cannot be fulfilled. Examples of irreducible systems are easily deduced by considering just the irreducible systems with inertialess objects given above in this chapter, by adding inertial "blocks" to them, for example, blocks of linear type.

The existence of irreducible systems is one further illustration of the fact that the feedback concept has absolute significance in a known sense. Processes arising in feedback systems are richer in their own contents than processes in open-loop systems, and have features that are not inherent in any open-loop systems.

We shall examine certain questions relating to the construction of block diagrams of controllers in irreducible systems. For concreteness let us consider one of the simplest examples of an irreducible system with the block diagram shown in Fig. 6.24. This circuit differs from the diagram of Fig. 6.8, where $x_s = \mu u_s$, only in the point of inclusion of purely random "noise." Instead of the independent random variables h_s in the feedback path (Fig. 6.8), the circuit depicted in Fig. 6.24 will contain the independent random variables ξ_s inside the object B. Let these variables be Gaussian, having zero mean value and the same probability density

$$P(\xi_s) = q(\xi_s) = \frac{1}{\sigma_\xi \sqrt{(2\pi)}} \exp \left\{ - \frac{\xi_s^{\,2}}{2\sigma_\xi^{\,2}} \right\}. \tag{6.373}$$

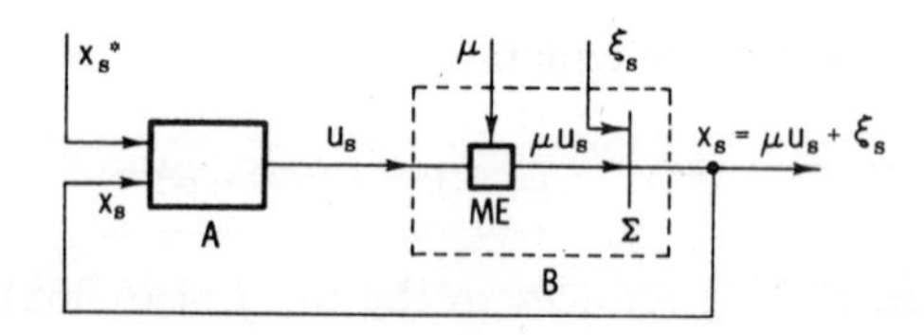

Fig. 6.24

The block diagram of the object B contains the multiplier element ME and the summing element Σ. Hence,

$$x_s = \mu u_s + \xi_s . \tag{6.374}$$

The "amplification coefficient" μ of the multiplier element ME, whose output variable is equal to μu_s, is a normally distributed random variable with *a priori* probability density

$$P_0(\mu) = \frac{1}{\sigma_\mu \sqrt{(2\pi)}} \exp \left\{ - \frac{(\mu - \mu_0)^2}{2\sigma_\mu^{\,2}} \right\}. \tag{6.375}$$

We will define the specific loss function in the following manner:

$$W_s = (x_s{}^* - x_s)^2, \tag{6.376}$$

where $x_s{}^*$ is an assigned constant which in general differs from zero.

If the random noise ξ_s were absent ($\xi_s = 0$), then, in order to ascertain the value of μ, one measurement of the variables u_s and x_s would suffice. Hence in this case the problem of determining the characteristic of the object B would prove to be trivial. On the other hand, if the variable μ is known and $\xi_s \neq 0$, then the characteristic of the object B experiences random measurements from time to time. But here the storage of information about the object characteristic is fundamentally impossible, since the individual values ξ_i and ξ_j are independent for $i \neq j$. Therefore in this case the study of the object characteristic likewise has no meaning. The problem becomes meaningful only when the quantity μ is unknown and in addition $\xi_s \neq 0$. In this case an exact measurement of μ after a finite number of times is impossible. Only by accumulating information about the data of the measurements of u_i and x_i can always newer *a posteriori* probability densities $P_{i+1}(\mu)$ be constructed with each new time, which permit a more and more precise estimate of the variable μ to be made, and hence a better and better control to the organized.

In previous sections problems with "internal noise" ξ_s in the object have not been considered. But the techniques of dual control theory presented above are applicable to such problems without appreciable changes. Let us first derive a formula for the conditional specific risk r_s. It is not necessary to write out the quantity $x_s{}^*$ in explicit form, since it is prescribed. From the circuit of Fig. 6.24 and formula (6.376) it follows that:

$$\begin{aligned} r_s &= M\{W_s \mid \mathbf{x}_{s-1}, \mathbf{u}_{s-1}\} \\ &= \int_{\Omega(x_s, u_s)} (x_s{}^* - x_s)^2 P_s(x_s \mid u_s) \cdot \Gamma_s(u_s \mid \mathbf{x}_{s-1}, \mathbf{u}_{s-1}) \, d\Omega. \end{aligned} \tag{6.377}$$

Here $P_s(x_s \mid u_s)$ is the *a posteriori* probability density of x_s for fixed u_s. This function characterizes the information about the object existing before the sth time. As in the previous account, the expression $\Gamma_s(u_s \mid \mathbf{x}_{s-1}, \mathbf{u}_{s-1})$ is used in general for denoting the random strategy of the controller A.

The function $P_s(x_s \mid u_s)$ is related to the *a posteriori* probability density $P_s(\mu)$ of the random variable μ by the obvious relation

$$P_s(x_s \mid u_s) = \int_{\Omega(\mu)} P(x_s \mid \mu, u_s) P_s(\mu) \, d\Omega, \tag{6.378}$$

where $P(x_s \mid \mu, u_s)$ is the conditional probability density of x_s for fixed μ and u_s, and $\Omega(\mu)$ is the region of possible values of μ. From Eq. (6.374) it follows that

$$\xi_s = x_s - \mu u_s . \tag{6.379}$$

Hence according to (6.373),

$$\begin{aligned} P(x_s \mid \mu, u_s) &= P(x_s \mid \mu \cdot u_s) = q(\xi_s = x_s - \mu u_s) \\ &= \frac{1}{\sigma_\xi \sqrt{(2\pi)}} \exp\left\{ -\frac{(x_s - \mu u_s)^2}{2\sigma_\xi^2} \right\}. \end{aligned} \tag{6.380}$$

Thus the first factor in the integrand of (6.378) is obtained.

We will find the expression for $P_s(\mu)$—the second factor in this expression—on the basis of Bayes' formula:

$$P_s(\mu) = P(\mu \mid \mathbf{u}_{s-1}, \mathbf{x}_{s-1}) = \frac{P(\mathbf{u}_{s-1}, \mathbf{x}_{s-1} \mid \mu) P_0(\mu)}{P(\mathbf{u}_{s-1}, \mathbf{x}_{s-1})} . \tag{6.381}$$

Let us examine the expression

$$\begin{aligned} P(u_i, x_i \mid \mu, \mathbf{u}_{i-1}, \mathbf{x}_{i-1}) &= P(x_i \mid \mu, u_i, \mathbf{u}_{i-1}, \mathbf{x}_{i-1}) P(u_i \mid \mu, \mathbf{u}_{i-1}, \mathbf{x}_{i-1}) \\ &= P(x_i \mid \mu, u_i) \cdot \Gamma_i(u_i \mid \mathbf{u}_{i-1}, \mathbf{x}_{i-1}). \end{aligned} \tag{6.382}$$

In the first factor on the right side $\mathbf{u}_{i-1}$, $\mathbf{x}_{i-1}$ are discarded, since fixing μ and u_i completely determines the conditional probability density for x_i, and adding $\mathbf{u}_{i-1}$, $\mathbf{x}_{i-1}$ does not give any new information because the object B does not have memory. The second factor is the strategy Γ_i of the controller A, depending only on $\mathbf{u}_{i-1}$ and $\mathbf{x}_{i-1}$. Therefore μ to the right of the line in this factor can be discarded.

From (6.382) it follows that

$$P(\mathbf{u}_{s-1}, \mathbf{x}_{s-1} \mid \mu) = \prod_{i=0}^{s-1} P(x_i \mid \mu, u_i) \cdot \prod_{i=0}^{s-1} \Gamma_i . \tag{6.383}$$

In addition, the mean specific risk is

$$R_s = \int_{\Omega(\mathbf{u}_{s-1}, \mathbf{x}_{s-1})} r_s P(\mathbf{u}_{s-1}, \mathbf{x}_{s-1}) \, d\Omega.$$

After substituting (6.383) into (6.381), and then (6.381) and (6.380) into (6.378), (6.378) into (6.377), and (6.377) into the expression for

R_s, with (6.375) taken into account we will obtain to within a proportionality coefficient

$$R_s \equiv \int_{\Omega(\mu, \mathbf{u}_s, \mathbf{x}_s)} (x_s^* - x_s)^2 \exp\left\{-\frac{1}{2\sigma_\xi^2}\sum_{i=0}^{s}(x_i - \mu u_i)^2 - \frac{(\mu - \mu_0)^2}{2\sigma_\mu^2}\right\} \cdot \prod_{i=0}^{s} \Gamma_i \, d\Omega. \quad (6.384)$$

The general method of finding the optimal strategy in this case is the same as in the ones considered earlier.

We introduce the function

$$\alpha_s(u_s, \mathbf{x}_{s-1}, \mathbf{u}_{s-1}) = \int_{-\infty}^{\infty}\int_{-\infty}^{\infty} (x_s^* - x_s)^2 \exp\left\{-\frac{1}{2\sigma_\xi^2}\sum_{i=0}^{s}(x_i - \mu u_i)^2 - \frac{(\mu - \mu_0)^2}{2\sigma_\mu^2}\right\} dx_s \, d\mu. \quad (6.385)$$

Further, we determine u_n^* minimizing α_n with respect to u_n; then we find γ_{n-1}. By minimizing γ_{n-1} with respect to u_{n-1}, we determine γ_{n-1}^*, and so on.

The function α_s can be found in final form. As a result of the integration of (6.385) we obtain to within a proportionality coefficient:

$$\alpha_s \equiv \exp\left\{-\frac{1}{2\sigma_\xi^2}\sum_{i=0}^{s-1} x_i^2\right\}\left[\sigma_\xi^2 + (x_s^*)^2 + \frac{u_s^2}{2r_{s-1}}\left(1 + \frac{2q_{s-1}^2}{r_{s-1}}\right) - 2u_s x_s^* \frac{q_{s-1}}{r_{s-1}}\right] \cdot \frac{1}{\sqrt{r_{s-1}}} \exp\left\{\frac{q_{s-1}^2}{r_{s-1}}\right\}, \quad (6.386)$$

where

$$q_{s-1} = \frac{1}{2\sigma_\xi^2}\sum_{i=0}^{s-1} x_i u_i + \frac{\mu_0}{2\sigma_\mu^2},$$
$$r_{s-1} = \frac{1}{2\sigma_\xi^2}\sum_{i=0}^{s-1} u_i^2 + \frac{1}{2\sigma_\mu^2}. \quad (6.387)$$

The function α_n assumes the minimal value when the derivative with respect to u_n vanishes in the brackets of expression (6.386) for $s = n$:

$$\frac{u_n}{r_{n-1}}\left(1 + \frac{2q_{n-1}^2}{r_{n-1}}\right) - 2x_n^* \frac{q_{n-1}}{r_{n-1}} = 0. \quad (6.388)$$

Hence it follows that

$$u_n^* = 2x_n^* \frac{q_{n-1}}{1 + 2q_{n-1}^2/r_{n-1}}. \tag{6.389}$$

Thus the best value u_n^* is defined by the expression $\mu' u_n^* = x_n^*$, where μ' is some "best" estimate of the variable μ. From (6.389) it follows that

$$\mu' = \frac{x_n^*}{u_n^*} = \frac{1}{2q_{n-1}}\left(1 + \frac{2q_{n-1}^2}{r_{n-1}}\right) = \frac{1}{2q_{n-1}} + \frac{q_{n-1}}{r_{n-1}}$$

$$= \frac{1}{\dfrac{\mu_0}{\sigma_\mu^2} + \dfrac{1}{\sigma_\xi^2}\sum_{i=0}^{n-1} x_i u_i} + \frac{(2\sigma_\xi^2)^{-1}\sum_{i=0}^{n-1} x_i u_i + \mu_0/2\sigma_\mu^2}{(2\sigma_\xi^2)^{-1}\sum_{i=0}^{n-1} u_i^2 + 1/2\sigma_\mu^2}. \tag{6.390}$$

With a large number of trials, when $\Sigma x_i u_i$ and Σu_i^2 become large numbers, this estimate approaches the value

$$\mu'' = \frac{\Sigma_{i=0}^{n-1} x_i u_i}{\Sigma_{i=0}^{n-1} u_i^2}. \tag{6.391}$$

If the relation $x_i = \mu u_i$ were fulfilled, then from (6.391) the equality $\mu'' = \mu$ would follow.

The quantity γ_{n-1} is determined from the expression

$$\gamma_{n-1} = \alpha_{n-1} + \int_{-\infty}^{\infty} \alpha_n^* \, dx_{n-1}. \tag{6.392}$$

The result of the integration is not expressed by a final formula. But the sequence of integrations and minimizations can be carried out by storing only functions of two variables, in the same way as in the related example analyzed in this chapter [see Eq. (6.141) and (6.145)]. As is seen from (6.387), the corresponding variables are the quantities

$$L_{s-1} = \sum_{i=0}^{s-1} x_i u_i, \qquad M_{s-1} = \sum_{i=0}^{s-1} u_i^2. \tag{6.393}$$

As in the example cited, the optimal control u_s^* in this case depends on the values of L_{s-1} and M_{s-1}. It is also useful to emphasize that in general u_s^* depends on the number of the time, more precisely on the number of times $n - s$ which have remained until the end of the process. Thus,

$$u_s^* = u_s^*[L_{s-1}, M_{s-1}, (n - s)]. \tag{6.394}$$

From Eq. (6.393) and (6.394) it follows that the block diagram of the optimal controller A can be depicted as in Fig. 6.25. The quantities $x_s u_s$ and u_s^2 are generated in the multiplier element ME and in the square-law function generator SL. These quantities enter the summing and storing blocks Σ_1 and Σ_2, which deliver the values of L_{s-1} and M_{s-1}. The time counter C gives the values of $n - s$. The quantities x_s^*, L_{s-1}, M_{s-1} and $n - s$ enter the unit A', which, for example, determines the value u_s^* from formulas computed in advance. Of course, this block can also be constructed in the form of a computer into which definite formulas for u_s^* are not inserted in advance. In such a case a series of integrations and minimizations in turn, necessary for the formation of formula (6.394), must be accomplished in the block A'.

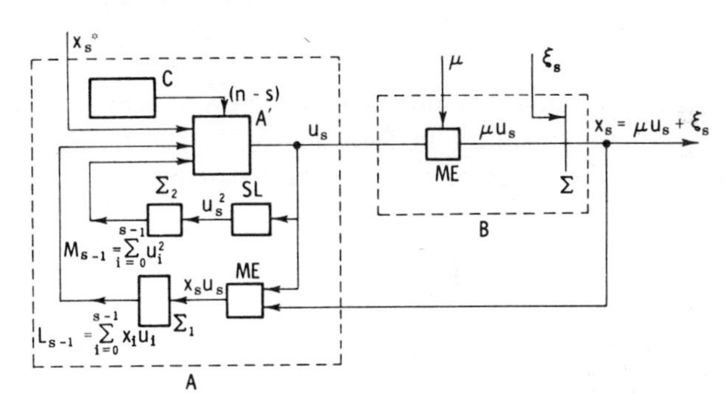

FIG. 6.25

From Fig. 6.25 it is seen that the quantities L_{s-1} and M_{s-1} are fed to the input of the device A' as sufficient characteristics. Indeed the values of these quantities reflect the information stored in the controller A about the properties of the controlled object B.

By another method we can also find grounds for selecting as sufficient characteristics just these quantities or others related to them. Let us examine the controlled object B in Fig. 6.26a by itself. As before let ξ_s be a purely random process characterized by the probability density $q(\xi_s)$, which is defined by Eq. (6.373). We will also assume that the *a posteriori* probability density $P_{s-1}(\mu)$ for the variable μ is defined by the expression

$$P_{s-1}(\mu) = \frac{1}{\sigma_{s-1}\sqrt{(2\pi)}} \exp\left\{-\frac{(\mu - m_{s-1})^2}{2\sigma_{s-1}^2}\right\}, \tag{6.395}$$

where m_{s-1} and σ_{s-1} are the mean value and standard deviation respectively. Let some action u_s be fed to the input of the object and some reaction x_s be obtained at the output. It is asked: What will be the new *a posteriori* probability density $P_s(\mu)$ found by taking into account the new observation made at the *s*th time? It is evident that $P_s(\mu)$ must depend both on the values σ_{s-1}, m_{s-1} characterizing the previous density $P_{s-1}(\mu)$, in this case playing the part of an *a priori* density, and on the values of u_s and x_s:

$$P_s(\mu) = P_s(\mu, \sigma_{s-1}, m_{s-1}, u_s, x_s) = P_s(\mu \mid u_s, x_s, \sigma_{s-1}, m_{s-1}). \tag{6.396}$$

It is not difficult to find an expression for $P_s(\mu)$ in explicit form. By representing $P(x_s, \mu \mid u_s)$ in the form of a product, it is not difficult to obtain the Bayes' formula:

$$P_s(\mu) = P(\mu \mid x_s, u_s) = \frac{P_{s-1}(\mu)P(x_s \mid \mu, u_s)}{P(x_s)}. \tag{6.397}$$

Since $P(x_s)$ is some quantity not depending on μ, which therefore does not play a fundamental part in formula (6.397), it can be set equal to a constant C_1. The value of this constant can be computed from the condition

$$\int_{-\infty}^{\infty} P_s(\mu)\, d\mu = 1. \tag{6.398}$$

After substituting $P_{s-1}(\mu)$ from expression (6.395) and $P(x_s \mid \mu, u_s)$ from (6.380), we find:

$$P_s(\mu) = C_2 \exp\left\{-\frac{(x_s - \mu u_s)^2}{2\sigma_\xi^2} - \frac{(\mu - m_{s-1})^2}{2\sigma_{s-1}^2}\right\}, \tag{6.399}$$

where C_2 is a constant. This formula can easily be represented in the form

$$P_s(\mu) = C_3 \exp\left\{-\frac{(\mu - m_s)^2}{2\sigma_s^2}\right\}, \tag{6.400}$$

where C_3 is a new constant, and m_s and σ_s are defined by the expressions

$$m_s = \frac{u_s x_s + (\sigma_\xi/\sigma_{s-1})^2 m_{s-1}}{u_s^2 + (\sigma_\xi/\sigma_{s-1})^2},$$
$$\sigma_s^2 = \frac{\sigma_{s-1}^2}{1 + (\sigma_{s-1}/\sigma_\xi)^2 u_s^2}. \tag{6.401}$$

A number of useful conclusions can be made from formulas (6.400) and (6.401). In the first place, it turns out that the *a posteriori* distribution characterized by the formula for $P_s(\mu)$ is also normal. Hence it follows that when the *a priori* distribution density $P_0(\mu)$ is prescribed in the form of the normal law (6.375) the subsequent distribution density $P_1(\mu)$, and after it $P_2(\mu)$, $P_3(\mu)$, and so on, are likewise described by normal laws. Hence any *a posteriori* probability density can be described in all by only two numbers—the mean value m_s and the dispersion σ_s^2. Further, from expressions (6.401) it is seen that the new mean value m_s is constructed from the previous mean value m_{s-1} by a linear transformation, where besides σ_{s-1} functions of the values u_s and x_s obtained at the sth step enter: u_s^2 and $u_s x_s$. Finally, the new dispersion σ_s^2 is found to be less than the previous dispersion σ_{s-1}^2, where σ_s^2 depends on the value of u_s^2 (but not on x_s).

The property of preserving the normal distribution law is inherent not only in the diagram depicted in Fig. 6.26a. For example, let us consider another circuit for the controlled object (see Fig. 6.26b). Here the "internal noise" ξ_s in the object is absent; but the measurement of the output variable x_s is realized with an error h_s. Therefore the variable

$$y_s = x_s + h_s = \mu u_s + h_s \tag{6.402}$$

is introduced into the controller A.

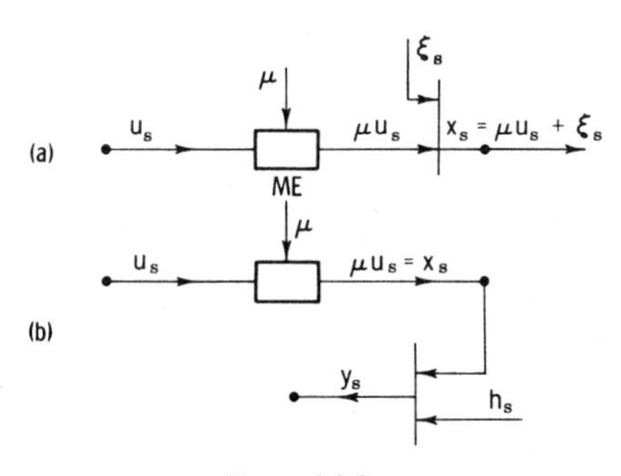

FIG. 6.26

Let $P_{s-1}(\mu)$ be the *a posteriori* probability density for μ after the $(s - 1)$st time. Let us find the subsequent *a posteriori* probability density $P_s(\mu)$ after that, when the values of u_s and y_s at the sth time are fixed. Evidently an equation analogous to (6.397) holds:

$$P_s(\mu) = P(\mu \mid y_s, u_s) = \frac{P_{s-1}(\mu)P(y_s \mid \mu, u_s)}{P(y_s)}. \tag{6.403}$$

We will assume that $P_{s-1}(\mu)$ is described by the normal distribution law (6.395). From formula (6.402) and the distribution law (6.121) for h_s it follows that:

$$P(y_s \mid \mu, u_s) = P(y_s \mid \mu \cdot u_s) = q(h_s = y_s - \mu u_s)$$
$$= \frac{1}{\sigma_h \sqrt{(2\pi)}} \exp\left\{-\frac{(y_s - \mu u_s)^2}{2\sigma_h^2}\right\}. \tag{6.404}$$

After substituting (6.395) and (6.404) into (6.403), we find:

$$P_s(\mu) = C_2 \exp\left\{-\frac{(y_s - \mu u_s)^2}{2\sigma_h^2} - \frac{(\mu - m_{s-1})^2}{2\sigma_{s-1}^2}\right\}$$
$$= C_3 \exp\left\{-\frac{(\mu - m_s)^2}{2\sigma_s^2}\right\}, \tag{6.405}$$

where C_2 and C_3 are constants. Expression (6.405) differs from (6.399) only in that x_s has been replaced by y_s, and σ_ξ by σ_h. Therefore the formulas for m_s and σ_s are obtained as ones analogous to (6.401):

$$m_s = \frac{u_s y_s + (\sigma_h/\sigma_{s-1})^2 m_{s-1}}{u_s^2 + (\sigma_h/\sigma_{s-1})^2},$$
$$\sigma_s^2 = \frac{\sigma_{s-1}^2}{1 + (\sigma_{s-1}/\sigma_h)^2 u_s^2}. \tag{6.406}$$

At each time instant $t = s$ the existing probability density $P_{s-1}(\mu)$ can be regarded as an *a priori* one. Therefore at $t = s$ the optimal control action u_s^* must be chosen depending on the parameters m_{s-1} and σ_{s-1} of the function $P_{s-1}(\mu)$, and also depending on how many times remained until the end of the process. The latter number is equal to $n - s$. In addition, u_s^* depends on x_s^*. Thus,

$$u_s^* = u_s^*[x_s^*, m_{s-1}, \sigma_{s-1}, (n - s)]. \tag{6.407}$$

Starting from formulas (6.401) and (6.407), we can construct the block diagram of the controller A for the object depicted in Fig. 6.26a, as is shown in Fig. 6.27. The values of u_s^2 and $u_s x_s$ are generated in the blocks *SL* and *ME* respectively. These values enter the unit Φ, in which the parameters m_s and σ_s are determined according to formulas (6.401). The preceding values m_{s-1} and σ_{s-1} of the parameters are stored in the memory cells R_1 and R_2 respectively, where they enter from the output of the block Φ through the time-delay cell τ. The values m_s, σ_s, x_s^*, and $n - s$ (the latter value is supplied from the time counter C)

act on the unit A', which delivers the optimal control action u_s^*. The similarity between the diagrams, which are equivalent to one another (Figs. 6.25 and 6.27), is evident.

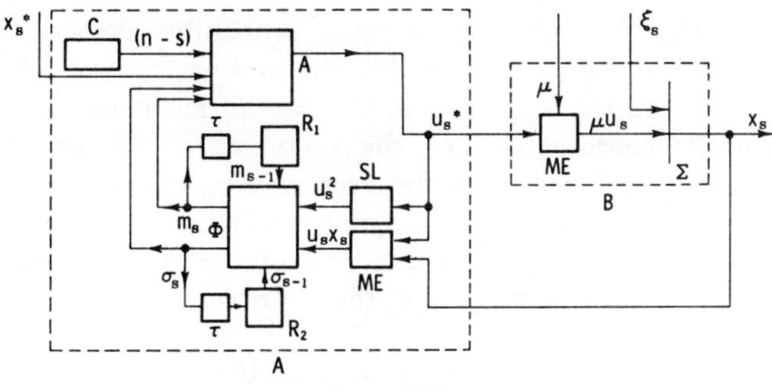

FIG. 6.27

If not x_s^* but y_s^*—the mixture of useful signal and noise—acts on the input to the controller A, then a filter must be added in the channel of y_s^*. Examples of such devices have been discussed in Chapter V; for certain very simple cases they have been considered in this chapter as well [see, for example, Eqs. (6.111) and (6.112)]. Therefore special consideration of such filters can be omitted here.

The preservation of the character of the distribution law after the transition from $P_{s-1}(\mu)$ to $P_s(\mu)$ can occur not only for the normal, but also for a number of other distribution laws (see [5.32, 5.33, 6.16]). In the more general case let the distribution density P_{s-1} of the vector $\bar{\mu}$ be a function of the parameters $a_{1,s-1}, \ldots, a_{q,s-1}$, combined into the vector $\bar{a}_{s-1}$:

$$P_{s-1}(\bar{\mu}) = P_{s-1}(\bar{\mu}, a_{1,s-1}, \ldots, a_{q,s-1}) = P_{s-1}(\bar{\mu}, \bar{a}_{s-1}). \qquad (6.408)$$

Further, we will assume that the observation of the values $\bar{u}_s$ and $\bar{y}_s$ at the output of the controller A and at the input of it, which is connected to the output of the controlled object, permits a new vector $\bar{a}_s$ to be obtained instead of the vector $\bar{a}_{s-1}$. This vector is obtained by some transformation $\bar{\Phi}$ of the previous vector $\bar{a}_{s-1}$, and certainly also depends on $\bar{u}_s$ and $\bar{y}_s$:

$$\bar{a}_s = \bar{\Phi}(\bar{a}_{s-1}, \bar{u}_s, \bar{y}_s). \qquad (6.409)$$

Expressions (6.406) are a special case of this relation.

The vector $\bar{a}_s$ can be called an information vector, since essentially all the *a posteriori* information about the characteristics of the controlled object (but not about its state) is concentrated in it. The function $P_s(\bar{\mu})$ depends on this vector:

$$P_s(\bar{\mu}) = P_s(\bar{\mu}, \bar{a}_s). \tag{6.410}$$

Considering that $P_{s-1}(\bar{\mu})$ is the *a priori* probability density before the sth time, it is not difficult to see that the optimal control $\bar{u}_s^*$ will be a function of the vector $\bar{a}_{s-1}$, and in the case of an inertial object also a function of its coordinates, i.e., the vector $\bar{x}_{s-1}$. (If the exact value $\bar{x}_{s-1}$ cannot be measured, then the vectors $\mathbf{y}_{s-1}$ and $\mathbf{u}_{s-1}$ are necessary for its estimation.) In addition, $\bar{u}_s^*$ depends on $\bar{x}_s^*$ and on the number of times $n - s$ remaining until the end of the process. Thus in the special case of $\bar{h}_s = 0$ and $\bar{x}_s = \bar{y}_s$ the expression

$$\bar{u}_s^* = \bar{u}_s^*[\bar{a}_{s-1}, \bar{x}_{s-1}, \bar{x}_s^*, (n - s)] \tag{6.411}$$

holds. The components of the vectors $\bar{a}_{s-1}$ and $\bar{x}_{s-1}$, reflecting the information about the characteristics and state of the object, can be designated as sufficient coordinates in this case. The application of sufficient coordinates has been recommended for the solution of statistical problems in optimal control, not relating to the field of dual control theory, in [5.33, 5.40, 6.10]. In [6.16] the techniques associated with sufficient coordinates have been used for solving an example of dual control theory.

We will examine a general diagram of a controller constructed on the basis of Eqs. (6.408), (6.409), (6.411). This diagram, shown in Fig. 6.28, is a generalization of the scheme depicted in Fig. 6.27. The channels G and H with noises $\bar{g}_s$ and $\bar{h}_s$ respectively are present in the diagram of Fig. 6.28. The value $\bar{a}_s$ is generated in the block $\bar{\Phi}$ from $\bar{a}_{s-1}$, $\bar{u}_s$, and $\bar{y}_s$, according to expression (6.409).

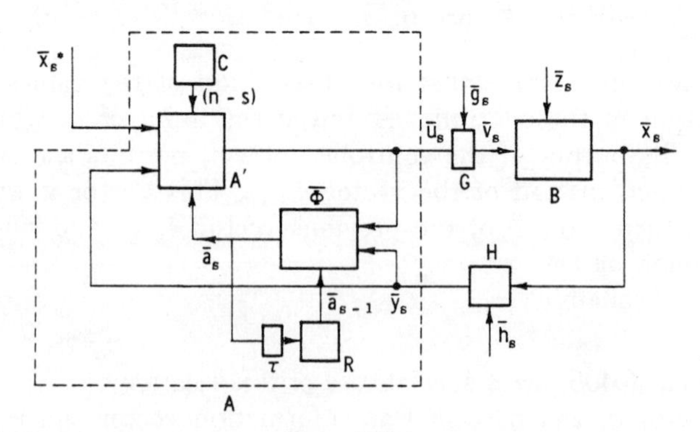

FIG. 6.28

The preceding value $\bar{a}_{s-1}$ is stored in the memory cell R, where it enters through the single-time delay element τ. The quantities $\bar{y}_s$, $\bar{a}_s$, and also $\bar{x}_s{}^*$ and $n - s$ act on the input to the block A', which can also contain delay elements τ inside it. The output variable $\bar{u}_s{}^*$ of this unit corresponds to expression (6.411).

Let us consider the example given in the paper [6.16] (Fig. 6.29). Here as before, the quantity μ and the sequence of independent variables ξ_s are subject to the distribution laws (6.375) and (6.374) respectively.

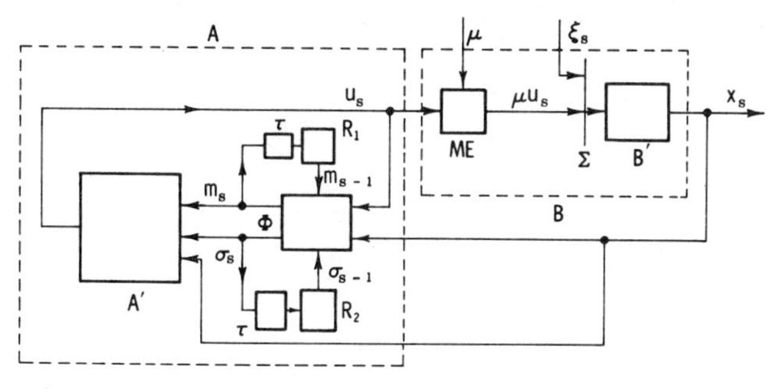

FIG. 6.29

The subunit B' is a discrete integrating element with input variable $\mu u_s + \xi_s$, and moreover the equation of the entire object B has the form

$$x_{s-1} - x_s = \mu u_s + \xi_s . \tag{6.412}$$

If the *a priori* probability density $P_s(\mu)$ for μ is subject to a normal law with mean value m_s and dispersion $\sigma_s{}^2$, then the *a posteriori* density obtained after measurement of the values u_s, x_s, and x_{s+1} will also be expressed by a normal law, with a mean value m_{s+1} and dispersion σ^2_{s+1}. This is easily proved by the same method as in the earlier examples. In fact, if we regard $P_s(\mu)$ as the "*a priori*" probability density of μ, then the equality

$$\begin{aligned} P(\mu, x_{s+1} \mid x_s, u_s) &= P_s(\mu) \cdot P(x_{s+1} \mid x_s, \mu, u_s) \\ &= P(x_{s+1} \mid x_s, u_s) \cdot P(\mu \mid x_s, x_{s+1}, u_s) \end{aligned} \tag{6.413}$$

holds. In view of the independence of $P(x_{s+1} \mid x_s, u_s)$ from μ we will replace this probability density by $(C_1)^{-1}$. Then

$$\begin{aligned} P_{s+1}(\mu) = P(\mu \mid x_s, x_{s+1}, u_s) &= \frac{P_s(\mu) P(x_{s+1} \mid x_s, \mu, u_s)}{P(x_{s+1} \mid x_s, u_s)} \\ &= C_1 \cdot P_s(\mu) \cdot P(x_{s+1} \mid x_s, \mu, u_s). \end{aligned} \tag{6.414}$$

From (6.412) and (6.373) we find:

$$P(x_{s+1} \mid x_s, \mu, u_s) = P(x_{s+1} \mid \mu u_s + x_s)$$
$$= q(\xi_s = x_{s+1} - x_s - \mu u_s)$$
$$= \frac{1}{\sigma_\xi \sqrt{(2\pi)}} \exp\left\{-\frac{(x_{s+1} - x_s - \mu u_s)^2}{2\sigma_\xi^2}\right\}. \tag{6.415}$$

Substituting this expression into (6.414) together with the formula for $P_s(\mu)$:

$$P_s(\mu) = \frac{1}{\sigma_s \sqrt{(2\pi)}} \exp\left\{-\frac{(\mu - m_s)^2}{2\sigma_s^2}\right\}, \tag{6.416}$$

after simple transformations we arrive at the expression

$$P_{s+1}(\mu) = C_2 \exp\left\{-\frac{(\mu - m_{s+1})^2}{2\sigma_{s+1}^2}\right\}, \tag{6.417}$$

where C_2 is a constant, and m_{s+1} and σ_{s+1} are obtained from the formulas

$$m_{s+1} = \frac{u_s(x_{s+1} - x_s) + (\sigma_\xi/\sigma_s)^2 m_s}{u_s^2 + (\sigma_\xi/\sigma_s)^2},$$
$$\sigma_{s+1}^2 = \frac{\sigma_s^2}{1 + (\sigma_s/\sigma_\xi)^2 u_s^2}. \tag{6.418}$$

Fig. 6.29 depicts the block Φ, whose operation is realized according to Eqs. (6.418).

We will now conduct the time reading not from the beginning, but from the end of the process. Thus u_1, x_1 will be the last values; u_2, x_2 the next to the last; and so on.

Let the loss function have the form

$$W_s = x_{s-1}^2 + u_s^2. \tag{6.419}$$

Thus W_s is determined by the control u_s and the output variable x_{s-1} which will appear at the following time.

Let us introduce the notation

$$S_r = \sum_{i=1}^{r} R_i = \sum_{i=1}^{r} M\{W_i\} = \sum_{i=1}^{r} M\{x_{i-1}^2 + u_i^2\}. \tag{6.420}$$

We agree to denote by S_r^* the optimal value of the function S_r attainable with the optimal controls u_r^*, u_{r-1}^*, ..., u_1^* belonging to the admissible region $\Omega(u_i)$:

$$S_r^* = \min S_r = \min \sum_{i=1}^{r} M\{x_{i-1}^2 + u_i^2\}. \tag{6.421}$$

The techniques used in [6.16] for deriving optimal control formulas for the example under consideration differ somewhat from those adopted in this chapter, although they are completely equivalent and are based on the method of dynamic programming. Here it is essential that the function S_r^* depend on the "initial" state of the object B (and for a first-order object this state is described by only one coordinate x_r), on the parameters m_r, σ_r of the "initial" probability density $P_r(\mu)$ for μ, and also on the number of times r until the end of the process. Thus,

$$S_r^* = S_r^*(r, x_r, m_r, \sigma_r) = S_r^*(x_r, m_r, \sigma_r).$$

The letter r need not be indicated in the parentheses, since it is shown in the index of S_r^*.

In the same way as in Chapter IV, from the principle of dynamic programming it follows that:

$$\begin{aligned} S_r^*(x_r, m_r, \sigma_r) &= \min_{u_r} M\left\{W_r + \min_{u_{r-1},\ldots,u_1} M\left\{\sum_{i=1}^{r-1} W_i\right\}\right\} \\ &= \min_{u_r} M\{x_{r-1}^2 + u_r^2 + S_{r-1}^*(x_{r-1}, m_{r-1}, \sigma_{r-1})\}. \end{aligned} \tag{6.422}$$

Moreover, considering the last time, we obtain the formula

$$S_1^*(x_1, m_1, \sigma_1) = \min_{u_1} M\{x_0^2 + u_1^2\}. \tag{6.423}$$

Here by x_0 is meant the state after supplying the control action u_1 at the last time, i.e., the state at the end of the entire process.

From (6.412) with a new order of time indexing it follows that

$$x_{r-1} = x_r + \mu u_r + \xi_r. \tag{6.424}$$

In (6.418) replacing the difference $x_{s+1} - x_s$, or with the new indexing $x_{r-1} - x_r$, by the value $\mu u_r + \xi_r$ equal to it according to (6.424), we arrive at the expressions

$$\begin{aligned} m_{r-1} &= \frac{\mu u_r^2 + m_r(\sigma_\xi/\sigma_r)^2 + u_r \xi_r}{u_r^2 + (\sigma_\xi/\sigma_r)^2}, \\ \sigma_{r-1}^2 &= \frac{\sigma_r^2}{1 + (\sigma_r/\sigma_\xi)^2 u_r^2}. \end{aligned} \tag{6.425}$$

Let us now examine the possibility of organizing a chain of computations of the values u_r^* from Eqs. (6.422)–(6.425). The determination of u_r^* from (6.422) requires the calculation of S_{r-1}^* as a preliminary. In addition, expressions (6.425) must appear as the arguments of S_{r-1}^*. But the value of μ in the first of them is not known. Therefore in [6.16] this expression is averaged with respect to μ, where it is assumed that μ has an *a priori* distribution density $P_0(\mu)$. This operation can be explained by the fact that after a large number of trials with the system the distribution of μ will have just such a form. Thus S_{r-1}^* will now be a function of the mean value m_{r-1} of the random variable. Analogous formulas can be derived with the aid of the techniques presented in the previous sections of this chapter.

The chain of computations starts from formula (6.423), where instead of x_0 its value from (6.424) is substituted:

$$\begin{aligned} S_1^*(x_1, m_1, \sigma_1) &= \min_{u_1} M\{x_0^2 + u_1^2\} \\ &= \min_{u_1} M\{u_1^2 + (x_1 + \mu u_1 + \xi_1)^2\} \\ &= \min_{u_1}[x_1^2 + (m_1^2 + \sigma_1^2)u_1^2 + \sigma_\xi^2 + 2m_1 x_1 u_1 + u_1^2]. \end{aligned} \tag{6.426}$$

Here after opening the brackets the averaging is performed, where as was indicated above the *a priori* normal distribution is taken for μ with mean value m_1 and dispersion σ_1^2. The averaging of the quantity ξ_1^2 leads to the appearance of the term σ_ξ^2.

Equating the derivative of the expression in the brackets in (6.426) to zero, we obtain u_1^* and then S_1^* as well:

$$\begin{aligned} u_1^* &= -\frac{m_1 x_1}{m_1^2 + \sigma_1^2 + 1}, \\ S_1^*(x_1, m_1, \sigma_1) &= \frac{x_1^2(1 + \sigma_1^2)}{m_1^2 + \sigma_1^2 + 1} + \sigma_\xi^2. \end{aligned} \tag{6.427}$$

We pass to the determination of S_2^* and u_2^*. From (6.422) it follows that:

$$S_2^*(x_2, m_2, \sigma_2) = \min_{u_2} M\{x_1^2 + u_2^2 + S_1^*(x_1, m_1, \sigma_1)\}. \tag{6.428}$$

Here if the expression for S_1^* from (6.427) and x_1 from (6.424) are introduced, then a complicated enough function of u_2 will be obtained and the determination of u_2^* by analytical means proves to be difficult. Therefore further computations were carried out on a digital computer. In programming the problem the function S_r^* was approximated by polynomials up to fourth degree. As a result of the computations and the determination of the dependences

$$u_i^* = u_i^*(x_i, m_i, \sigma_i), \tag{6.428'}$$

it turned out that, as indicated in [6.16], by now starting from the fourth time from the end, these dependences gain a steady-state character, and with an accuracy acceptable in practice can be approximated by the equations

$$u_i^* \cong \frac{-m_i x_i}{m_i^2 + \frac{1}{2}\sigma_i^2 + 1}. \tag{6.429}$$

This law can be put into the subunit A' shown in Fig. 6.29.

If the control were performed without a parallel study of the object, then as is not difficult to show, the values of u_i would be smaller in absolute value. A study of the object simultaneous with the control requires increases in the values of u_i, since the larger the u_i are, the larger the quantities μu_i will also be in comparison with the noise ξ_i, and hence the more precise the estimate of the unknown amplification coefficient μ will turn out to be. Thus the larger $|u_i|$ is, the more intensively the process of studying the characteristic of the controlled object B will take place. But the form of the loss function W_i (6.419) indicates that an excessively large value of u_i is undesirable. First, this increases the "cost" of control, i.e., the term u_i^2 in the formula for W_i. Second, such an increase can give rise to a growth in the output variable at subsequent times, which likewise will cause an increase in the loss function. The optimal control u_i^* represents a compromise between the tendencies indicated above which conflict with one another.

Conclusion

What are the trends of development and immediate perspectives of the theory of optimal systems?

It is difficult to give a satisfactorily complete answer to this question, since the rapid growth of the science of control will possibly give rise in the near future to unexpected turns and new situations. We can only make certain statements whose correctness is highly probable.

First, much work is necessary in order to build a bridge between theory and practice. The practical applications of optimal system theory are comparatively few at present, but their number is rapidly growing. The development of specific types of optimal systems with chemical, metallurgical, and other objects and the design of optimal control strategies for complex transportation and power systems are revealing a series of new problems whose solution will be useful in practice and will serve as a new impetus for the development of the theory. The problem of developing systems that are close to optimal, and the determination of quasi-optimal algorithms represent a great, relatively independent, and exceedingly urgent theoretical problem (see, for example, [3.67]). The solution to this problem can be related to the use of the abundant arsenal of general methods of modern analysis—for example, functional analysis—and the powerful tools of computing technology. The questions of the practical synthesis of systems also enter into this range of problems—now not the synthesis of algorithms or block diagrams of abstract type, but the synthesis of workable optimal controller circuits and their units.

Second, the development of optimal system theory will lead to its extension to the case when certain *a priori* information is given not only about the object, but also about the controller. Thus in this case known constraints will also be imposed on the freedom of choice of the controller. Similar questions, still weakly developed, have hardly been raised in this book. But the importance of such an approach is certain. As a matter of fact, an actual controller is constructed from actual elements having limited possibilities. Noise acts on them; these elements have a limited frequency transmission band, or in general a limited capacity in an information-theoretic sense.

Further, these elements are not sufficiently reliable; they can depart from design, and meanwhile it is required to provide reliable enough system operation as a whole. Finally, economic considerations dictate the choice of a variant of the system that will ensure the best advantage in one sense or another, for example, in the sense of minimum initial expenditures and operating cost necessary for a definite quantity of production. Here optimal system theory is linked to reliability theory, and the problems of this theory touch on economic and other problems, in general belonging to the realm of the so-called theory of compound systems.

Progress in the field of the concept of uncertainty will emerge as the third important direction in the development of optimal system theory. The situation of uncertainty, the absence of complete *a priori* information about the object, and all the more about the controller is generally typical for control theory. This has been repeatedly referred to in the text of this book.

This book has examined two main ranges of problems.

1. Systems with complete information about the controlled object.

2. Systems with partial information about the object, in which, however, *a priori* information of a probabilistic character is prescribed ("Bayes'" approach).

In reality, however, the *a priori* probabilistic characteristics are very frequently either completely lacking or are known only partially, or finally are measured in a way not expected beforehand. In such cases information about the controlled object is even less complete than in the systems considered in this book. The extension of optimal system theory to these cases is very urgent from a practical point of view. Part of R. Bellman's book is devoted to such an extension [2.38]. Of course, the generalization of the theory can be made in various ways. Apparently the game-theoretic approach will be used all the more widely in this field [5.10–5.14, 5.32–5.35]. Probably this approach itself will be modified as a development of game theory and the inclusion of new fields in its realm, for example, many-person games where the players are not necessarily hostile to one another, and can temporarily cooperate and help one another for reasons of final gain. The development of the theory of games of automata with one another also offers a very perspective direction [6.18]. But in a "game" with nature—being the central process in automatic systems—the game-theoretic approach does not always prove to be sufficiently justified (see Chapter V). Therefore other approaches will also be developed, for example, the combination of the minimax game-theoretic approach with the taking into account of the information obtained in the process of operating the system (see, in particular, [5.36]). Such an approach utilizes all the information about the object that is obtained at the given current instant. We will assume that this information is insufficient for uniquely determining the probabilistic characteristics, and some set of possible forms of characteristics corresponds to it. But this set is limited by the information having entered in the process of system operation, and its extent is less than at the beginning of operation. The greater the information that has entered the controller, the more limited is the indicated set. By choosing from this set the "worst" characteristics, the most risky from the positions of the selected optimality criterion, i.e., by applying the minimax approach, the minimax optimal strategy can be found for this case. According to the extent of information storage about the object, this strategy will be closer and closer to the optimal strategy corresponding to complete or maximal information about the object.

Another approach, apparently not very distant from the previous one, consists of the application of the concept of inductive probability. The sources of this concept go back even to the founders of probability theory, J. Bernoulli and Laplace. In our time this concept has been examined in the work of R. Carnap [5.37]. Certain problems related to the concept of inductive probability were analyzed in R. Bellman's book [2.38] and in the paper [5.38].

In the simplest modification the idea of inductive probability consists of the following. We assume that nothing at all is known about a certain random event. In such a case there are no grounds for considering its probability greater than or less than $1/2$. Therefore it is recommended to regard it equal to $1/2$. On the other hand, if n trials are performed and of them the event appeared in m cases, then for large n the frequency m/n can be taken as a measure of the probability of the event. The simplest formula, giving a result equal to $1/2$ for $n = m = 0$ and a result equal to m/n for large m and n, has the form

$$p_i = \frac{1 + m}{2 + n}.$$

This is the simplest kind of formula for inductive probability. Such a formula gives a known measure for the probability of future occurrences of the random event, based on past trial, where the number of trials n performed in the past can be arbitrary—large or small. Of course, the questions of justifying formulas of such a type lie outside the scope of probability theory [5.37].

In general, the development of methods of investigating indeterminate situations has primary significance both for understanding the operation of complex organized systems and the theory of self-organizing and self-adaptive systems, and by the same token for all of cybernetics as well.

The growing universality of problem formulation and the development of all the more general methods of their solution are an important trend in the development of optimal system theory. A wider and wider range of problems is spanned by a unified point of view, by a unity of the methods of approach. This trend apparently will last into the future, and as a consequence it will be an ever-growing bond, increasing the coalescence of optimal system theory with the other directions of engineering cybernetics.

Such a coalescence is extremely important, since it will emerge as the guarantee of the formation of a clear, unified, scientific-theoretical basis for cybernetics as a whole. Apparently, such a unified basis and general laws, formulating what is possible and what is impossible in cybernetic systems, will be constructed on an information-theoretic foundation. In fact, the concepts related to information and its processing are fundamental to cybernetics.

C. Shannon's brilliant and profound theory of information, with its modern ramifications, represents only one of the first steps in the formation of the scientific fundamentals of cybernetics. The concept of precision of information, introduced in Shannon's theory, suffices for the isolated examination of questions of information transmission in a communication channel. But the theory of processes of information processing in closed-loop systems requires a broadening of the range of information theory concepts, and the inclusion in it of at least the concepts of meaning and value of information. A discussion of these concepts falls outside the scope of this book. The necessity of introducing such concepts can only be illustrated by the example of the operation of biological systems. Man receives a vast amount of various stimuli, and if he were to react to all these stimuli, his actions would be a sequence of very diverse, chaotic motions. In reality man, acting single-mindedly, makes a selection of the information coming to him according to its meaning, i.e., according to its relation to real objects with which he interacts, and according to its value, i.e., according to the degree of utility of the information for those actions he undertakes.

The optimal processes of information selection, the perception processes, the formations of general conceptions, and finally the planning processes, the preparations for the direct realization of the actions—in other words, the simulation of thought in a definite realm—all this must find its place in a general theory of optimal systems in the future.

Then the processes of learning, self-organization, and other forms of adaptation could also be considered from a unified point of view, classifying them according to the degree of effectiveness for achieving the stated purpose.

The more complex and branching automatic systems become, the more abundant the assortment of possible actions on them, the more flexible the structure of systems, then the more necessary the development of unified methods of approach, a unified point of view and general conceptions of the theory become. Meanwhile the logic of automation will lead not only to an unlimited expansion of the operating realm of automatic systems, but also to a qualitative complication of the problems arising.

Actually, the automation of routine labor processes, the replacement by automata of workers of lower and middle qualification, gives rise to a concentration of people at more qualified forms of activity, frequently bearing an auxiliary character. Mainly these activities are control, adjustment, design, programming, analysis, and so on. Therefore further progress in automation will be confined to a considerable degree to the replacement and complementation by automata of such forms of man's mental activity as bear a high intellectual character. Side by side with technological automata effecting the basic industrial process, controlling adjusting, designing, and programming automata will be developed; there will also be automata facilitating research activity and replacing man at some of its stages.

Of course, the process of the growing replacement of man's individual functions by automata does not mean at all his exclusion from productive activity in general. Man always remains as the commander of the automatic system, and with the complication and development of the system his activity as the leader will only rise to a higher level.

The stages in the replacement of man's functions by the operation of automata can be traced from the diagram depicted in Fig. C.1. We assume that the automatic system is designed and realized with a controlled object B and controller A, on which the noises z_B and z_A act respectively. To construct this system it is necessary as a preliminary to effect the study of the object (the operation C) and the development of a control strategy, which is put into the controller (the operation D).

At the first stage of automation these operations are carried out by man, but later on they can be automated. If the study of the object and the correction of the strategy of the controller are performed by automata C and D not once, but continuously or episodically, then depending on the character of the processes in such a system it is called self-adapting or self-adjusting. But this system is also connected to man, if only because the information about the method of developing the control strategy and about the method of studying the object enters from man (E in Fig. C.1).

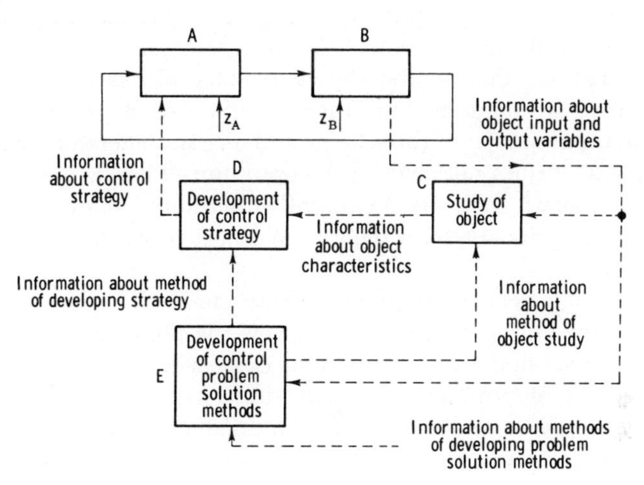

FIG. C.1

The following stage of automation can consist of replacing E by an automaton. But in this case the connection to man also remains, if only because the information about the methods of developing control problem solution techniques enters the automaton from man. Such a process can be continued as far as desired, but always there remains a connection to man in any automaton.

The problems of optimal systems arise at each stage of automation. At the first stage such a problem is posed as the problem of developing an optimal strategy for the controller A. At the following stage of automation the system A–B can be regarded as invariable (in one version of the problem formulation); then the development of an optimal control system reduces to an optimal strategy for studying the object and the control (the optimal strategy of the blocks C and D). The solution to this problem reduces to the construction of an optimal self-adapting or self-adjusting system. In another version of the problem formulation the blocks A, C, D are combined into a single device controlling the object B. It is required to find the optimal controller, where only information about the methods of solving control problems acts on it from E. Analogous problems also arise at later stages of automation.

Automata take upon themselves the fulfillment of all the more complex functions, also including the functions in developing control systems. A question arises in connection with this: What will be the role of theory, including optimal system theory, at the new stages of automation? Will not new trends in the field of automation lead to the fact that the development of optimal strategy by man will be replaced by the work of automata?

We will assume that at first the elementary blocks of the controller are combined at random, and then in the process of learning and adaptation the character of the scheme and its parameters vary automatically as long as it does not achieve the optimum. Then is optimal system theory necessary? Will its "twilight" therefore follow the present flourishing of optimal system theory? Won't it prove to be unnecessary in the future?

This question relates in equal measure to each theory in general. It seems to the author that only one answer is possible: Theory cannot be replaced by blind experimentation, even when it is performed not by people, but by automata. Human ideas, hypotheses, constructions, and hence theory as well will be of even greater necessity to the science and technology of the future than to today's. Therefore the judgment that the theory will "die off" is incorrect. But theory itself will necessarily change its character. If the automaton finds the needed result more simply, more rapidly, or more exactly than man making calculations on paper, then of course such calculations should be entrusted to the automaton. In this case the work of the theoretician is considered complete, if he brings the solution of the problem to the stage of programming it on the machine. But a large field of activity still remains for theory up to this stage.

Besides, new theoretical problems arise in the interaction with machines. For example, what must be the elementary blocks of a learning automaton in order to solve a sufficiently large or the largest possible volume of problems with its aid? Under what conditions will the learning process converge? What is the dependence of the indices of the learning process on the initial state of the automaton? How close does it come to the optimal as a result of learning? Finally, how is the learning process to be organized in some optimal way, for example, to make its duration minimal or its result the most correct?

Thus new theoretical problems, more general than the previous ones, arise at the new stage of automation. Solutions for them will require new methods, new approaches. Moreover, the application of machines expands the possibilities of theory, making practical the formulation of problems that were not conceived of previously. As the final result, machines and theory, influencing one another, promote mutual perfection and development.

As man's practical activity, guiding an intricate complex of new automata, rises to a new, higher level, the problems of theory also rise to a new, higher level. This rise is the indispensable condition for mastering the technology of the present and creating the technology of the future.

Bibliography

Chapter I

1.1 Wiener, N., "Cybernetics," 1948, 150 pp. New York, John Wiley & Sons.

1.2 Fel'dbaum, A. A., Optimal systems, *in* Peschon, J. (ed.), "Disciplines and Techniques of Systems Control," 1963, 547 pp. Boston, Mass., Ginn and Company.

1.3 Kharkevich, A. A., On a special frequency characteristic for estimating the distortion in the shape of a curve. *J. Theoretical Physics*, 1936, vol. 6, no. 9, pp. 1551 ff (Russian).

1.4 Krasovskii, A. A., "Integral Estimates of the Quality of a Control Process" (Lecture in the seminar on automatic control theory, All-Union Scientific Engineering and Technical Society-Assembly), 1949, 24 pp. Moscow, State Press for Machinery (Russian).

1.5 Newton, G. C., Jr., Gould, L. A., and Kaiser, J. F., "Analytic Design of Linear Feedback Controls," 1957, 419 pp. New York, John Wiley & Sons.

1.6 Fel'dbaum, A. A., Integral criteria for the quality of control. *Automation and Remote Control*, 1948, vol. 9, no. 1, pp. 3–19 (Russian).

1.7 Gibson, G., How to specify the performance of closed loop systems. *Control Engineering*, 1956, vol. 3, September, pp. 122–129.

1.8 Shultz, W. C., and Rideout, V. C., Performance measures: past, present and future. *IRE Trans. on Automatic Control*, 1961, vol. AC-6, no. 1, pp. 22–35.

1.9 Fel'dbaum, A. A., "Electrical Automatic Control Systems," 1st ed., 1954, 736 pp., 2nd ed., 1957, 807 pp. Moscow, State Press for Defense (Russian).

1.10 Pugachev, V. S., "Theory of Random Functions and Its Application to Automatic Control Problems," 1st ed., 1957, 659 pp., 2nd ed., 1960, 883 pp. Moscow, State Press for Physico-Mathematical Literature (Russian).

1.11 Laning, J. H., Jr., and Battin, R. H., "Random Processes in Automatic Control," 1956, 434 pp. New York, McGraw-Hill Book Company.

1.12 Davenport, W. B., Jr., and Root, W. L., "An Introduction to the Theory of Random Signals and Noise," 1958, 393 pp. New York, McGraw-Hill Book Company.

1.13 Lévin, B. R., "Theory of Random Processes and Its Application in Radio Engineering, 1st ed., 1957, 495 pp., 2nd ed., 1960, 662 pp. Moscow, Soviet Radio (publisher) (Russian).

1.14 Kolmogorov, A. N., Interpolation and extrapolation of stationary random sequences. *News of the Academy of Sciences USSR, Mathematical Series,* 1941, vol. 5, no. 1, pp. 3–14 (Russian).

1.15 Wiener, N., "Extrapolation, Interpolation and Smoothing of Stationary Time Series," 1949, 163 pp. New York, John Wiley & Sons.

1.16 Zadeh, L. A., and Ragazzini, T. R., Extension of Wiener's theory of prediction. *J. Appl. Phys.*, 1950, vol. 21, no. 7, pp. 645–655.

1.17 Zadeh, L. A., Optimum nonlinear filters. *J. Appl. Phys.*, 1953, vol. 24, no. 4, pp. 396–404.

1.18 Lubbock, J. K., The optimization of a class of non-linear filters. *Proc. IEE* (Great Britain), 1960, vol. 107C, no. 11, pp. 60–74.

1.19 Khinchin, A. Ya., Correlation theory of stationary stochastic processes. *Progress of Mathematical Sciences*, 1938, no. 5, pp. 42–51 (Russian).

1.20 Seifert, W. W., and Steeg, C. W. (eds.), "Control Systems Engineering," 1960, 944 pp. New York, McGraw-Hill Book Company.

1.21 Lee, Y. W., "Statistical Theory of Communication," 1960, 528 pp., New York, John Wiley & Sons.

1.22 Chang, S. S. L., "Synthesis of Optimum Control Systems," 1961, 381 pp. New York, McGraw-Hill Book Company.

1.23 Solodov, A. V., "Linear Automatic Control Systems with Variable Parameters," 1962, 324 pp. Moscow, State Press for Physico-Mathematical Literature (Russian).

1.24 Westcott, J. H., Synthesis of optimum feedback systems satisfying a power limitation, *ASME Frequency Response Symposium, New York,* 1953, Paper no. 53-A-17.

1.25 Pellegrin, M., "Statistical Design of Tracking Systems," 1957, 223 pp. Moscow, Publisher of Foreign Literature (Russian).

1.26 Andreev, N. I., General extremum condition for a prescribed function of the mean square error and the square of the mathematical expectation of the error in a dynamical system. *Automation and Remote Control*, 1959, vol. 30, no. 7, pp. 833–838 (Russian).

1.27 Westcott, J. H., Design of multivariable optimum filters. *Trans. ASME*, 1958, vol. 80, pp. 463–467.

1.28 Kalman, R. E., and Bucy, R. S., New results in linear filtering and prediction theory. *J. Basic Eng. Trans. ASME, Ser. D*, 1961, vol. 83, March, pp. 95–108.

Chapter II

2.1 Gnedenko, B. V., "Course in Probability Theory," 2nd ed., 1954, 412 pp., 3rd ed., 1961, 406 pp. Moscow, State Press for Physico-Mathematical Literature (Russian).

2.2 Venttzel', E. S., "Theory of Probability," 1st ed., 1958, 464 pp., 2nd ed., 1962, 564 pp. Moscow, State Press for Physico-Mathematical Literature (Russian).

2.3 Feller, W., "An Introduction to Probability Theory and Its Applications," 1st ed., 1950, 419 pp., 2nd ed., 1957, 461 pp. New York, John Wiley & Sons.

2.4 Smirnov, N. V., and Dunin-Barkovskii, I. V., "Short Course in Mathematical Statistics for Engineering Applications," 1959, 436 pp. Moscow, State Press for Physico-Mathematical Literature (Russian).

2.5 Smirnov, V. I., Krylov, V. I., and Kantorovich, L. V., "Calculus of Variations," 1933, 204 pp. Leningrad, Kubuch (publisher) (Russian).

2.6 Lavrent'ev, M. A., and Lyusternik, L. A., "Course in the Calculus of Variations," 2nd ed., 1950, 296 pp. Moscow, Leningrad, State Press for Technical and Theoretical Literature (Russian).

2.7 El'sgol'ts, L. E., "Calculus of Variations," 1952, 167 pp. Moscow, Leningrad, State Press for Technical and Theoretical Literature (Physico-Mathematical Library of the Engineer) (Russian).

2.8 Gel'fand, I. M., and Fomin, S. V., "Calculus of Variations," 1961, 228 pp. Moscow, State Press for Physico-Mathematical Literature (Russian).

2.9 Bellman, R., "Dynamic Programming," 1957, 342 pp. Princeton, N. J., Princeton University Press.

2.10 Bellman, R., and Dreyfus, S., Functional approximation and dynamic programming. *Mathematical Tables and Other Aids to Computation*, 1959, vol. 13, no. 68, pp. 247–251.

2.11 Boltyanskii, V. G., Gamkrelidze, R. V., and Pontryagin, L. S., Toward the theory of optimal processes. *Reports of the Academy of Sciences USSR*, 1956, vol. 110, no. 1, pp. 7–10 (Russian).

2.12 Gamkrelidze, R. V., Theory of processes which are optimal in speed of response in linear systems. *News of the Academy of Sciences USSR, Mathematical Series*, 1958, vol. 22, no. 4, pp. 449–474 (Russian).

2.13 Boltyanskii, V. G., The maximum principle in the theory of optimal processes. *Reports of the Academy of Sciences USSR*, 1958, vol. 119, no. 6, pp. 1070–1073 (Russian).

2.14 Gamkrelidze, R. V., Toward the general theory of optimal processes. *Reports of the Academy of Sciences USSR*, 1958, vol. 123, no. 2, pp. 223–226 (Russian).

2.15 Rozonoer, L. I., On the sufficient conditions for optimality. *Reports of the Academy of Sciences USSR*, 1959, vol. 127, no. 3, pp. 520–523 (Russian).

2.16 Pontryagin, L. S., Optimal control processes. *Progress of Mathematical Sciences*, 1959, vol. 14, no. 1, pp. 3–20 (Russian).

2.17 Rozonoer, L. I., The maximum principle of L. S. Pontryagin in the theory of optimal systems. *Automation and Remote Control*, 1959, vol. 20, no. 10, pp. 1320–1334; no. 11, pp. 1441–1458; no. 12, pp. 1561–1578 (Russian).

2.18 Gamkrelidze, R. V., Processes which are optimal in speed of response with bounded phase coordinates. *Reports of the Academy of Sciences USSR*, 1959, vol. 125, no. 3, pp. 475–478 (Russian).

2.19 Rozonoer, L. I., On variational methods of investigating the quality of automatic control systems, *in* "Automatic and Remote Control," 1960, vol. IV, pp. 1621–1624. London, Butterworths Scientific Publications.

2.20 Desoer, C. A., Pontriagin's maximum principle and the principle of optimality. *J. Franklin Inst.*, 1961, vol. 271, no. 5, pp. 361–367.

2.21 Pontryagin, L. S., Boltyanskii, V. G., Gamkrelidze, R. V., and Mishchenko, E. F., "The Mathematical Theory of Optimal Processes," 1962, 360 pp. New York, Interscience Publishers (Translation).

2.22 Krasovskii, N. N., Toward the theory of optimal control. *Automation and Remote Control*, 1957, vol. 18, no. 11, pp. 960–970 (Russian).

2.23 Krasovskii, N. N., On a problem of optimal control. *Applied Mathematics and Mechanics*, 1957, vol. 21, no. 5, pp. 670–677 (Russian).

2.24 Akhiezer, N., and Krein, M. G., "On Certain Questions in the Theory of Moments," 1938, 254 pp. Chapter 4: The *L*-problem in an abstract linear normed space. Khar'kov, State Scientific and Technical Institute (Ukr. SSR) (Russian).

2.25 Kulikowski, R., Concerning a class of optimum control systems. *Bulletin of the Polish Academy of Sciences*, 1960, vol. 8, no. 10, pp. 595–600.

2.26 Kulikowski, R., On optimum automatic control networks. *Depository of Automation and Remote Control*, 1961, vol. 6, no. 2–3, pp. 236–296 (Polish).

2.27 Kirillova, L. S., Certain questions in the theory of optimal control. *Bulletin of Higher Educational Institutions, Mathematics*, 1962, vol. 5, no. 3, pp. 48–58 (Russian).

2.28 Gamkrelidze, R. V., On sliding optimal conditions. *Reports of the Academy of Sciences USSR*, 1962, vol. 143, no. 6, pp. 1243–1245 (Russian).

2.29 Krotov, V. F., Methods of solving variational problems on the basis of sufficient conditions for an absolute minimum, pt. 1. *Automation and Remote Control*, 1962, vol. 23, no. 12, pp. 1571–1583 (Russian).

2.30 Shreider, Yu. A., The problem of dynamic planning and automata. *Problems of Cybernetics, Moscow*, 1961, no. 5, pp. 31–48 (Russian).

2.31 Boltyanskii, V. G., Sufficient conditions for optimality. *Reports of the Academy of Sciences USSR*, 1961, vol. 140, no. 5, pp. 994–997 (Russian).

2.32 Miele, A., General variational theory of the flight paths of rocket powered aircraft, missiles and satellite carriers. *Astronautica Acta*, 1958, vol. 4, no. 4, pp. 264–288.

2.33 Berkovitz, L. D., Variational methods in problems of control and programming. *J. Mathematical Analysis and Applications*. 1961, vol. 3, no. 1, pp. 145–169.

2.34 Troitskii, V. A., Variational problems in the optimization of control processes in systems with bounded coordinates. *Applied Mathematics and Mechanics*, 1962, vol. 26, no. 3, pp. 431–443 (Russian).

2.35 Bellman, R., and Dreyfus, S., "Applied Dynamic Programming," 1962, 363 pp. Princeton, N. J., Princeton University Press.

2.36 Desoer, C. A., The bang-bang servo problem treated by variational techniques. *Information and Control*, 1959, vol. 2, no. 4, pp. 333–348.

2.37 Chan Sien, Toward the theory of optimal control. *Applied Mathematics and Mechanics*, 1961, vol. 25, no. 3, pp. 413–419 (Russian).

2.38 Bellman, R., "Adaptive Control Processes: a Guided Tour," 1961, 256 pp. Princeton, N. J., Princeton University Press.

2.39 Chang, S. S. L., Minimal time control with multiple saturation limits. *IEEE Trans. on Automatic Control*, 1963, vol. AC-8, no. 1, pp. 35–42.

2.40 Krylov, N. A., and Chernous'ko, F. L., On the method of successive approximations for solving optimal control problems. *J. Computer Mathematics and Mathematical Physics*, 1962, vol. 2, no. 6, pp. 1132–1139 (Russian).

2.41 Katz, S., A discrete version of Pontrjagin's maximum principle. *J. Electronics and Control, Ser. 1*, 1962, vol. 13, no. 2, pp. 179–184.

2.42 Gabasov, R., Toward the question of uniqueness of optimal control in discrete systems. *News of the Academy of Sciences USSR, Department of Technical Sciences, Power Engineering and Automation*, 1962, vol. 4, no. 5, pp. 99–106 (Russian).

2.43 Jurovics, S. A., and McIntyre, J. E., The adjoint method and its application to trajectory optimization. *ARS Journal*, 1962, vol. 32, no. 9, September, pp. 1354–1358.

2.44 Leitmann, G. (ed.), "Optimization Techniques with Applications to Aerospace Systems," 1962, 453 pp., New York, Academic Press.

2.45 Chang, S. S. L., Optimal control in bounded phase space. *Automatica*, 1962, vol. 1, pp. 55–67. New York, Pergamon Press.

2.46 Chang, S. S. L., Minimal time control with multiple saturation limits. *IEEE Trans. on Automatic Control*, 1963, Vol. AC-8, no. 1, pp. 35–42.

2.47 Butkovskii, A. G., The extended maximum principle for optimal control problems. *Automation and Remote Control*, 1963, vol. 24, no. 3, pp. 314–327 (Russian).

2.48 Dubovitskii, A. Ya., and Milyutin, A. A., Problems on the extremum in the presence of constraints. *Reports of the Academy of Sciences USSR*, 1963, vol. 149, no. 4, p. 759 (Russian).

Chapter III

3.1 Mar'yanovskii, D. I., and Svecharnik, D. V., Patent no. 77023, claim no. 181007 of February 25, 1935 (Russian).

3.2 Fel'dbaum, A. A., The simplest relay systems of automatic control. *Automation and Remote Control*, 1949, vol. 10, no. 4, pp. 249–266 (Russian).

3.3 Hopkin, A. M., A phase-plane approach to the compensation of saturating servomechanisms. *Trans. AIEE*, 1951, pt. 1, vol. 70, pp. 631–639.

3.4 Lerner, A. Ya., Improvement of the dynamical properties of automatic compensators with the aid of nonlinear couplings. *Automation and Remote Control*, 1952, vol. 13, no. 2, pp. 134–144; no. 4, pp. 429–444 (Russian).

3.5 Fel'dbaum, A. A., Optimal processes in automatic control systems. *Automation and Remote Control*, 1953, vol. 14, no. 6, pp. 712–728 (Russian).

3.6 Lerner, A. Ya., On the limiting speed of response of automatic control systems. *Automation and Remote Control*, 1954, vol. 15, no. 6, pp. 461–477 (Russian).

3.7 Bogner, I., and Kazda, L. F., An investigation of the switching criteria for higher order contactor servomechanisms. *Trans. AIEE*, 1954, pt. 2, vol. 73, pp. 118–127.

3.8 Silva, L. M., Predictor servomechanisms. *IRE Trans. on Circuit Theory*, 1954, vol. CT-1, no. 1, pp. 36–70.

3.9 LaSalle, J. P., Basic principle of the bang-bang servo. *Bull. AMS*, 1954, vol. 60, no. 2, p. 154.

3.10 Fel'dbaum, A. A., Toward the question of the synthesis of optimal automatic control systems. *Trans. of the 2nd All-Union Conference on Automatic Control Theory, 1953, Moscow,* 1955, pp. 325–360. Moscow, Academy of Sciences USSR (Russian).

3.11 Fel'dbaum, A. A., On the synthesis of optimal systems with the aid of phase space. *Automation and Remote Control*, 1955, vol. 16, no. 2, pp. 120–149 (Russian).

3.12 Hopkin, A. M., and Iwama, M., A study of a predictor type "air-frame" controller designed by phase-space analysis. *AIEE Applications and Industry*, 1956, no. 23, pp. 1–9.

3.13 Kuba, R. E., and Kazda, L. F., A phase-space method for the synthesis of nonlinear servomechanisms. *AIEE Applications and Industry*, 1956, no. 27, pp. 282–289.

3.14 Bass, R. W., Equivalent linearization, nonlinear circuit synthesis and the stabilization and optimization of control systems. *Proc. of the Symposium on Nonlinear Circuit Analysis, Polytech. Inst. of Brooklyn*, 1957, pp. 163–198.

3.15 Tsien Siu Sen, "Engineering Cybernetics," 1956, 462 pp. Moscow, Publisher of Foreign Literature (Russian).

3.16 Rosenman, E. A., On optimal transient processes in systems with a power constraint. *Automation and Remote Control*, 1957, vol. 18, no. 6, pp. 497–513 (Russian).

3.17 Bushaw, D. W., Optimal discontinuous forcing terms, *in* "Contributions to Nonlinear Oscillations," (S. L. Lefschetz, ed.) vol. IV, *Annals of Mathematics Study*, 1958, vol. 41. Princeton, N. J., Princeton University Press.

3.18 LaSalle, J. P., On time-optimal control systems. *Proc. National Acad. Sci.*, 1959, vol. 45, pp. 573–577.

3.19 Novosel'tsev, V. N., The optimal process in a second order relay-impulse system. *Automation and Remote Control*, 1960, vol. 21, no. 5, pp. 569-574 (Russian).

3.20 Desoer, C. A., and Wing, J., An optimal strategy for a saturating sampled data system. *IRE Trans. on Automatic Control*, 1961, vol. AC-6, no. 1, pp. 5–15.

3.21 Smith, F. B., Jr., Time-optimal control of higher-order systems. *IRE Trans. on Automatic Control*, 1961, vol. AC-6, no. 1, pp. 16–21.

3.22 Gosiewski, A., An optimum second order relay servomechanism, *Polish Academy of Sciences, Electrotechnical Reports*, 1961, vol. 7, no. 1, pp. 17–67 (Polish).

3.23 Chandaket, P., Leondes, C. T., and Deland, E. C., Optimum nonlinear bang-bang control systems with complex roots, I–II. *AIEE Applications and Industry*, 1961, no. 54, pp. 82–102.

3.24 Desoer, C. A., and Wing, J., A minimal time discrete system. *IRE Trans. on Automatic Control*, 1961, vol. AC-6, no. 2, pp. 111–124.

3.25 Fel'dbaum, A. A., "Computers in Automatic Systems," 1959, 800 pp. Moscow, State Press for Physico-Mathematical Literature (Russian).

3.26 Kulikowski, R., Concerning a class of optimum control systems. *Bull.*

Polish Academy of Sciences, Scientific Technical Series, 1960, vol. 8, no. 10, pp. 595–600.

3.27 Kipiniak, W., "Dynamic Optimization and Control, a Variational Approach," 1961, 233 pp. Cambridge, Mass., Technology Press of MIT.

3.28 Letov, A. M., Analytical design of regulators, 1–3. *Automation and Remote Control*, 1960, vol. 21, no. 4, pp. 436–441; no. 5, pp. 561–568; no. 6, pp. 661–665 (Russian).

3.29 Ostrovskii, G. M., On a method of solving variational problems. *Automation and Remote Control*, 1962, vol. 23, no. 10, pp. 1284–1289 (Russian).

3.30 Tsypkin, Ya. Z., On optimal processes in automatic impulse systems. *Reports of the Academy of Sciences USSR*, 1960, vol. 134, no. 2, pp. 308–310 (Russian).

3.31 Tsypkin, Ya. Z., Optimal processes in automatic impulse systems. *News of the Academy of Sciences USSR, Department of Technical Sciences, Power Engineering and Automation*, 1962, no. 4, pp. 74–93 (Russian).

3.32 Eckman, D. P., and Lefkowitz, J., A report on optimizing control of a chemical process. *Control Engineering*, 1957, vol. 4, no. 9, pp. 197–204.

3.33 Gould, L. A., and Kipiniak, W., Dynamic optimization and control of a stirred-tank chemical reactor. *AIEE Communication and Electronics*, 1961, no. 52, pp. 734–743.

3.34 Merriam, C. W., III, "Synthesis of Adaptive Controls," Report 7793–R-3, February, 1959, Servomechanisms Lab., Cambridge, Mass., Massachusetts Inst. of Technology.

3.35 Merriam, C. W., III, A class of optimum control systems. *J. Franklin Inst.*, 1959, vol. 267, no. 4, pp. 267–281.

3.36 Peterson, E. L., "Statistical Analysis and Optimization of Systems," 1961, 190 pp. New York, John Wiley & Sons.

3.37 Soong Tsien, Synthesis of an optimal control system on the basis of the isochrone field. *News of the Academy of Sciences USSR, Department of Technical Sciences, Power Engineering and Automation*, 1960, no. 5, pp. 96–103 (Russian).

3.38 Zadeh, L. A., Optimal control problems in discrete-time systems, *in* Leondes, C. T. (ed.), "Computer Control Systems Technology," 1961, pp. 389–414. New York, McGraw-Hill Book Company.

3.39 Tsien Shen-vee, A problem in the synthesis of optimal systems from the maximum principle. *Automation and Remote Control*, 1961, vol. 22, no. 10, pp. 1302–1308 (Russian).

3.40 Soong Tsien, Optimal control in a nonlinear system. *Automation and Remote Control*, 1960, vol. 21, no. 1, pp. 3–14 (Russian).

3.41 Bor-Ramenskii, A. E., and Soong Tsien, An optimal tracking servo with two control parameters. *Automation and Remote Control*, 1961, vol. 22, no. 2, pp. 157–170 (Russian).

3.42 Butkovskii, A. G., and Lerner, A. Ya., On optimal control by distributed parameter systems. *Automation and Remote Control*, 1960, vol. 21, no. 6, pp. 682–691 (Russian).

3.43 Butkovskii, A. G., Optimal processes in distributed parameter systems. *Automation and Remote Control*, 1961, vol. 22, no. 1, pp. 17–26 (Russian).

3.44 Butkovskii, A. G., The maximum principle for optimal systems with distributed parameters. *Automation and Remote Control*, 1961, vol. 22, no. 10, pp. 1288–1301 (Russian).

3.45 Bellman, R., On the application of the theory of dynamic programming to the study of control processes. *Proc. of the Symposium on Nonlinear Circuit Analysis, Polytech. Inst. of Brooklyn*, 1957, pp. 199–213.

3.46 Bellman, R., Glicksberg, I., and Gross, O., On the bang-bang control problem. *Quarterly of Applied Mathematics*, 1956, vol. 14, no. 1, pp. 11–18.

3.47 Kalman, R. E., and Koepcke, R. W., Optimal synthesis of linear sampling control systems using generalized performance indexes. *Trans. ASME*, 1958, vol. 80, pp. 1820–1826.

3.48 Ho Yu-Chi, Solution space approach to optimal control problems, ASME Paper no. 60-JAC-11, 1960, *Joint Automatic Control Conference, Cambridge, Mass, Sept. 7-9, 1960.*

3.49 Kramer, J. D. H., On control of linear systems with time lags. *Information and Control*, 1960, vol. 3, pp. 299–326.

3.50 Smith, F. B., Jr., Time-optimal control of higher-order systems. *IRE Trans. on Automatic Control*, 1961, vol. AC-6, no. 1, pp. 16–21.

3.51 Chang, S. S. L., Computer optimization of nonlinear control systems by means of digitized maximum principle. *IRE International Convention Record*, 1961, vol. 9, pt. 4, pp. 48–55.

3.52 Aronovich, G. V., On the effect of wave phenomena in a pressure pipe-line on optimal control. *Bull. of Higher Educational Institutions, Radiophysics*, 1962, vol. 5, no. 2, pp. 362–369 (Russian).

3.53 Krasovskii, N. N., and Letov, A. M., Toward the theory of the analytical design of regulators. *Automation and Remote Control*, 1962, vol. 23, no. 6, pp. 713–720 (Russian).

3.54 Gabasov, R., Toward optimal processes in coupled systems of the discrete type. *Automation and Remote control*, 1962, vol. 23, no. 7, pp. 872–880 (Russian).

3.55 Balakirev, V. S., The maximum principle in the theory of second order optimal systems. *Automation and Remote Control*, 1962, vol. 23, no. 8, pp. 1014–1022 (Russian).

3.56 Kranc, G. M., and Sarachik, P. E., An application of functional analysis to the optimal control problem, *in* "Joint Automatic Control Conference" (ASME, ISA, AIEE, IRE, AIChE), 1962, no. 8-2, pp. 1–8.

3.57 Kazda, L. F., Application of state variables to optimal control system problems, *in* "Joint Automatic Control Conference" (ASME, ISA, AIEE, IRE, AIChE), 1962, no. 11-2, pp. 1–5.

3.58 Flügge-Lotz, I., and Marbach, H., The optimal control of some attitude control systems for different performance criteria, *in* "Joint Automatic Control Conference" (ASME, ISA, AIEE, IRE, AIChE), 1962, no. 12-1, pp. 1–12.

3.59 Neustadt, L., Time-optimal control systems with position and integral limits. *J. Mathematical Analysis and Applications*, 1961, vol. 3, no. 3, pp. 406–427.

3.60 Friedland, B., The structure of optimum control systems. *J. Basic Eng. Trans. ASME, Ser. D*, 1962, vol. 84, no. 1, pp. 1–13.

3.61 Isaev, V. K., and Sonin, V. V., On a nonlinear optimal control problem. *Automation and Remote Control*, 1962, vol. 23, no. 9, pp. 1117–1129 (Russian).

3.62 Nguyen Chan Bang, Toward the solution of certain problems in the theory of dynamic programming with the aid of electronic simulation systems. *Automation and Remote Control*, 1962, vol. 23, no. 9, pp. 1130–1140 (Russian).

3.63 Paraev, Yu. I., On a particular control in optimal processes which are linear with respect to the control actions. *Automation and Remote Control*, 1962, vol. 23, no. 9, pp. 1202–1209 (Russian).

3.64 Savvin, A. B., Toward the theory of processes which are optimal in speed of response in second order systems. *News of the Academy of Sciences USSR, Department of Technical Sciences, Power Engineering and Automation*, 1960, vol. 2, no. 6, pp. 162–164 (Russian).

3.65 Litvin-Cedoi, M. Z., and Savvin, A. B., On the synthesis of second-order automatic control systems with bounded transient processes. *Herald of Moscow University, Ser. 1, Mathematics and Mechanics*, 1961, vol. 4, no. 5, pp. 77–83 (Russian).

3.66 Brennan, P. J., and Roberts, A. P , Use of an analog computer in the application of Pontryagin's maximum principle to the design of control systems with optimum transient response. *J. Electronics and Control, Ser. 1*, 1962, vol. 12, no. 4, pp. 345–352.

3.67 Pavlov, A. A., On the increase in speed of response of certain third order relay systems. *News of the Academy of Sciences USSR, Department of Technical Sciences, Power Engineering and Automation*, 1962, vol. 4, no. 2, pp. 59–71 (Russian).

3.68 Athanassiades, M., and Smith, O. J. M., Theory and design of higher order bang-bang control systems. *IRE Trans. on Automatic Control*, 1961, vol. AC-6, no. 2, pp. 125–134.

3.69 Friedland, B., A minimum response-time controller to amplitude and energy constraints, *IRE Trans. on Automatic Control*, 1962, vol. AC-7, 1, pp. 73–74.

3.70 Nelson, W. L., Optimal control methods for on-off sampling systems, *J. Basic Eng. Trans. ASME, Ser. D*, 1962, vol. 84, March, pp. 91–99.

3.71 Fel'dbaum, A. A., On the automatic synthesis of optimal automatic control systems, *in* "Composite Computers," B. Ya. Kogan (ed.) (Conference Transactions of the Seminar on the Theory and Methods of Mathematical Modelling, Moscow, 1961), 1962, 294 pp. Moscow, Publisher of the Academy of Sciences USSR (Russian).

3.72 Kallay, N., An example of the application of dynamic programming to the design of optimal control programs. *IRE Trans. on Automatic Control*, 1962, vol. AC-7, no. 3, p. 10–21.

3.73 Wazewski, T., On problems of optimum guidance in the nonlinear case. *Depository of Automation and Remote Control*, 1962, vol. 7, no. 1–2, pp. 19–32 (Polish).

3.74 Kirillova, L. S., The problem of the optimization of the final state of a control system. *Automation and Remote Control*, 1962, vol. 23, no. 12, pp. 1584–1594 (Russian).

3.75 Salukvadze, M. E., Toward the problem of the synthesis of an optimal controller in linear systems with delay, subject to continuously acting disturbances. *Automation and Remote Control*, 1962, vol. 23, no. 12, pp. 1595–1601 (Russian).

3.76 Utkin, V. I., The design of a class of optimal automatic control systems without using "pure" undistorted derivatives in the control law. *Automation and Remote Control*, 1962, vol. 23, no. 12, pp. 1631–1642 (Russian).

3.77 Lerner, A. Ya., "Design Principles for Rapid-Acting Tracking Systems and Controllers," 1961, 151 pp. Moscow, Leningrad, State Press for Power Engineering (Russian).

3.78 Johnson, C. D., and Gibson, J. E., Singular solutions in problems of optimal control. *IEEE Trans. on Automatic Control*, 1963, vol. AC-8, no. 1, pp. 4–15.

Chapter IV

4.1 Roitenberg, Ya. N., On certain indirect methods of obtaining information about the position of a control system in phase space. *Applied Mathematics and Mechanics*, 1961, vol. 25, no. 3, pp. 440–444 (Russian).

4.2 Fuller, A. T., Optimization of non-linear control systems with random inputs. *J. Electronics and Control, Ser. 1*, 1960, vol. 9, no. 1, pp. 65–80.

4.3 Florentin, J. J., Optimal control of continuous time, Markov, stochastic systems. *J. Electronics and Control, Ser. 1*, 1961, vol. 10, no. 6, pp. 473–488.

4.4 Aoki, M., Stochastic-time optimal control systems. *AIEE Applications and Industry*, 1961, no. 54, pp. 41–44, Disc., pp. 44–46.

4.5 Krasovskii, N. N., and Lidskii, E. A., Analytical design of controllers in systems with random properties. *Automation and Remote Control*, 1961, vol. 22, no. 9, pp. 1145–1150; no. 10, pp. 1273–1278; no. 11, pp. 1425–1431 (Russian).

4.6 Novosel'tsev, V. N., Optimal control in a second order relay-impulse system in the presence of random disturbances. *Automation and Remote Control*, 1961, vol. 22, no. 7, pp. 865–875 (Russian).

4.7 Howard, R. A., "Dynamic Programming and Markov Processes," 1960, 136 pp. Cambridge, Mass., Technology Press of MIT.

4.8 Bellman, R., A Markovian decision process. *J. Mathematics and Mechanics*, 1957, vol. 6, pp. 679–684.

4.9 Huggins, W. H., Signal-flow graphs and random signals. *Proc. IRE*, 1957, vol. 45, no. 1, pp. 74–86.

4.10 Sittler, P. W., System analysis of discrete Markov processes. *IRE Trans. on Circuit Theory*, 1956, vol. CT-3, no. 4, pp. 257–265.

4.11 Fel'dbaum, A. A., The steady-state process in the simplest discrete extremal system in the presence of random noise. *Automation and Remote Control*, 1959, vol. 20, no. 8, pp. 1056–1070 (Russian).

4.12 Eaton, J. H., and Zadeh, L. A., Optimal pursuit strategies in discrete state probabilistic systems. *J. Basic Eng. Trans. ASME, Ser. D*, 1962, vol. 84, March, pp. 23–29.

4.13 Romanovskii, V. I., "Discrete Markov Chains," 1949, 436 pp. Moscow, State Press for Technical and Theoretical Literature (Russian).

4.14 Feller, W., "An Introduction to Probability Theory and Its Applications," 1st ed., 1950, 419 pp., 2nd ed., 1957, 461 pp. New York, John Wiley & Sons.

4.15 Gel'fond, A. O., "The Calculus of Finite Differences," 1952, 480 pp. Moscow, State Press for Technical and Theoretical Literature (Russian).

4.16 Tsypkin, Ya. Z., "Theory of Impulse Systems," 1958, 724 pp. Moscow, State Press for Physico-Mathematical Literature (Russian).

4.17 Bush, R. R., and Mosteller, F., "Stochastic Models for Learning," 1955, 365 pp. New York, John Wiley & Sons.

4.18 Bush, R. R., and Estes, W. (eds.), "Studies in Mathematical Learning Theory," 1959, 432 pp. Stanford, Calif., Stanford University Press.

4.19 Novosel'tsev, V. N., Control systems which are optimal in speed of response in the presence of random noise. *Automation and Remote Control*, 1962, vol. 23, no. 12, pp. 1620–1630 (Russian).

4.20 Bellman, R., Dynamic programming and stochastic control processes. *Information and Control*, 1958, vol. 1, no. 3, pp. 228–239.

4.21 Florentin, J. J., Optimal control of systems with generalized Poisson inputs, *in* "Joint Automatic Control Conference" (ASME, ISA, AIEE, IRE, AIChE), N. Y. U., 1962, no. 14-2, pp. 1–5.

4.22 Katz, S., Best endpoint control of noisy systems. *J. Electronics and Control*, 1962, vol. 12, no. 4, pp. 323–343.

4.23 Andronov, A. A., Vitt, A. A., and Pontryagin, L. S., On the statistical analysis of dynamical systems. *J. Experimental and Theoretical Physics*, 1933, vol. 3, no. 3, pp. 165–180; Andronov, A. A., "Collection of Works," 1956, 533 pp. Moscow, Publisher of the Academy of Sciences USSR (Russian).

4.24 Barrett, J. F., Application of Kolmogorov's equations to randomly disturbed automatic control systems, *in* "Automatic and Remote Control," 1960, vol. II, pp. 724–733. London, Butterworths Scientific Publications.

4.25 Chen, P. P., An algorithm for stochastic control through dynamic programming techniques. *IRE International Convention Record*, 1962, vol. 10, pt. 2, pp. 112–124.

4.26 Stratonovich, R. L., and Shmal'gauzen, V. I., Certain stationary problems in dynamic programming. *News of the Academy of Sciences USSR, Department of Technical Sciences, Power Engineering and Automation*, 1962, vol. 4, no. 5, pp. 131–139.

4.27 Chen, P. P., An algorithm for stochastic control through dynamic programming, *IRE Trans. on Automatic Control*, 1962, vol. AC-7, no. 5, pp. 110–118.

Chapter V

5.1 Kulebakin, V. S., High-quality invariant control systems, *in* "Transactions of the Conference on The Theory of Invariance and Its Application to Automatic Systems, Kiev, 1958," 1959, pp. 11–39. Moscow, Published for Academy of Sciences Ukr. SSR (Russian).

5.2 Petrov, B. N., The invariance principle and the conditions for its application in the design of linear and nonlinear systems, *in* "Automatic and Remote Control," 1960, vol. II, pp. 1123–1128. London, Butterworths Scientific Publications.

5.3 Smith, O. J. M., "Feedback Control Systems," 1958, 694 pp. New York, McGraw-Hill Book Company.

5.4 Meerov, M. V., "Synthesis of the Structures of Automatic Control Systems of High Accuracy," 1959, 284 pp. Moscow, State Press for Physico-Mathematical Literature (Russian).

5.5 Kogan, B. Ya., "Electronic Simulators and Their Application to the Investigation of Control Systems," 1959, 432 pp. Moscow, State Press for Physico-Mathematical Literature (Russian).

5.6 Shchipanov, G. V., Theory and design methods for automatic controllers. *Automation and Remote Control*, 1939, no. 1, pp. 49–66 (Russian).

5.7 Cramer, H., "Mathematical Methods of Statistics," 1946, 575 pp. Princeton, N. J., Princeton University Press.

5.8 Smirnov, N. V., and Dunin-Barkovskii, I. V., "Short Course of Mathematical Statistics for Engineering Applications," 1959, 436 pp. Moscow, State Press for Physico-Mathematical Literature (Russian).

5.9 Linnik, Yu. V., "The Method of Least Squares and the Fundamentals of the Mathematical-Statistical Theory of the Processing of Observations," 1st ed., 1958, 333 pp., 2nd ed., 1962, 349 pp. State Press for Physico-Mathematical Literature (Russian). Translation: Linnik, Yu. V., "Method of Least Squares and Principles of the Theory of Observations," 1st ed., 1961, 360 pp. New York, Pergamon Press.

5.10 Williams, J. D., "The Compleat Strategyst (or a Primer on the Theory of Games of Strategy)," 1954, 234 pp. New York, McGraw-Hill Book Company.

5.11 Venttzel', E. S., "Elements of the Theory of Games," 1959, 67 pp. Moscow, State Press for Physico-Mathematical Literature (Popular Lectures on Mathematics) (Russian).

5.12 Vajda, S., "An Introduction to Linear Programming and the Theory of Games," 1960, 76 pp. New York, John Wiley & Sons.

5.13 McKinsey, J. C. C., "Introduction to the Theory of Games," 1952, 371 pp. New York, McGraw-Hill Book Company.

5.14 Luce, R. D., and Raiffa, H., "Games and Decisions," 1957, 509 pp. New York, John Wiley & Sons.

5.15 Wald, A., "Statistical Decision Functions," 1950, 179 pp. New York, John Wiley & Sons.

5.16 Kotel'nikov, V. A., "Theory of Potential Noise Stability in the Presence of Fluctuation Noise," Doctoral Dissertation, 1946, 364 pp. Moscow, Molotov Energy Institute; Kotel'nikov, V. A., "The Theory of Potential Noise Stability," 1956, 151 pp. Moscow, Leningrad, State Press for Power Engineering (Russian); Translation: Kotel'nikov, V. A., "The Theory of

Optimum Noise Immunity," 1959, 140 pp. New York, McGraw-Hill Book Company.

5.17 Vainshtein, L. A., and Zubakov, V. D., "The Extraction of Signals from a Background of Random Noise," 1960, 447 pp. Moscow, Soviet Radio (publisher) (Russian); Translation: Wainstein, L. A., and Zubakov, V. D., "Extraction of Signals from Noise", 1962, 382 pp. Englewood Cliffs, N.J., Prentice-Hall.

5.18 "The Reception of Signals in the Presence of Noise," Collection of Translations under the editorship of L. S. Gutkin, 1960, 343 pp. Moscow, Publisher of Foreign Literature (Russian).

5.19 "The Reception of Impulse Signals in the Presence of Noise," Collection of Translations under the editorship of A. E. Basharinov and M. S. Aleksandrov, 1960, 283 pp. Moscow, Leningrad, State Press for Power Engineering (Russian).

5.20 Fal'kovich, S. E., "The Reception of Radar Signals from a Background of Fluctuation Noise," 1961, 311 pp. Moscow, Soviet Radio (publisher) (Russian).

5.21 Gutkin, L. S., "The Theory of Optimal Methods of Radio Reception in the Presence of Fluctuation Noise," 1961, 488 pp. Moscow, Leningrad, State Press for Power Engineering (Russian).

5.22 Middleton, D., "An Introduction to Statistical Communication Theory," 1960, 1140 pp. New York, McGraw-Hill Book Company.

5.23 Chow, C. K., An optimum character recognition system using decision functions. *IRE Trans. on Electronic Computers*, 1957, vol. EC-6, no. 4, pp. 247–254.

5.24 Fel'dbaum, A. A., Theory of dual control, II. *Automation and Remote Control*, 1960, vol. 21, no. 11, pp. 1453–1464 (Russian).

5.25 Truxal, J. G., and Padalino, I. I., Decision theory, pt. 4, sec. 5, *in* Mishkin, E., and Braun, L. (eds.), "Adaptive Control Systems," 1961, 533 pp. New York, McGraw-Hill Book Company.

5.26 Ryzhik, I. M., and Gradshtein, I. S. "Tables of Integrals, Sums, Series and Products," 3rd ed., 1948, 464 pp. Moscow, State Press for Technical and Theoretical Literature (Russian).

5.27 Chernoff, H., and Moses, L. E., "Elementary Decision Theory," 1959, 364 pp. New York, John Wiley & Sons.

5.28 Tarasenko, V. P., Toward the question of optimal methods for extremal control in the presence of noise, *in* "Computer Engineering, Automation, Information Theory," 1961, issue no. 40, pp. 47–57. Tomsk, Tomsk University (Russian).

5.29 Hsu, J. C., and Meserve, W. E., Decision-making in adaptive control systems. *IRE Trans. on Automatic Control*, 1962, vol. AC-7, no. 1, pp. 24–32.

5.30 Wald, A., "Sequential Analysis," 1947, 212 pp., New York, John Wiley & Sons.

5.31 Basharinov, A. E., and Fleishman, B. S., "Methods of Statistical Sequential Analysis and Their Radio-Engineering Applications," 1962, 352 pp. Moscow, Soviet Radio (publisher) (Russian).

5.32 Bellman, R., and Kalaba, R., On adaptive control processes. *IRE National Convention Record*, 1959, vol. 7, pt. 4, pp. 3–11.

5.33 Freimer, M. A., Dynamic programming approach to adaptive control processes. *IRE National Convention Record*, 1959, vol. 7, pt. 4, pp. 12–17.

5.34 Kelendzheridze, D. L., On a problem of optimal pursuit. *Automation and Remote Control*, 1962, vol. 23, no. 8, pp. 1008–1013 (Russian).

5.35 Gadzhiev, M. Yu., Application of game theory to certain problems of automatic control. *Automation and Remote Control*, 1962, vol. 23, no. 8, pp. 1023–1036; no. 9, pp. 1144–1153 (Russian).

5.36 Repin, V. G., On the solution of the filtering problem with a limited knowledge of *a priori* statistics. *Automation and Remote Control*, 1962, vol. 23, no. 8, pp. 1108–1111 (Russian).

5.37 Carnap, R., "Logical Foundations of Probability," 1950, 2nd ed., 1962, 640 pp., Chicago, Ill., University of Chicago Press.

5.38 Schwartz, L. S., Harris, B., and Hauptschein, A., Information rate from the viewpoint of inductive probability. *IRE National Convention Record*, 1959, vol. 7, pt. 4, pp. 102–112.

5.39 Kiefer, J., and Wolfowitz, J., Optimum design in regression problems. *Annals of Mathematical Statistics*, 1959, vol. 30, pp. 271–294.

5.40 Bellman, R., and Kalaba, R., The theory of dynamic programming and feedback control systems, *in* "Automatic and Remote Control," 1960, vol. I, pp. 122–125. London, Butterworths Scientific Publications.

5.41 Kushner, H. J., Optimal stochastic control. *IRE Trans. on Automatic Control*, 1962, vol. AC-7, no. 5, pp. 120–122.

5.42 Rozonoer, L. I., A variational approach to the problem of invariance. *Automation and Remote Control*, 1963, vol. 24, no. 6, pp. 744–752; no. 7, pp. 861–870 (Russian).

Chapter VI

6.1 Fel'dbaum, A. A., Theory of dual control. *Automation and Remote Control*, 1960, vol. 21, no. 9, pp. 1240–1249; no. 11, pp. 1453–1464; 1961, vol. 22, no. 1, pp. 3–16; no. 2, pp. 129–142 (Russian).

6.2 Kazakevich, V. V., Extremal control systems and certain methods of improving their stability, *in* "Automatic Control and Computer Engineering," V.V. Solodovnikov (ed.), 1958, pp. 66–96. Moscow, State Press for Machinery (Russian).

6.3 Morosanov, I. S., Methods of extremal control. *Automation and Remote Control*, 1957, vol. 18, no. 11, pp. 1029–1044 (Russian).

6.4 Fel'dbaum, A. A., On the application of computers to automatic systems. *Automation and Remote Control*, 1956, vol. 17, no. 11, pp. 1046–1056 (Russian).

6.5 Fitsner, L. N., On the design principles and analysis methods for certain types of extremal systems. *Trans. of the Conference on the Theory and Application of Discrete Automatic Systems (Institute of Automation and*

Telemechanics), Moscow, 1960, pp. 480–504. Moscow, Publisher of the Academy of Sciences USSR (Russian).

6.6 Fel'dbaum, A. A., An automatic optimizer. *Automation and Remote Control*, 1958, vol. 19, no. 8, pp. 731–743 (Russian).

6.7 Fel'dbaum, A. A., On information storage in closed-loop automatic control systems. *News of the Academy of Sciences USSR, Department of Technical Sciences, Power Engineering and Automation*, 1961, vol. 3, no. 4, pp. 107–119 (Russian).

6.8 van der Waerden, B. L., "Mathematical Statistics," 1957, 360 pp. Berlin, Springer Publishing House (German).

6.9 Cramer, H., "Mathematical Methods of Statistics," 1946, 575 pp., Princeton, N. J., Princeton University Press.

6.10 Stratonovich, R. L., Toward the theory of optimal control, sufficient coordinates. *Automation and Remote Control*, 1962, vol. 23, no. 7, pp. 910–917 (Russian).

6.11 Fel'dbaum, A. A., On optimal control by Markov objects. *Automation and Remote Control*, 1962, vol. 23, no. 8, pp. 993–1007 (Russian).

6.12 Fel'dbaum, A. A., On problems in dual control theory, *in* "Automatic and Remote Control", 1963. London, Butterworths Scientific Publications.

6.13 Pike, E. W., and Silverberg, T. R., Designing mechanical computers. *Machine Design*, 1952, vol. 24, no. 7, pp. 131–137; no. 8, pp. 159–163.

6.14 Eterman, I. I., "Continuously Operating Mathematical Machines," 1957, 236 pp. Moscow, State Press for Machinery (Russian).

6.15 Zhigulev, V. N., Synthesis of a single class of optimal systems. *Automation and Remote Control*, 1962, vol. 23, no. 11, pp. 1431–1438 (Russian).

6.16 Florentin, J. J., Optimal, probing, adaptive control of a simple Bayesian system. *J. Electronics and Control, Ser. 1*, 1962, vol. 13, no. 2, pp. 165–177.

6.17 Fel'dbaum, A. A., On the application of statistical decision theory to open-loop and closed-loop automatic control systems. *News of the Academy of Sciences USSR, Department of Technical Sciences, Series "Engineering Cybernetics,"* 1963, vol. 1, no. 1 (Russian).

6.18 Florentin, J. J., Partial observability and optimal control. *J. Electronics and Control, Ser. 1*, 1962, vol. 13, no. 3, pp. 263–279.

6.19 Fuller, A. T., Bibliography of optimum non-linear control of determinate and stochastic-definite systems, *J. Electronics and Control, Ser. 1*, 1962, vol. 13, no. 6, pp. 589–611.

Author Index

Numbers in parentheses are reference numbers and indicate that an author's work is referred to although his name is not cited in the text. Numbers in italics show the page on which the complete reference is listed.

Subject Index